Mally, Scott 981-6845

D0849219

Structural Design in Architecture

SECOND EDITION

MARIO SALVADORI
Columbia University

MATTHYS LEVY
Columbia University

With example and problem solutions by
HOWARD H. M. HWANG

Prentice-Hall, Inc., Englewood Cliffs, N.J. 07632

Library of Congress Cataloging in Publication Data

Salvadori, Mario George, 1907–
Structural design in architecture.

Includes index.
1. Structural design. I. Levy, Matthys, joint author. II. Title.
TH845.S32 1981 729'.3 80-1150
ISBN 0-13-853473-X

Editorial supervision by STEVEN BOBKER
Interior design by VIRGINIA RUBENS
Manufacturing buyer: TONY CARUSO

© 1967, 1981 by Prentice-Hall, Inc., Englewood Cliffs, N.J. 07632

All rights reserved. No part of this book may be reproduced in any form or by any means without permission in writing from the publisher.

Printed in the United States of America

10 9 8 7 6 5 4 3 2

Prentice-Hall International, Inc., *London*
Prentice-Hall of Australia Pty. Limited, *Sydney*
Prentice-Hall of Canada, Ltd., *Toronto*
Prentice-Hall of India Private Limited, *New Delhi*
Prentice-Hall of Japan, Inc., *Tokyo*
Prentice-Hall of Southeast Asia Pte. Ltd., *Singapore*
Whitehall Books Limited, Wellington, *New Zealand*

Contents

7. FRAMES 139

8. TORSION 177

9. CABLES 193

10. ARCHES 215

11. PLANE GRID SYSTEMS 234

Preface to the Second Edition

The favor encountered by the first edition of *Structural Design in Architecture*, including several translations, and recent new international developments in structural design convinced us of the advisability of preparing this revised edition in order to make the book more useful to practicing engineers and architects as well as to engineering and architectural students.

The most obvious revision of the present edition is the adoption of SI units, followed by English units in all the examples and problems. It is apparent that the International System of Units must gradually be adopted in the United States because we are at present using a system of units acceptable only to a few small countries throughout the world.

A few sections in this edition did not appear in the first edition. They are concerned mostly with approximate methods of evaluation of stresses in complex systems, which allow preliminary design and estimation in high-rise frames of steel and concrete, in column-slab systems of concrete, in fabric structures, and in steel domes, all structures which have acquired greater importance during the last decade.

Minor inaccuracies have been corrected and clarification of the presentation has been improved or simplified wherever possible.

The problems have been revised for greater clarity and so as to conform with changes in the text.

As was the case for the first edition, we extend deep gratitude to our students in the United States and abroad for their constructive criticism and suggestions. Many of our colleagues were also kind enough to write us to propose modifications here and there.

We are grateful to our publishers for their interest in our book and for their acceptance of the introduction of SI units.

We express our thanks to Dr. Howard H. M. Hwang who painstakingly solved all examples and problems in both SI and English units and to Mr. Alexander G. Simpson who prepared the new index.

If this second edition is of greater use to our young colleagues and our future students, our many hours of work will have been abundantly rewarded.

Mario Salvadori
Matthys Levy

Preface to the First Edition

This book is concerned with simplified methods of analysis required for preliminary design of structures, and is dedicated to the practicing engineer and the practicing architect, as well as the architectural and the engineering student. The methods presented, while obviously needed by engineers, are of particular value to architects because their interest in structures is directed essentially toward preliminary design.

The architect may approach structures from two different viewpoints: a fairly rigorous mathematical viewpoint, akin to that used by the engineer, or an intuitive viewpoint based on physical experience and the use of models. After a combined experience of over 40 years in teaching structures to architects and engineers, and over 50 years in structural consulting practice, we have come to the conclusion that these two exclusive approaches are unsatisfactory both educationally and professionally. Educationally, it seems improper to load the already crammed architectural curriculum with unnecessary analytical courses and with delicate experimental techniques that often hide the structural woods because of the electronic trees. Professionally, it appears obvious that the architect will never engage in refined structural analysis, since it is more efficient and practically unavoidable that the services of a specialist be utilized.

The philosophy of the structural approach we have adopted at Columbia University to solve this dilemma can be simply stated, and should be congenial to both the architect and the engineer.

We believe on the one hand that the architect should be able to conceive his buildings correctly and, on the other, that the architectural student will never become interested in structures unless he is duly motivated.

Hence, we have tried to indoctrinate students with the importance of structures at the very beginning of their careers, eschewing at this time mathematical tools, but we present the subject again, later, making use of mathematics to obtain quantitative answers. Both times structures are considered from the exclusive viewpoint of preliminary design.

Our approach implies that structures be taught at first by appealing to those intuitions we all gain from our daily experiences with forces and materials, that is, from a purely physical viewpoint. But it also implies that later on the student must acquire enough *manipulative* knowledge of mathematics to be able to "figure out" quantitative answers, within wide margins.

We believe this approach to be valid for both the engineering and the architectural student.

The first exposure of our students to structures occurs at the freshmen level, through experiments, models, slides, motion pictures, and words. No mathematics beyond arithmetic is used at this stage. *Structure in Architecture* by the senior author and Robert Heller is the basis for this exposure.

The present volume translates into quantitative terms the qualitative statements of that previous book and makes use of a variety of methods we have found particularly useful in our consulting practice. In order to use it profitably the reader need only be familiar with the mathematical knowledge summarized in its entirety on page xvii of this book. The reader should be able to use arithmetic, algebra, and trigonometry but does not need to have more than an understanding of the two basic concepts of the calculus (the integral as limit of a sum and the derivative as limit of increment ratios). This much mathematical knowledge can be mastered by any student within a few weeks, often by self-study of a programmed book on elementary mathematics.

As to our choice of subject matter, we feel that a rudimentary knowledge of loads and materials is a prerequisite to structures. Hence, these matters are concisely covered in the first two chapters of the book. The next four chapters deal with the use of elementary states of stress and their combinations and cover tensed and compressed elements, beams in bending, shear and torsion, beam-columns and their buckling, trusses and frames. The chapter on beams is necessarily long, although elementary; it covers continuous beams and gives a hint of beams on elastic supports, a topic of importance both in foundation design and in shell theory.

The following two chapters deal with one-dimensional curved elements: cables and arches. Plane two-dimensional structures are considered in two more chapters on grid systems and plates; plates are treated both as two-dimensional elastic bodies and as bents in the manner of most codes.

A chapter on plasticity, stress redistribution, and ultimate design introduces the reader to new concepts and to new methods of analysis, at times mathematically simpler than those of elastic analysis.

Two-dimensional curved structures are covered in four chapters on the geometry of surfaces, membranes, membrane stresses, and bending stresses in shells.

The last chapter covers plane and curved space-frames. General formulas to check the strength and stability of the most commonly used space-frames are presented, with elementary derivations wherever possible.

To enhance basic understanding of the material, each section of the book is cross referenced to those sections of Structure in Architecture, *where the same subject matter is presented in a purely intuitive manner.*

This book would be of little practical use if principles and formulas were not made clear through their application to actual structural situations. To this end practically all of the 95 sections of the book contain illustrative examples worked out to the last significant figure or conclusion. By comparing the situations that will be encountered in practice with those considered in the over 150 examples in the book, the reader should be able to acquire, without undue hardship, that "structural sense" that is the goal of every practitioner of the art.

The teacher of structures will also find, at the end of the text, problems subdivided by chapters, with answers to approximately half of them. There are over 400 problems in the book.

All these features should make the book useful to architects and engineers in professional practice, as well as to students.

We are deeply indebted to:

our students and our architectural friends for the insight they have given us in the structural needs of the profession;

Mr. John J. Farrell, who checked all the illustrative examples and most of the problems;

our partner Paul Weidlinger for innumerable discussions on structures, the fruits of which pervade this book;

Dean K. A. Smith of the School of Architecture at Columbia, an old friend and an engineer, for his encouragement in pursuing this approach;

Miss Katherine Anner for carefully typing the manuscript;

Mr. Nicholas Romanelli and Mr. Gregory Hubit of Prentice-Hall, who used their exceptional talents to give this book an appearance conducive to a clear understanding of our ideas;

our families for their patience during the laborious months dedicated to the writing of this book;

those students, architects, engineers, and teachers who will read our book with enough interest to feel they should communicate to us their dissatisfaction so that we may try to improve it through well-directed criticism.

Mario Salvadori
Matthys Levy

Mathematical Formulas

The following formulas represent the entire mathematical knowledge required by the reader for an understanding of the subject matter covered in this book.

1. $(a \pm b)^2 = a^2 \pm 2ab + b^2$ **2.** $a^2 - b^2 = (a + b)(a - b)$

3. $\frac{1}{1 + x} \approx 1 - x \quad (x \ll 1)$ **4.** $\sqrt{1 + x} \approx 1 + \frac{1}{2}x \quad (x \ll 1)$

5. $\frac{1}{\sqrt{1 + x}} \approx 1 - \frac{1}{2}x \quad (x \ll 1)$

6. $x^2 + bx + c = 0; \quad x_{1,2} = -\frac{b}{2} \pm \sqrt{\left(\frac{b}{2}\right)^2 - c}$

7. $\begin{matrix} ax + by = c \\ dx + ey = f \end{matrix} \quad x = \frac{ce - bf}{ae - bd}; \quad y = \frac{af - cd}{ae - bd}$

8. $\sin (\alpha + 90°) = \cos \alpha$ **9.** $\cos (\alpha + 90°) = -\sin \alpha$

10. $2 \sin \alpha \cos \alpha = \sin 2\alpha$ **11.** $\sin \alpha \approx \alpha - \frac{\alpha^3}{6} \quad (\alpha \ll 1)$

12. $\cos \alpha \approx 1 - \frac{\alpha^2}{2} \quad (\alpha \ll 1)$

13. $\frac{dx^n}{dx} = nx^{n-1}$ **14.** $\frac{de^{ax}}{dx} = ae^{ax}$

15. $\frac{d \sin ax}{dx} = a \cos ax$ **16.** $\frac{d \cos ax}{dx} = -a \sin ax$

17. $\int x^n \, dx = \frac{1}{n + 1} x^{n+1} + C \quad (n \neq -1)$

18. $\int \frac{dx}{x} = \ln x + C$

19. $\int \sin ax\, dx = -\frac{1}{a} \cos ax + C$

20. $\int \cos ax\, dx = \frac{1}{a} \sin ax + C$

21. $\int e^{ax}\, dx = \frac{1}{a} e^{ax} + C$

Mathematical Symbols and Abbreviations

LIST OF MATHEMATICAL SYMBOLS

$=$	equal to
$\equiv$	equal by definition to
$\approx$	approximately equal to
$\neq$	unequal to
$<$	less than
$>$	greater than
$\ll$	much less than
$\gg$	much greater than
$\therefore$	therefore
$\| \ \|$	absolute value (positive value of a number)

LIST OF ABBREVIATIONS

English Units

in. = inch
ft = foot
yd = yard (3 ft)
mi = mile (5,280 ft)
in.^{2} or sq in. = square inches
ft^2 or sq ft = square feet
in.^{3} or cu in. = cubic inches
ft^3 or cu ft = cubic feet
cu yd = cubic yards (27 cu ft)
lb = pound
k = kip (1,000 lb)
t = ton (2,000 lb)
pli = pounds per lineal inch
plf = pounds per lineal foot

SI Units

mm = millimeter
m = meter
km = kilometer
mm^2 = square millimeter
m^2 = square meter
mm^3 = cubic millimeter
m^3 = cubic meter
kg = kilogram
N = newton (0.102 kg force)
kN = kilonewton (1 000 newtons)
kg/m = kilograms per lineal meter
kN/m = kilonewtons per lineal meter
kg/m^2 = kilograms per square meter

klf = kips per lineal foot
psi = pounds per square inch
psf = pounds per square foot
ksi = kips per square inch
ksf = kips per square foot
lb/cu in. = pounds per cubic inch
lb/cu ft = pounds per cubic foot
k/cu yd = kips per cubic yard
in·lb = inch-pounds
ft·lb = foot-pounds
ft·k = foot-kips
sec = second (of time)
°F = degrees Fahrenheit
′ = minutes of a degree (angle)
″ = seconds of a degree (angle)
o.c. = on center
kN/m^2 = kilonewtons per square meter
Pa = pascal (N/m^2)
kPa = kilopascal
MPa = megapascal (N/mm^2)
kg/m^3 = kilograms per cubic meter
kN/m^3 = kilonewtons per cubic meter
kN·m = kilonewton-meter
°C = degrees Celsius (degrees centigrade)

Conversion Tables

SI TO ENGLISH UNITS

kilometer	= 0.62 mile
meter	= 3.28 feet
millimeter	= 0.0394 inch
newton	= 0.225 pounds (force)
kilogram	= 2.2 pounds (mass)
square meter	= 10.76 square feet
square millimeter	= 0.00155 square inch
cubic meter	= 35.3 cubic feet = 1.3 cubic yard
cubic millimeter	= 0.000061 cubic inch
kilonewton per meter	= 68.53 pounds per foot
kilonewtons per square meter	= 20.88 pounds per square foot
megapascal	= 145 pounds per square inch
kilonewton per cubic meter	= 6.365 pounds per cubic foot
kilonewton-meter	= 738 foot-pounds
°Celsius, °C	= $\frac{5}{9}$(°F − 32)

ENGLISH TO SI UNITS

mile	= 1.61 kilometers
foot	= 0.305 meter
inch	= 25.38 millimeters
pound (force)	= 4.44 newtons
pound (mass)	= 0.45 kilogram
square foot	= 0.093 square meter
square inch	= 645.2 square millimeters
cubic foot	= 0.014 6 cubic meter
cubic yard	= 0.769 cubic meter
cubic inch	= 16 393 cubic millimeters

pounds per foot	= 0.014 6 kilonewton per meter
pounds per square foot	= 0.047 9 kilonewton per square meter
pounds per square inch	= 0.006 9 megapascal
pounds per cubic foot	= 0.157 kilonewton per cubic meter
foot-pound	= 0.001 355 kilonewton-meter

The SI System of Units

The units used in this book are primarily those of the SI system, the system of measurements adopted worldwide (with the major exception of the United States). Although also known as the "metric system," the SI, or International System of Units, differs in one major respect from the old European metric system in that it uses different units for *mass* and for *force*.

The primary SI unit for *mass* is the kilogram (kg); the SI unit for *force* is the newton (N). The introduction of the newton assures internal consistancy for all the SI units and adds clarity of concept to all the well-known advantages of the decimal measurement base inherent in the metric system.

Throughout this book multiples and submultiples are given in powers of 1 000. For instance, the preferred units of length are the millimeter, the meter, and the kilometer. The prefixes used throughout the book for the following three multiples are:

mega (M)	1 000 000
kilo (k)	1 000
milli (m)	0.001

(This approach eliminates the centimeter as a common unit of measurement.)

In this book, the following standard SI conventions have been adopted:

(a) For designating thousands or multiples of thousands, a space is used rather than a comma. For instance, 1,005,010 is written 1 005 010.

(b) Decimal points are used to indicate fractional numbers, and numbers smaller than unity are always preceded by a zero: 74.26, 0.15.

Since many construction industry standards in the United States have not yet been converted to the SI system, a "soft conversion," i.e., an approximate conversion as used for instance, as in Table A.1, Steel Wide Flange Sections given in Appendix A. Elsewhere, "hard conversions," i.e., exact conversions from given SI units have been shown, as for instance, as in Table 2.4.1, Properties of Concrete, which shows 20 MPa (2,900 psi), rather than 20.7 MPa (3,000 psi). The hard conversions adopted in this book are based on the standards developed by other English-speaking countries that have recently changed to the SI system, such as Canada and Australia.

One of the difficulties inherent in changing from one measurement system to another is the loss of immediate recognition of physical magnitudes, which comes about through familiarity. It is hoped that by being shown SI units followed by English units in parentheses the reader will gain familiarity with magnitudes represented by the SI numbers and begin to recognize them without resorting to the mental exercise of translation.

1. Loads

1.1 LOAD CHARACTERISTICS [2.2, 2.5, 2.6]*

The primary purpose of a structure is to enclose or define a space. In so doing, the structure becomes subjected first to its own weight, the *dead load*, and secondly to loads due to other natural forces or conditions (such as wind and snow, earthquakes, or differences in temperature) and to loads imposed by its occupants, the *live loads*. In considering a load, one must not only define its intensity, in kilograms (pounds), but state how the load is applied to the specific structure under study.

Due to the elastic behavior of structural materials (see Section 2.1), every structure deflects under load and returns to its original shape when the load is removed. Hence, structures have a tendency to oscillate. The time it takes a structure to perform a complete oscillation is called its *period*. A structure usually has any number of periods, the longest of which is called its *fundamental period*. Fundamental periods of buildings vary from 0.2 sec for low, stiff buildings to as much as 9 or more seconds for tall, flexible buildings. When a load is applied to a structure in a time *much longer* than its fundamental period, the load is said to act *statically* on the structure; the load is then completely defined by its intensity, direction, and location. The pressure of a wind gust growing from zero to its maximum in 3 sec is a static force for a short, stiff building. When the time of application of the load is *short* in comparison with its fundamental period, the load is said to act *dynamically*. The pressure of the same wind gust acts dynamically on a skyscraper with a period of 9 sec.

Dynamic loads must be defined in terms of their maximum intensity and

*Numbers in square brackets following section titles refer to sections of *Structure in Architecture: The building of buildings* (2nd Edition) by M. Salvadori and R. Heller, Prentice-Hall, Inc., Englewood Cliffs, N.J., where the same topic is treated on a nonmathematical, intuitive basis.

their time variation. Two variations in time are particularly important. A load which grows to its maximum intensity very rapidly is called an *impact load* and may be equivalent to a static load of much greater intensity. A load which increases and decreases in time periodically is called an *oscillatory load* and may be particularly dangerous when its period of oscillation coincides with one of the periods of the structure. In this case, the oscillatory load is said to be *in resonance* with the structure, and, if it is applied to the structure for a long enough time, it may produce deflections comparable to those due to a static load of much greater intensity. The weight of an elevator stopping rapidly imparts to the supporting structures an impact load which may equal twice the weight of the elevator, thus having a dynamic *magnification factor* of 2. A rotating piece of machinery may impart both vertical and horizontal oscillatory loads to a structure. Such oscillations are particularly dangerous if they are in resonance with the structure.

The dynamic loads produced by rapid movements of the soil, such as those due to earthquakes, depend essentially on the elasticity and the weight of the structure. They are dealt with in Section 1.7.

Temperature differences between various parts of a structure produce relative expansions (or contractions), which result in loads only if they are partially or totally prevented. The stresses developed when thermal expansions are prevented by the stiffness of the structure may require at times a modification of the structure. Examples of such conditions are considered throughout this book.

1.2 DEAD LOAD [2.3]

The weight of the materials used to construct a building is called its *dead load.* It includes the weight of the structural frame, the weight of fixed mechanical equipment and of distribution systems as well as that of all finishes, partitions, and walls. (Partitions are often included among the live loads by building codes, and a uniformly distributed load is designated to approximate the actual partition loads.) The dead load must be estimated before the design of a structure can be started. Such an estimate can be made with the aid of Table 1.2.1. For example, if it is required to design the floor of an office building, consisting of beams and a concrete slab, the following loads on the beams are estimated from the table:

	N/m²	lb/sq ft	
Vinyl tile finish	50	1	
75 mm (3 in.) lightweight concrete fill	510	27	(for electrical
25 mm (1 in.) plaster ceiling	480	10	distribution ducts)
150 mm (6 in.) concrete slab	3 450	75	
Subtotal	4 490	113	
Partitions	1 000	20	
Total	5 490	133	

Table 1.2.1 Weights of Building Materials

Materials	N/m²	lb/sq ft
Ceilings:		
Channel suspended system	50	1
Lathing and plastering	*See* Partitions	
Floors:		
Steel deck	100–500	2–10
Concrete—plain, 10 mm		
Stone	230	5
Lightweight	50–170	1–4
Finishes		
Terrazzo, 10 mm	240	5
Ceramic or quarry tile, 20 mm	500	10
Plastic tile, 6 mm	50	1
Wood, 20 mm	100–200	2–4
Roofs:		
Copper, tin, or steel	50–240	1–5
5-ply felt and gravel	290	6
Shingles		
Wood	100	2
Asphalt, cement asbestos	140	3
Clay tile	430–670	9–14
Slate, 6 mm	480	10
Sheathing		
Wood or gypsum, 20 mm	150	3
Insulation, 10 mm		
Rigid	30	0.6
Partitions:		
Clay tile		
100 mm	860	18
200 mm	1 630	34
Gypsum block		
50 mm	455	9
100 mm	600	12
150 mm	885	18
Wood studs 50 × 100 mm (400 mm o.c.)	100	2
Steel partitions	200	4
Plaster, 10 mm		
Cement	190	4
Gypsum	95	2
Lathing		
Metal	25	0.5
Gypsum board, 10 mm	75	1.5
Walls:		
Brick		
100 mm	1 950	40
Hollow concrete block (heavy aggregate)		
150 mm	2 100	43
200 mm	2 700	55
300 mm	3 900	80
Hollow concrete block (light aggregate)		
150 mm	1 500	30
200 mm	1 900	38
300 mm	2 700	55
Stone, 100 mm	2 680	55
Glass block, 100 mm	880	18
Windows (glass, frame, and sash)	400	8

Materials	Weight		
	kg/m³	kN/m³	lb/cu ft
Aluminum	2 770	27.2	173
Brick	1 920	18.9	120
Steel	7 850	77.0	490
Wood	640	6.3	40
Concrete	2 245	22.0	140
Sand	1 539	15.0	96
Glass	2 646	25.9	165

1.3 USE OR OCCUPANCY LOADS [2.4]

The primary purpose of a floor structure is to support the weight of people, furniture, machines, or stores. These loads are essentially variable: a person can stand anywhere in a room, can move about, can stand in a group, furniture is movable and can be placed anywhere in a room, etc. It is obviously impossible to consider separately all the conceivable loading

Table 1.3.1 MINIMUM UNIFORMLY DISTRIBUTED LIVE LOADS

Occupancy or use	Live load kN/m²	Live load lb/sq ft
Armories and drill rooms	7.2	150
Assembly halls and other places of assembly:		
Fixed seats	2.9	60
Movable seats	4.8	100
Balcony (exterior)	4.8	100
Bowling alleys, poolrooms, and similar recreational areas	3.6	75
Corridors:		
First floor	4.8	100
Other floors, same as occupancy served except as indicated		
Dance halls	4.8	100
Dining rooms and restaurants	4.8	100
Garages (passenger cars)	2.4	50
Gymnasiums, main floors, and balconies	4.8	100
Hospitals:		
Operating rooms, laboratories	3.6	75
Private rooms and wards	1.9	40
Libraries:		
Reading rooms	2.9	60
Stack rooms	7.2	150
Manufacturing	6.0	125
Office buildings:		
Offices	3.8	80
Lobbies	4.8	100
Penal institutions:	4.8	100
Cell blocks	1.9	40
Corridors	4.8	100
Residential:		
Multifamily houses:		
Private apartments	1.9	40
Public rooms	4.8	100
Corridors	2.9	60
Dwellings	1.9	40
Hotels:		
Guest rooms	1.9	40
Public rooms	4.8	100
Corridors serving public rooms	4.8	100
Private corridors	1.9	40
Reviewing stands and bleachers	4.8	100
Schools:		
Classrooms	1.9	40
Corridors	4.8	100
Sidewalks, vehicular driveways, and yards, subject to trucking	12.0	250
Skating rinks	4.8	100
Stairs, fire escapes, and exitways	4.8	100
Storage warehouse, light	6.0	125
Storage warehouse, heavy	12.0	250
Stores:		
Retail:		
First-floor, rooms	4.8	100
Upper floors	3.6	75
Wholesale	6.0	125
Theaters:		
Aisles, corridors, and lobbies	4.8	100
Orchestra floors	2.9	60
Balconies	2.9	60
Stage floors	7.2	150
Yards and terraces, pedestrians	4.8	100

conditions. Hence, a statistical approach is used to define a uniformly distributed static load which is safely equivalent to the weight of the maximum concentration of occupants expected for a particular use. These loads are defined in codes, e.g., the ANSI Code*, as equivalent distributed live loads and are given in Table 1.3.1. In addition to these distributed loads, concentrated loads are defined for certain occupancies and are given in Table 1.3.2.

Table 1.3.2 CONCENTRATED DESIGN LOADS FOR FLOORS

Use of area of floor	Minimum concentrated design load	
	kN	lb
Floors and areas used by passenger cars	11	2,500
Office space	9	2,000
Driveways or sidewalks adjacent to driveways over basements, cellars, or other open areas	54	12,000

Moreover, certain structural elements are required to resist horizontal as well as vertical loads due to their use. Bleachers, for instance, are subject to lateral motions due to the swaying together of a group of spectators: a lateral force of 0.36 kN/m (24 lb/ft) of seats must thus be considered. Similarly, handrails and balcony railing must be designed for a horizontal force perpendicular to the top edge of 0.75 kN/m (50 lb/ft).

In a multistory building, it is unlikely that all floors will be fully loaded by a specified live load. Similarly, there is little likelihood of a structural element supporting a large floor or roof area having to carry the total live load specified for that area. Due to the low probability of these occurrences, codes generally permit live load reductions to be used in the design of columns in multistory buildings and girders supporting large floor areas. For instance, the ANSI recommendations specify:

> For live loads 4.8 kN/m² (100 psf) or less, the design live load on any member supporting 14 m² (150 sq ft) or more may be reduced at the rate of 0.86% per m² (0.08% per sq ft) of area supported. The reduction shall not exceed 60%, or:
>
> $$R = 23\left(1 + \frac{D}{L}\right), \tag{1.3.1}$$
>
> where R = % reduction,
> D = dead load per m² (sq ft),
> L = live load per m² (sq ft).

Minimum Design Loads in Buildings and Other Structures, ANSI A58.1, American National Standards Institute, Inc., New York, 1972.

1.4 SNOW LOADS [2.4]

Snow loads on the roofs of buildings vary greatly depending on geographic location and elevation above sea level. In the United States, for example, the snow load at low elevations ranges from 0 to 1.9 kN/m² (40 psf), while at high elevations a load of 4.8 kN/m² (100 psf) may be encountered. The snow map of Fig. 1.4.1 is based on reports accumulated over a period of many years. Special local conditions, particularly in mountainous areas, can be evaluated by securing maximum snow accumulation from local weather bureaus. This accumulation can be converted to load by using a weight of 0.1 kN/m² per meter (0.5 psf per inch) of snow depth. [This weight has been found to vary from 0.08 to 0.18 kN/m² (0.4 to 0.9 psf), depending on the density of the snow, with the lighter snow occurring at higher elevations.]

The snow load is usually measured on the horizontal projection of the roof. The slope of the roof obviously affects the retention of snow; the greater the slope, the lesser the retention. Figure 1.4.2 shows the reduction in load per unit of horizontal projection from the basic snow load, depending on the slope of the roof.

The shape of the roof, that is, whether it is flat, curved, hipped, etc., must also be considered in determining the snow load. Accumulations in valleys, around monitors, and at parapets will require special consideration, since snow will always drift where an obstruction is present. In this case, the direction of prevailing winds will often indicate to the designer where to expect the maximum accumulation.

Example 1.3.1. In a 14-story office building in Washington, D.C., with a column spacing of 4 m (13 ft) in both directions, what is the total live load for which a column must be designed at the ground floor? (Assume equation (1.3.1) does not control.)

From Fig. 1.4.1, the snow load on the roof is 1 kN/m² (20 psf). The design live load for a typical floor is 3.8 kN/m² (80 psf), from Table 1.3.1. This live load is reduced in accordance with the provisions of the ANSI recommendation and results in the following:

Floor	Roof	14	13	12	11	10	9	8	7	6	5	4	3	2	Total
% Reduction:	13.8	27.5	41.3	55.0	60	60	60	60	60	60	60	60	60	60	
Full live load (kN):	16	60.8	60.8	60.8	60.8	60.8	60.8	60.8	60.8	60.8	60.8	60.8	60.8	60.8	806.4
Reduced live load (kN):	13.8	44.1	35.7	27.4	24.3	24.3	24.3	24.3	24.3	24.3	24.3	24.3	24.3	24.3	364

The resulting design live load of 364 kN represents a 55% reduction from the full live load specified for the floors of the building.

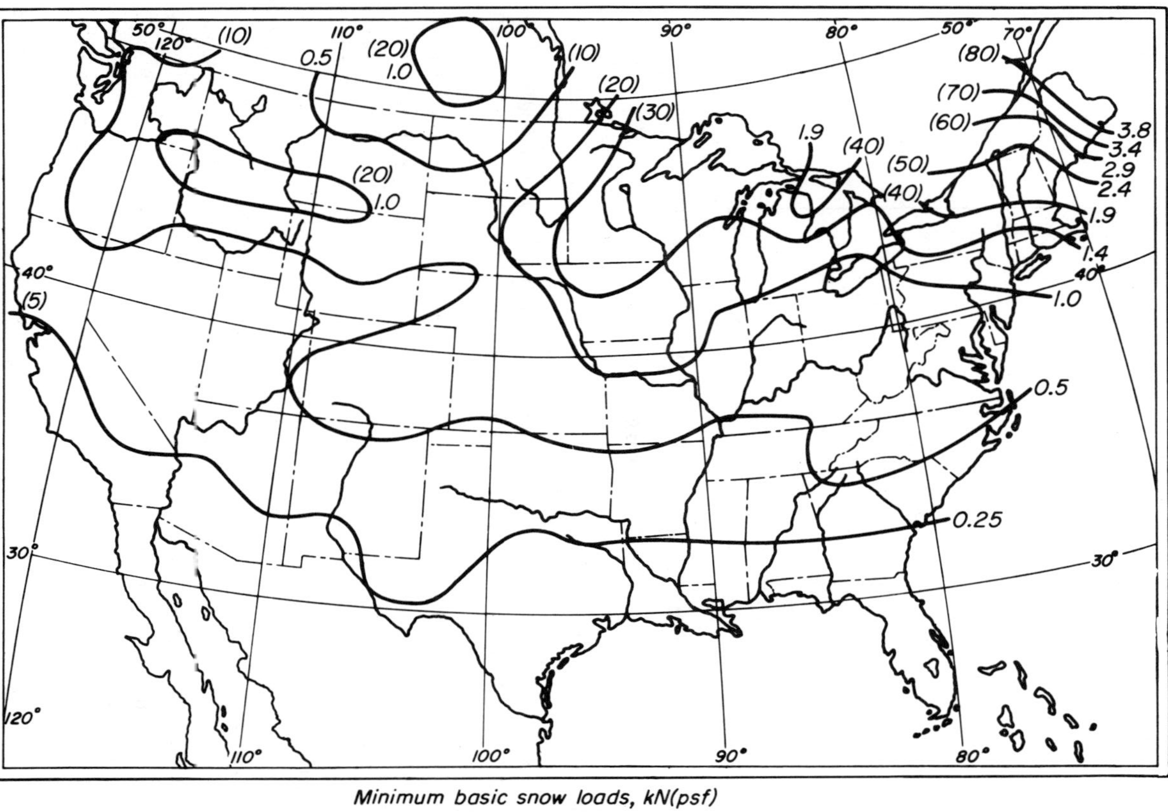

Minimum basic snow loads, kN(psf)

FIGURE 1.4.1

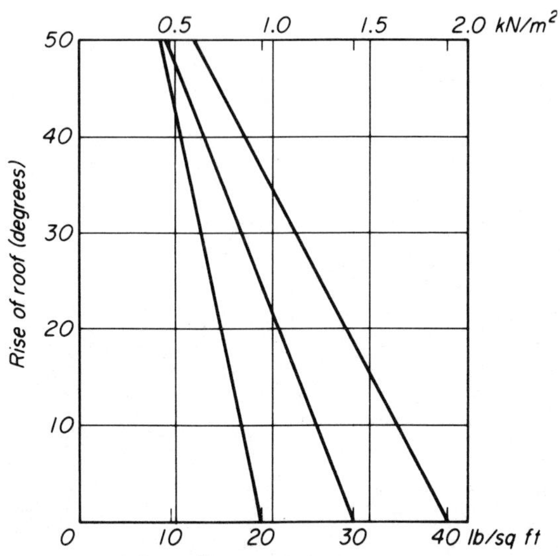

FIGURE 1.4.2

1.5 WIND LOADS [2.4]

Wind loads depend on the velocity of the wind, on the shape of the building, its height, its geographic location, the quality of its surface, and the stiffness of its structure.

Much of what is known today about the response of structures to wind loads is the result of investigations conducted into the failure of both buildings and bridges. This points out the importance of the effect of wind forces. Basic wind forces are obtained by considering the maximum wind velocity in a particular locality. The map shown in Fig. 1.5.1 presents contours of the maximum wind velocity 10 m (30 ft) above ground with a 100-year period of recurrence. This basic velocity increases with the height above ground, as shown in Fig. 1.5.2 for the case of a suburban area. A similar but somewhat lesser increase is experienced in centers of cities and a larger increase in open country, but the values shown in the figure may always be used for preliminary design. In most municipal building codes the continuous change in wind pressure with height is usually simplified into a stepped change (Fig. 1.5.3).

Knowing the velocity V, the force acting on a structure is obtained from the equation:*

$$p = 0.000\,047\,3\; C_D V^2, \qquad (1.5.1)$$

*"Wind Forces on Structures," American Society of Civil Engineers, *Transactions*, Vol. 126, Part II, 1961.

km/hr	mph
97	60
108	67
121	75
129	80
137	85
145	90
153	95
161	100
177	110
193	120

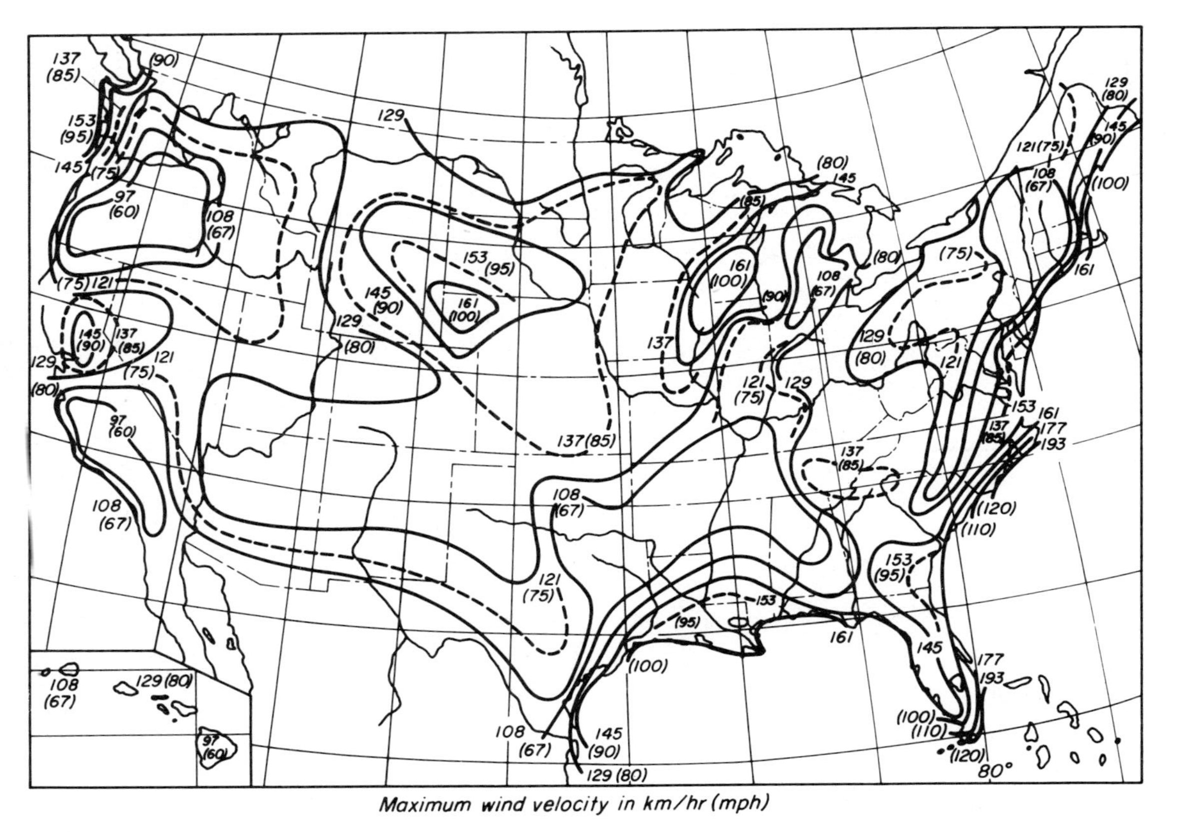

Maximum wind velocity in km/hr (mph)

FIGURE 1.5.1

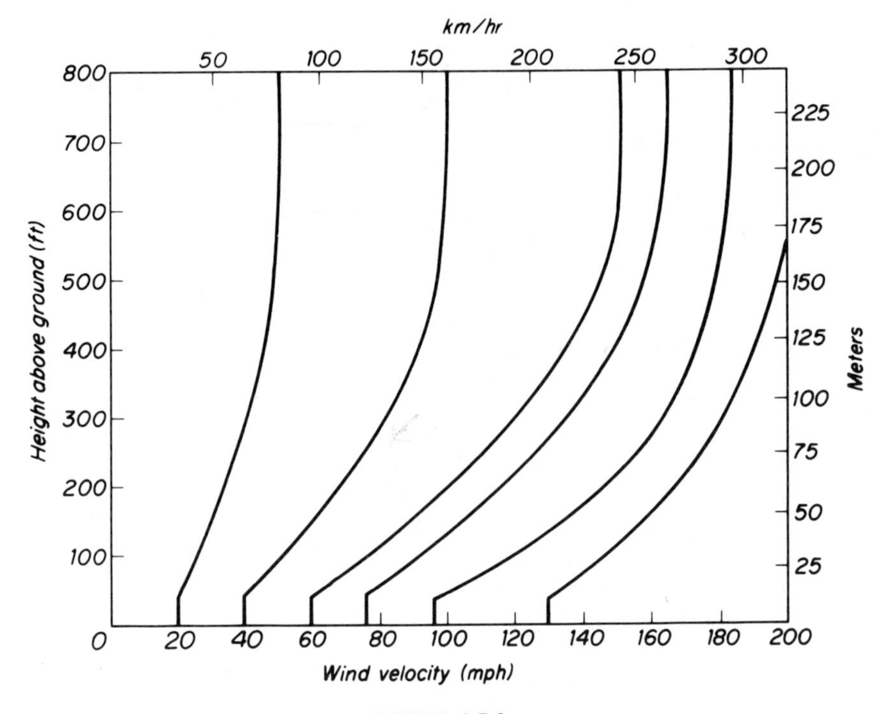

FIGURE 1.5.2

where p = pressure in kN/m² of vertical projection,
C_D = shape coefficient,
V = velocity in km/hr,

or, from the equation:

$$p = 0.002\,558\ C_D V^2, \tag{1.5.1a}$$

where p is measured in psf and V in mph.

Since both pressure and suction forces are exerted by the wind on most structures, pressure and suction shape coefficients have been determined experimentally for various types of buildings, roofs, and shelters. For example, for a box-shaped structure the pressure coefficient on the windward side is 0.8 and the suction coefficient on the leeward side is 0.6, with a combined shape coefficient of 1.4. For a basic wind velocity of 129 km/hr (80 mph), (1.5.1) or (1.5.1a) indicates a pressure of 0.78 kN/m² (16.4 psf). For the same basic velocity, but at an altitude of 100 m (300 ft) above ground, Fig. 1.5.2 shows a velocity of 240 km/hr (150 mph), which corresponds to a pressure of 2.72 kN/m² (57 psf). Multiplying these values by the shape factor,

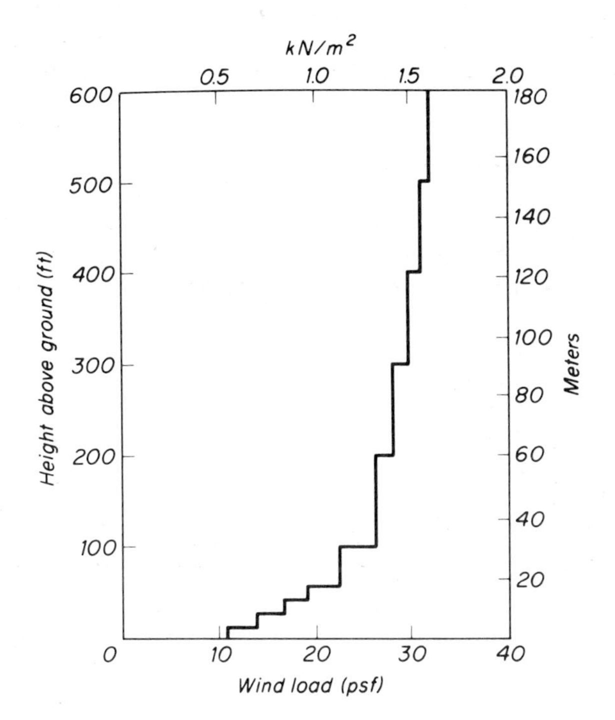

FIGURE 1.5.3

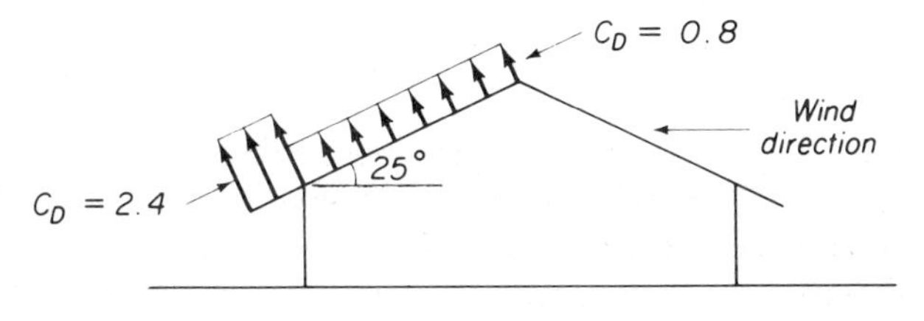

FIGURE 1.5.4

we obtain the design wind pressures on a building of 1.09 kN/m² at 10 m and 3.81 kN/m² at 100 m (23 psf at 30 ft and 80 psf at 300 ft).

Roofs of buildings are also subject to wind pressures or suctions, depending on the shape of the roof. For instance, a shallow (25° angle) gabled roof has an external suction coefficient of 0.8, whereas an overhanging eave has a suction coefficient of 2.4 (Fig. 1.5.4).

In addition to the purely static effects described above, light, elastic structures such as stacks, towers, and cable-supported structures exhibit a tendency to oscillate or "flutter" under certain wind conditions. The nature of this problem is too complex to be treated in a simplified manner; the reader is referred to works in this field for the analysis of such structures.

1.6 EARTH AND WATER LOADS

Structures below ground are required to resist the lateral pressure of earth, which for dry soils is equal to approximately 4.7 kN/m² for every meter (30 psf per foot) of depth below ground. For a basement wall which extends

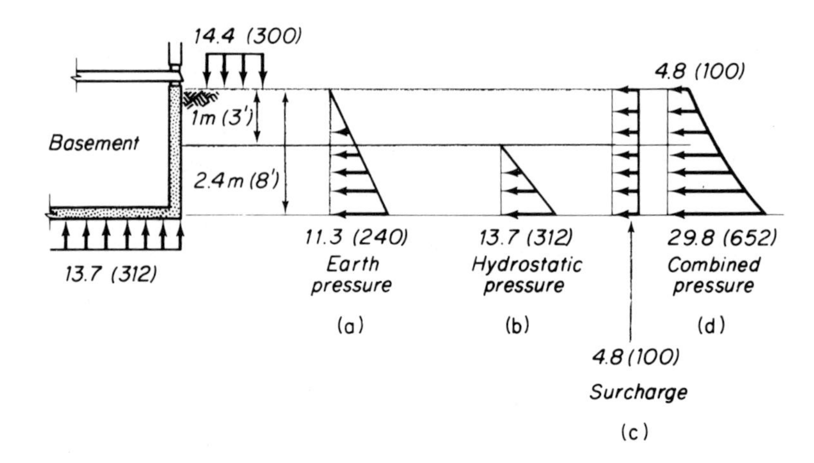

FIGURE 1.6.1

2.4 m (8 ft) below grade, this means a pressure distribution which varies from zero at the top to 11.3 kN/m² (240 psf) at the bottom [Fig. 1.6.1(a)].

If the structure is submerged partially or wholly in water, it must resist, in addition, the hydrostatic pressure of water, which is 9.8 kN/m² per meter (62.4 psf per foot) of depth. If, in the example above, the ground water table lies 1 m (3 ft) below grade, the basement wall must resist a water pressure which varies from zero at a point 1 m (3 ft) below grade to 13.7 kN/m² (312 psf) at the bottom [Fig. 1.6.1(b)]. Furthermore, the basement slab will also have to resist an upward pressure equal to 13.7 kN/m² (312 psf) [Fig. 1.6.1(e)].

If the ground surface adjacent to the basement wall is surcharged by, for instance, vehicles on a street, additional lateral wall pressures will result. These are often approximated by assuming that the additional lateral pressure is one-third of the surcharge loading [see Fig. 1.6.1(c)].

1.7 EARTHQUAKES [2.6]

The devastating effects of earthquakes have been felt since earliest recorded time. No area of the earth is totally free of earthquakes, but certain regions are subject to periodic tremors of destructive intensity. Earthquakes are classified according to their destructive intensity by the Modified Mercalli scale (MMI) with a range from 1 to 12 which is related to observable phenomena such as:

MMI 6

Everybody runs outdoors; damage to buildings varies depending on quality of construction; noticed by drivers of autos.

For the United States, the regions or zones of equal risk of earthquake occurrence of a particular intensity are shown in Fig. 1.7.1. The five zones which are correlated to the MMI scale range from 0 for intensity less than MMI 5, to 4 for intensity greater than MMI 8. These zone designations have been arrived at by an analysis of both recorded tremors and site geology.

Earthquakes are, by their very nature, random dynamic phenomena. The record of ground shaking during an earthquake, when measured in terms of ground acceleration versus time, shows a seemingly random trace (Fig. 1.7.2). At times, the force felt by a building can be approximated by multiplying the maximum ground acceleration by the mass of the structure:

$$F = Ma, \tag{1.7.1}$$

where $M = W/g$ (W = weight of structure, g = acceleration of gravity).

However, this is only true for an infinitely rigid building resting on a rigid foundation. Since all structures are flexible to some degree, the actual force may be more or less than that defined by equation (1.7.1). This relationship between structures and earthquakes can be clarified by constructing a *response spectrum.*

To understand the concept of a response spectrum, imagine lining up a series of oscillators, consisting of a mass on a spring, in order of increasing natural periods [Fig. 1.7.3(a)], and subjecting these to shaking by a seismic force. Each oscillator will respond in its own specific manner, which can be measured in terms of its maximum acceleration, velocity, or displacement versus its period. The graph of the accelerator responses versus their periods constitutes a so-called response spectrum [Fig. 1.7.3(b)]. Note that a simple oscillator with a zero period (infinitely rigid spring) has the same response as the ground motion. For higher period oscillators, the response is either greater or less than that of the ground motion.

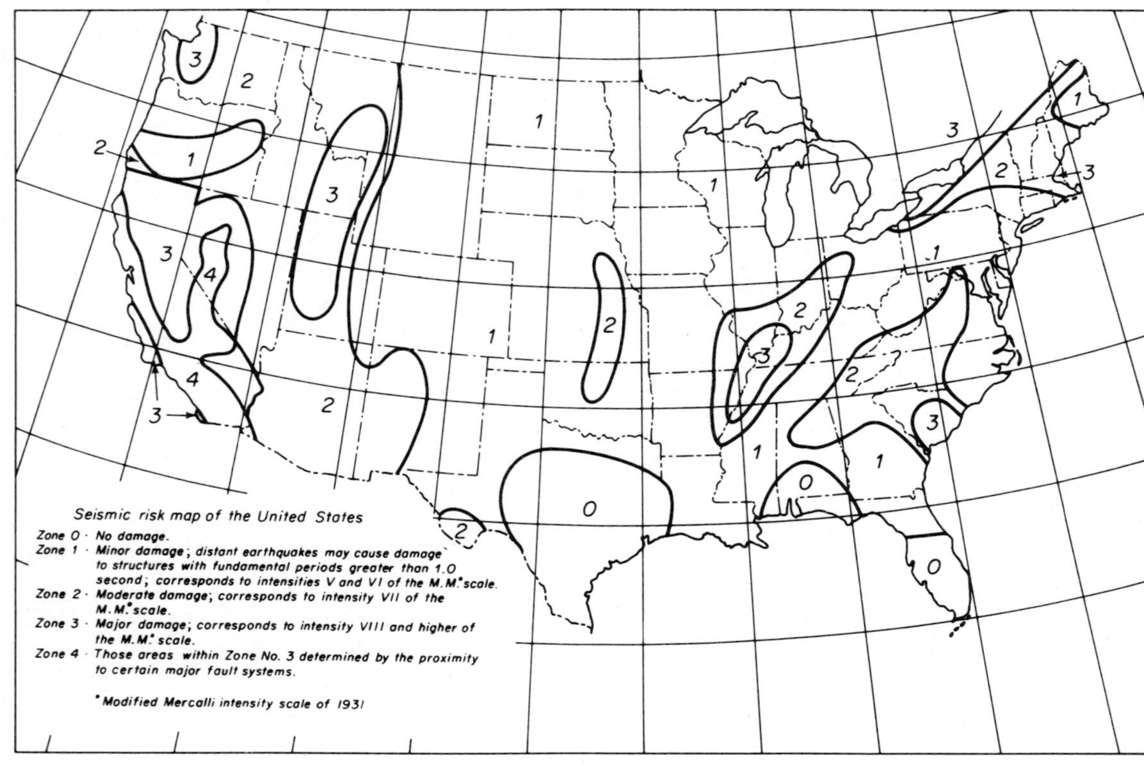

Seismic zone map of the United States

FIGURE 1.7.1

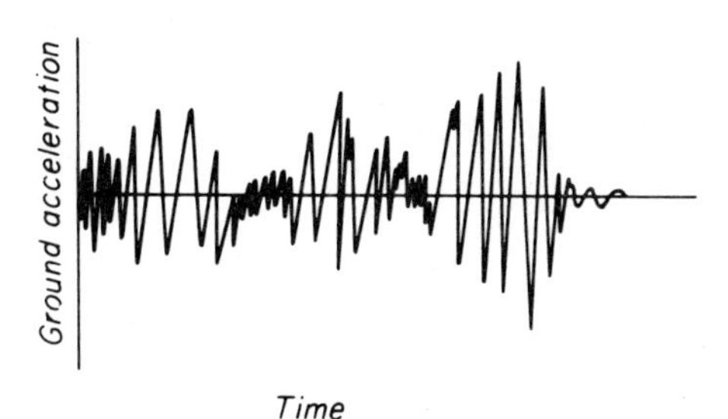

FIGURE 1.7.2

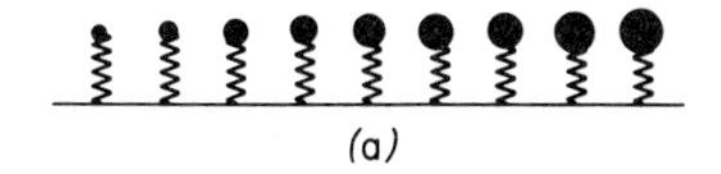

(a)

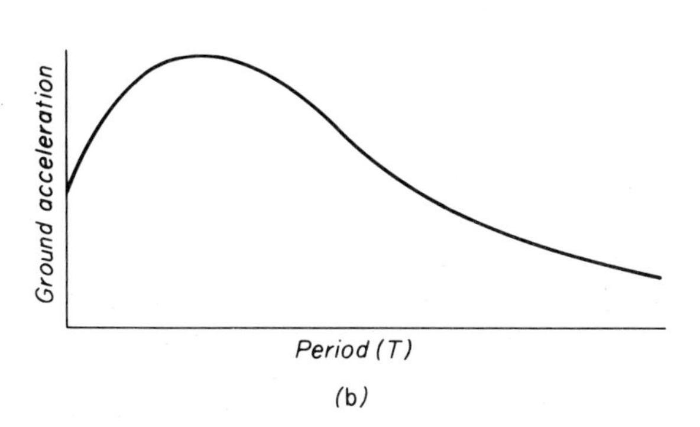

(b)

FIGURE 1.7.3

A real building is a complex structure that vibrates simultaneously at various periods. Its lowest period of vibration is called its *fundamental period*. The deformed shape of the structure, which corresponds to the fundamental period, is called the *first mode*. Each higher mode has a distinct mode shape (Fig. 1.7.4) and contributes to the bending, shear, and axial forces experienced by the structure. For a complex structure, a dynamic analysis must be performed to determine these mode shapes which must be added (in a complex manner) in order to provide a good respresentation of the behavior of the structure. In many cases, however, a simplified approach can be taken, which recognizes that the primary contribution to the response of the structure is fairly well represented by the first mode. The Uniform Building Code, for

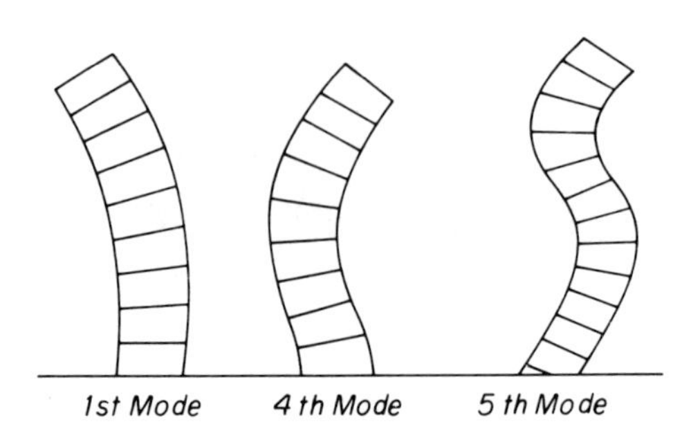

FIGURE 1.7.4

instance, defines the horizontal force acting on a building [in the same form as (1.7.1)] as:

$$F = WZKCSI, \qquad (1.7.2)$$

where W = total dead load,
Z = seismic probability zone factor,
= 0 in zone 0,
= 3/16 in zone 1,
= 3/8 in zone 2,
= 3/4 in zone 3,
= 1 in zone 4.

The factor K depends on the type of structure and varies from 0.67 for ductile frames to 1.33 for shear walls. A *ductile frame* is one in which the element resisting the lateral force is a rectangular grid of beams and columns that are rigidly connected to each other but which can deform without failure beyond the yield point of the material (see Chapter 2). In a shear wall type of structure, the horizontal forces are carried either by walls extending the full height of the structure, essentially without openings, or by vertical trusses.

The factor C is a function of the fundamental period of the structure (T), which itself depends on the stiffness and the weight of the structure. For a ductile moment resisting frame, T can be approximated by:

$$T = 0.1N \quad \text{sec}, \qquad (1.7.3)$$

where N is the total number of floors above grade.

For all other structures, T is given approximately by:

$$T = \frac{0.09h}{\sqrt{D}}, \qquad (1.7.4)$$

where h = height of building above grade (m),
D = depth of the building in the direction of the earthquake force (m),

$$\text{or:} \quad T = \frac{0.05h}{\sqrt{D}}, \quad \text{if } h \text{ and } D \text{ are given in feet.} \tag{1.7.4a}$$

The seismic coefficient C is then expressed as a function of T as:

$$C = \frac{1}{15\sqrt{T}}. \tag{1.7.5}$$

The factor S depends on the soil conditions at the site and varies from 1.0 for rock or stiff soils less than 200 feet deep overlying rock, to 1.5 for soft to medium stiff clays and sands.

The factor I is an importance factor, which is 1.0 for ordinary buildings, 1.25 for places of assembly, and 1.5 for essential facilities like hospitals.

The horizontal force F given by (1.7.2) is distributed over the height of the building in two parts. First, a concentrated force F_1, equal to 0.07 TF, but not larger than $0.25F$, is applied at the top of the building. The balance of the force, $F - F_1$, is distributed as a triangular load diminishing to zero at its bottom (Fig. 1.7.5).

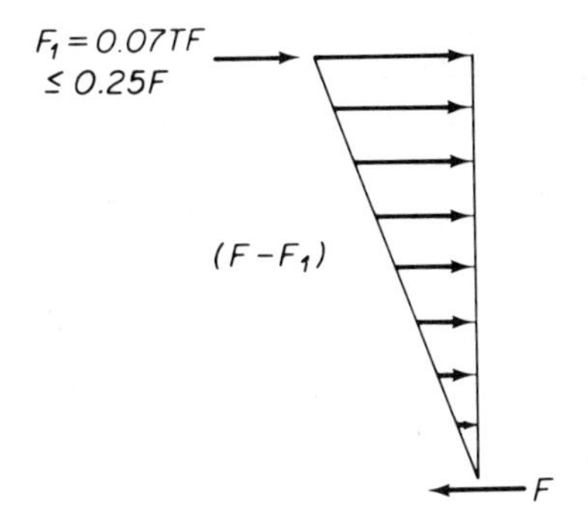

FIGURE 1.7.5

PROBLEMS

1.1 A 4-story office building is to be designed in northern Texas. List all the superimposed loads that must be considered in its design.

1.2 The basement of a building is 3.5 m (12 ft) below grade. Ground water is located 2.75 m (9 ft) below grade. What thickness concrete slab is required to exactly balance the hydrostatic uplift?

1.3 The inside dimensions of a concrete pit below the basement of Problem 1.2 are 1.2 m × 1.2 m × 1.2 m (4 ft × 4 ft × 4 ft). What are the outside dimensions necessary to exactly balance the hydrostatic uplift, if its walls are 150 mm (6 in.) thick?

1.4 An apartment tower to be built in southern Florida is 75 m (250 ft) high and has plan dimensions of 20 m × 40 m (70 ft × 140 ft). What are the wind shears and moments at the base in the direction of each tower wall? (Use 25 m stepped increments.)

1.5 What is the minimum snow load to be considered for the design of a barn to be built in northern Michigan, if the slope of its roof is 20°?

1.6 A steel television tower 120 m (400 ft) high and pyramidal in shape is to be built in the northwest corner of Washington State. The base dimensions of the tower are 15 m × 15 m (50 ft × 50 ft) and its dimensions at the top are 3 m × 3 m (10 ft × 10 ft). What are the wind shears and moments at the base? What is the maximum tension or compression in each of the four legs? (*Note:* Since the tower faces are trussed and not solid, it is sufficiently accurate for preliminary design to assume a ratio of solid to total area between 20% and 40%, say 30%.) (Use 30 m stepped increments.)

1.7 What are the base shears due to an earthquake for the tower of Problem 1.6? The assumed average weight of the tower is 5 kN/m (350 lb per foot) of height. Assume foundations on rock and an ordinary building.

1.8 Compute the total weight of a concrete office building 35 m × 40 m (120 ft × 140 ft) in plan dimension, 110 m (360 ft) and 30 stories high, with 10% corridor area and 3% column area, assuming an average slab thickness of 200 mm (8 in.). The building is to be built in northern New York state on a 250 kN/m² (2½ ton/sq ft) soil. Determine the width of its foundation mat, assuming the length to be 42 m (147 ft) and the thickness to be 600 mm (2 ft). (*Note:* The contributary area for the columns is 18 m² (200 sq ft) for LL reduction.)

2. Structural Materials

2.1 MATERIAL PROPERTIES* [3.1]

The structural behavior of materials can be studied in terms of a set of idealized properties. These are homogeneity and isotropy, elasticity, plasticity, hardness, brittleness, stiffness, and ductility.

A material whose properties are identical throughout is called *homogeneous*. Wood, in which the density varies between heartwood and sapwood and in the neighborhood of knots, is obviously nonhomogenous. (However, it is often idealized as a homogenous material for the purpose of defining its structural behavior.) If the properties of a material are identical in all directions, the material is called *isotropic*. The fibrous nature of wood categorizes it as a nonisotropic material.

All materials deform under load. If the deformation disappears when the load is removed, a material is said to be *elastic*. If the deformation remains after removal of the load, the material is said to behave *plastically*. Most materials have both an elastic and a plastic range, depending on the load intensity. Most materials are used structurally within their elastic range, in order to avoid permanent deformations. If the deformations of an elastic material are proportional to the applied load, its behavior is termed *linearly elastic*. The stiffness of a material is defined by its *elastic modulus*, which is the ratio of the *stress* f or force per unit of area to the corresponding *strain* ϵ or deformation per unit length:

$$E = \frac{f}{\epsilon}. \qquad (2.1.1)$$

This relationship is known as *Hooke's law*.

**Mechanics of Deformable Solids*, Irving Shames, Robert Krieger Press, New York, N.Y., 1979.

Ductile materials deform plastically before breaking, while *brittle* materials have no plastic range. Steel is ductile; cast iron is brittle. Ductility is a basic requirement of structural materials. It permits readjustment of stress with elimination of stress concentrations (see Chapter 13), and gives warning of impending failure through large deformations in the plastic range. One measure of the structural efficiency of a material is given by its *strength-density ratio* or *specific strength:*

$$k_1 = \frac{f}{\gamma},$$

where f = ultimate stress, MPa (lb/sq in.),
γ = specific gravity, kg/m^3 (lb/cu in.).

The higher the specific strength, the stronger the material is on a kilogram per kilogram basis. The deformation of a structure under its own weight can be measured by its *specific elasticity:*

$$k_2 = \frac{E}{\gamma}.$$

Two materials with equal specific elasticity will deform identically under their own weight. For superimposed loads, however, the deformations will be in the ratio of the moduli of elasticity of the two materials. For instance, consider a steel and an aluminum structure which are geometrically identical and identically loaded. The specific elasticities of steel and aluminum are approximately equal; the elastic modulus of aluminum is one-third that of steel (Fig. 2.1.1). The deformations due to the weight of the two structures will be equal, but under superimposed loads the aluminum structure will deform three times as much as the steel structure.

2.2 STEEL [3.3]

Steel is possibly the most efficient and certainly one of the most used structural materials. It can be formed into structural shapes, such as wide flange beams, or into plates or sheets by rolling. It can be cast into complex shapes, like those of bridge bearings. It can be bolted, riveted, or welded. It can be alloyed with other metals, such as chromium, nickel, or copper, to obtain an increased resistance to corrosion, or with metals like manganese or silicon to obtain an increase in strength.

Steel is one of the few structural materials which demonstrates a well defined *yield stress*, i.e., a stress above which it yields or flows with almost no increase in stress, as shown in Fig. 2.2.1. The small increase in stress required to produce a large increase in deformation above the yield point

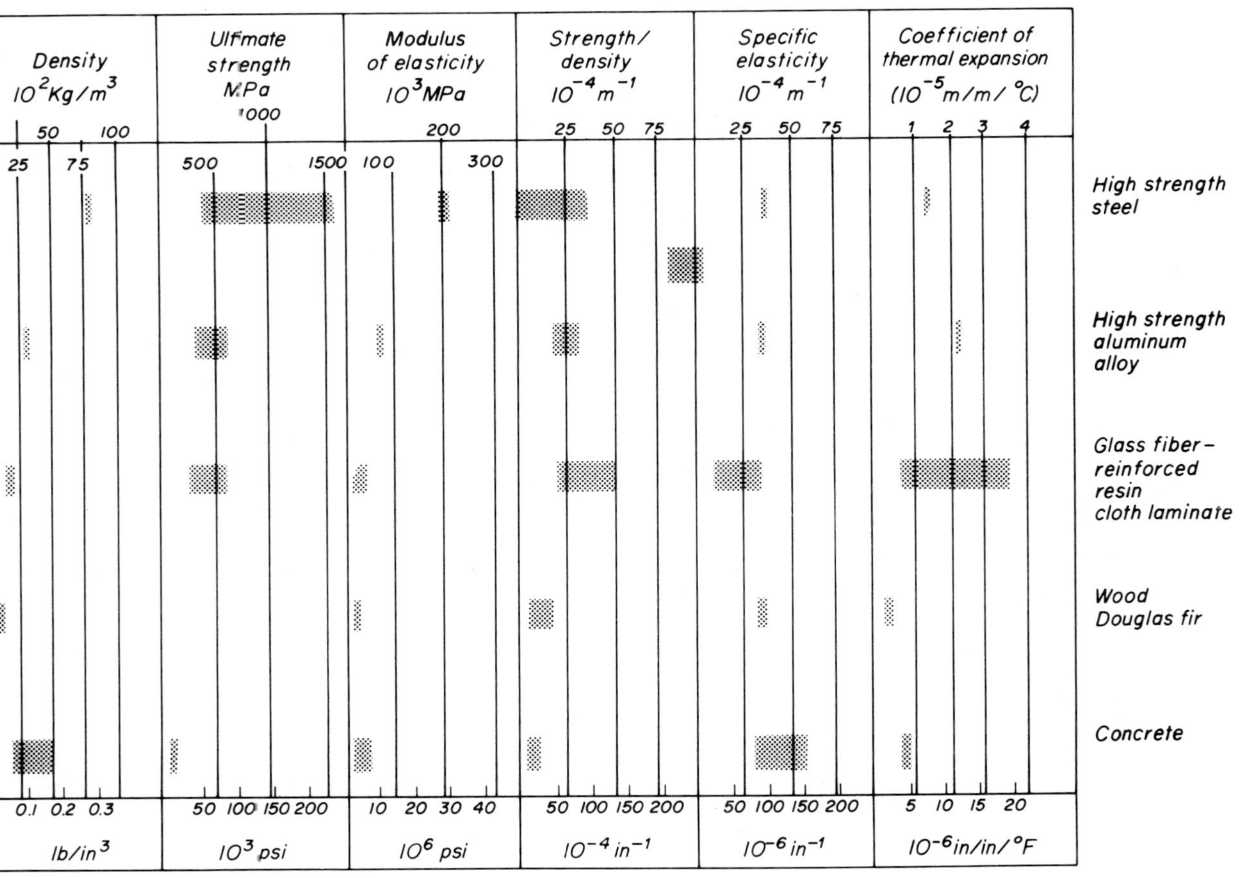

Density
10^2 Kg/m^3
25 50 75 100
0.1 0.2 0.3
lb/in^3
Ultimate strength MPa
500 1000 1500
50 100 150 200
10^3 psi
Modulus of elasticity
10^3 MPa
100 200 300
10 20 30 40
10^6 psi
Strength/density
10^-4 m^-1
25 50 75
50 100 150 200
10^-4 in^-1
Specific elasticity
10^-4 m^-1
25 50 75
50 100 150 200
10^-6 in^-1
Coefficient of thermal expansion
(10^-5 m/m/°C)
1 2 3 4
5 10 15 20
10^-6 in/in/°F
High strength steel
High strength aluminum alloy
Glass fiber–reinforced resin cloth laminate
Wood Douglas fir
Concrete

FIGURE 2.1.1

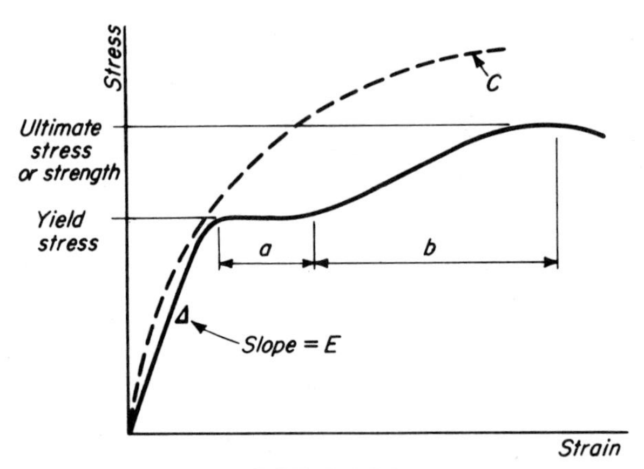

FIGURE 2.2.1

shows that steel *strain hardens* above the yield stress before reaching its *ultimate strength,* at which it fractures (Fig. 2.2.1). The modulus of elasticity of steel is measured by the slope of the elastic portion of its stress-strain curve, and equals 200 000 MPa (29,000,000 psi) for practically all grades of steel. Steel is one of the densest structural materials and weighs 7 850 kg/m^3 (490 lb/cu ft). Its coefficient of thermal expansion is approximately 1×10^{-5}/°C (0.65×10^{-5}/°F).

The mechanical properties of various grades of structural steel are given in Table 2.2.1. The ductility of steel is measured by the elongation at fracture on a 200 mm (8 in.) length; Table 2.2.1 shows that ductility decreases as strength

Table 2.2.1 PROPERTIES OF STRUCTURAL STEEL

	Minimum yield stress		Minimum ultimate tensile stress		Elongation in 200 mm (8 in.)
ASTM grade	MPa	ksi	MPa	ksi	%
A36	248	36	414	60	20
A572 Grade 50	345	50	448	65	18
A588*	345	50	483	70	19

*Corrosion resistant steel.

increases. Heat treated alloy steels with a yield stress of 690 MPa (100,000 psi) are also produced, with reduced ductility but with a resistance to corrosion of up to four times that of carbon steels. Since the limiting strength of steel is controlled by the bonding forces between iron atoms, which is over 27 600 MPa (4×10^6 psi), higher strength steels will continue to be developed as the chemistry of steel and quenching and tempering processes are improved. "Whiskers" of steel with a strength of 6 900 MPa (10^6 psi) have already been manufactured. However, the modulus of elasticity of steel remains practically constant, so that the relative stiffness of a structural member made out of steel decreases with increasing strength.

Steel is also used as reinforcing bars in concrete. Steel rods in diameters varying from 10 to 50 mm ($\frac{1}{4}$ in. to $2\frac{1}{4}$ in.) are produced with "deformations" on their surface to increase their bond to concrete. The properties of the various grades of steel used as reinforcing bars are shown in Table 2.2.2. In addition to bars, welded fabrics made with cold drawn wires are used as reinforcing for concrete slabs and shells. This wire has an ultimate strength of 517 MPa (75,000 psi) and a yield point of 448 MPa (65,000 psi).

Table 2.2.2 PROPERTIES OF STEEL REINFORCING BARS

ASTM grade	Minimum yield stress		Allowable tensile stress		Minimum ultimate tensile stress		Elongation in 200 mm (8 in.)
	MPa	ksi	MPa	ksi	MPa	ksi	%
A615	276	40	138	20	483	70	7–11
A615	414	60	165	24	621	90	7–9
A615	517	75	165	24	689	100	5

Table 2.2.3 gives the basic allowable stresses for structural steel in tension, compression, bending, and shear.* Steel loses its strength rapidly above 400°C (700°F) and becomes brittle at −34°C (−30°F).

Table 2.2.3 ALLOWABLE STRESSES IN STRUCTURAL STEEL

ASTM grade	Yield stress		Allowable stress, MPa (ksi)			
	MPa	ksi	Tension	Compression*	Flexure†	Shear
A36	248	(36)	151 (22)	149–25.7 (21.56–3.73)	165 (24)	100 (14.5)
A572	345	(50)	207 (30)	206–25.7 (29.94–3.73)	228 (33)	138 (20)

*Decreases with increasing slenderness (see Chapter 5).
†Decreases for members not adequately braced and members with thin outstanding flanges.

**Steel Design Handbook*, American Institute of Steel Construction, New York, 1980.

Steel wire strands and rope are used in suspended roofs, cable-stayed or suspension bridges, fabric roofs, and in other structural applications. A *strand* is a helical arrangement of wires about a center wire. A rope consists of multiple strands helically wound around a central plastic core; it is more flexible, has a lower elastic modulus, and lower carrying capacity than a strand. Ultimate tensile strength of up to 1 725 MPa (250,000 psi) are common for *stress-relieved* wire. The yield characteristics of ropes and strands are different from those of steel bars because of the composite action of the helically wound wires around the core. Typical properties of zinc-coated wire rope are shown in Table 2.2.4.

Table 2.2.4 PROPERTIES OF ZINC-COATED STEEL STRUCTURAL WIRE ROPE

Modulus of elasticity: $E = 138\ 000$ MPa (20,000,000 psi)
Ultimate strength: $F_u = 1\ 517$ MPa (220,000 psi)

Nominal diameter		Minimum breaking strength		Area,	Weight,
mm	in.	10^3 kg	10^3 lb	mm^2	kg/m
9.53	($\frac{3}{8}$)	5.9	13.0	41.9	0.36
12.70	($\frac{1}{2}$)	10.4	22.8	76.8	0.62
15.88	($\frac{5}{8}$)	16.3	36.0	117.4	0.97
19.05	($\frac{3}{4}$)	23.6	52.0	172.9	1.41
22.23	($\frac{7}{8}$)	31.8	70.0	232.9	1.90
25.40	(1)	41.5	91.4	303.9	2.48
31.75	($1\frac{1}{4}$)	65.5	144.4	480.7	3.93
38.10	($1\frac{1}{2}$)	94.3	208.0	694.2	5.68
44.45	($1\frac{3}{4}$)	130.0	286.0	948.4	7.80
50.80	(2)	169.0	372.0	1 238.8	10.19

For prestressing concrete, strands, wire, and rods are used with ultimate strengths of up to 1 860 MPa (270,000 psi) for ASTM A416 strand.

2.3 ALUMINUM [3.3]

Aluminum is often used as a structural material whenever light weight combined with strength is an important factor. Together with its superior corrosion resistance, these properties have made it useful in airplane structures, light roof framing, and bridges.* Although half of the earth's crust contains sources of aluminum, it was a rare metal until 1886, when Charles

***Aluminum in Modern Architecture*, Vol. II, Paul Weidlinger, Reynolds Metals Co., Louisville, Ky., 1956.

Hall discovered an electrolytic reduction process to obtain metallic aluminum from aluminum oxide.

Since pure aluminum is extremely soft and ductile, the addition of alloying elements and heat treatment or cold working are used to impart to it the strength required by structural members. The ultimate tensile strength of pure aluminum is only 90 MPa (13,000 psi) but it goes up to 690 MPa (100,000 psi) for the highest strength alloys. Magnesium, silicon, zinc, and copper are used either singly or in combination in aluminum alloys. The properties of the most common structural aluminum alloys are shown in Table 2.3.1. It should be noted that welding results in a significant reduction

Table 2.3.1 PROPERTIES OF ALUMINUM ALLOYS

Alloy	Major alloying constituents	MPa (ksi) Minimum UTS	MPa (ksi) Minimum TYS	Minimum elongation unwelded, %
6061T6	Mg, Si	290 (42)	255 (37)	12
7001T6*	Zn, Cu, Mg	676 (98)	627 (91)	9
7075T6*	Mg, Si, Zn, Cu	572 (83)	503 (73)	11

*Not weldable.

Mg = magnesium
Si = silicon
Cu = copper
Zn = zinc

TYS = tensile yield strength
UTS = ultimate tensile strength

of strength for all heat treated alloys. Aluminum exhibits a high degree of corrosion resistance through the formation of an oxide film on its surface. Although alloying reduces this resistance, special protection is not normally required by aluminum structures. For severe exposure, their surface can be treated by coating with a film of pure aluminum, by anodic oxidation, or by painting. When aluminum is in contact with other metals in the presence of an electrolyte, galvanic corrosion may cause damage. For this reason, steel and aluminum must be carefully separated in a structure, either by means of painting or through the use of a nonconductive material.

Aluminum is one of the most workable of the common metals. As a result, it can be formed into foil and plates, drawn into wire and rods, extruded at high speeds into simple I-beams or intricate shapes. Although extrusions are limited to members under 90 cm (36 in.) in maximum dimension, hot rolling can produce heavier structural shapes. Aluminum members can be joined by riveting, bolting, or, to a lesser extent, by welding.

The modulus of elasticity of structural aluminum is 69 000 MPa (10,000,000 psi); its coefficient of thermal expansion is 2.4×10^{-5}/°C (1.3×10^{-5}/°F); its unit weight is 2 770 kg/m³ (173 lb/cu ft).

Table 2.3.2 gives the allowable stresses in tension, compression, bending, and shear for the most commonly used aluminum alloy. Aluminum loses strength rapidly above 100°C (212°F) but exhibits an increase in strength without loss of ductility at low temperatures.

Table 2.3.2 ALLOWABLE STRESSES FOR RIVETED OR BOLTED STRUCTURES, 6061-T6 IN MPa (ksi)

Tension	131 (19)
Compression	131 (19) (maximum; decreases for increasing slenderness)
Shear	83 (12) (decreases for slender webs)
Bearing	235 (34) on bolts and rivets
	159 (23) on milled surfaces

Note: At welds the allowable stresses decrease by over 40%.

2.4 CONCRETE [3.3]

Concrete is a mixture of Portland cement, water, and aggregates (usually sand and crushed stone). An ideal mixture is one in which:

1. a minimum amount of cement-water paste is used to fill the interstices between the particles of aggregate;
2. a minimum amount of water is provided to complete the chemical reaction with the cement.

In such an ideal mixture, the volume of aggregates is about three-fourths of the total volume of concrete, and the cement paste, which acts as a binding agent, fills one-quarter of the volume. Since the types of aggregate components used (e.g., gravel, crushed stone, manufactured lightweight stone, sea shells, sands) vary greatly, a correct mixture must be custom designed for each particular concrete. The aggregate itself must be properly graded, that is, it must consist of a uniform distribution of particles varying in size from the largest to the smallest. The smaller particles, up to 6 mm ($\frac{1}{4}$ in.) in size, are called *fine aggregates*, and the larger particles, from 6 mm ($\frac{1}{4}$ in.) up, *coarse aggregates*. The maximum particle size that can be used depends on the size of the concrete member but will rarely exceed 50 mm (2 in.) except for massive structures such as dams.

Portland cement is a mixture of calcareous and argillaceous materials which are calcined in a kiln and then pulverized. When mixed with water, cement first sets, i.e., becomes firm, and then hardens for an indefinite period of time, through a process called *hydration*. This material, first discovered in 1824 by an English mason, is today one of the most widely used construction materials.

The mechanical properties of concrete depend on numerous factors: the strength of the aggregate, the ratio of water to cement, the ratio of the paste to the aggregate, the curing conditions, and age. The standard measure of concrete strength is the breaking strength of a cylinder 150 mm (6 in.) in diameter and 300 mm (12 in.) high at 28 days of age. Strengths in excess of 70 MPa (10,000 psi) are possible, although most commercial concrete is produced with strengths under 40 MPa (6,000 psi). A decrease in the water-cement ratio normally results in an increase in strength. Curing of concrete in a moist atmosphere will result in higher strengths than curing in a dry atmosphere. The modulus of elasticity of concrete depends on its weight and strength. Representative values are 23 000 MPa (3.3×10^6 psi) for stone concrete and 14 000 MPa (2×10^6 psi) for lightweight concrete. A summary of the mechanical properties of concrete is given in Table 2.4.1.

Table 2.4.1 PROPERTIES OF CONCRETE

Aggregate type	Strength at 28 days		Modulus of elasticity		Unit weight		Average values at 1 year for stress-strength ratio of 0.5		Coefficient of thermal expansion ($\times 10^{-5}$)	
	MPa	psi	MPa	psi	kg/m^3	lb/cu ft	Shrinkage %	Creep %	per °C	per °F
Lightweight	20	2,900	12 400	1.8×10^6	1 600	100	0.07	0.14	0.7	0.45
	50	7,250	22 000	3.2×10^6	1 760	110				
Stone	20	2,900	21 700	3.15×10^6	2 320	145	0.05	0.10	1.0	0.65
	50	7,250	33 800	4.9×10^6	2 480	155				

In *reinforced concrete*, concrete is combined with steel rods to create a new material that has the advantage of the tensile strength of steel and the compressive strength of concrete. This combination is possible because steel and concrete have approximately the same coefficient of thermal expansion [about 1×10^{-5}/°C (0.65×10^{-5}/°F)]. If this were not the case, the bond between steel and concrete would be broken by a change in temperature, since the two materials would have different elongations. Since both steel and concrete are available in a variety of strengths, innumerable combinations of these two materials can be created to meet specific design requirements.

Reinforced concrete can be poured into an infinite variety of shapes to create beams, columns, slabs, and shells. Concrete elements may be *pretensioned* or *post tensioned* to put them in a state of compression which eliminates in part or totally the tensile stresses due to applied loads. In this manner, the weakness of concrete in tension is practically eliminated.

When concrete is poured into a mold or form, it contains free water (not involved in the cement chemical reaction) which evaporates. As the concrete sets and cures, it releases that moisture over a period of time and shrinks. The amount of shrinkage depends on the conditions of curing (dry or moist), the amount of reinforcing, and the type of aggregate and mixture used. For plain concrete, shrinkage after one year is about 0.05%, so that, for example, a 30 m (100 ft) long member shrinks about 16 mm ($\frac{5}{8}$ in.). In lightweight concrete, shrinkage is 40% greater. Unless limited, shrinkage produces cracks in the concrete. The addition of reinforcing bars with an area equal to 2% of the concrete section reduces shrinkage by about 75% and causes most of the shrinkage to occur within one month.

Shrinkage may also be reduced by using a so-called "gap graded" mix, in which the coarse aggregates touch each other because aggregates of intermediate size are omitted. Shrinkage is thus reduced because the movement of the large aggregate is restricted to a rearrangement of particles instead of consisting of a free movement of particles in the cement matrix.

Concrete under water has a tendency to expand by absorbing moisture. The amount of this expansion is very slight but points to the fact that the longer concrete can be kept moist, the less it will shrink. To counter the effects of shrinkage, *expansive* cements have been developed by the addition of calcium sulfoaluminate to Portland cement. This kind of concrete, expanding rather than shrinking, makes possible prestressing of concrete by chemical means, since the reinforcing rods restrain the expansion of the concrete, which thus becomes compressed.

Since concrete is a plastic material, it will deform in time. This property of flowing under load is called *creep*. The amount of creep depends on the load, the strength and type of aggregate used, and the age of the concrete. Creep continues indefinitely in time but at a decreasing rate. The total creep after 20 years is approximately twice the creep after 1 year. At the end of 1 year, the creep for stone concrete is about 0.1%, so that an unreinforced column in a 150 m (500 ft) high building would shorten by 150 mm (6 in.). For lightweight concrete, the same column would shorten 50% more, or by 225 mm (9 in.). The addition of steel reinforcing bars to the concrete has the effect of reducing the amount of creep. With 2% reinforcing, for instance, creep is reduced by 50%. With 5% reinforcing, creep is reduced by 70%.

Another way of reducing creep is to reduce the concrete *stress-strength ratio*, i.e., the ratio of the actual stress to the ultimate strength of the concrete. Creep is also reduced if concrete is kept moist; this is apparently due to the fact that concrete, when kept moist, increases in strength more than when exposed to air. Table 2.4.2 indicates allowable stresses for concretes of different strengths.*

Table 2.4.2 ALLOWABLE CONCRETE STRESSES, MPa (psi)

Ultimate concrete strength at 28 days (f_c')	Compressive stress in flexure (f_c)	Shear stress in beams (v_c)*	Tensile stress†
20 (2,900)	9 (1,305)	0.41 (59.2)	2.24 (323)
30 (4,350)	13.5 (1,957)	0.50 (72.5)	2.74 (395)
40 (5,800)	18.0 (2,610)	0.58 (83.8)	3.16 (457)
50 (7,250)	22.5 (3,263)	0.64 (93.7)	3.54 (511)

*10% higher in concrete joists.
†In prestressed concrete members.

2.5 MASONRY [3.1]

Masonry consists of either natural materials, such as stone, or of manufactured products, such as brick and concrete blocks, stacked and bonded together with a mortar. It has been used as a construction material since ancient times: mud bricks were used in the city of Babylon for secular buildings, and stone masonry was used for the great temples of the Nile valley. The igloo, built of ice blocks by the Eskimos, is also a good example of masonry construction. Modern structural masonry materials fall into three general categories: burnt clay or slate bricks, concrete bricks or blocks, and stone masonry. They are all basically compression materials and have low tensile strength (1/10 of their compressive strength).

Masonry construction is made up of two materials, the masonry unit and the mortar used in joining the units. Mortar used in masonry is a mixture of sand, masonry cement, and either Portland cement or hydrated lime. Portland cement mortars have greater strength and durability. The allowable strengths of various masonry units, shown in Table 2.5.1, depend essentially on the strength of the mortar used in joining them.†

***Building Code Requirements for Reinforced Concrete ACI 318*, American Concrete Institute, Detroit, 1977.
†*National Building Code*, National Board of Fire Underwriters.

Table 2.5.1 ALLOWABLE COMPRESSIVE STRENGTH ON THE GROSS AREA OF MASONRY, MPa (psi)

Masonry unit	Mortar strength	
	17 MPa (2,500 psi)	5 MPa (750 psi)
Solid brick*	1.2 (175)	1.0 (140)
Solid concrete block*	1.2 (175)	1.0 (140)
Hollow concrete block†	1.17 (170)	1.0 (140)
Stone:		
Granite	5.6 (800)	4.4 (640)
Sandstone	2.8 (400)	2.2 (320)
Rubble stone	1.0 (140)	0.7 (100)

*On net area.
†Minimum strength 17 MPa (2,500 psi).

2.6 TIMBER [3.1]

Timber is one of the earliest construction materials and one of the few natural materials with good tensile properties. Of over 600 species of wood found in the United States, only 20 are used structurally. Each species exhibits different characteristics.*

Lumber is graded according to use, i.e., post and timber, beam and stringer, joist and plank, and allowable stresses are given within these classifications. Lumber gradations depend primarily on the location, size, and number of defects, such as knots, checks, shakes, warp, and bow (Fig. 2.6.1).

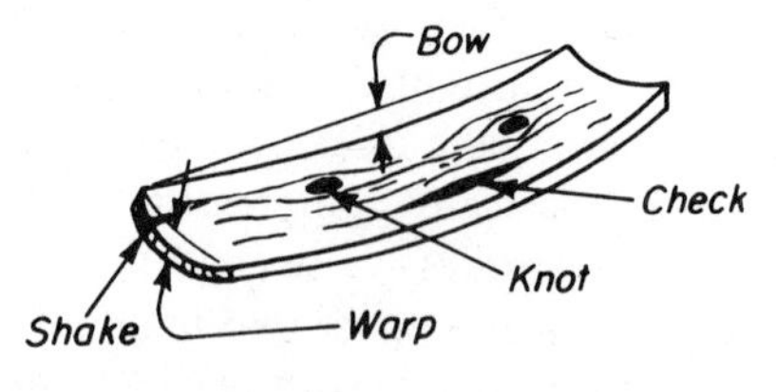

FIGURE 2.6.1

Lumber is subject to seasoning or drying. Green lumber has a moisture content of 35 to 40%. From green to oven dried lumber, shrinkage occurs of up to 21% by volume. This shrinkage occurs primarily across the grain (up to 14%); longitudinal shrinkage is at most 0.3%. Once dry, lumber will reabsorb water from the atmosphere and the final moisture content may be

*AITC, Timber Construction Manual, John Wiley & Sons, N.Y., 1974.

5% at 20% relative humidity and 23% at 90% relative humidity. Lumber is *nonisotropic*, i.e., its properties vary in different directions. In a direction parallel to the grain, the tensile strength of timber is about three times greater than its compressive strength. But timber is rarely used in pure tension because tensile connections involve shear along the grain or compression across the grain, with ultimate strengths much smaller than the tensile strength.

The strength of timber is time dependent: its sustained load strength is one-half of its short-time load strength. Timber is resilient, i.e., it is a good shock absorber: its strength-stiffness ratio is greater than for steel. Its strength decreases with increasing moisture content. Wood chars at 150°C (300°F) and ignites at 200°C (400°F).

The most common species of timber used in construction are Douglas fir, southern pine, hemlock, and larch. Representative values of allowable stresses are shown for Douglas fir in Table 2.6.1. Allowable stresses for the various wood species are given in design specifications.* The unit weight of lumber at 12% moisture content varies from 384 kg/m³ (24 lb/cu ft) for spruce to 832 kg/m³ (52 lb/cu ft) for hickory.

Table 2.6.1 ALLOWABLE STRESSES IN MPa (psi) FOR DOUGLAS FIR

Grade	Extreme fiber in bending	Tension parallel to grain	Horizontal shear	Compression perpendicular to grain	Compression parallel to grain	Modulus of elasticity
Construction	8.3 (1,200)	4.3 (625)	0.66 (95)	2.7 (385)	7.9 (1,150)	10 340 (1,500,000)
Dense #1	14.1 (2,050)	8.3 (1,200)	0.66 (95)	3.1 (455)	10 (1,450)	13 100 (1,900,000)
Laminated	16.5 (2,400)	11.0 (1,600)	1.1 (165)	—	10.4 (1,500)	12 400 (1,800,000)

Laminated members are made of strips of wood, at most 5 cm (2 in.) thick, glued together to form either straight or curved structural elements. Generally, the laminations are horizontal, i.e., the strips of wood are stacked vertically one on top of the other. The strength of a glued-laminated member may be greater than the strength of a virgin timber member because of the greater selectivity and quality control which can be exercised on the individual laminations. Bending strengths of up to 16.5 MPa (2,400 psi) are common for these members. Also, members larger than those obtained from wood

***National Design Specifications for Stress Grade Lumber and Its Fastenings*, National Lumber Manufacturers Association, Washington, D.C., 1979.

in its natural state can be fabricated. With the introduction of synthetic resin glues, glued-laminated members can be used outdoors as well as indoors.

Plywood is a glued wood panel in which the grain of the adjacent layers are perpendicular to each other. There are generally an odd number of layers, so that the grain direction is the same for both faces. Because of the alternating grain directions, a plywood panel has almost equal strength in the two orthogonal directions. Bending strengths may vary according to species of wood and glues used, from 6.9 MPa (1,000 psi) to 13.8 MPa (2,000 psi).

2.7 STRUCTURAL PLASTICS [3.3]

Plastics are important construction materials because of their great variety and their inherent strength, durability, and light weight.

A plastic is a synthetic material or resin which can be molded into a desired shape, using an organic substance as a binder. Organic plastics may be divided into two general groups: thermosetting and thermoplastic compositions. The thermoplastic group remains soft at elevated temperatures and must be cooled before becoming rigid; plastics in this group can be shaped and molded by the application of heat. The thermosetting group becomes rigid through a chemical change which takes place when heat is applied; once set, they can no longer be remolded. There are hundreds of separate compositions which fall into each of these two groups, but only those of structural interest and, particularly, reinforced plastics are considered here.

Glass fibers are the most common reinforcing material added to the resin to improve its mechanical properties. Glass fibers are made by forcing molten glass through fine orifices and blowing jets of high pressure steam or air against them. The material is then spun into filaments and strands of exceptionally high strength. Single filaments have been produced with tensile strengths of 4 140 MPa (600,000 psi), and strands with tensile strengths of 2 070 MPa (300,000 psi). Filament winding, in which strands are wound around a cylindrical or spherical surface and then impregnated with epoxy resin, results in a reinforced plastic with tensile strengths of up to 1 725 MPa (250,000 psi). This process is used in missile bodies and pressure tanks to provide the highest strength composition produced to date, but new advances are made constantly in this field.

Woven yarn of glass, and particularly glass fabrics, offer great potential for structural use when used to reinforce epoxy or polyester resins. The reinforced plastic thus created is used to produce structural panels, and shapes such as I-beams.* Glass fiber strands have been used experimentally as prestressing tendons for concrete beams. Protective coatings are required for these strands, since they decompose when left in contact with concrete.

**Fiberglas Reinforced Plastics*, Owens Corning Fiberglas Co., Toledo, Ohio, 1962.

Typical properties of fabric-reinforced polyester resins are shown in Table 2.7.1. Both polyesters and epoxies are thermosetting resins. In these resins the rate of flow, or *creep*, under load decreases rapidly to almost zero as the resin sets, or is polymerized.

Table 2.7.1 TYPICAL ENGINEERING PROPERTIES OF FIBERGLAS-REINFORCED PLASTIC LAMINATES

Type	Direction of stress, deg	Modulus of elasticity, MPa ($\times 10^{-6}$ psi)	Strength, MPa (psi)		
			Tension	Compression	Flexure
Mat	0, 45, 90	5 520 (0.8)	76 (11,000)	152 (22,000)	141 (20,500)
Woven	0, 90	16 560 (2.4)	228 (33,000)	117 (17,000)	214 (31,000)
Roving	45	2 760 (0.4)	69 (10,000)	69 (10,000)	128 (18,500)
Unidirectional	0	34 500 (5.0)	448 (65,000)	269 (39,000)	483 (70,000)

Another class of plastics finds structural uses as foams. Urethane and polystyrene foams are used in laminated panels between two layers of a surface material, such as plywood, aluminum, or unreinforced and reinforced plastics. These foams have very low compressive and tensile strengths but sufficient shear strength to allow the manufacturing of panels with high stiffness.

Fabrics are available in a variety of combinations, with plastic coatings ranging from the short-life vinyls to the permanent Teflon. These fabrics are used in both air-supported roof structures and in tension, or tent-like, structures. Teflon coated fiberglas (with a life expectancy of over 30 years) is used to economically cover spans of 50 to 300 m (150–900 ft). It is fire-resistant, will not degrade under ultraviolet radiation, and has a highly reflective surface. Typical properties of various fabrics are shown in Table 2.7.2.

Table 2.7.2 PROPERTIES OF PLASTIC FABRICS

	Total weight		Tensile strength	
	g/m^2	oz/sq yd	N/mm	lb/in.
Vinyl coated fiberglas	881	26	70	400
Teflon coated fiberglas	1 271	37.5	91	520
Teflon coated fiberglas	1 526	45	122	700
Vinyl coated nylon	949	28	70	400
Neoprene coated nylon	678	20	57	325
Vinyl coated polyester	949	28	82	470
Vinyl coated polyester	1 220	36	122	700

2.8 SOILS [2.5]

Although not ordinarily considered as such, soils are nevertheless basic structural materials since all structures rest on them. The design of a foundation depends on many soil factors: the type of soil, the stratification and thickness of the soil layers, their compaction, the ground water conditions, proximity to adjacent structures, etc.

Soils are rarely encountered in nature as materials of a single composition, but appear as mixtures and in layers of varying thickness. For the purpose of their evaluation, soils are graded according to particle size. The particle size increases from silt to clay, to sand, to gravel, and to rock. In a well graded mixture, the voids between the larger particles are entirely filled by smaller and smaller particles; i.e., sand fills the space between particles of gravel, clay between particles of sand, and silt between particles of clay.

The finer grained soils, i.e., silt and clay, become fluid when mixed with water, and exhibit spongy and, at times, slippery characteristics when wet. In a dry condition, clay becomes hard and impenetrable; silt becomes powdery.

All soils beneath grade are in a state of compaction. For instance, at a point 3 m (10 ft) below natural grade, the soil is under a vertical pressure equal to the weight of the soil above it: 3 m (10 ft) at 1 600 kg/m^3 (100 lb/cu ft) or about 4 800 kg/m^2 (1,000 psf). An excavation to this depth removes this pressure.

Many soils, except most sands and gravels, exhibit elastic properties;

Table 2.8.1 ALLOWABLE BEARING PRESSURES FOR SOIL AND ROCK

Type and condition of soil or rock	Maximum allowable bearing pressure	
	kPa	tons/sq ft
Dense sand, dense sand, and gravel	300	3
Compact sand, compact sand, and gravel	150	1.5
Loose sand, loose sand and gravel	50	0.5
Dense silt	150	1.5
Compact silt	100	1.0
Very stiff clay	300	3.0
Stiff clay	150	1.5
Firm clay	75	0.75
Soft clay	40	0.40
Till, dense or hard	400	4.0
Till, compact or firm	150	1.5
Cemented sand, and gravel	500	5.0
Clay shale	300	3.0
Sound rock	500–10 000	5.0 –100

i.e., they deform when compressed under load and rebound when the load is removed. The elasticity of soils is often time dependent; i.e., the deformations occur over a length of time which may vary from minutes to years. Because of these properties, a building which imposes on the soil a load greater than the natural compaction weight of the soil may settle in time. Conversely, a building in a deep excavation, which imposes on the soil loads smaller than the natural compaction weight, may heave in time. Because such movements can occur both during and after construction, careful analysis of the behavior of soils under a structure is important. In preliminary designs of foundations, allowable bearing values may be assigned to a soil in accordance with Table 2.8.1. In order to safely estimate these values, preliminary explorations by means of borings or test pits must be made to determine the types of soils on which the structure will rest and the depth of the water table.

PROBLEMS

2.1 Compare the elongation due to a 30°C (54°F) rise in temperature in four beams 12 m (40 ft) long, made of steel, aluminum, concrete, and wood.

2.2 What is the stress in the beams of Problem 2.1 if the elongation is prevented?

2.3 The safety factor is defined here as the ratio of ultimate to working stress. What is the apparent safety factor in flexure for steel ASTM A36, aluminum 6061-T6, concrete $f'_c = 30$ MPa (4,350 psi)?

2.4 Compare the unit strains of short columns made out of wood, steel, aluminum, and concrete loaded to their allowable compressive stress.

2.5 Glass is a brittle material which fails suddenly in tension, with no plastic deformation. Following the principle that makes reinforced concrete possible, what material can be bonded to glass to increase its flexural strength? The properties of glass are:

Ultimate tensile strength	69 MPa (10,000 psi)
Working stress in tension	6.9 MPa (1,000 psi)
Modulus of elasticity	69 000 MPa (10,000,000 psi)
Coef. of thermal expansion	8.1×10^{-6} mm/mm/°C
Density	25.9 kN/m^3 (165 lb/cu ft)

2.6 A W310 × 97 (W12 × 65) column is made out of A36 steel. Determine its total working load and the load at which it would fail, if prevented from buckling both locally and as a whole.

3. Straight Structural Elements Under Direct Stress

3.1 SIMPLE TENSION [5.1]

Simple tension is a structural mechanism that transfers loads along a line. In a straight structural element the tensile load is transferred along the axis of the element, and every unit area of its cross section carries the same amount of load. Hence, the material of the element is *uniformly* stressed and every one of its fibers can be stressed up to the allowable stress.

The most commonly used tensile materials are metals like steel, the least expensive structural material per unit weight, or aluminum, one of the lightest materials. Metals are not only strong, but stiff, i.e., they have large moduli of elasticity; hence, elastic deflections are usually very small in elements under tension, like bars, structural shapes, and pipes.

Limitations in the use of tensile elements come from two sources:

1. A tensile element must "hang" from some other nontensile element, since all loads must eventually reach the earth (unless they are suspended from lighter-than-air elements or magnetically supported).

2. Tensile elements may be so flexible that they may tend to oscillate, unless stiffened or "stabilized."

3.2 STRESS, STRAIN, AND ELONGATION IN SIMPLE TENSION [5.1]

If T (N, lb) is the tensile force to be carried by an element of cross-sectional area A (mm², sq in.), the *tensile stress* f_t (MPa, psi) is given by:

$$f_t = \frac{T}{A} \quad \text{(MPa, psi).} \qquad (3.2.1)$$

Calling F_t (MPa, psi) the *allowable stress* in tension of the material, the minimum area A_{min} needed to carry a force T is given by:

$$A_{min} = \frac{T}{F_t} \quad (\text{mm}^2, \text{sq in.}). \tag{3.2.2}$$

If the tensile modulus of elasticity of the material is E_t (MPa, psi), the elongation per unit length or *strain* ϵ_t (mm/mm, in./in.) is given by Hooke's law, (2.1.1):

$$\epsilon_t = \frac{f_t}{E_t} \quad (\text{mm/mm, in./in.}). \tag{3.2.3}$$

If L (mm, in.) is the length of the element, its *elongation* ΔL equals:

$$\Delta L = \epsilon_t L \quad (\text{mm, in.}), \tag{3.2.4}$$

provided the stress, and hence the strain, are constant along the length of the element. When the strain equals $\epsilon_1, \epsilon_2, \epsilon_3, \ldots$ over the partial lengths $L_1, L_2, L_3, \ldots$ the elongation equals:

$$\Delta L = \epsilon_1 L_1 + \epsilon_2 L_2 + \epsilon_3 L_3 + \ldots. \tag{3.2.4a}$$

The reduction in width in a direction at right angles to the axis of the element, the so-called *Poisson's phenomenon*, equals:

$$\Delta d = \nu \epsilon_t d \quad (\text{mm, in.}), \tag{3.2.5}$$

where ν is *Poisson's ratio*, equal to 0.3 for most metals, and d (mm, in.) is a lateral dimension of the element (the diameter for circular cross sections, the width or depth for rectangular cross sections, etc.).

The stress conditions at the ends of the tensile element are usually complicated, due to the connections with adjoining elements, but de St. Venant's theorem guarantees that simple tension will exist over most of the element's length [9.1].

Example 3.2.1. A chandelier weighing 500 kg (1,000 lb) hangs from the dome of a theater. The rod from which it hangs is 10 m (30 ft) long. We wish to determine the cross section of the rod, comparing its cost in steel and aluminum.

Neglecting the weight of the rod itself in comparison with the weight of the chandelier, the rod must support a force $T = 500 \times 9.8$ kN. From Table 2.2.3, an ASTM A36 structural steel with $F_t = 151$ MPa (22,000 psi) (equal to six-tenths of its *yield stress* F_y) requires by (3.2.2) an area:

$$A = \frac{500 \times 9.8}{151} = 32.5 \text{ mm}^2 \text{ (0.05 sq in.)}.$$

From the AISC manual we find that a round bar of 8 mm diameter with $A = 50.0$ mm², or a square rod of side 6 mm with $A = 36.5$ mm² are the nearest available sizes. We choose the rod. The actual stress in the rod is:

$$f_t = \frac{500 \times 9.8}{36.5} = 134 \text{ MPa (20,500 psi)}.$$

With $E = 200\,000$ MPa, the corresponding strain is, by (3.2.3):

$$\epsilon_t = \frac{134}{200\,000} = 0.000\,67 \text{ mm/mm.}$$

The bar elongation, by (3.2.4), equals:

$$\Delta L = 0.000\,67 \times (10 \times 1\,000) = 6.7 \text{ mm (0.26 in.).}$$

The weight per unit length of the rod is 0.282 kg/m; its total weight, 0.282 × 10 = 2.82 kg, is negligible in comparison with the weight of the chandelier, as assumed in the calculations. At 70 ¢/kg the steel rod costs $1.98. A 6063-T6 aluminum bar with $F_t = 131$ MPa (15 000 psi) requires an area:

$$A = \frac{500 \times 9.8}{131} = 37.4 \text{ mm}^2.$$

From the *Alcoa Structural Handbook*, e.g., the nearest available bar is a 6 mm square bar with $A = 36$ mm². The stress in the bar is:

$$f_t = \frac{500 \times 9.8}{36} = 136 \text{ MPa (19,720 psi)}$$

and, with $E = 69\,000$ MPa, the strain is:

$$\epsilon_t = \frac{136}{69\,000} = 0.001\,97 \text{ mm/mm.}$$

Hence, the aluminum bar elongates by:

$$\Delta L = 0.001\,97 \times (10 \times 1\,000) = 19.7 \text{ mm (0.776 in.)}$$

or three times as much as the steel rod. The weight of the aluminum bar is 0.10 kg/m. Its total weight is 1.0 kg, and its cost at $4.40/kg is $4.40. The steel rod is cheaper but will require maintenance; the aluminum bar is more expensive but will not require maintenance.

Example 3.2.2. A 20-passenger elevator in a 20-story building is lifted by a steel rope. We wish to design the rope according to the provisions of the New York City Code, which requires a factor of safety of 11 against the ultimate force T_u of the rope.

The rope length L is 3 m (10 ft) per story times 19 stories plus 3 m (10 ft) for the pulley housing; i.e., $L = 60$ m (200 ft).

From Table 2.2.4, for example, we choose a 25.4 mm (1 in.) diameter rope with an ultimate force $T_u = 41\,500$ kg (54 000 lb). The loads on the rope are:

20 people at 70 kg each	1 400 kg	(3,000 lb)
20-passenger cab	680 kg	(1,500 lb)
60 m of rope at 2.2 kg/m	132 kg	(300 lb)
	$T =$ 2 212 kg	(4,800 lb)

The allowable load on the rope with a factor of safety of 11 is:

$$T = \frac{T_u}{11} = \frac{41\,500}{11} = 3\,772 \text{ kg (4,900 lb).}$$

The net resisting area of a rope of diameter d is about two-thirds of its gross area $\pi d^2/4$:

$$A = \left(\frac{2}{3}\right)\left(\frac{\pi}{4}\right)(25.4)^2 = 337 \text{ mm}^2 \text{ (0.5 sq in.)}.$$

The stress in the steel rope is:

$$f_t = \frac{2\,212 \times 9.8}{337} = 64.3 \text{ MPa (9,800 psi)};$$

its strain is:

$$\epsilon_t = \frac{64.3}{200\,000} = 0.000\,32 \text{ mm/mm*}$$

The rope elongation equals:

$$\Delta L = 0.000\,32 \times (60 \times 100) = 19.2 \text{ mm (0.72 in.)}.$$

Example 3.2.3. What is the longest 25.4 mm (1 in.) diameter steel rope that will not snap under its own weight of 2.48 kg/m (1.45 lb/ft)?

Defining L_u as the length of the rope in feet, its weight must equal its ultimate force:

$$2.48 \times L_u = 41\,500 \text{ kg},$$

from which:

$$L_u = \frac{41\,500}{2.48} = 16\,733 \text{ m} = 16.733 \text{ km (10.40 miles)}.$$

It is apparent that the limit length is independent of the rope diameter. In fact, the volume of a unit length of rope is $A \times 1 = A$, and, indicating by ρ the weight per unit volume of the material, the weight per unit length of rope w is:

$$w = \rho \times A \times 1 = \rho A.$$

The weight of a length L is thus:

$$W = wL = \rho AL.$$

Calling F_u the ultimate strength of the rope, its ultimate force is given by:

$$T_u = F_u \times A,$$

and equating W to T_u:

$$L_u = \frac{F_u}{\rho}.$$

*The same result is obtained by using the nominal area of the rope:

$$A_n = \frac{\pi}{4}(25.4)^2 = 506.7 \text{ mm}^2 \text{ (0.8 sq in.)},$$

which gives a nominal stress:

$$f_{t,n} = \frac{2\,212 \times 9.8}{506.7} = 42.8 \text{ MPa (6,125 psi)},$$

and by using the nominal rope modulus $E_n = 138\,000$ MPa, by means of which:

$$\epsilon_t = \frac{42.8}{138\,000} = 0.000\,32.$$

L_u is thus independent of the area A and, hence, of its diameter. As a check, for the 25.4 mm rope:

$$F_u = \frac{T_u}{A} = \frac{41\,500 \times 9.8}{337} = 1\,207 \text{ MPa (108,000 psi)},$$

$$\rho = 64 \text{ kN/m}^3 \text{ (0.24 lb/cu in.)},$$

$$L_u = \frac{1\,207 \times 1\,000}{64} = 18\,859 \text{ m (37,000 ft)}.$$

For a structural steel bar of unit weight 77 kN/m³ (0.29 lb/cu in.) and ultimate strength 414 MPa (60,000 psi), the ultimate length is:

$$L_u = \frac{414 \times 1\,000}{77} = 5\,377 \text{ m (17,000 ft)}.$$

Example 3.2.4. A 3 m × 6 m (10 ft × 20 ft) hotel marquee hangs from two rods inclined at an angle of 30° (Fig. 3.2.1). The dead load and snow load on the marquee add up to 4.8 kN/m² (100 psf). We wish to design the two rods out of A36 steel with $F_t = 151$ MPa (22,000 psi).

The contributory area to one rod is 3 × 3 = 9 sq m, the corresponding load 9 × 4.8 = 43.2 kN. The inclined tension T in the rod must provide a vertical reaction R of 21.6 kN, since the remaining 21.6 kN are supported by the wall. Hence:

$$R = T \sin 30° = \tfrac{1}{2}T = 21.6$$

$$\therefore \quad T = 43.2 \text{ kN (10,000 lb)};$$

$$A = \frac{T}{F_t} = \frac{43.2 \times 1\,000}{151} = 286 \text{ mm}^2 \text{ (0.46 sq in.)}.$$

A circular rod 20 mm in diameter, with an area of 314 mm² and a weight of 2.466 kg/m is the nearest available standard size. The stress in the rod is:

$$f_t = \frac{43.2 \times 1\,000}{314} = 137.6 \text{ MPa (19,200 psi)};$$

its strain is:

$$\epsilon_t = \frac{137.6}{200\,000} = 0.000\,688.$$

The length of the rod is 3/cos 30° = 3/0.87 = 3.448 m, and its elongation:

$$\Delta L = 0.000\,688 \times (3.448 \times 1\,000) = 2.4 \text{ mm (0.1 in.)}.$$

The reduction in the diameter of the bar is:

$$\Delta d = \nu \epsilon_t d = \tfrac{3}{10} \times 0.000\,688 \times 20 = 0.004\,1 \text{ mm (0.000 16 in.)}.$$

The weight of the rod is 3.448 × 2.466 = 8.5 kg (20 lb) and can be neglected in comparison with the weight of the marquee.

Example 3.2.5. In contemporary reinforced concrete construction the spandrel beam of a floor slab is often outside the glass-line of the building. In winter, when the building is heated and the spandrel is at the temperature of the outside air, a difference of as much as 55°C may exist between the spandrel and the floor slab. Since these two elements are rigidly connected, the spandrel cannot reduce its

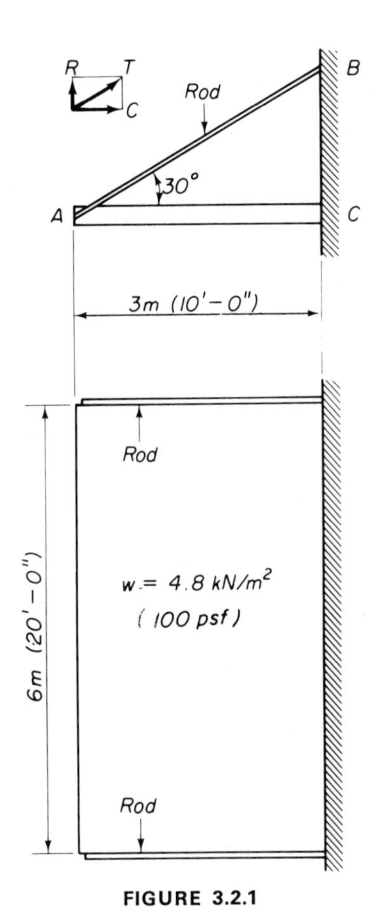

FIGURE 3.2.1

length as it would if it were free to shrink, and high tensile stresses are set up in the spandrel reinforcement, which we wish to evaluate.

The thermal expansion coefficient of steel and of concrete, α, is approximately 1.0×10^{-5} mm/mm/°C. Hence, for a temperature difference $\Delta t = -55$°C the shrinkage strain in the free spandrel would be:

$$\epsilon_t = \alpha \Delta t = -0.000\,55 \text{ mm/mm}.$$

Continuity with the slabs makes it practically impossible for the spandrel to shrink, since the slab is very rigid in its own plane. Hence, the slab exerts a tensile force T on the spandrel capable of stretching it to its original length by means of a tensile strain $\epsilon_t = +0.000\,55$. Neglecting the tensile strength of the concrete and indicating the steel modulus by E, we obtain a tensile stress f_s in the steel:

$$f_s = \epsilon_t E = 0.000\,55 \times 200\,000 = 110 \text{ MPa } (15{,}950 \text{ psi}).$$

Since the steel reinforcing rods are also stressed by the vertical loads, the total stress may be above the allowable stress for A615 steel (see Table 2.2.2) with some flow in the steel and cracking of the concrete (Fig. 3.2.2). For a 300 mm × 600 mm (12 in. × 24 in.) spandrel with 8% reinforcement, i.e., with:

$$A_s = \tfrac{8}{100} \times (300 \times 600) = 14\,400 \text{ mm}^2 \text{ (23 sq in.)},$$

the total tensile force developed in the spandrel is:

$$T = f_s A_s = 14\,400 \times \frac{110}{1\,000} = 1\,584 \text{ kN (414 k)}.$$

This tensile force in the spandrel produces compressive stresses in the slab that are essentially concentrated at the slab corners and introduce in it a radial compression (Fig. 3.2.2).

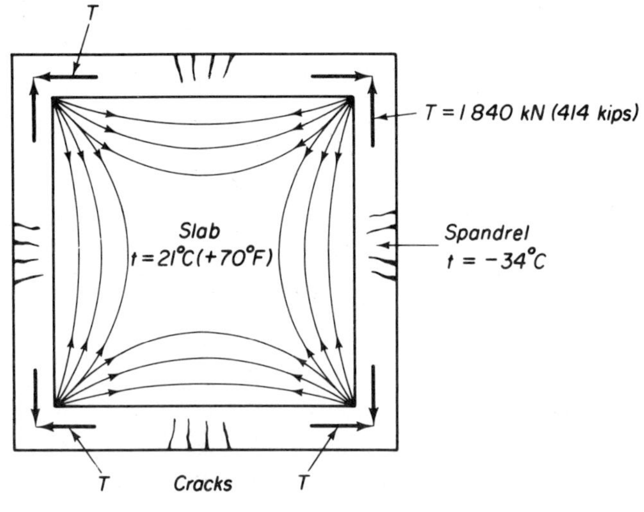

FIGURE 3.2.2

3.3 SIMPLE COMPRESSION AND ITS LIMITATIONS [5.2]

All structural materials are capable of withstanding some compression. Since, moreover, all loads must eventually reach the earth, compressive elements like columns, pillars, or buttresses are found in almost all structural systems. When the compressive force acts along the axis of the element, *uniform* compressive stresses are developed all over its cross section, and the compression mechanism of force transfer is then most efficient. Depending on whether the material is stronger in compression or in shear, the ultimate load under compression may occur either because the material is crushed or because the shear resistance at 45° to the direction of compression (Fig. 3.3.1) is overcome (see Section 13.1).

With the advent of materials like metals, which are exceptionally strong in compression, the cross section required to withstand a given compressive load has become progressively smaller. Consequently, the ratio of the length to the lateral dimensions of the element has increased and modern compressive elements have become *slender*. Just as the slenderness of tensile elements is limited by their flexibility, compression in unbraced slender elements is limited by their tendency to bend out, or *buckle* (Fig. 3.3.2) as soon as the compressive load reaches a specific value, called its *critical value*. The danger of buckling of a slender element or of any thin part of an element under compression, is related to a bending behavior. Its quantitative analysis appears in Section 5.3.

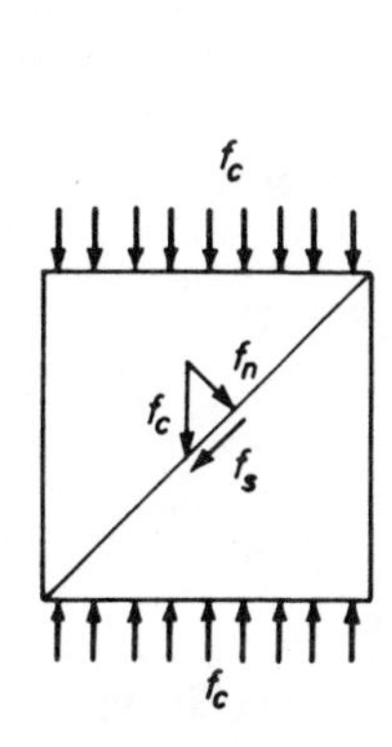

FIGURE 3.3.1

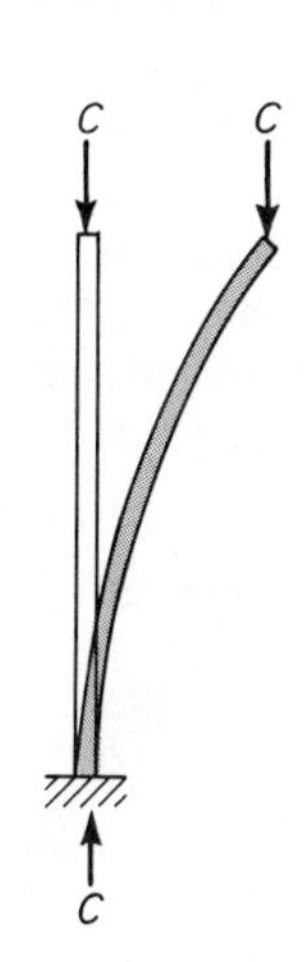

FIGURE 3.3.2

3.4 STRESS, STRAIN, AND ELONGATION IN COMPRESSION [5.2]

Calling C (N, lb) the axial compressive force carried by an element of cross-sectional area A (mm², sq in.), the *compressive stress* f_c (MPa, psi) equals:

$$f_c = \frac{C}{A} \quad \text{(MPa, psi).} \tag{3.4.1}$$

Calling F_c the allowable stress of the material in compression, the minimum area $A_{\min}$ required to carry a force C is:

$$A_{\min} = \frac{C}{F_c} \quad (\text{mm}^2, \text{sq in.}). \tag{3.4.2}$$

The compressive strain ϵ_c (mm/mm, in./in.) is given by:

$$\epsilon_c = \frac{f_c}{E_c}, \tag{3.4.3}$$

where E_c is the compressive elastic modulus.

The negative elongation (shortening) of an element of length L is:

$$\Delta L = -\epsilon_c L \quad \text{(mm, in.)}, \tag{3.4.4}$$

and its Poisson expansion equals:

$$\Delta d = \nu \epsilon_c d \text{ (mm, in.)}, \tag{3.4.5}$$

where d is a lateral dimension of the element and ν is Poisson's ratio (0.30 for most metals and 0.10 to 0.15 for concrete).

Example 3.4.1. A one-story factory building with 6 m × 9 m (20 ft × 30 ft) bays has a roof supported by 3.5 m (12 ft) high steel columns. Design the columns for a roof dead load of 5.0 kN/m² (100 psf), a snow load of 2.0 kN/m² (40 psf), and a live load of 2.5 kN/m² (50 psf).

Each column supports a bay and, hence, neglecting the weight of the column, we find a total load:

$$C = (5.0 + 2.0 + 2.5)(6 \times 9) = 513 \text{ kN (114 k)}.$$

With $F_c = 69$ MPa (a low value to avoid the danger of excessive slenderness), the column area is:

$$A = \frac{513 \times 1\,000}{69} = 7\,435 \text{ mm}^2 \text{ (11.4 sq in.)}.$$

From Appendix A, one may choose a standard 310 × 60.7 beam weighing 60.7 kg/m. The total weight of the column is 212.5 kg and is negligible in comparison with the applied load. It is shown in Section 5.4 how such columns are checked against buckling.

Example 3.4.2. The two 3.5 m (12 ft) high brick walls of a garage support a roof slab 6 m (20 ft) wide. The roof weighs 5.0 kN/m² (100 psf) and carries a snow load of 2.5 kN/m² (50 psf). Determine the thickness of the brick walls.

Each meter of wall carries (5.0 + 2.5) × 3 = 22.5 kN of roof load. Assuming the walls to be 100 mm thick and the brick to weigh 18.9 kN/m³ (see Table 1.2.1), each meter of wall weighs $3.5 \times 18.9 \times \frac{1}{10} = 6.59$ kN. The total load on the wall is 29.09 kN/m. Hence, the compressive stress at the foot of the wall is:

$$f_c = \frac{29.1 \times 1\,000}{100 \times 1\,000} = 0.291 \text{ MPa (41 psi)}.$$

From Table 2.5.1, F_c for brick varies from 1.0 to 1.2 MPa. Hence, the design is safe.

With an ultimate strength in compression $F_u = 2$ MPa for the brick-mortar wall, the highest brick wall of thickness d capable of supporting its own weight of

w kN/m³ would have a height H such that

$$(H \times d \times 1) \times w = F_u \times (d \times 1);$$

thus:

$$H_u = \frac{F_u}{w} = \frac{2\,000 \times 1\,000}{1\,920 \times 9.807} = 106.2 \text{ m (360 ft)}. \tag{3.4.6}$$

Example 3.4.3. A square concrete column of side a carries the load of a 3 m × 3 m (10 ft × 10 ft) contributory area in a 12-story apartment building with floor heights of 3 m (10 ft). It is reinforced with 1% steel. Determine the column size at the ground floor of the building.

With floor slabs weighing 5.0 kN/m² (100 psf) and a live load of 2.0 kN/m² (40 psf), each floor contributes 3 × 3 × 7.0 = 63.0 kN (14,000 lb) of load to the column, and the load at the ground floor is 63 × 12 = 756 kN (170,000 lb).

The area of the steel replaces 1% of the concrete area and is equivalent to the replaced concrete area multiplied by the ratio n of the moduli E_s and E_c of the steel and the concrete*:

$$n = \frac{E_s}{E_c} = \frac{200\,000}{20\,700} \simeq 10.$$

Hence, the *transformed* column area is:

$$A_t = A - \tfrac{1}{100}A + 10\tfrac{1}{100}A = 1.09A.$$

With an F_c for concrete of 7 MPa (1,000 psi) and $A = a^2$:

$$F_c = 7 = \frac{756}{1.09A},$$

$$A = \frac{756 \times 1\,000}{1.09 \times 7} = 99\,082 \text{ mm}^2,$$

$$a = \sqrt{A} = 315 \text{ mm (12.5 in.)}.$$

With a concrete weight of 23 kN/m³ (150 pcf), the 12-story column weighs:

$$\left(\frac{315}{1\,000}\right)^2 \times 3 \times 12 \times 23 \simeq 82 \text{ kN (21,000 lb)}.$$

The total load on the column is 756 + 82 = 838 kN, or 11% more than assumed so far in the calculations. Hence, the stress in the concrete is $f_c = 7 \times 1.11 = 7.77$ MPa, an acceptable working stress for a 35 MPa (5,000 psi) concrete. The stress in the steel is $f_s = nf_c = 77.7$ MPa.

By (3.4.6), a column made out of concrete with an ultimate strength of 35 MPa could be as high as:

$$H_u = \frac{35 \times 1\,000}{23} = 1\,522 \text{ m (4,750 ft)}.$$

*This equivalence stems from the assumption that plane sections remain plane, i.e., that the strains in the steel and the concrete are equal, so that:

$$\epsilon = \frac{f_s}{E_s} = \frac{f_c}{E_c} \quad \text{or} \quad \frac{f_s}{f_c} = \frac{E_s}{E_c}.$$

If the column of this example is to be supported on a soil with a bearing capacity of 400 kN/m^2, the area of its footing must be:

$$A = \frac{838}{400} = 2.10 \text{ sq m } (22.6 \text{ sq ft}),$$

or about 1.45 m square.

Example 3.4.4. A reinforced concrete building covers an area A and has N stories. The weight of 1 sq m (sq ft) of floor, inclusive of the weight of columns, is w. Determine the ratio of the total column area A_c at the ground floor to the floor area A, if the allowable stress in compression is F_c, and determine the number of stories for which one-tenth of the area is occupied by columns.

The weight of N stories is $N \times wA$. Hence:

$$A_c = \frac{NwA}{F_c} \quad \therefore \quad \frac{A_c}{A} = \frac{Nw}{F_c}.$$

With $w = 15$ kN/m^2 and $F_c = 7$ MPa,

$$N = \frac{F_c}{w}\frac{A_c}{A} = \left(\frac{7 \times 1\,000}{15}\right)\frac{A_c}{A} \simeq 470\frac{A_c}{A}.$$

Thus:

$$N = 47 \text{ for } \frac{A_c}{A} = \frac{1}{10}.$$

With a floor height of 3.5 m (12 ft), the building is 164.5 m high. For $N = 80$, i.e., a height of 280 m,

$$\frac{A_c}{A} = \frac{80}{470} \times 100 = 17\%.$$

For a steel building with $F_c = 138$ MPa:

$$N = \frac{138 \times 1\,000}{15}\frac{A_c}{A} = 9\,200\frac{A_c}{A}.$$

A building with 100 stories and a height of 370 m (1,200 ft) has $A_c/A = 1.09\%$. The area occupied by the columns is actually greater, due to the I shape of the steel columns, which must be enclosed in concrete to fireproof them.

PROBLEMS

3.1 A 20-story office building is to be built with a concrete central core and perimeter walls suspended by hangers 3 m (10 ft) on centers from cantilevered roof girders. The floor construction consists of a steel deck with 75 mm (3 in.) of lightweight concrete fill, beams with an average weight of 0.3 kN/m^2 (6 psf), and a hung acoustic tile ceiling. The span from the core to the perimeter is 10.5 m (35 ft), and the floor-to-floor height is 3.5 m (12 ft). Compute the total load of each floor supported by one hanger. Determine the sizes of A36 steel angles to be used as hangers, assuming their size to be increased only every second floor. Compute their total elongation due to dead and live loads. (Note that this represents the deflection of the lowest floor at the perimeter with respect to the roof.)

3.2 A glass pane is to be held by clamping its upper edge. What is the highest pane of glass that can be safely suspended in this manner, neglecting the bending induced by lateral loads? (Refer to Problem 2.5.)

3.3 A 25.4 mm (1 in.) diameter rod of A36 steel, 150 m (500 ft) long is suspended from the top of a building. What is the decrease in the diameter of the rod at the top, and what is the ratio of the area of the rod at the top to the area of the rod at the bottom?

3.4 A steel tie rod is required to prevent the base of an arch spanning 150 m (500 ft) from spreading. If the force required is 1780 kN (400 k), what elongation of the A36 tie rod is required to produce this force?

3.5 A baldachin 7.5 m $\times$ 15 m (24 ft $\times$ 48 ft) is to be suspended over an orchestra stage. The unit weight of the baldachin is 2 kN/m^2 (40 psf); the live load on it is 2 kN/m^2 (40 psf). If it is suspended at the four corner points, what is the size of the required A36 steel rods?

3.6 A grandstand is braced against lateral forces by diagonal tension rods. Each of these rods is 15 m (50 ft) long and at 30° to the ground. If each rod must resist the forces due to sidesway of 15 rows of bleachers, each 14 m (45 ft) long, what is the size of each A36 rod and the tension in it?

3.7 If in Problem 3.6 the wind force against the stands introduces an additional horizontal force in each rod of 35 kN (8,000 lb) what, if any, is the change in the size of the rod?

3.8 A block of concrete 3.5 m (12 ft) square and 2.5 m (8 ft) high is pulled by a steel cable attached to a tractor. If the coefficient of friction between the concrete and the ground is 0.3, what is the tension in the cable and the required size of the cable?

3.9 What is the maximum load W that can be carried by a nominal 50 mm $\times$ 100 mm (2 in. $\times$ 4 in.) wood stud [40 mm $\times$ 90 mm ($1\frac{5}{8}$ in. $\times$ $3\frac{5}{8}$ in.) actual dimensions] in a stud wall construction, assuming it is properly braced against buckling? What floor area can be carried by this stud in a house? Assume the floor construction to consist of a 25 mm (1 in.) parquet, 15 mm ($\frac{5}{8}$ in.) subfloor, 50 mm $\times$ 300 mm (2 in. $\times$ 12 in.) joists at 400 mm (16 in.) o.c., and a 25 mm (1 in.) plaster ceiling.

3.10 A column made out of A36 structural steel pipe with an outside diameter of 150 mm (6 in.) supports a statue weighing 40 kN (4 tons). What is the required pipe thickness?

3.11 In a 50-story concrete office building the exterior columns support a floor area of 3 m $\times$ 6 m (10 ft $\times$ 20 ft). If the weight of the floor is 6 kN/m^2 (120 psf), compute the required size of a square column at the base. Assume

$F_c = 7$ MPa (1,000 psi) and 3% reinforcing. Note that the live load in a multistory building may be reduced by assuming that the top 10 stories carry the full live load and the balance of the floors carry half the live load.

3.12 If in the building of Problem 3.11 the floor-to-floor height is 3.8 m (12 ft 6 in.) and the columns are fully stressed, what is the anticipated elastic shortening in a floor height? What is the anticipated creep shortening?

3.13 A multistory apartment building consists of 150 mm (6 in.) concrete slabs spanning 4.5 m (15 ft) between 200 mm (8 in.) brick walls. How many floors can be supported by these walls if the floor-to-floor height is 2.5 m (8 ft 6 in.)?

3.14 A pyramidal circus tent 45 m (150 ft) on a side and 25 m (80 ft) high is supported by a central aluminum pole. The weight of the fabric is 0.005 kN/m^2 (1 lb/sq yd) and the tension in each of the four guy wires at the corners is 18 kN (4,000 lb). Compute the thickness of the pole if it has an outside diameter of 600 mm (24 in.), assuming a basic snow load in horizontal projection of 1.5 kN/m^2 (30 psf). If the pole is to be supported on a 100 kN/m^2 (1 ton/sq ft) soil, what are the dimensions of its footing?

3.15 The circular grid carrying lights at the center of a circular arena weighs 400 kN (40 tons) and is suspended from cables anchored in the perimeter of the ceiling, 9 m (30 ft) above the grid, every 15°. If the diameter of the grid is 15 m (50 ft) and the diameter of the arena is 90 m (300 ft), determine the size of the cables, assuming a working stress of 690 MPa (100,000 psi).

3.16 The arch of Problem 3.4 carries a total load of 48 kN/m (3.33 k/ft) in horizontal projection, producing a thrust of 1 780 kN (400 k). If the tie-less arch is supported on rock at the springing, determine the inclination and the magnitude of the force exerted by the arch on the rock. Determine the minimum rock area required to support the arch.

3.17 Determine the limiting height of columns made of granite, 30 MPa concrete, glass, A36 steel, 6061-T6 aluminum, and douglas fir, if the columns are prevented from buckling. Base analysis on the allowable strength of the material.

3.18 Determine the limiting length of hangers (based on allowable stresses) made out of granite, concrete, glass, steel, aluminum, and wood. Assume the allowable tensile strength of granite to be 1.7 MPa (250 psi) and of concrete to be 1 MPa (150 psi).

4. Beams

4.1 BENDING AND SHEAR [7.1, 7.2]

Tension and compression are structural mechanisms used to transfer loads *along* the line of action of the loads. Bending-and-shear is a mechanism for transferring loads at right angles to their line of action. As such, it is the basic mechanism used to transfer gravity loads horizontally, and to solve the floor and roof problems. Because beams are basic structural elements working mostly in bending and shear, bending-and-shear is often referred to as *beam action.*

The following fundamental assumptions are the basis of simple beam theory and allow the derivation of the corresponding stress and deflection formulas for *prismatic beams*, i.e., beams of constant cross section.

1. The beam material is elastic, i.e., obeys Hooke's law (2.1.1); hence, the bending stresses f_b are proportional to the strains ϵ_b:

$$f_b = E\epsilon_b. \tag{a}$$

2. Plane sections remain plane; i.e., a cross section of a beam perpendicular to the undeflected beam axis remains perpendicular to the axis of the deflected beam. This assumption implies that the deformations due to shear (in which a section "slides" with respect to an adjoining section) are negligible.

3. The distribution of the bending stresses f_b across the depth of the section is linear, with maximum and minimum (algebraic) values at the bottom and top fibers of the beam (Fig. 4.1.1).

4. The bending stress f_b at a section x (Fig. 4.1.1) is proportional to the distance y of the beam fiber from the neutral axis of the beam (which is its

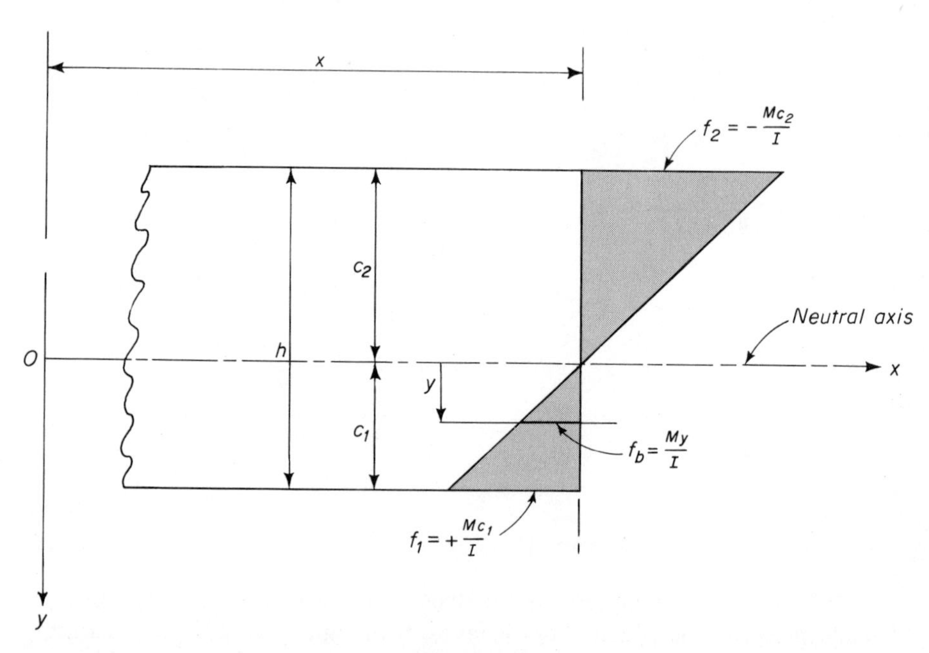

FIGURE 4.1.1

centroidal axis for homogeneous sections), and to the bending moment M*, and is inversely proportional to a characteristic of the beam cross section called its *moment of inertia* I:

$$f_b = \frac{My}{I}. \tag{4.1.1}$$

Indicating by $c_{1,2}$ the maximum positive and negative values of y, the maximum (tensile) and minimum (compressive) stresses are given by:

$$f_{1,2} = \frac{Mc_{1,2}}{I} = \frac{M}{S_{1,2}}, \tag{4.1.2}$$

where the *section moduli* $S_{1,2}$ are given by: .

$$S_{1,2} = \frac{I}{c_{1,2}}. \tag{4.1.3}$$

5. The bending moment is proportional to the curvature C of the deflected beam axis $y(x)$:

*The bending moment M at a section x is the moment about the centroid of the section of all the applied loads and reactions from either beam end up to that section x.

$$M = -EIC = -EI\frac{d^2y}{dx^2}^*, \tag{4.1.4a}$$

where EI, the product of the modulus of elasticity and the moment of inertia, is called the *flexural rigidity* of the beam. Hence, the bending moment is zero where the deflected beam has no curvature, e.g., at points of inflection of its axis (Fig. 4.1.2).

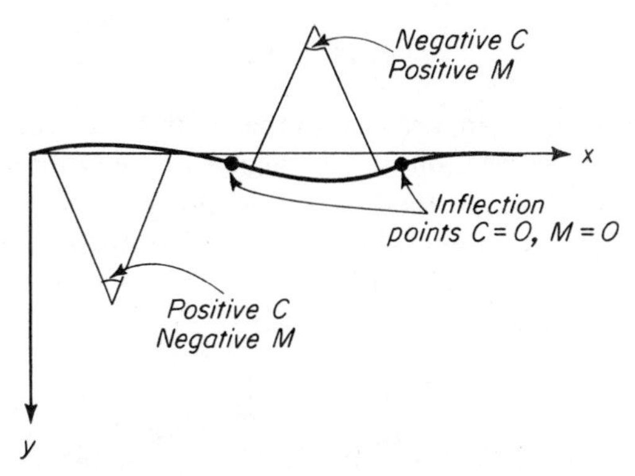

FIGURE 4.1.2

The shear V is equal to the rate of change of the moment:

$$V = \frac{dM}{dx} = -EI\frac{d^3y}{dx^3}. \tag{4.1.4b}$$

The distributed load q is equal and opposite to the rate of change of the shear:

$$q = -\frac{dV}{dx} = EI\frac{d^4y}{dx^4}. \tag{4.1.4c}$$

Equations (4.1.4b) and (4.1.4c) state that an element of the beam of length dx is in equilibrium in rotation and in the y-direction under the load q, the shear V, and the moment M.

6. The shear stress f_v on a beam section x perpendicular to the beam axis [and hence, for rotational equilibrium, also on a section parallel to the

*Equation (4.1.4a) is derived in Section 4.3. The curvature C is the rate of change of the slope of the deflected beam axis $y(x)$. For deflections y small compared to the beam span:

$$C(x) \approx \frac{d^2y}{dx^2}. \tag{b}$$

The minus sign in (4.1.4a) is due to the fact that, usually, a bending moment is called positive when it induces tension at the bottom and compression at the top of the beam, and therefore produces an upward curvature, which is a geometrically negative curvature when the y-axis is positive downward (Fig. 4.1.2).

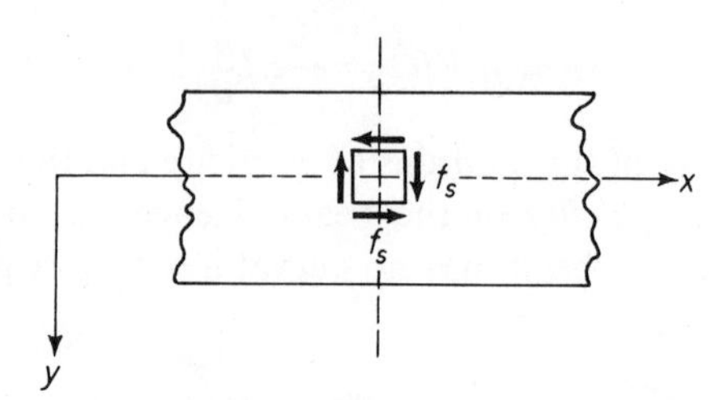

FIGURE 4.1.3

beam axis (Fig. 4.1.3)] is proportional to the shear V, i.e., the resultant of all the loads (including the reaction) perpendicular to the beam from one end of the beam up to the section x. Its distribution along the beam depth depends on the shape of the cross section. For rectangular cross sections the distribution of shear stress is parabolic with the maximum value of f_s occurring at the neutral axis and equal to:

$$f_{s,\max} = 1.5\frac{V}{A}, \tag{4.1.5}$$

where A is the area of the cross section.

For I- and wide-flange beams f_s is approximately constant for the full depth of the section and equals:

$$f_s = \frac{V}{A_w} \tag{4.1.6}$$

where A_w is the area of the web.

Beam dimensions are usually determined by bending stresses but must also be checked for shear, since sometimes shear is the governing factor in beam design. This is particularly true for sections near the supports of beams made out of materials with low tensile strength or thin sections.

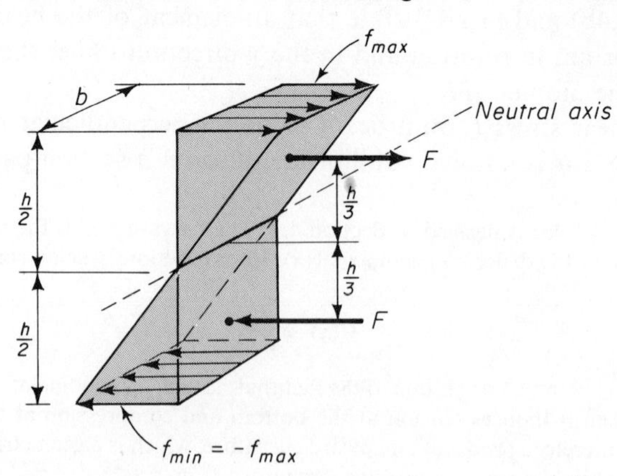

FIGURE 4.1.4

The bending stress formula (4.1.2) states the equilibrium in rotation between the moment M of the applied loads and the resisting couple M_f of the bending stresses. For example, given a homogeneous rectangular cross section of depth h and width b (Fig. 4.1.4), the resultants of the stresses on the upper and lower halves of the section:

$$F = \frac{1}{2} f_{max} \frac{bh}{2} \tag{c}$$

are located at the centroid of their triangular distribution, i.e., at a distance $\frac{2}{3}(h/2) = h/3$ from the neutral axis. The couple of the F equals:

$$M_f = 2\left(\frac{1}{2} f_{max} \frac{bh}{2}\right)\frac{h}{3} = \frac{1}{6} h^2 b f_{max}. \tag{d}$$

Equating M_f to the bending moment M of the applied loads and remembering that for a rectangular cross section:

$$I = \frac{bh^3}{12}, \qquad c = \frac{h}{2}, \qquad S_{1,2} = \frac{I}{c} = \frac{bh^2}{6},$$

it is seen that (d) is identical with (4.1.2).

The cross-sectional characteristics I, c, S, A, and $r = \sqrt{I/A}$ (the so-called *radius of gyration*, which plays an important role in buckling theory) are given in most manuals on standard steel, aluminum, and wood sections. Table 4.1.1 shows A, I, c, and S for the most commonly encountered symmetrical cross-sectional shapes, including some hollow sections and some thin, curved sections.

Table 4.1.2 shows the moment and shear diagrams for the most commonly encountered loading and support conditions of structural beams.

In view of the assumption of linearly elastic behavior, the moments and shears due to a combination of loads may be obtained by *superposition* of the moments and shears due to each load. For example, the bending moment at midspan of the fixed-end beam of Fig. 4.1.5 on page 61 is obtained from Table 4.1.2 (12), (13), and (14):

$$M\left(\frac{l}{2}\right) = \frac{5}{32} Pl + \frac{1}{16} w_B l^2 + \frac{7}{240} w'_A l^2$$

$$= \frac{5}{32}(130 \times 6) + \frac{1}{16}(150 \times 6^2) + \frac{7}{240}(300 \times 6^2) = 775 \text{ kN·m } (577 \text{ k·ft}).$$

Example 4.1.1. A 4.5 m (15 ft) span, simply supported wood beam with $F_b =$ 6.2 MPa (900 psi) carries the contributory load of a width of 3 m (10 ft) of floor weighing 4.8 kN/m² (100 psf) plus a live load of 2.4 kN/m² (50 psf). Design the beam with a depth of not more than 450 mm (18 in.).

The distributed load on the beam is:

$$w = (4.8 + 2.4) \times 3 = 21.6 \text{ kN/m}.$$

Table 4.1.1

No.	Section	A	C_x	I_x	S_x
1	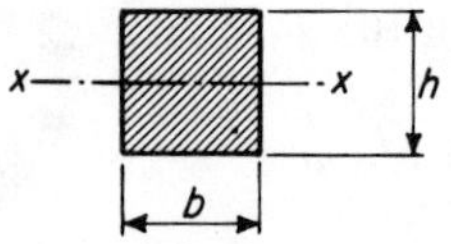	bh	$\frac{h}{2}$	$\frac{bh^3}{12}$	$\frac{bh^2}{6}$
2	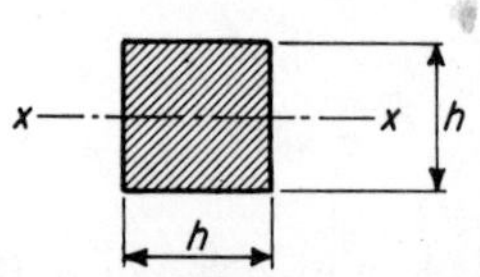	h^2	$\frac{h}{2}$	$\frac{h^4}{12}$	$\frac{h^3}{6}$
3	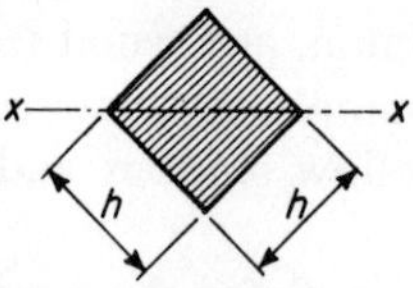	h^2	$\frac{h}{\sqrt{2}}$	$\frac{h^4}{12}$	$\frac{\sqrt{2}}{12}h^3 = 0.12h^3$
4	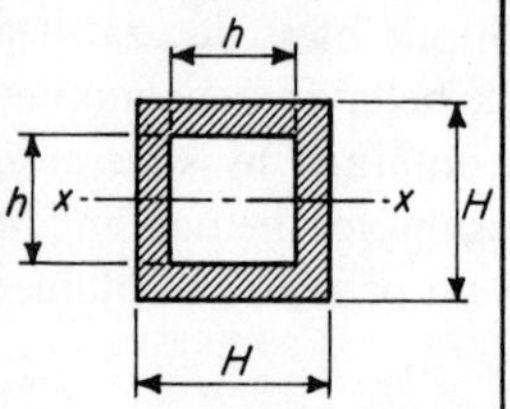	$H^2 - h^2$	$\frac{H}{2}$	$\frac{H^4 - h^4}{12}$	$\frac{H^4 - h^4}{6H}$
5	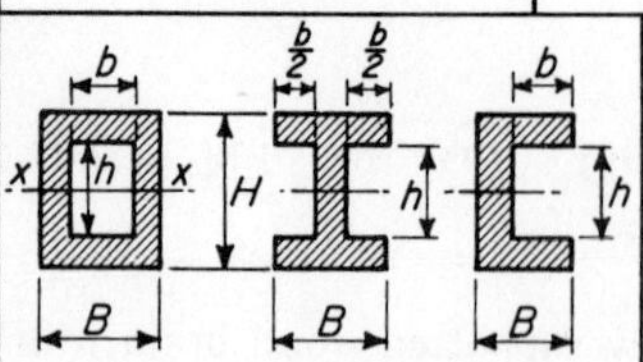	$BH - bh$	$\frac{H}{2}$	$\frac{BH^3 - bh^3}{12}$	$\frac{BH^3 - bh^3}{6H}$

Table 4.1.1. Continued.

No.	Section	A	C_x	I_x	S_x
6		$b(H-h) = 2bt$	$\frac{H}{2}$	$\frac{b}{12}(H^3-h^3)$; for $t \ll H$ $\frac{1}{2}bth^2 = \frac{1}{4}Ah^2$	$\frac{b}{6}\frac{H^3-h^3}{H}$; for $t \ll H$ $bth = \frac{1}{2}Ah$
7		$BH + bh$	$\frac{H}{2}$	$\frac{BH^3+bh^3}{12}$	$\frac{BH^3+bh^3}{6H}$
8		$\frac{bh}{2}$	$C_2 = \frac{2}{3}h$ $C_1 = \frac{1}{3}h$	$\frac{bh^3}{36}$	$S_2 = \frac{bh^2}{24}$ $S_1 = \frac{bh^2}{12}$
9		$\frac{\pi}{4}D^2 = 0.79D^2 = 3.14R^2$	$\frac{D}{2} = R$	$\frac{\pi D^4}{64} \approx 0.05D^4 \approx 0.79R^4$	$\frac{\pi D^3}{32} \approx 0.1D^3 \approx 0.79R^3$
10	$(t \ll D)$	$\pi Dt = 6.28Rt$	$\frac{D}{2} = R$	$\frac{\pi}{8}D^3t = 0.39D^3t = 3.14R^3t$	$\frac{\pi}{4}D^2t = 0.79D^2t = 3.14R^2t$

Table 4.1.1. Continued.

No.	Section	A	C_x	I_x	S_x
11		$\frac{\pi}{2}R^2 = 1.57R^2$	$C_2 = 0.58R$ $C_1 = 0.42R$	$0.11R^4$	$S_2 = 0.19R^3$ $S_1 = 0.26R^3$
12	$t \ll R$	πRt	$C_2 = 0.36R$ $C_1 = 0.64R$	$0.30R^3t$	$S_2 = 0.83R^2t$ $S_1 = 0.47R^2t$

Its maximum moment by Table 4.1.2(5) equals:

$$M = \tfrac{1}{8}(21.6)(4.5)^2 = 54.68 \text{ kN}\cdot\text{m } (42.2 \text{ k}\cdot\text{ft}).$$

Hence, with $S = bh^2/6$ and $h = 450$ mm (18 in.):

$$6.2 = \frac{54.68 \times 1\,000^2}{\tfrac{1}{6}(450)^2 b} \quad \therefore \quad b = \frac{54.68 \times 1\,000^2 \times 6}{6.2 \times (450)^2} \approx 260 \text{ mm } (\approx 12 \text{ in.}).$$

The weight of the beam, which was neglected, is for a timber of specific weight 6.3 kN/m³ (40 lb/cu ft):

$$260 \times 450 \times 6.3 = 0.737 \text{ kN/m } (60 \text{ lb/ft}).$$

The actual maximum bending stress in the beam is thus:

$$f_b = \frac{1}{8}\frac{(22.337 \times 4.5^2)1\,000^2}{\tfrac{1}{6}(260 \times 450^2)} = 6.44 \text{ MPa } (710 \text{ psi}).$$

The maximum shear in the beam is:

$$V = \tfrac{1}{2}(22.337)(4.5) = 50.26 \text{ kN},$$

and by (4.1.5) the maximum shear stress:

$$f_{s,\max} = 1.5\frac{50.26}{260 \times 450} = 0.000\,644 \text{ kN/mm}^2 = 0.644 \text{ MPa } (93 \text{ psi})$$

is smaller than the allowable shear stress for wood, which is about 0.66 MPa (see Table 2.6.1).

Example 4.1.2. A steel beam is fixed at one end A, simply supported at the other end B, and has a span of 10 m (30 ft). It carries a uniform load of 45 kN/m (3 k/ft) and a triangular load with a peak intensity of 60 kN/m (4 k/ft) at the fixed

Table 4.1.2

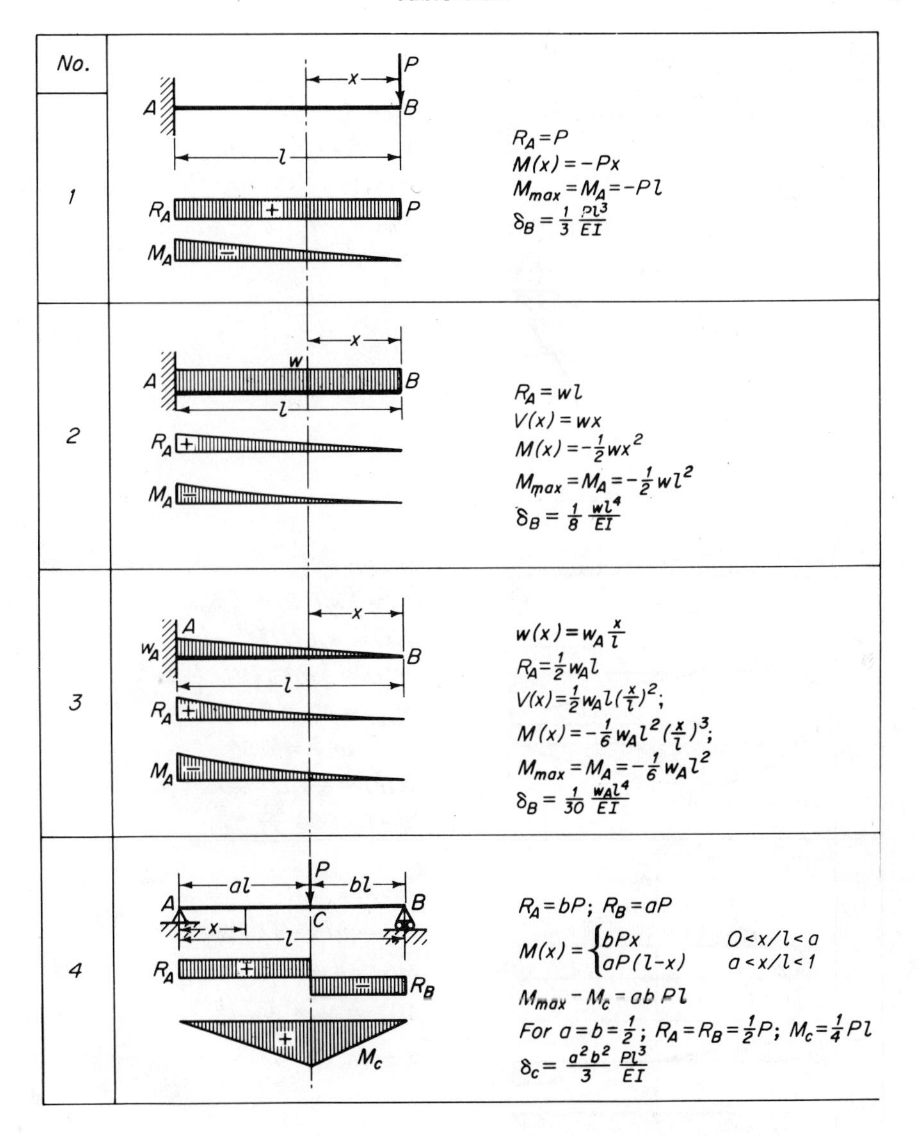

No.	Formulas
1	$R_A = P$ $M(x) = -Px$ $M_{max} = M_A = -Pl$ $\delta_B = \frac{1}{3}\frac{Pl^3}{EI}$
2	$R_A = wl$ $V(x) = wx$ $M(x) = -\frac{1}{2}wx^2$ $M_{max} = M_A = -\frac{1}{2}wl^2$ $\delta_B = \frac{1}{8}\frac{wl^4}{EI}$
3	$w(x) = w_A\frac{x}{l}$ $R_A = \frac{1}{2}w_A l$ $V(x) = \frac{1}{2}w_A l(\frac{x}{l})^2;$ $M(x) = -\frac{1}{6}w_A l^2(\frac{x}{l})^3;$ $M_{max} = M_A = -\frac{1}{6}w_A l^2$ $\delta_B = \frac{1}{30}\frac{w_A l^4}{EI}$
4	$R_A = bP;\ R_B = aP$ $M(x) = \begin{cases} bPx & 0 < x/l < a \\ aP(l-x) & a < x/l < 1 \end{cases}$ $M_{max} = M_c = ab\,Pl$ For $a = b = \frac{1}{2};\ R_A = R_B = \frac{1}{2}P;\ M_c = \frac{1}{4}Pl$ $\delta_c = \frac{a^2b^2}{3}\frac{Pl^3}{EI}$

Table 4.1.2. Continued.

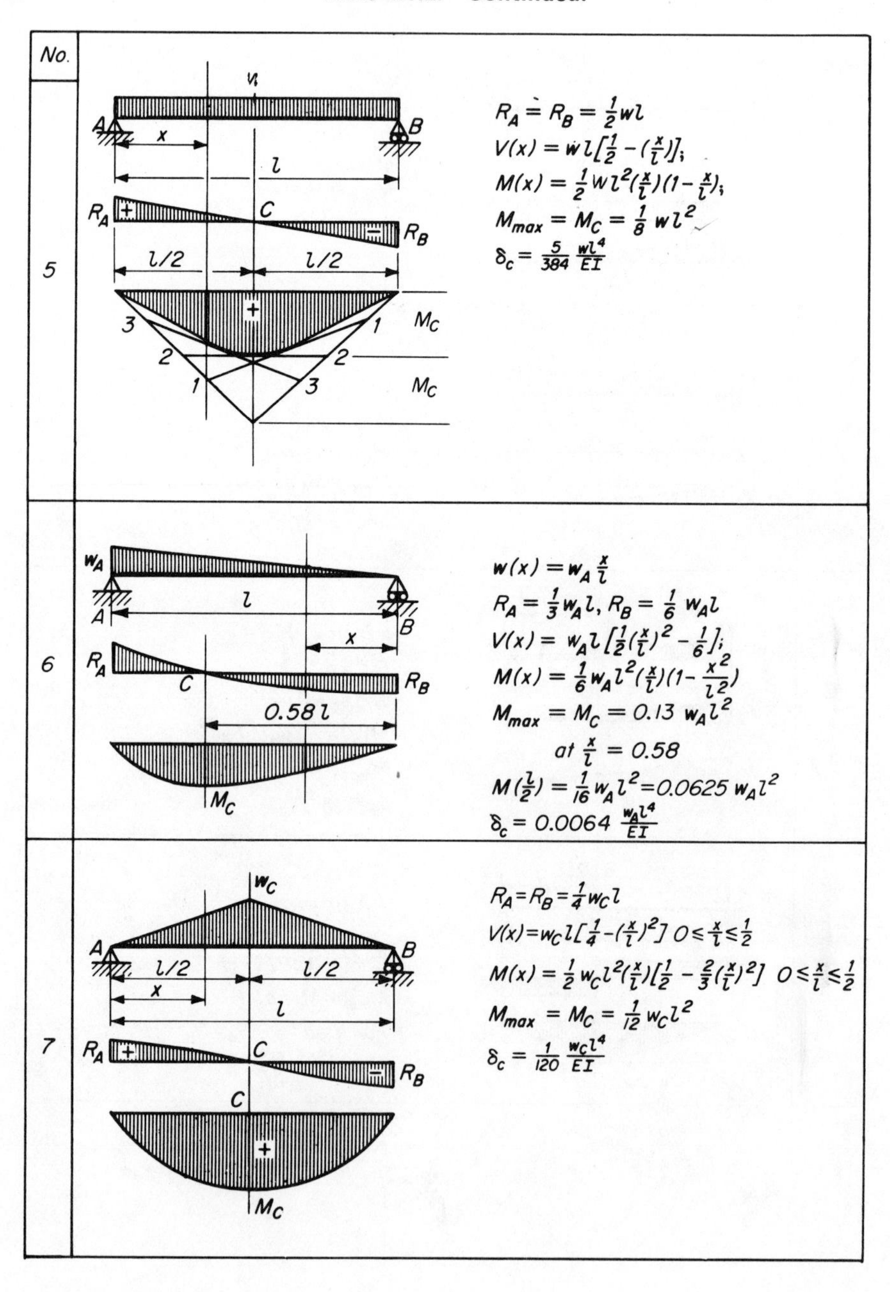

No.	Formulas
5	$R_A = R_B = \frac{1}{2}wl$ $V(x) = wl[\frac{1}{2} - (\frac{x}{l})]$; $M(x) = \frac{1}{2}wl^2(\frac{x}{l})(1-\frac{x}{l})$; $M_{max} = M_C = \frac{1}{8}wl^2$ $\delta_c = \frac{5}{384}\frac{wl^4}{EI}$
6	$w(x) = w_A\frac{x}{l}$ $R_A = \frac{1}{3}w_A l,\ R_B = \frac{1}{6}w_A l$ $V(x) = w_A l[\frac{1}{2}(\frac{x}{l})^2 - \frac{1}{6}]$; $M(x) = \frac{1}{6}w_A l^2(\frac{x}{l})(1-\frac{x^2}{l^2})$ $M_{max} = M_C = 0.13\ w_A l^2$ at $\frac{x}{l} = 0.58$ $M(\frac{l}{2}) = \frac{1}{16}w_A l^2 = 0.0625\ w_A l^2$ $\delta_c = 0.0064\ \frac{w_A l^4}{EI}$
7	$R_A = R_B = \frac{1}{4}w_C l$ $V(x) = w_C l[\frac{1}{4} - (\frac{x}{l})^2]\ 0 \le \frac{x}{l} \le \frac{1}{2}$ $M(x) = \frac{1}{2}w_C l^2(\frac{x}{l})[\frac{1}{2} - \frac{2}{3}(\frac{x}{l})^2]\ 0 \le \frac{x}{l} \le \frac{1}{2}$ $M_{max} = M_C = \frac{1}{12}w_C l^2$ $\delta_c = \frac{1}{120}\frac{w_C l^4}{EI}$

Table 4.1.2. Continued.

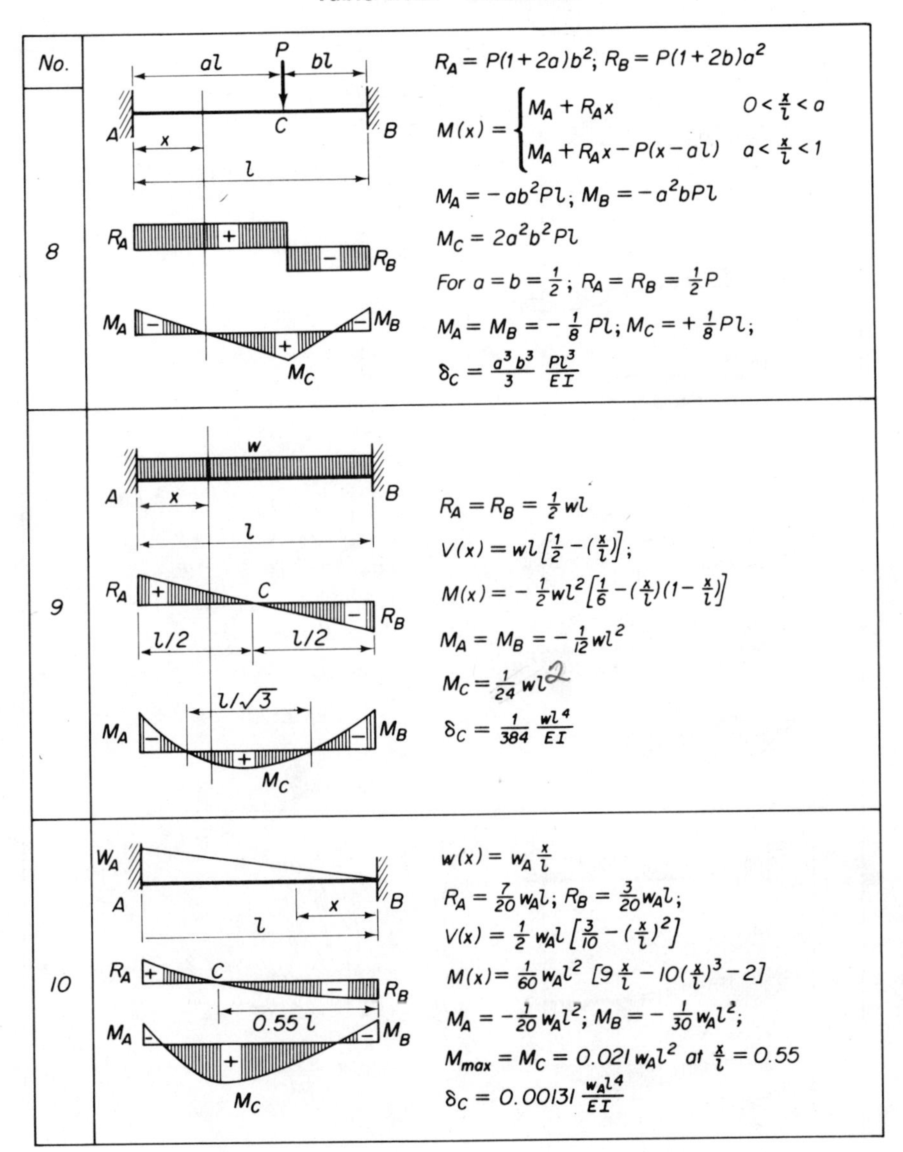

No.	Formulas
8	$R_A = P(1+2a)b^2;\ R_B = P(1+2b)a^2$ $M(x) = \begin{cases} M_A + R_A x & 0 < \frac{x}{l} < a \\ M_A + R_A x - P(x - al) & a < \frac{x}{l} < 1 \end{cases}$ $M_A = -ab^2Pl;\ M_B = -a^2bPl$ $M_C = 2a^2b^2Pl$ For $a = b = \frac{1}{2};\ R_A = R_B = \frac{1}{2}P$ $M_A = M_B = -\frac{1}{8}Pl;\ M_C = +\frac{1}{8}Pl;$ $\delta_C = \frac{a^3b^3}{3}\frac{Pl^3}{EI}$
9	$R_A = R_B = \frac{1}{2}wl$ $V(x) = wl\left[\frac{1}{2} - \left(\frac{x}{l}\right)\right];$ $M(x) = -\frac{1}{2}wl^2\left[\frac{1}{6} - \left(\frac{x}{l}\right)\left(1 - \frac{x}{l}\right)\right]$ $M_A = M_B = -\frac{1}{12}wl^2$ $M_C = \frac{1}{24}wl^2$ $\delta_C = \frac{1}{384}\frac{wl^4}{EI}$
10	$w(x) = w_A\frac{x}{l}$ $R_A = \frac{7}{20}w_A l;\ R_B = \frac{3}{20}w_A l;$ $V(x) = \frac{1}{2}w_A l\left[\frac{3}{10} - \left(\frac{x}{l}\right)^2\right]$ $M(x) = \frac{1}{60}w_A l^2\left[9\frac{x}{l} - 10\left(\frac{x}{l}\right)^3 - 2\right]$ $M_A = -\frac{1}{20}w_A l^2;\ M_B = -\frac{1}{30}w_A l^2;$ $M_{max} = M_C = 0.021\,w_A l^2$ at $\frac{x}{l} = 0.55$ $\delta_C = 0.00131\frac{w_A l^4}{EI}$

Table 4.1.2. Continued.

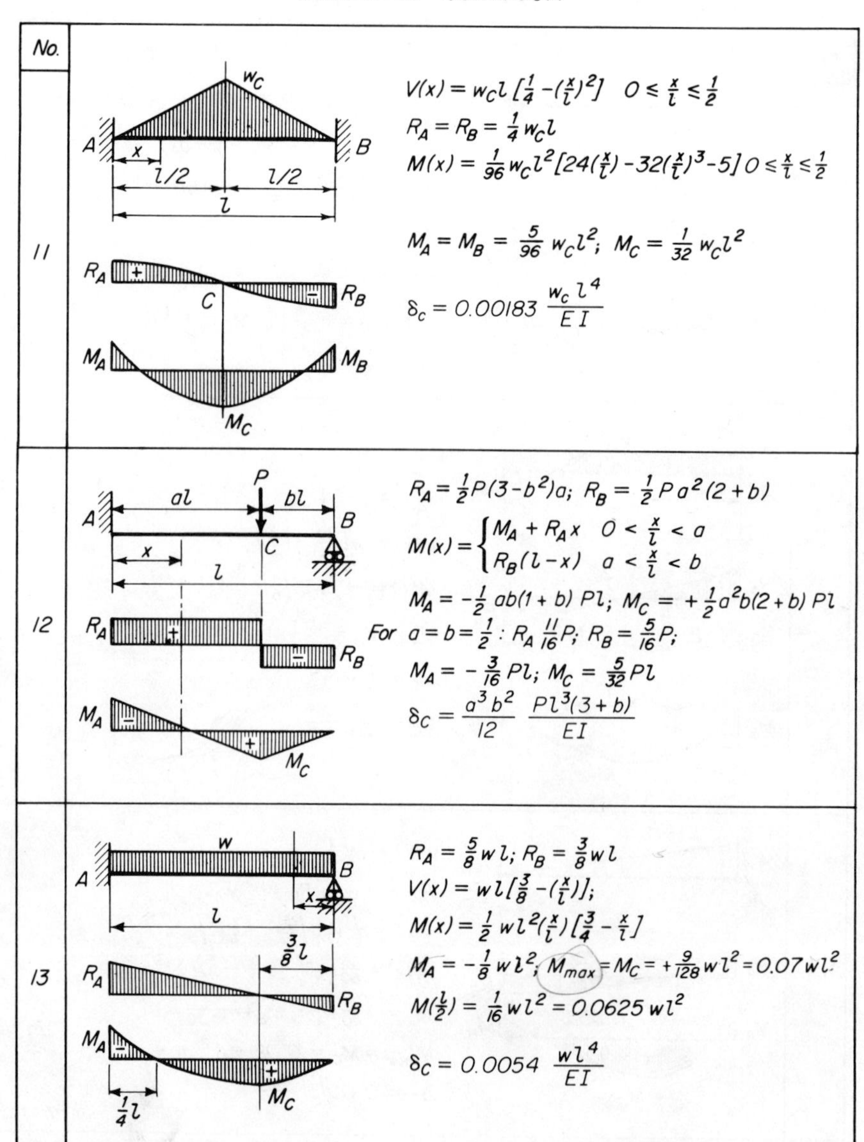

No.	
11	$V(x) = w_C l\,[\frac{1}{4} - (\frac{x}{l})^2] \quad 0 \le \frac{x}{l} \le \frac{1}{2}$ $R_A = R_B = \frac{1}{4} w_C l$ $M(x) = \frac{1}{96} w_C l^2 [24(\frac{x}{l}) - 32(\frac{x}{l})^3 - 5] \quad 0 \le \frac{x}{l} \le \frac{1}{2}$ $M_A = M_B = \frac{5}{96} w_C l^2;\ M_C = \frac{1}{32} w_C l^2$ $\delta_c = 0.00183 \frac{w_c l^4}{EI}$
12	$R_A = \frac{1}{2} P(3-b^2)a;\ R_B = \frac{1}{2} P a^2 (2+b)$ $M(x) = \begin{cases} M_A + R_A x & 0 < \frac{x}{l} < a \\ R_B(l-x) & a < \frac{x}{l} < b \end{cases}$ $M_A = -\frac{1}{2} ab(1+b) Pl;\ M_C = +\frac{1}{2} a^2 b(2+b) Pl$ For $a = b = \frac{1}{2}$: $R_A \frac{11}{16} P;\ R_B = \frac{5}{16} P;$ $M_A = -\frac{3}{16} Pl;\ M_C = \frac{5}{32} Pl$ $\delta_C = \frac{a^3 b^2}{12} \frac{Pl^3 (3+b)}{EI}$
13	$R_A = \frac{5}{8} wl;\ R_B = \frac{3}{8} wl$ $V(x) = wl[\frac{3}{8} - (\frac{x}{l})];$ $M(x) = \frac{1}{2} wl^2 (\frac{x}{l}) [\frac{3}{4} - \frac{x}{l}]$ $M_A = -\frac{1}{8} wl^2;\ M_{max} = M_C = +\frac{9}{128} wl^2 = 0.07 wl^2$ $M(\frac{l}{2}) = \frac{1}{16} wl^2 = 0.0625 wl^2$ $\delta_C = 0.0054 \frac{wl^4}{EI}$

Table 4.1.2. Continued.

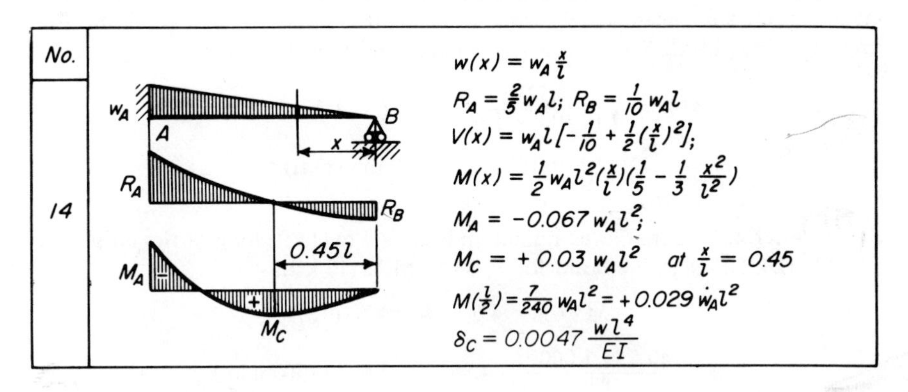

end. Design the beam for $F_b = 165$ MPa (24 ksi) and $F_s = 100$ MPa (14.5 ksi). By Table 4.1.2 (13) and (14):

$$M_{\max} = M_A = -0.125 \times 45 \times 10^2 - 0.067 \times 60 \times 10^2 = -965 \text{ kN}\cdot\text{m}.$$

$$S = \frac{M}{F_b} = \frac{965 \times 1\,000 \times 1\,000}{165} = 5\,848 \times 10^3 \text{ mm}^3 \text{ (347 cu in.)}.$$

From Appendix A, the minimum weight beam is a W760 × 185 (W30 × 124) with $S = 5\,820 \times 10^3$ mm³. A W610 × 216 (W24 × 145) with $S = 6\,080 \times 10^3$ mm³ will give an acceptable shallower but heavier beam. The maximum shear in the beam equals, by Table 4.1.2 (13) and (14):

$$V = \tfrac{3}{8} \times 45 \times 10 + \tfrac{2}{5} \times 60 \times 10 = 521.3 \text{ kN}.$$

The web of a W760 × 185 is 766 mm high and 14.9 mm wide; hence, by (4.1.6):

$$f_{s,\max} = \frac{521.3}{766 \times 14.9} = 0.045\,7 \text{ kN/mm}^2 = 45.7 \text{ MPa (5.9 ksi)},$$

or well below an F_s of 100 MPa.

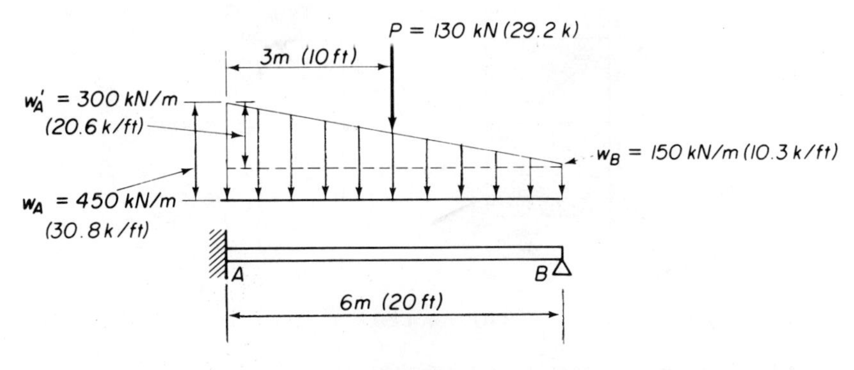

FIGURE 4.1.5

Example 4.1.3. Evaluate the *longest span* l_u of a simply supported W920 × 446 (W36 × 300) steel beam barely capable of supporting its dead load elastically. With $F_y = 414$ MPa, $S = 18\,100 \times 10^3$ mm³, and $w = 4.37$ kN/m:

$$414 = \frac{1}{8}\frac{(4.37)l_u^2}{18\,100 \times 10^3} \quad \therefore \quad l_u^2 = 1.37 \times 10^{10};$$

$$l_u = 117.1 \times 10^5 \text{ mm} = 117 \text{ m (383 ft).*}$$

Example 4.1.4. A cantilever aluminum beam 4.5 m (15 ft) long carries a tip load of 9 kN (2 k). Design the beam for $F_b = 83$ MPa (12 ksi).

$$M = 9 \times 4.5 = 40.5 \text{ kN·m};$$

$$S = \frac{40.5 \times 1\,000^2}{83} = 4.88 \times 10^5 \text{ mm}^3 \text{ (30 cu in.)}.$$

From the *Alcoa Structural Handbook*, for example, a W200 × 19 (W8 × 13) with $S = 495 \times 10^3$ mm³ is chosen with a web area of 200 × 12.7 = 2 540 mm². The maximum shear stress in the beam is

$$f_s = \frac{9 \times 1\,000}{2\,450} = 3.54 \text{ MPa (510 psi)}.$$

4.2 REINFORCED CONCRETE BEAMS [7.2]

The resisting moment of a rectangular concrete section is based on the assumption that the section is cracked by tension either below or above the neutral axis and that, therefore, the steel reinforcement carries all the tensile stress (Fig. 4.2.1). The total tensile and compressive forces F in the steel and the concrete are:

$$F = A_s f_s = \tfrac{1}{2}b(kd)f_c, \tag{a}$$

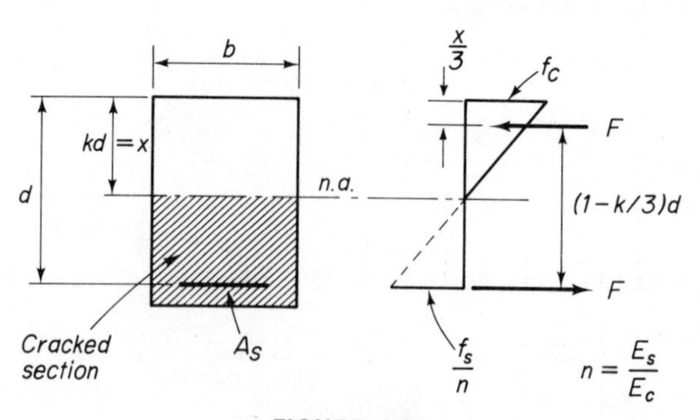

FIGURE 4.2.1

*It will be seen in Example 5.4.4 that the ultimate span is limited by buckling considerations, if the beam is not supported laterally.

in which f_s and f_c are the maximum steel and concrete stresses, respectively, and kd is the depth of the compressed part of the section. The resisting moment M_R of the section is the couple of the internal forces F:

$$M_R = A_s f_s\left(1 - \frac{k}{3}\right)d = \frac{1}{2}bd^2k\left(1 - \frac{k}{3}\right)f_c. \tag{b}$$

The value of k for sections with tensile reinforcement only is approximately $\frac{3}{8}$. A good approximation for M_R is thus:

$$M_R = \tfrac{1}{6}f_c bd^2. \tag{b'}$$

Equating M_R to the maximum moment in a uniformly loaded, simply supported beam, we obtain:

$$\frac{wl^2}{8} = \frac{1}{6}f_c bd^2 \quad \therefore \quad \left(\frac{d}{l}\right)^2 = \frac{3}{4}\frac{w}{f_c b}. \tag{c}$$

The depth-span ratio d/l varies between $\frac{1}{10}$ and $\frac{1}{20}$ for beams and may be as low as $\frac{1}{30}$ for slabs. The curves of Fig. 4.2.2 give *w in kN/m (lb/ft) versus b in mm (in.)* for various ratios d/l and values of the *ultimate concrete stress* f'_c. The maximum allowable compressive stress f_c is taken from Table 2.4.2.

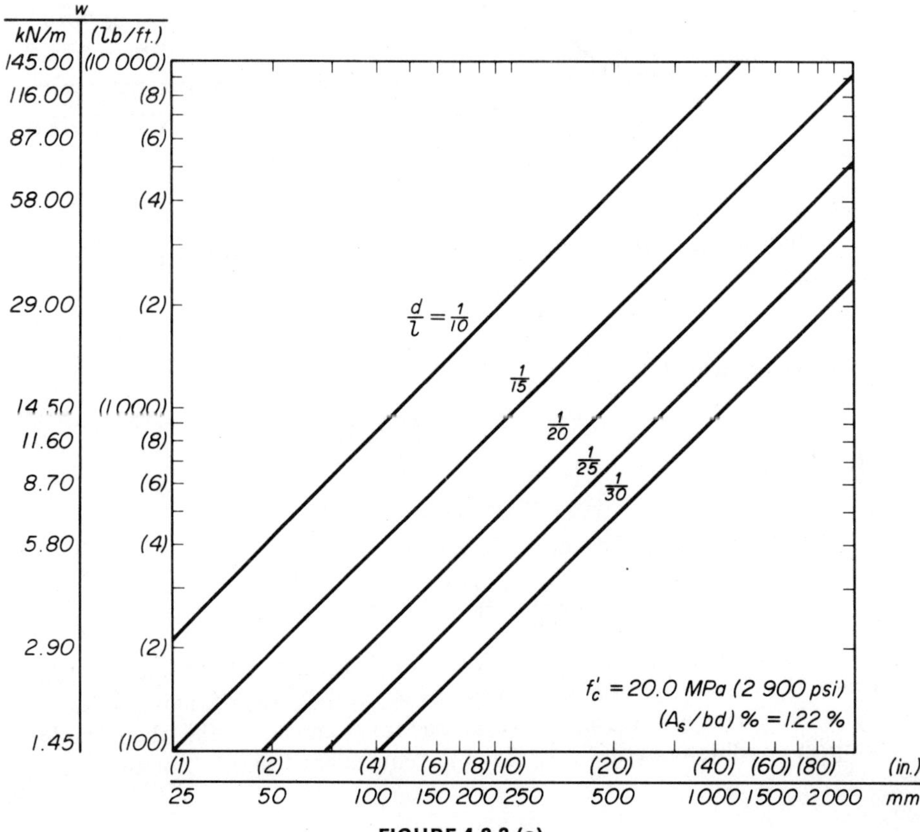

FIGURE 4.2.2 (a)

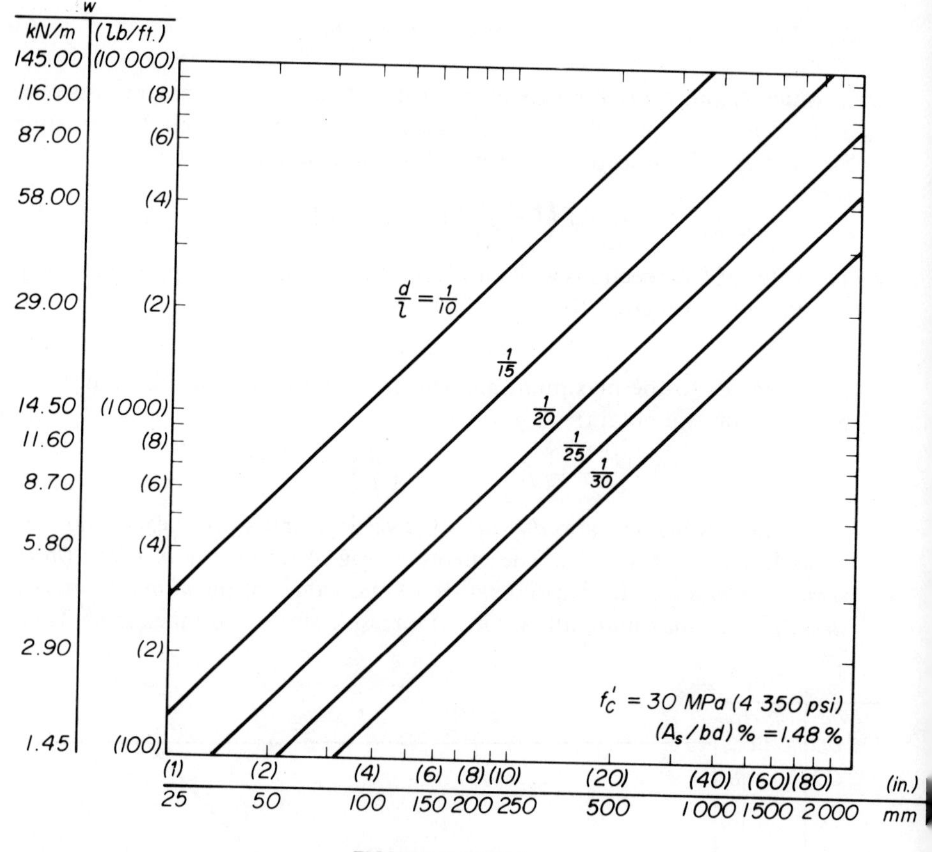

FIGURE 4.2.2 (b)

Indicating by n the ratio E_s/E_c, the stress f_s/n is expressed in terms of f_s (Fig. 4.2.1) by:

$$\frac{f_s/n}{f_c} = \frac{(1-k)d}{kd},$$

from which, with $k = \frac{3}{8}$:

$$f_s = n\frac{1-k}{k}f_c = \frac{5}{3}nf_c. \tag{d}$$

By (d) and (a),

$$A_s f_s = \tfrac{5}{3}A_s n f_c = \tfrac{1}{2}k(bd)f_c,$$

and the percentage of steel area is given by:

$$100\frac{A_s}{bd} = \frac{3}{5}\frac{k}{2n}100 = \frac{45}{4n}. \tag{e}$$

The ratio n takes the values 9.2, 7.6, 6.6 for f'_c equal to 20, 30, and 40 MPa hence, the percentage of steel is, respectively, 1.22%, 1.48%, and 1.70%.

The curves of Figs. 4.2.2 may be used for beams with other than simple support conditions by noticing that the "equivalent length" l_e of such uni-

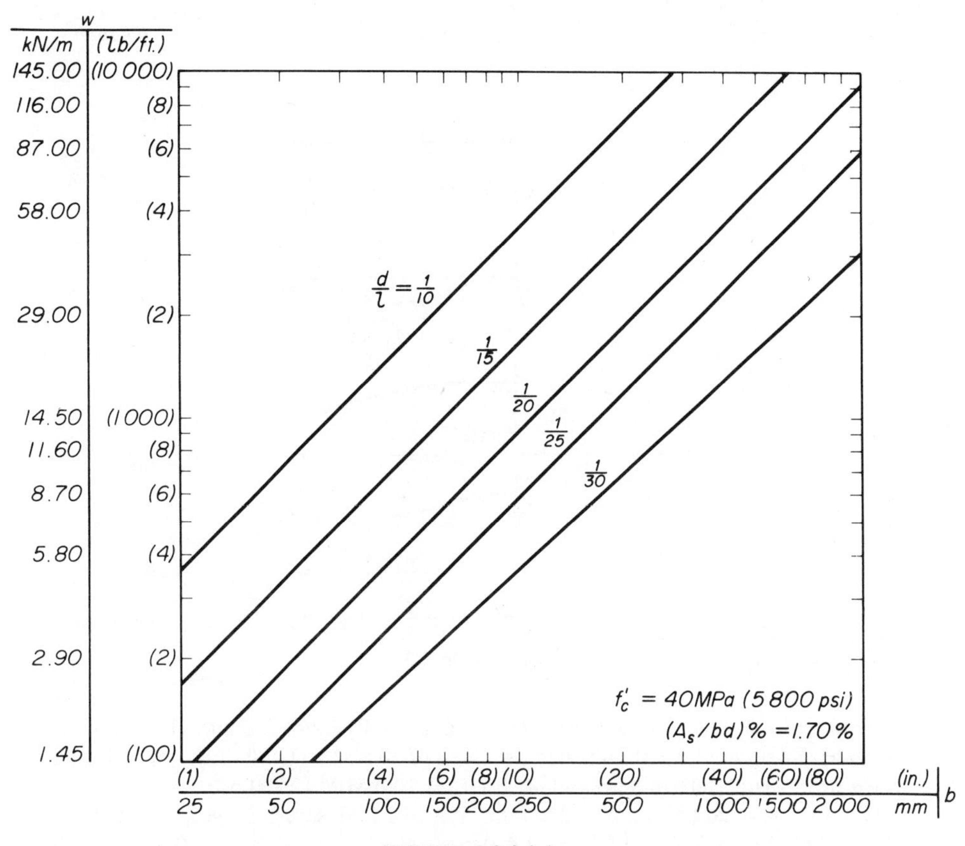

FIGURE 4.2.2 (c)

formly loaded beams (see Sections 4.4 and 4.5) is the length between their inflection points (Fig. 4.2.3):

$l_e = 0.58l$ for fixed-fixed beams,

$l_e = 2l$ for cantilever beams,

$$l_e = \begin{cases} 0.75l \text{ for the positive moment for fixed, simply supported beams,} \\ 1.00l \text{ for the negative moment for fixed, simply supported beams.} \end{cases}$$

The shear in a concrete beam is checked at a distance from a support equal to the distance d of the tensile steel from the extreme compressive fibers. The maximum allowable stress v_c is determined by the formula:

$$v_c = \frac{V}{bd},$$

and is limited by code to the values shown in Table 2.4.2. Since the average shear in reinforced concrete beams cannot be higher than a value established by code (of the order of 0.50 MPa, 70 psi), any shear above this value must be absorbed either by stirrups or by inclined bars in the direction of the equivalent tensile stress (see Section 13.2).

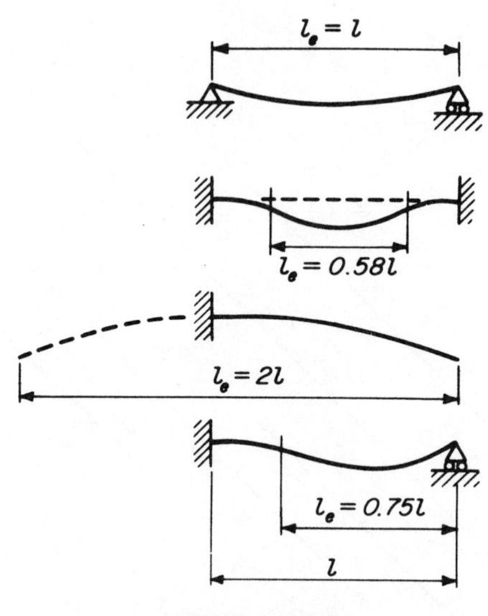

FIGURE 4.2.3

Example 4.2.1. A fixed concrete beam, 6 m (20 ft) long, 250 mm (10 in.) wide, with tensile reinforcement only, carries a load of 20 kN/m (1,500 lb/ft), including its own dead load. Determine d and A_s for $f_c' = 20.0$ MPa (2,900 psi).

For $w = 20$ kN/m, $b = 250$ mm, Fig. 4.2.2(a) gives $d/l_e \approx \frac{1}{12}$ and, with $l_e = 0.58l = 0.58 \times 6 \times 1\,000 = 3\,480$ mm, $d = \frac{1}{12} \times 3\,480 = 290$ mm ≈ 300 mm, $A_s = 0.0122 \times 250 \times 300 = 915$ mm^2 = two 26 mm (#8) bars and $h = 300 + 50 = 350$ mm.

With 50 mm of concrete to cover the steel bars, the beam depth is 350 mm and the beam dead load is $(250 \times 350/1\,000^2) \times 22.8 = 2$ kN/m, or 10% of the total load.

The shear stress in the beam measured at a distance $d = 300$ mm from each end is:

$$v_c = \frac{20 \times (3 - 0.3)}{250 \times 300} = 0.000\,7 \text{ kN/mm}^2 = 0.7 \text{ MPa (101.5 psi).}$$

This exceeds the allowable stress of 0.41 MPa (60 psi), which implies that either stirrups will have to be provided or the area of the beam will have to be increased to:

$$A_{\text{req.}} = \frac{20 \times (3 - 0.3)10^3}{0.41} = 132\,000 \text{ mm}^2 \text{ (205 sq in.),}$$

e.g., $b = 132\,000/300 \simeq 440$ mm (17 in.).

Example 4.2.2. A cantilevered concrete beam 2 m (7 ft) long with $d = 300$ mm (12 in.) carries a live load of 15 kN/m (1,000 lb/ft). Determine b and A_s for $f_c = 30$ MPa (4,350 psi).

Assuming the beam's dead load to be 10% of its live load, its total load is

$w = 16.5$ kN/m. Its d/l_e equals $300/(2 \times 2\,000) \simeq \frac{1}{13}$. With $f_c' = 30$ MPa, Fig. 4.2.2(b) gives:

$$b = 200 \text{ mm}; \quad A_s = 0.014\,8 \times 300 \times 200 = 888 \text{ mm}^2 \text{ (1.38 sq in.)},$$

or two 24 mm bars. The shear stress 300 mm from the support is:

$$v_c = \frac{16.5(2 - 0.3)}{200 \times 300} = 0.000\,47 \text{ kN/mm}^2 = 0.47 \text{ MPa (68 psi)},$$

which is less than the allowable stress of 0.5 MPa.

Example 4.2.3. A concrete beam fixed at one end and simply supported at the other has a span of 4.5 m (15 ft) and a cross section of 250 mm × 400 mm (10 in. × 15 in.). Determine the maximum live load it can carry if $f_c' = 30$ MPa (4,350 psi).

For the largest moment, which is negative, with $l_e = l = 4.5 \times 1\,000 = 4\,500$ mm, $d/l_e = \frac{400}{4\,500} \simeq \frac{1}{11} = \frac{1}{12}$, and $b = 250$ mm, Fig. 4.2.2(b) gives $w = 36$ kN/m. Since the dead load of the beam is $250 \times 400/[(1\,000)^2 \times 22.8] = 2.28$ kN/m, the maximum live load equals 33.72 kN/m.

The maximum shear at a distance of 400 mm (15 in.) from the fixed support is, from Table 4.1.2 (13):

$$V = \tfrac{5}{8}wl - w\tfrac{400}{1\,000} = 2.41w.$$

With an allowable shear stress F_s of 0.50 MPa (70 psi) (for $f_c' = 30$ MPa), the maximum load which can be carried without using stirrups is:

$$w = \frac{F_s bd}{2.41} = \frac{0.50 \times 250 \times 400}{2.41(1\,000)} \simeq 20.75 \text{ kN/m (1.42 k/ft)},$$

or a maximum live load of $20.75 - 2.28 = 18.47$ kN/m (1.27 k/ft).

4.3 STATICALLY DETERMINATE BEAM DEFLECTIONS [7.1, 7.2]

Although deflections and rotations of beams due to bending will be shown to be very small, their knowledge is essential to the solution of beam problems in which the equations of statics are insufficient for stress-determination, i.e., in the solution of *statically indeterminate beam* problems. Deflections due to shear are smaller than those due to bending in beams with $h/l < 3$ and are usually neglected (see Section 4.9).

Given a small element dx of a bent beam, the rotation $d\varphi$ of its right section with respect to its left section (Fig. 4.3.1) is:

$$d\varphi \approx \tan d\varphi = \frac{\epsilon\, dx}{h/2} = \frac{f_b/E}{h/2}\, dx, \tag{a}$$

and by (4.1.2), with $c = h/2$, i.e., with $f_b = M(h/2)/I$:

$$d\varphi = \frac{M\, dx}{EI}. \tag{b}$$

Since dx is an arc of radius r (*radius of curvature*) and angle $d\varphi$:

$$dx = r\, d\varphi \quad \therefore \quad \frac{dx}{d\varphi} = r, \tag{c}$$

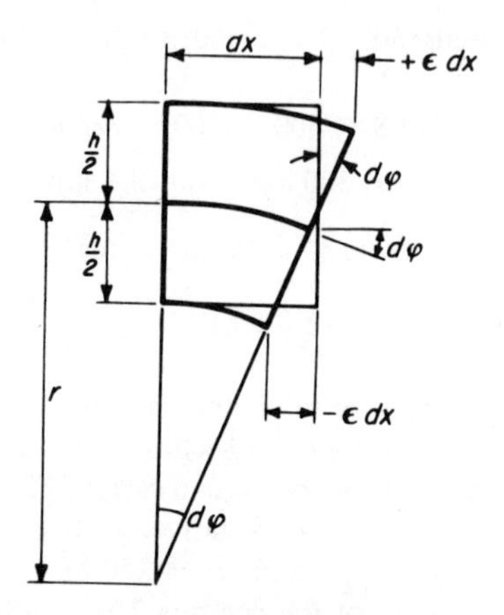

FIGURE 4.3.1

by means of which (b) becomes:

$$\frac{1}{r} = \frac{M}{EI}. \tag{d}$$

The reciprocal of the radius of curvature is called the *curvature* C and is assumed positive when downward, while M is positive when producing tensile stresses at the bottom of the beam. Thus:

$$C = -\frac{M}{EI}, \tag{e}$$

which is the basic *beam equation* (4.1.4a).

The rotation $\phi_{B/A}$ of a section B of a beam with respect to another section A is obtained by adding up all the small rotations $d\varphi$ of (b) from A to B. In mathematical terms:

$$\phi_{B/A} = \int_A^B d\varphi = \int_A^B \frac{M(x)\,dx}{EI}. \tag{4.3.1}$$

By (4.3.1) the rotations of the free end B of a cantilever of length l and flexural rigidity EI due to (a) a moment M at B, (b) a concentrated load P at B and (c) a uniform load w (Fig. 4.3.2) are:

(a) $M(x) = \text{const.}, \quad \phi_{B/A} = \int_A^B \frac{M\,dx}{EI} = \frac{M}{EI}\int_0^l dx = \frac{Ml}{EI},$ (4.3.2)

(b) $M(x) = Px, \quad \phi_{B/A} = \frac{P}{EI}\int_0^l x\,dx = \frac{1}{2}\frac{Pl^2}{EI},$ (4.3.3)

(c) $M(x) = \frac{wx^2}{2}, \quad \phi_{B/A} = \frac{w}{2EI}\int_0^l x^2\,dx = \frac{1}{6}\frac{wl^3}{EI}.$ (4.3.4)

To derive the end rotations of a simply supported beam under (a) equal end moments M, (b) a concentrated load P at midspan, and (c) a distributed load w (Fig. 4.3.3), we notice that the midspan section of the beam

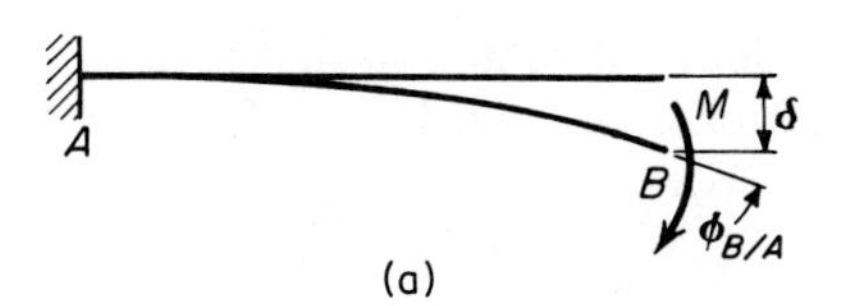

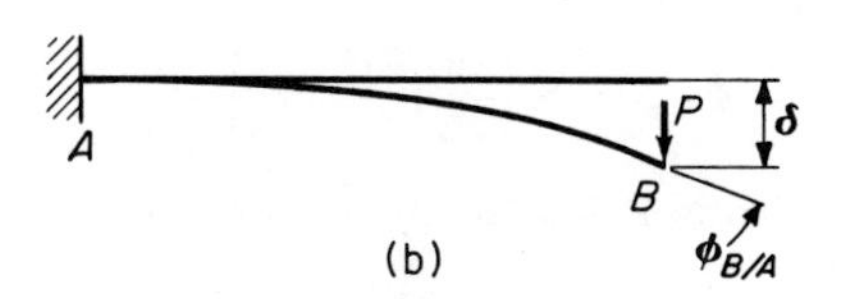

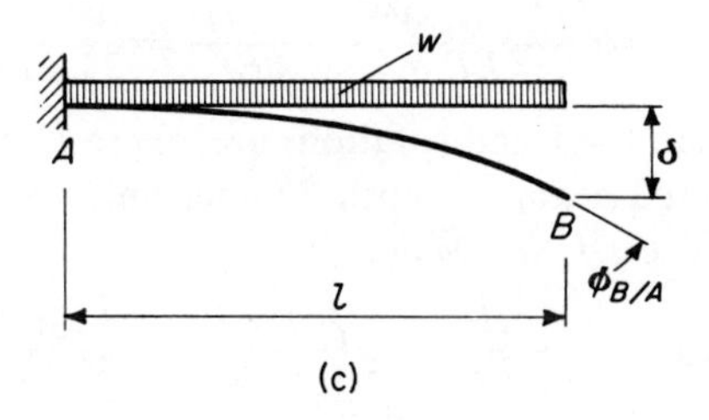

FIGURE 4.3.2

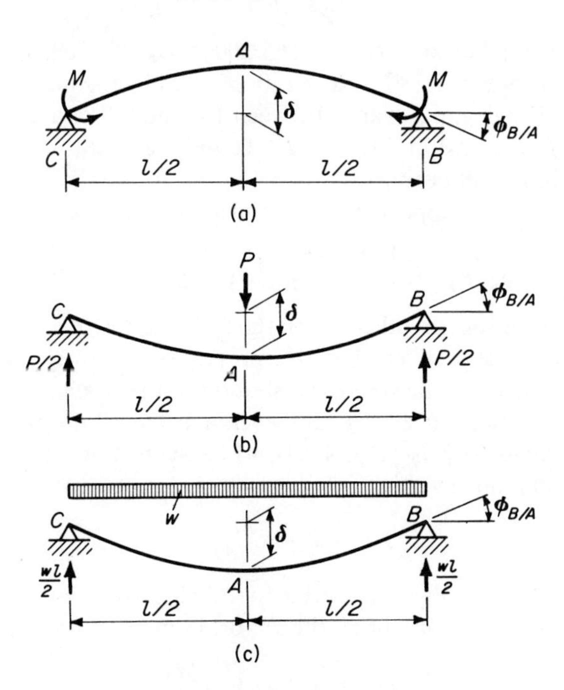

FIGURE 4.3.3

remains vertical, i.e., does not rotate, so that each half of the beam behaves like a cantilever beam of length $l/2$. Hence, for case (a):

$$\text{(a)} \qquad \phi_{B/A} = \frac{1}{2}\frac{Ml}{EI}. \tag{4.3.5}$$

For case (b), each half of the beam behaves like a cantilever of span $l/2$ under a tip load $P/2$; hence:

$$\text{(b)} \qquad \phi_{B/A} = \frac{(P/2)(l/2)^2}{2EI} = \frac{1}{16}\frac{Pl^2}{EI}. \tag{4.3.6}$$

For case (c), the half beam is loaded by a distributed load w down and a tip load $wl/2$ up; hence:

$$\text{(c)} \qquad \phi_{B/A} = \frac{(wl/2)(l/2)^2}{2EI} - \frac{w(l/2)^3}{6EI} = \frac{1}{24}\frac{wl^3}{EI}. \tag{4.3.7}$$

To see how small these end rotations are, express, e.g., the end rotation (4.3.4) of a cantilevered beam of depth h under uniform load in terms of its maximum bending stress $f_1 = (M_A h/2)/I$:

$$\phi_{B/A} = \frac{1}{6}\frac{wl^3}{EI} = \frac{1}{3}\left(\frac{wl^2}{2}\right)\frac{l}{EI} = \frac{1}{3}\frac{M_A l}{EI}$$

$$= \frac{1}{3}\frac{M_A h/2}{EI}\frac{l}{h/2} = \frac{2}{3}\left(\frac{f_1}{E}\right)\left(\frac{l}{h}\right). \tag{4.3.8}$$

For an A36 steel beam, $f_1/E = 165/200\,000 = 1/1\,200$, $l/h = 10\text{–}20$, and $\phi_{B/A}$ varies between 1/180 and 2/180 of a radian. Since a radian equals $180/3.14 = 57.3$ deg, $\phi_{B/A}$ varies between 1/3 and 2/3 of a degree. For a concrete beam $f_1/E = 9.0/21\,700 = 1/2\,400$, and the elastic end rotations are 1/2 as large as in a steel beam.

For a simply supported beam under uniform load.

$$\phi_{B/A} = \frac{1}{24}\frac{wl^3}{EI} = \frac{1}{3}\left(\frac{wl^2}{8}\right)\frac{l}{EI} = \frac{2}{3}\frac{Mh/2}{EI}\left(\frac{l}{h}\right) = \frac{2}{3}\left(\frac{f_1}{E}\right)\left(\frac{l}{h}\right),$$

and the end rotations are identical to the end rotation of a cantilevered beam.

It will be seen later that the prevention of these minute rotations substantially changes the stress distribution in the beam.

If only an element dx of a cantilever, a distance x from its free end, were elastic and rotated by $d\varphi$ (Fig. 4.3.4), while the rest of the beam is rigid, the free end would move vertically by $d\delta = x\,d\varphi$, or, by (b):

$$d\delta = x\frac{M\,dx}{EI}. \tag{f}$$

The vertical displacement δ due to the elastic rotation of all the elements dx of the beam is the sum of all the $d\delta$, or, in mathematical terms:

$$\delta = \int_A^B d\delta = \int_0^l x\frac{M\,dx}{EI}. \tag{4.3.9}$$

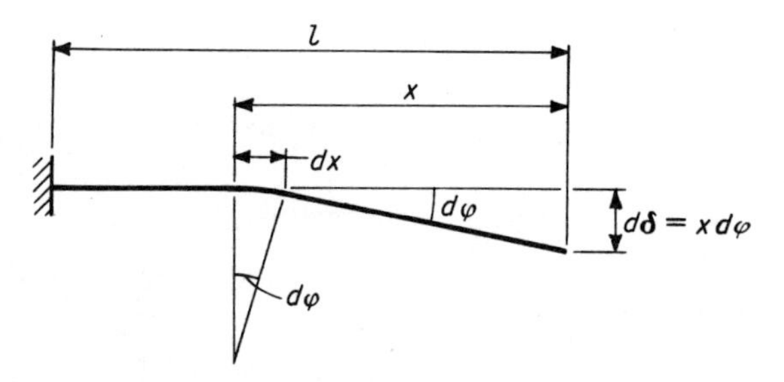

FIGURE 4.3.4

By (4.3.9) the deflection δ of the cantilevers of Figs. 4.3.2(a), (b), and (c) are:

(a) $$\delta = \frac{M}{EI}\int_0^l x\,dx = \frac{1}{2}\frac{Ml^2}{EI}, \tag{4.3.10}$$

(b) $$\delta = \frac{P}{EI}\int_0^l x^2\,dx = \frac{1}{3}\frac{Pl^3}{EI}, \tag{4.3.11}$$

(c) $$\delta = \frac{w}{2EI}\int_0^l x^3\,dx = \frac{1}{8}\frac{wl^4}{EI}. \tag{4.3.12}$$

For the simply supported beams of Figs. 4.3.3(a), (b), and (c), we obtain, by considering each half beam as a cantilever:

(a) $$\delta = \frac{M(l/2)^2}{2EI} = \frac{1}{8}\frac{Ml^2}{EI}, \tag{4.3.13}$$

(b) $$\delta = \frac{1}{3}\frac{(P/2)(l/2)^3}{EI} = \frac{1}{48}\frac{Pl^3}{EI}, \tag{4.3.14}$$

(c) $$\delta = \frac{1}{3}\frac{(wl/2)(l/2)^3}{EI} - \frac{1}{8}\frac{w(l/2)^4}{EI} = \frac{5}{384}\frac{wl^4}{EI}. \tag{4.3.15}$$

To obtain the end rotations of a simply supported beam under an end moment M_B (Fig. 4.3.5), notice, first, that for vertical equilibrium the beam reactions must be equal and opposite, and that for rotational equilibrium their couple must equal M_B, so that $R = M_B/l$. Drawing the tangent AC at A to the deflected beam axis, we see that the simply supported beam AB behaves like the cantilevered beam AC built-in at A and acted upon by both a moment M_B and a tip load R. Hence, by (4.3.10), (4.3.11), (4.3.2), and (4.3.3), the deflection δ and rotation $\phi_{B/A}$ referred to AC are:

$$\delta = \frac{1}{2}\frac{M_Bl^2}{EI} - \frac{1}{3}\frac{(M_B/l)l^3}{EI} = \frac{1}{6}\frac{M_Bl^2}{EI},$$

$$\phi_{B/A} = \frac{M_Bl}{EI} - \frac{1}{2}\frac{(M_B/l)l^2}{EI} = \frac{1}{2}\frac{M_Bl}{EI}.$$

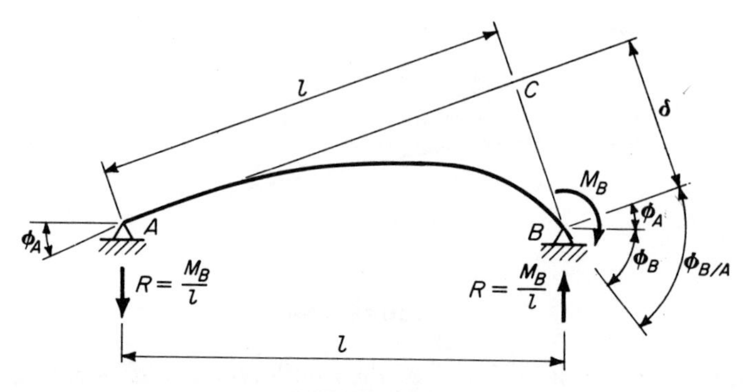

FIGURE 4.3.5

To obtain the end rotations ϕ_A and ϕ_B referred to the horizontal axis, we notice that ϕ_A is very small and, hence:

$$\phi_A \approx \frac{\delta}{l} = \frac{1}{6}\frac{M_B l}{EI},$$
$$\phi_B = \phi_{B/A} - \phi_A = \frac{1}{2}\frac{M_B l}{EI} - \frac{1}{6}\frac{M_B l}{EI} = \frac{1}{3}\frac{M_B l}{EI}. \tag{4.3.16}$$

Equations (4.3.16) allow the evaluation of the end rotations of a beam acted upon by two end moments M both rotating in the *same* sense (Fig. 4.3.6), by adding *algebraically* the rotations due to each moment:

$$\phi_A = -\frac{1}{6}\frac{Ml}{EI} + \frac{1}{3}\frac{Ml}{EI} = \frac{1}{6}\frac{Ml}{EI},$$
$$\phi_B = \frac{1}{6}\frac{Ml}{EI} - \frac{1}{3}\frac{Ml}{EI} = -\frac{1}{6}\frac{Ml}{EI}. \tag{4.3.17}$$

The end rotations are equal and opposite, the midspan deflection is zero, and there is an inflection point (zero curvature) with zero bending moment at midspan. Such a deflection is called *antisymmetrical.*

Expressing the cantilever displacement δ under a uniform load (4.3.12), in terms of the maximum bending stress, we obtain:

$$\delta = \frac{1}{8}\frac{wl^4}{EI} = \frac{1}{4}\left(\frac{wl^2}{2}\right)\frac{l^2}{EI} = \frac{1}{4}\left(\frac{M_A h/2}{EI}\right)\frac{2}{h}l^2,$$

from which:

$$\frac{\delta}{l} = \frac{1}{2}\left(\frac{f_1}{E}\right)\left(\frac{l}{h}\right).$$

For a cantilevered steel beam:

$$\frac{\delta}{l} = \frac{1}{2}\frac{165}{200\,000}\left(\frac{l}{h}\right),$$

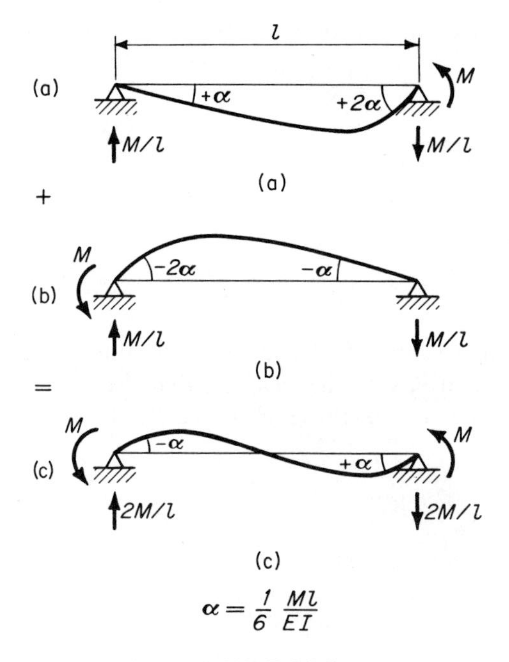

FIGURE 4.3.6

and with $10 \leq l/h \leq 20$:

$$\frac{1}{240} \leq \frac{\delta}{l} \leq \frac{1}{120}. \tag{g}$$

For an aluminum cantilever, $f_1 = 131$ MPa (19 ksi), $E = 69\,000$ MPa (10×10^3 ksi) and with $10 \leq l/h \leq 20$:

$$\frac{1}{105} \leq \frac{\delta}{l} \leq \frac{1}{53}. \tag{h}$$

For a wood cantilevered beam with $f_1 = 14.1$ MPa (2,050 psi), $E = 13\,100$ MPa (1.9×10^3 ksi) and with $10 \leq l/h \leq 20$:

$$\frac{1}{186} \leq \frac{\delta}{l} \leq \frac{1}{93}. \tag{i}$$

For uniformly loaded, simply supported beams:

$$\delta = \frac{5}{384}\frac{wl^4}{EI} = \frac{5 \times 2}{48}\left(\frac{wl^2}{8}\right)\frac{h/2}{EI}\left(\frac{l}{h}\right)l$$

or:

$$\frac{\delta}{l} = \frac{5}{24}\left(\frac{f_1}{E}\right)\left(\frac{l}{h}\right).$$

For a simply supported steel beam with $10 \leq l/h \leq 20$:

$$\frac{1}{580} \leq \frac{\delta}{l} \leq \frac{1}{290}; \tag{j}$$

for an aluminum beam:

$$\frac{1}{254} \leq \frac{\delta}{l} \leq \frac{1}{127}; \tag{k}$$

for a wood beam:

$$\frac{1}{446} \leq \frac{\delta}{l} \leq \frac{1}{223}. \tag{l}$$

It is interesting to compare the *bending* deflection of a cantilever under a concentrated tip load P, with the *compressive* deflection of the same beam under the same load acting axially (Fig. 4.3.7). The horizontal displacement equals:

$$\delta_h = \epsilon l = \frac{f_c}{E} l = \frac{Pl}{EA};$$

the vertical displacement equals:

$$\delta_v = \frac{1}{3} \frac{Pl^3}{EI}.$$

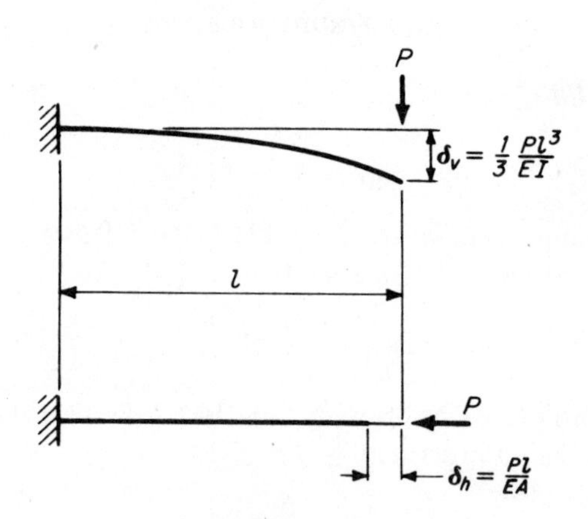

FIGURE 4.3.7

Hence:

$$\frac{\delta_v}{\delta_h} = \frac{1}{3} \frac{Pl^3}{EI} \bigg/ \frac{Pl}{EA} = \frac{l^2}{3I/A} \tag{m}$$

For a rectangular cross-sectional beam:

$$I = \frac{1}{12} bh^3, \qquad A = bh, \qquad \frac{I}{A} = \frac{h^2}{12}$$

and:

$$\frac{\delta_v}{\delta_h} = 4\left(\frac{l}{h}\right)^2. \tag{n}$$

For l/h between 10 and 20, δ_v is 400 to 1,600 times δ_h.

For an ideal I beam:

$$I = 2\left(\frac{A}{2}\right)\left(\frac{h}{2}\right)^2 = \frac{1}{4}Ah^2; \qquad \frac{I}{A} = \frac{1}{4}h^2$$

and:

$$\frac{\delta_v}{\delta_h} = \frac{4}{3}\left(\frac{l}{h}\right)^2, \tag{o}$$

so that δ_v is 133 to 533 times δ_h.

These results indicate how much stiffer a beam is under direct stress than in bending. The same result applies to all structural systems.

Table 4.3.1 gives the rotations and deflections derived in this section for statically determinate beams.

4.4 FIXED-END BEAMS [7.3]

The ends of simply supported beams are entirely free to rotate. The ends of *fixed-end* (or *built-in*) beams are completely *restrained from rotating*. (It is essential to notice that for both types of beams one end is assumed free to move with respect to the other along the beam axis, so that the beam neutral axis is neither stretched nor compressed.)

The end supports of a fixed-end beam develop bending moments M_A, M_B capable of wiping out the end rotations ϕ_A, ϕ_B that would be induced by the loads if the ends were simply supported. Thus, by Table 4.3.1 (a), (c), for the case of a uniform load w with rotations $\phi_A = \phi_B = \frac{1}{24}(wl^3/EI)$ due to the load and rotations $\frac{1}{2}(Ml/EI)$ due to the end moments:

$$\frac{1}{24}\frac{wl^3}{EI} + \frac{1}{2}\frac{Ml}{EI} = 0 \quad \therefore \quad M_A = M_B = -\frac{1}{12}wl^2. \tag{4.4.1}$$

The bending moment diagram for the fixed-end beam [Table 4.1.2 (9)] is obtained by subtracting from the diagram for the simply supported beam the ordinate $-(\frac{1}{12})wl^2$ of the constant diagram due to the end moments. Therefore, the midspan moment M_C is:

$$M_C = \tfrac{1}{8}wl^2 - \tfrac{1}{12}wl^2 = \tfrac{1}{24}wl^2. \tag{4.4.2}$$

Writing (4.4.2) in the form:

$$M_C = \frac{1}{8}w\left(\frac{l}{\sqrt{3}}\right)^2, \tag{a}$$

Table 4.3.1

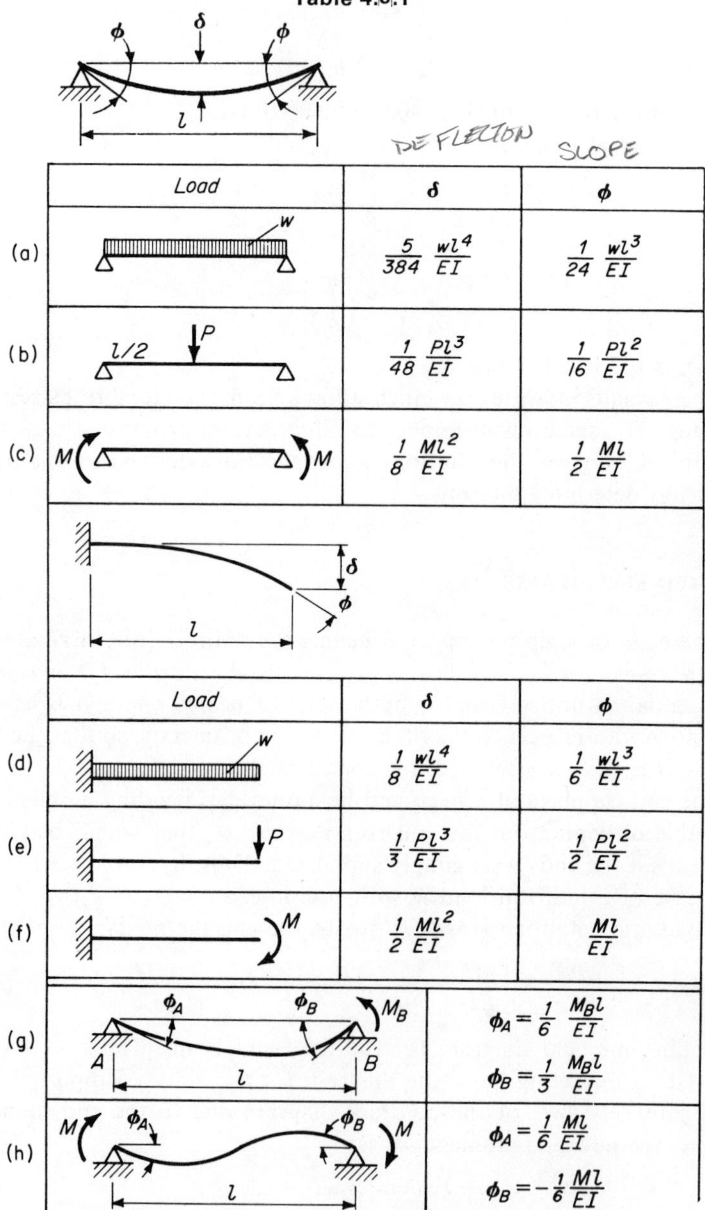

	Load	δ	ϕ
(a)	w	$\frac{5}{384}\frac{wl^4}{EI}$	$\frac{1}{24}\frac{wl^3}{EI}$
(b)	$l/2$, P	$\frac{1}{48}\frac{Pl^3}{EI}$	$\frac{1}{16}\frac{Pl^2}{EI}$
(c)	M, M	$\frac{1}{8}\frac{Ml^2}{EI}$	$\frac{1}{2}\frac{Ml}{EI}$

	Load	δ	ϕ
(d)	w	$\frac{1}{8}\frac{wl^4}{EI}$	$\frac{1}{6}\frac{wl^3}{EI}$
(e)	P	$\frac{1}{3}\frac{Pl^3}{EI}$	$\frac{1}{2}\frac{Pl^2}{EI}$
(f)	M	$\frac{1}{2}\frac{Ml^2}{EI}$	$\frac{Ml}{EI}$

	Load	Slopes
(g)	ϕ_A, ϕ_B, M_B, A, B, l	$\phi_A = \frac{1}{6}\frac{M_B l}{EI}$ $\phi_B = \frac{1}{3}\frac{M_B l}{EI}$
(h)	M, ϕ_A, ϕ_B, M, l	$\phi_A = \frac{1}{6}\frac{Ml}{EI}$ $\phi_B = -\frac{1}{6}\frac{Ml}{EI}$

we see that the center portion of the beam, between the inflection points, has an equivalent length $l_e = l/\sqrt{3} = 0.58l$ and hangs, so to say, from two cantilevers of length $(\sqrt{3} - 1)l/(2\sqrt{3}) = 0.21l$.

Similarly, for a load P concentrated at midspan, Table 4.3.1 (b) gives:

$$\frac{1}{16}\frac{Pl^2}{EI} + \frac{1}{2}\frac{Ml}{EI} = 0 \quad \therefore \quad M_A = M_B = -\frac{1}{8}Pl,$$

$$M_C = \tfrac{1}{4}Pl - \tfrac{1}{8}Pl = \tfrac{1}{8}Pl. \tag{4.4.3}$$

The midspan deflection of the fixed-end beam is the algebraic sum of the deflections due to the load and to the end moments. By Table 4.3.1 (a), (c), one obtains for the uniform load:

$$\delta_f = \frac{5}{384}\frac{wl^4}{EI} - \frac{1}{8}\frac{(\frac{1}{12}wl^2)l^2}{EI} = \frac{1}{384}\frac{wl^4}{EI} = \frac{1}{5}\delta_{ss}, \tag{4.4.4}$$

and for the midspan concentrated load:

$$\delta_f = \frac{1}{48}\frac{Pl^3}{EI} - \frac{1}{8}\frac{(\frac{1}{8}Pl)l^2}{EI} = \frac{1}{192}\frac{Pl^3}{EI} = \frac{1}{4}\delta_{ss}. \tag{4.4.5}$$

It is interesting to compare the fixed-end beam to the simply supported beam in terms of deflections, stresses, and influence of foundation settlements.

The fixed-end beam, depending on the load distribution, is between four and five times stiffer than an identical simply supported beam. The minimum and maximum bending moments in the fixed-end, uniformly loaded beam equal $-\frac{2}{3}$ and $+\frac{1}{3}$ of the midspan moment in the simply supported beam: hence, a fixed *uniform* beam requires a section modulus at most $\frac{2}{3}$ that of the simply supported beam. This means that for rectangular cross-section beams of equal width with $S = bh^2/6$, i.e., $h = \sqrt{6S/b}$, the depth of the fixed-end beam may be $\sqrt{\frac{2}{3}} = 0.82$ of the depth of the simply supported beam, and that the cross-section areas bh are in the same ratio. Since the maximum shear is the same for both conditions of support, if the shear *stress* in the simply supported beam is equal to its maximum allowable value F_s, it will be $(1/0.82)F_s = 1.22F_s$, or 22% too high in the fixed-end beam. This overstress occurs in nonuniform beams also and may be eliminated by increasing the cross-section area toward the supports through *haunches* (Fig. 4.4.1), but it must be remembered that the added bending stiffness due

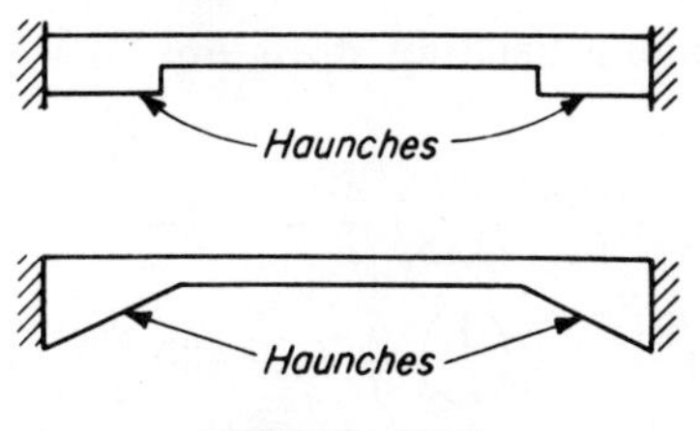

FIGURE 4.4.1

to the haunches increases the end moments and decreases the midspan moment. For the type of haunches used in reinforced concrete beams, the end moments increase by approximately 20%.

When an end support B of a simply supported beam settles by δ with respect to the other support A [Fig. 4.4.2(a)], the beam adapts to this condition by end rotations $\phi_A = \phi_B = \delta/l$ without developing bending stresses, as shown by the fact that it remains straight.

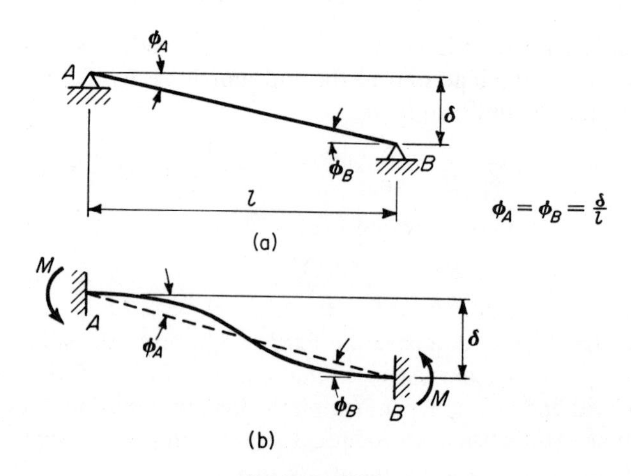

FIGURE 4.4.2

When a beam is fixed-ended, i.e., its ends are not free to rotate, its supports must develop equal and opposite bending moments capable of wiping out the rotations δ/l. By Table 4.3.1 (h), these antisymmetrical moments $M\delta$ are given by:

$$\frac{\delta}{l} = \frac{1}{6}\frac{M_\delta l}{EI} \quad \therefore \quad M_\delta = \pm\frac{6EI}{l^2}\delta, \tag{4.4.6}$$

and are proportional to the settlement δ and to the flexural rigidity of the beam EI. The ratio of the end moments M_δ to the end moments M_w due to a uniform load w in the same beam equals:

$$\frac{M_\delta}{M_w} = \pm\frac{6EI\delta/l^2}{M_w} = \pm\frac{[6E(\delta/l)(h/2]/l}{(M_w h/2)/I} = \pm 3\left(\frac{E}{f_1}\right)\left(\frac{h}{l}\right)\left(\frac{\delta}{l}\right), \tag{4.4.7}$$

where f_1 is the maximum bending stress due to w. With $l/h = 15$, (4.4.7) gives:

$$\frac{M_\delta}{M_w} = \pm 3\left(\frac{200\,000}{165}\right)\left(\frac{1}{15}\right)\frac{\delta}{l} \approx \pm 240\frac{\delta}{l} \quad \text{for steel,}$$

$$\frac{M_\delta}{M_w} = \pm 3\left(\frac{21\,700}{9}\right)\left(\frac{1}{15}\right)\frac{\delta}{l} = \pm 480\frac{\delta}{l} \quad \text{for concrete,} \tag{b}$$

and shows that (a) concrete beams are twice as sensitive to foundation settlements as steel beams, (b) a settlement equal to only $\frac{1}{240}$ to $\frac{1}{480}$ of the span doubles one of the end moments due to the load, (c) the advantages due to stiffness in the load carrying capacity of a fixed-end beam are somewhat countered by its sensitivity to foundation settlements.

These results may be generalized: the stiffer a structural system the smaller its deflections, and the more sensitive it is to imposed deformations, such as differential settlements and temperature deflections.

4.5 OTHER STATICALLY INDETERMINATE BEAMS [7.3]

The moment and shear diagrams for other statically indeterminate beams may be obtained by superposition, using Tables 4.1.2 and 4.3.1.

For example, a uniformly loaded beam fixed at its left support *A* and simply supported at its right support *B* (Fig. 4.5.1) may be analyzed as a cantilever with an upward force *R* at *B* capable of lifting the cantilever tip to the level of *A*. By Table 4.3.1 (d), (e):

$$-\frac{1}{3}\frac{Rl^3}{EI}+\frac{1}{8}\frac{wl^4}{EI}=0 \quad \therefore \quad R=\frac{3}{8}wl. \tag{a}$$

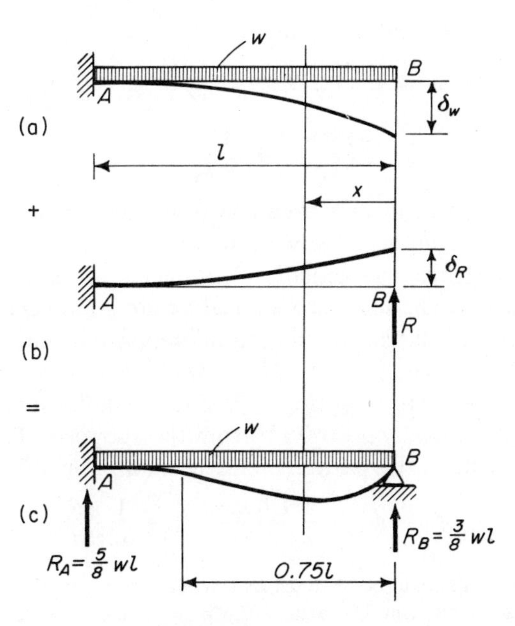

FIGURE 4.5.1

The bending moment, as shown in Table 4.1.2 (13), equals:

$$M(x) = -\frac{wx^2}{2} + \frac{3}{8}wlx, \qquad M_A = -\frac{wl^2}{2} + \frac{3}{8}wl^2 = -\frac{1}{8}wl^2.*$$

The inflection point occurs where $M(x) = 0$, i.e., where:

$$-\frac{wx^2}{2} + \frac{3}{8}wlx = 0 \quad \therefore \quad x = \frac{3}{4}l. \tag{b}$$

To analyze the influence of a settlement δ of B with respect to A, we notice that if δ is larger than the deflection of a cantilevered beam $\delta_0 = \frac{1}{8}wl^4/EI$, the beam will act as a cantilever. For $\delta < \delta_0$, the reaction R_B must lift B by $\delta_0 - \delta$, so that:

$$\frac{1}{3}\frac{R_B l^3}{EI} = \delta_0 - \delta = \frac{1}{8}\frac{wl^4}{El} - \delta,$$

from which:

$$R_B = \frac{3}{8}wl - \frac{3EI}{l^3}\delta,$$

$$M(x) = -\frac{wx^2}{2} + \frac{3}{8}wlx - \frac{3EI}{l^3}\delta x,$$

and the change in $M(x)$ due to δ is:

$$M_\delta(x) = -\frac{3EI}{l}\left(\frac{x}{l}\right)\left(\frac{\delta}{l}\right). \tag{4.5.1}$$

The ratio of M_δ at A to M_w at A equals:

$$\left.\frac{M_\delta}{M_w}\right]_{x=l} = \frac{(3EI/l)(\delta/l)}{M_w} = \frac{3E}{(M_w h/2)/I}\left(\frac{h/2}{l}\right)\left(\frac{\delta}{l}\right)$$

$$= \frac{3}{2}\left(\frac{E}{f_1}\right)\left(\frac{h}{l}\right)\left(\frac{\delta}{l}\right), \tag{4.5.2}$$

and is one-half as large as in a fixed-end beam, as shown by (4.4.7).

In practice, no support freely allows or totally prevents the rotation of the beam ends. Hence, the actual end moments are always less than the fixed-end moments. The most economical bending-moment diagram is one with equal values of the maximum and minimum moments. For a uniform load w (Fig. 4.5.2) this implies that $M_A + M_C = 2M_A = wl^2/8$, so that $M_A = -wl^2/16$, $M_C = +wl^2/16 = \frac{1}{8}w(l/\sqrt{2})^2$, and the inflection points occur at a distance $(\sqrt{2} - 1)l/2\sqrt{2} = 0.15l$ from the supports. The rotations due to the end moments in this *partially* fixed beam are, by Table 4.3.1 (c):

$$\phi_A = \phi_B = \frac{1}{2}\frac{(\frac{1}{16}wl^2)l}{EI} = -\frac{1}{32}\frac{wl^3}{EI},$$

*The same result could be obtained by starting with a simply supported beam and applying to it a moment M_A capable of wiping out the rotation ϕ_A:

$$\frac{1}{24}\frac{wl^3}{EI} + \frac{1}{3}\frac{M_A l}{EI} = 0 \quad \therefore \quad M_A = -\frac{1}{8}wl^2.$$

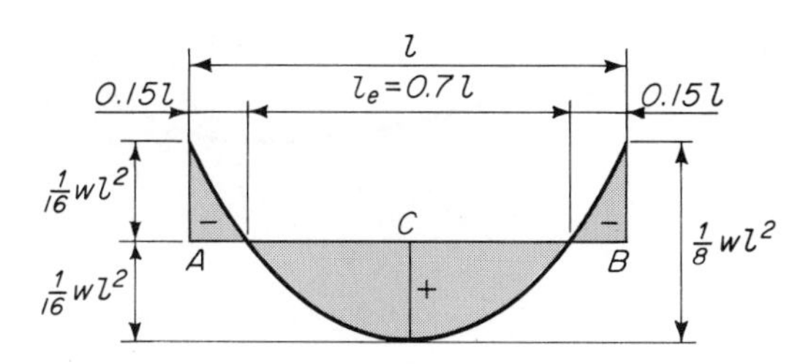

FIGURE 4.5.2

and the total rotations, obtained by subtraction from the simply supported rotations are:

$$\phi_A = \phi_B = \left(\frac{1}{24} - \frac{1}{32}\right)\frac{wl^3}{EI} = \frac{1}{96}\frac{wl^3}{EI}, \tag{4.5.3}$$

or one-quarter of the simply supported rotations.

4.6 CONTINUOUS BEAMS [7.3]

Continuous beams are beams on more than two supports (Fig. 4.6.1), in which the successive spans are rigidly connected to one another over the supports, so that under load the rotation $\phi_{A,l}$ of the right end of the span to the *left* of the support is equal *and opposite* to the rotation $\phi_{A,r}$ of the left end of the span to the *right* of the support (Fig. 4.6.1):

$$\phi_{A,l} = -\phi_{A,r}. \tag{4.6.1}$$

In a continuous beam *all spans* deflect even if only *one* of them is loaded. Hence, all spans help carry the load, but, as will be seen through the following examples, the carrying action of the adjoining spans peters out rapidly as one moves away from the loaded span.

In order to establish this characteristic behavior of continuous beams, we need to determine, first, the moment M_A at the fixed end of the beam of Fig. 4.6.2(a) due to a moment M_B applied to its simply supported end. M_A must wipe out the rotation ϕ_A due to M_B in a simply supported beam; hence, by Table 4.3.1(g):

$$\frac{1}{3}\frac{M_A l}{EI} + \frac{1}{6}\frac{M_B l}{EI} = 0, \qquad M_A = -\frac{1}{2}M_B. \tag{4.6.2}$$

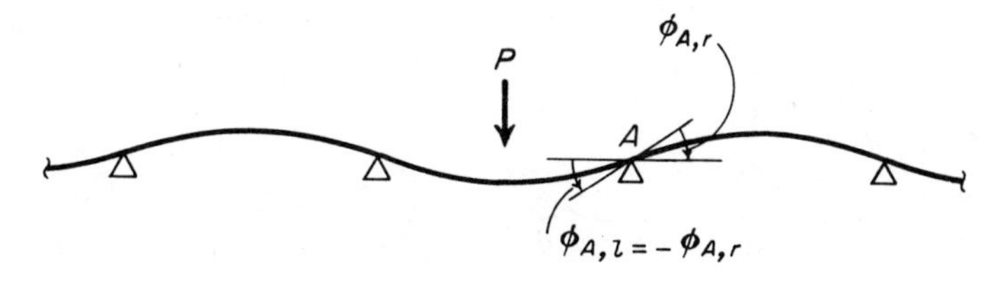

FIGURE 4.6.1

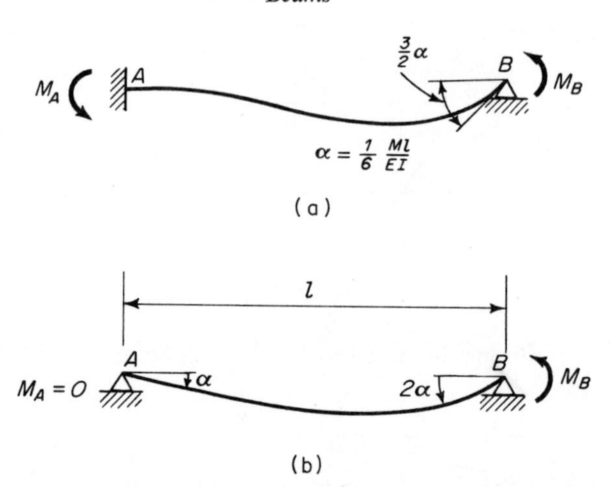

FIGURE 4.6.2

Equation (4.6.2) indicates that the beam "carries" the moment M_B from B to A by changing its sign and cutting it in half. A beam with a simple support at A, instead, carries no moment from B to A [Fig. 4.6.2(b)].

Example 4.6.1. Determine the moment over the support B of a two-span beam due to a moment M_C applied at C when the end A is (a) simply supported, and (b) fixed [Figs. 4.6.3(a), (b)].

Calling l_1, l_2 the two spans, EI_1, EI_2 their flexural rigidities, and M_A, M_B, M_C the moments over the supports, we obtain:

Case (a) $$\phi_{B,l} = \frac{1}{3}\frac{M_B l_1}{EI_1}; \quad \phi_{B,r} = \frac{1}{3}\frac{M_B l_2}{EI_2} + \frac{1}{6}\frac{M_C l_2}{EI_2},$$

and, by (4.6.1):

$$\frac{1}{3}\frac{M_B l_1}{EI_2} = -\left(\frac{1}{3}\frac{M_B l_2}{EI_2} + \frac{1}{6}\frac{M_C l_2}{EI_2}\right),$$

from which:

$$M_B = -\frac{1}{2}M_C\frac{1}{1 + (l_1/l_2)(I_2/I_1)}. \tag{4.6.3}$$

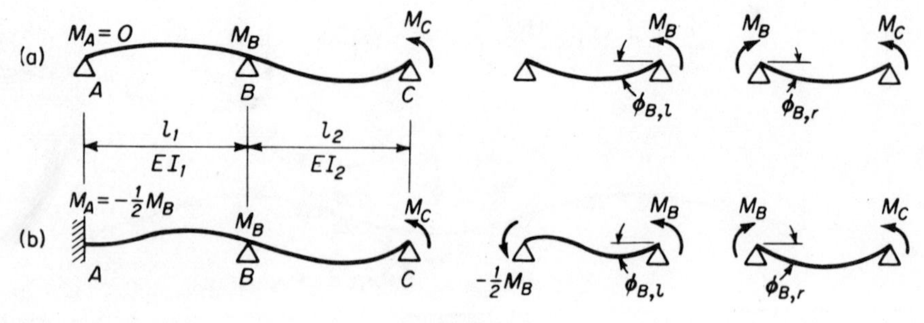

FIGURE 4.6.3

Case (b) Since, by (4.6.2), $M_A = -\frac{1}{2}M_B$, we obtain:

$$\phi_{B,l} = \frac{1}{3}\frac{M_B l_1}{EI_1} + \frac{1}{6}\frac{(-\frac{1}{2}M_B)l_1}{EI_1}, \qquad \phi_{B,r} = \frac{1}{3}\frac{M_B l_2}{EI_2} + \frac{1}{6}\frac{M_C l_2}{EI_2},$$

and, by (4.6.1):

$$M_B\left(\frac{l_1}{I_1} - \frac{1}{4}\frac{l_1}{I_1}\right) = -\left(M_B\frac{l_2}{I_2} + \frac{1}{2}\frac{M_2 l_2}{I_2}\right),$$

from which:

$$M_B = -\frac{1}{2}M_C\frac{1}{1 + \frac{3}{4}(l_1/l_2)(I_2/I_1)}. \tag{4.6.4}$$

Both (4.6.3) and (4.6.4) give (1) $M_B = -\frac{1}{2}M_C$ for $l_1 = 0$ or $I_1 =$ infinity, i.e., for an infinitely rigid left beam, which prevents rotation at A and is equivalent to a fixed-end support, and (2) $M_B = 0$ for $l_1 =$ infinity or $I_1 = 0$, i.e., for a perfectly flexible left beam incapable of preventing rotations.

For $l_1 = l_2$ and $I_1 = I_2$, (4.6.3) and (4.6.4) give:

$$M_B]_{\text{s.s.}} = \tfrac{1}{2}(-\tfrac{1}{2}M_C) = -\tfrac{1}{4}M_C, \qquad M_B]_{\text{f.e.}} = \tfrac{4}{7}(-\tfrac{1}{2}M_C) = -0.286M_C,$$

showing that the support condition at A makes *at most* a difference of $(\frac{4}{7} - \frac{1}{2})/\frac{1}{2} \approx 14\%$ on the value of M_B. For case (b) and for identical spans:

$$M_A = -\tfrac{1}{2}M_B = +\tfrac{1}{7}M_C = 0.144M_C,$$

indicating the damping out of M_C when carried over two spans.

Example 4.6.2. Determine the moments over the supports of a three-span beam with identical spans due to a moment M_D at D, when A is (a) a simple support and (b) a fixed support (Fig. 4.6.4). From the results of Example 4.6.1:

Case (a) $\qquad M_B = -\frac{1}{4}M_C, \qquad M_A = 0$

Case (b) $\qquad M_B = -\frac{2}{7}M_C, \qquad M_A = \frac{1}{7}M_C.$

Hence, applying (4.6.1) at C, and dropping the common factor $\frac{1}{6}(l/EI)$:

Case (a) $\qquad M_D + 2M_C = -[2M_C + (-\frac{1}{4}M_C)]$

$$\therefore \quad M_C = -\tfrac{4}{15}M_D = -0.267M_D, \quad M_B = \tfrac{1}{15}M_D = 0.067M_D, \quad M_A = 0.$$

Case (b) $\qquad M_D + 2M_C = -[2M_C + (-\frac{2}{7}M_C)]$

$$\therefore \quad M_C = -\tfrac{7}{26}M_D = -0.269M_D, \quad M_B = \tfrac{2}{26}M_D = 0.077M_D,$$

$$M_A = -\tfrac{1}{26}M_D = -0.038M_D.$$

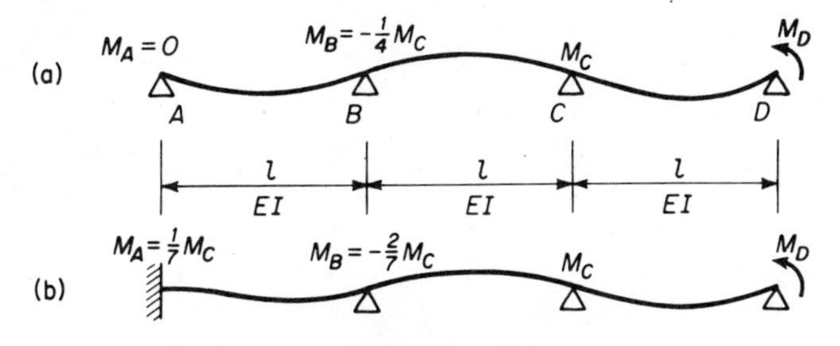

FIGURE 4.6.4

Comparison of the results of Examples 4.6.1 and 4.6.2 show that the influence on M_B of the third span is practically negligible and that a moment M_D is reduced by 96% when carried over three identical spans.

The results of Example 4.6.2 show that the values of the moments over the supports of an *unloaded* continuous beam oscillate in sign and are rapidly "damped out" as one moves away from an applied end moment.

Example 4.6.3. Draw the bending moment and shear diagrams for a two-span beam with identical spans, loaded uniformly on the right span, when the left support is (a) a simple support and (b) fixed (Fig. 4.6.5).

Applying (4.6.1) at B, ignoring the common factor l/EI, and noticing that the rotation of beam BC at B is the sum of the rotation due to the load and of that due to M_B, we obtain:

Case (a) $\quad \frac{1}{3}M_B l = -(\frac{1}{3}M_B l + \frac{1}{24}wl^3) \quad \therefore \quad M_B = -\frac{1}{16}wl^2,$

$$R_A = -\frac{1}{l}\left(\frac{wl^2}{16}\right) = -\frac{wl}{16}, \qquad R_B = \frac{1}{l}\left(\frac{wl^2}{16}\right) + \frac{1}{l}\left(\frac{wl^2}{16}\right) + \frac{wl}{2} = \frac{10wl}{16},$$

$$R_C = -\frac{1}{l}\left(\frac{wl^2}{16}\right) + \frac{wl}{2} = \frac{7wl}{16}.$$

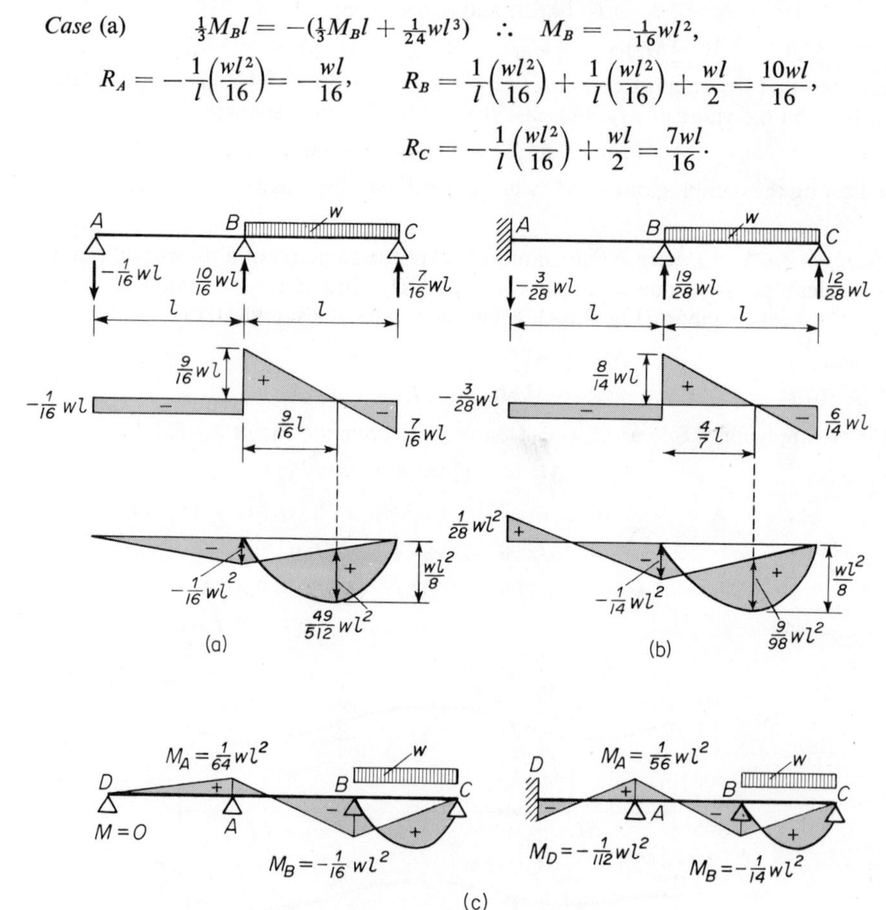

FIGURE 4.6.5

Case (b)
$$\tfrac{1}{3}M_B l + \tfrac{1}{6}(-\tfrac{1}{2}M_B l) = -(\tfrac{1}{3}M_B l + \tfrac{1}{24}wl^3)$$
$$\therefore \quad M_B = -\tfrac{1}{14}wl^2, \qquad M_A = \tfrac{1}{28}wl^2,$$
$$R_A = -\frac{1}{l}\left(\frac{wl^2}{28} + \frac{wl^2}{14}\right) = -\frac{3wl}{28}, \quad R_B = \frac{1}{l}\left(\frac{wl^2}{28} + \frac{2wl^2}{14}\right) + \frac{wl}{2} = \frac{19wl}{28},$$
$$R_C = -\frac{1}{l}\left(\frac{wl^2}{14}\right) + \frac{wl}{2} = \frac{12wl}{28}.$$

Here, again, the minor influence of the support conditions at A is worth noticing.

On the basis of the results of Examples 4.6.1 and 4.6.3, the moments M_A, M_B, M_C and the shears in a three-span beam uniformly loaded on the right span [Fig. 4.6.5(c)] may be approximated by the M_B in the two-span beam, "damped" to $-\frac{1}{4}$ of its value to give $M_A = -\frac{1}{4}M_B$, and to $M_D = 0$ for a simple left support, and to $M_D = -\frac{1}{2}M_A = +\frac{1}{8}M_B$ for a fixed support.

Example 4.6.4. Determine the moment and shear diagrams in a three-span beam with equal side spans, uniformly loaded on the middle span, with (a) simple supports and (b) fixed-end supports (Fig. 4.6.6).

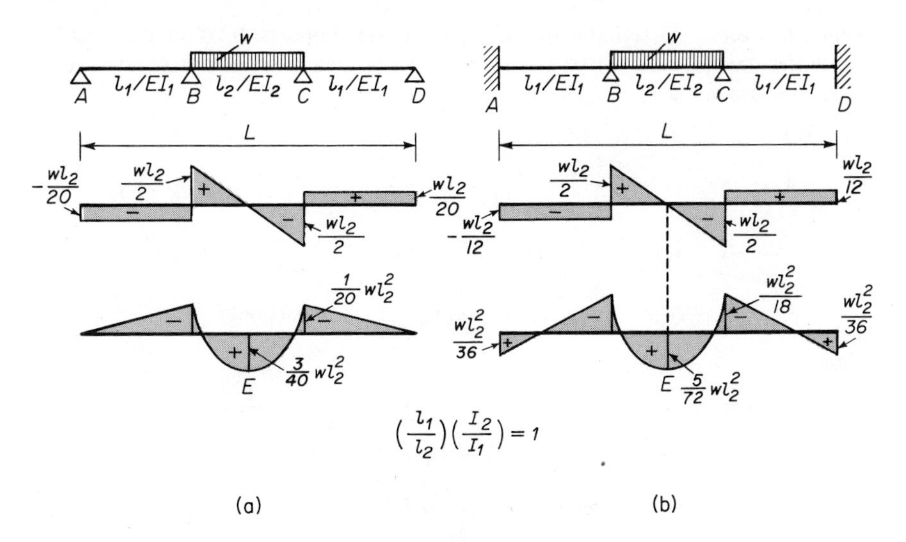

FIGURE 4.6.6

Noticing that, because of symmetry, $M_B = M_C$, we obtain:

Case (a)
$$\frac{1}{3}\frac{M_B l_1}{EI_1} = -\left(\frac{1}{3}\frac{M_B l_2}{EI_2} + \frac{1}{6}\frac{M_B l_2}{EI_2} + \frac{1}{24}\frac{wl_2^3}{EI_2}\right)$$
$$\therefore \quad M_B = M_C = -\frac{1}{12}wl_2^2\frac{1}{1 + \frac{2}{3}(l_1/l_2)(I_2/I_1)}, \tag{a}$$

and for $(l_1/l_2)(I_2/I_1) = 1$:
$$M_B = M_C = -\tfrac{1}{20}wl^2.$$

Noticing that $M_A = M_D = -\frac{1}{2}M_B$, we obtain:

Case (b) $$\frac{1}{6}\frac{(-\frac{1}{2}M_B)l_1}{EI_1} + \frac{1}{3}\frac{M_B l_1}{EI_1} = -\left(\frac{1}{3}\frac{M_B l_2}{EI_2} + \frac{1}{6}\frac{M_B l_2}{EI_2} + \frac{1}{24}\frac{wl_2^3}{EI_2}\right)$$

$$\therefore \quad M_B = -\frac{1}{12}wl_2^2 \frac{1}{1 + \frac{1}{2}(l_1/l_2)(I_2/I_1)},$$

and for $(l_1/l_2)(I_2/I_1) = 1$:

$$M_B = M_C = -\tfrac{1}{18}wl_2^2, \qquad M_A = M_D = \tfrac{1}{36}wl_2^2.$$

Example 4.6.5. Determine the ratio I_2/I_1 in the beam of Example 4.6.4(a) so as to equalize the maximum negative and positive moments in beam BC, when $l_1 = l_2 = l$.

If M_B must equal $-M_E$, $M_B = |-M_E| = wl_2^2/16$ and, by (a):

$$\frac{1}{16}wl^2 = \frac{1}{12}wl^2 \frac{1}{1 + \frac{2}{3}(I_2/I_1)}, \qquad \frac{12}{16} = \frac{1}{1 + \frac{2}{3}(I_2/I_1)}$$

$$\therefore \quad \frac{I_2}{I_1} = \frac{1}{2}.$$

Example 4.6.6. Determine the location of the supports B, C in the beam of Examples 4.6.4(a) and (b) so as to obtain equal maximum positive and negative moments, assuming $I_1 = I_2$.

Case (a)

$$M_B = M_C = -\frac{1}{16}wl_2^2 = -\frac{1}{12}wl_2^2 \frac{1}{1 + \frac{2}{3}(l_1/l_2)}, \qquad \frac{4}{3} = 1 + \frac{2}{3}\left(\frac{l_1}{l_2}\right)$$

$$\therefore \quad l_2 = 2l_1, \qquad 2l_1 + l_2 = 4l_1 \equiv L, \qquad l_1 = \frac{L}{4}, \qquad l_2 = \frac{L}{2},$$

$$M_B = -\frac{1}{16}wl_2^2 = -\frac{1}{16}w\left(\frac{L}{2}\right)^2 = -0.016wL^2.$$

Case (b)

$$M_B = M_C = -\frac{1}{16}wl_2^2 = -\frac{1}{12}wl_2^2 \frac{1}{1 + \frac{1}{2}(l_1/l_2)}, \qquad \frac{8}{3} = 2 + \left(\frac{l_1}{l_2}\right)$$

$$\therefore \quad l_2 = \tfrac{3}{2}l_1, \qquad 2l_1 + l_2 = \tfrac{7}{2}l_1 \equiv L, \qquad l_1 = \tfrac{2}{7}L, \; l_2 = \tfrac{3}{7}L,$$

$$M_B = -\tfrac{1}{16}wl_2^2 = -\tfrac{1}{16}w\left(\frac{3}{7}L\right)^2 = -0.011\,5\; wL^2.$$

4.7 MOMENT DISTRIBUTION

It was shown in Section 4.6 [see (4.6.2)] that when a moment M_B is applied at the simply supported end B of a beam AB of constant cross section fixed at A, a moment $M_A = -\frac{1}{2}M_B$ develops at the fixed end A (Fig. 4.6.2). If moments are considered positive when acting clockwise and negative when acting counterclockwise *on the beam ends*, (4.6.2) becomes:

$$M_A = \tfrac{1}{2}M_B. \tag{4.7.1}$$

The factor $\frac{1}{2}$ is called the *carry-over* factor for the beam.

By Table 4.3.1 (g), the rotation due to M_B in the simply supported fixed beam is:

$$\phi_B = \frac{1}{3}\frac{M_B l}{EI} + \frac{1}{6}\frac{(-\frac{1}{2})M_B l}{EI} = \frac{l}{4EI}M_B. \tag{a}$$

Equation (a) shows that the moment M_B due to a rotation ϕ_B is given by:

$$M_B = K\phi_B, \tag{b}$$

where the quantity:

$$K = \frac{4EI}{l} \tag{4.7.2}$$

is called the *beam stiffness.*

Given a two-span beam, *built-in* at the end supports A and C and simply supported at the middle support B, with stiffnesses K_l and K_r in the left and right spans, the moments $M_{B,l}$ and $M_{B,r}$ due to a rotation ϕ_B of the middle support are, by (b),

$$M_{B,l} = K_l\phi_B, \qquad M_{B,r} = K_r\phi_B,$$

so that:

$$\frac{M_{B,r}}{M_{B,l}} = \frac{K_r}{K_l}. \tag{4.7.3}$$

Equations (4.7.1) and (4.7.3) are the basis for the evaluation of bending moments in continuous beams by the Cross method of *moment distribution,* which consists of considering, first, all spans as built-in at the supports and in allowing, then, each support to rotate in succession until the beam is in equilibrium in its correct deflected shape. The physical meaning and mathematical simplicity of moment distribution have made this method popular among engineers and architects.

To illustrate the method, consider the beam of Fig. 4.7.1, which is uniformly loaded and has span stiffnesses in the ratio $K_1/K_2 = 2/1$ and spans in the ratio $l_2/l_1 = 2/1$ (and, hence, identical I's). If the two spans were fixed, their end moments would be:

$$M_1 = \tfrac{1}{12}wl_1^2, \qquad M_2 = \tfrac{1}{12}wl_2^2 = \tfrac{1}{12}w(2l_1)^2 = 4M_1. \tag{c}$$

These moments are considered *positive when acting clockwise* on the span ends; hence, they have the values shown in line (1) of Fig. 4.7.1(b).

The beam is not in equilibrium, since the moments at the supports A, B, and C do not each add up to zero. The equilibrate the *unbalanced moments* $-M_1$ at A, $M_1 - 4M_1 = -3M_1$ at B and $4M_1$ at C, the supports should be capable of exerting on the beam equal and opposite *balancing moments* M_1 at A, $3M_1$ at B, and $-4M_1$ at C. Since the simple supports cannot exert external moments on the beam, the beam will rotate at the supports and thus develop the required balancing moments.

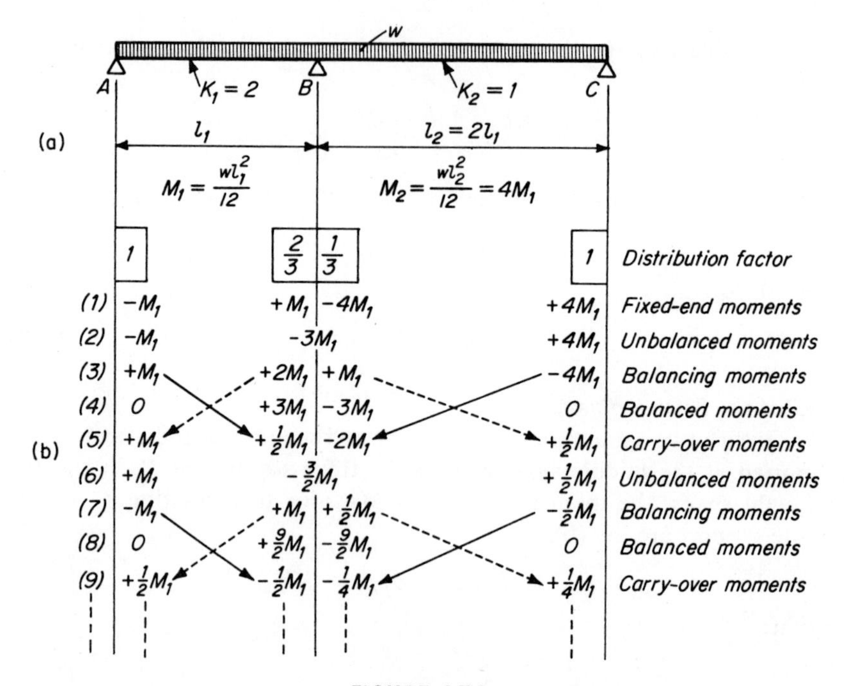

FIGURE 4.7.1

Let the beam rotate at A (by ϕ_A) but remain fixed at all other supports, until the balancing moment M_1 develops at A. Then, let the beam rotate at C only (by ϕ_C), until the balancing moment $-4M_1$ develops at C, and subsequently let it only rotate at B (by ϕ_B) until the balancing moment $3M_1$ develops at B. The balancing moment $3M_1$ at B must be *distributed*, i.e., split, between the two spans according to (4.7.3), so that:

$$M_{B,l} = \frac{K_l}{K_l + K_r}(3M_1) = \frac{2}{2+1}(3M_1) = 2M_1,$$

$$M_{B,r} = \frac{K_r}{K_l + K_r}(3M_1) = \frac{1}{2+1}(3M_1) = M_1.$$

The ratios $K_l/(K_l + K_r) = 2/3$, $K_r/(K_l + K_r) = 1/3$ are called the *distribution factors* at the joint B. The distribution factors at A and C are obviously equal to 1. The balancing moments at the span ends appear on line (3) of Fig. 4.7.1(b). After balancing, the end moments are those of line (4), and the beam is in equilibrium.

But it must be remembered that by (4.7.1) the balancing moment M_1 applied at A *while B is fixed* develops at B a *carry-over moment* $M_{B,l} = \frac{1}{2}M_1$. Similarly, the balancing moment $-4M_1$ at C develops at B a carry-over moment $M_{B,r} = \frac{1}{2}(-4M_1) = -2M_1$, and the balancing moments at B develop carry-over moments M_1 at A and $\frac{1}{2}M_1$ at C.

The carry-over moments of line (5) disturb the equilibrium reached at line (4) and must be distributed as was done with the original fixed-end moments. This is done in lines (5) to (9). The balancing and distributing process is continued until the unbalanced moments are negligible. The final moments are the sum of the fixed-end moments, the balancing moments, and the carry-over moments of lines (1), (3), (5), (7), (9), ... Obviously, lines (2), (4), (6), . . . need not be written.

When one end of a member is hinged, it is possible to accelerate the convergence of the process by using for such members a reduced stiffness factor K^R which may be shown to equal:

$$K^R = \tfrac{3}{4}K. \tag{4.7.4}$$

In this case, once the first balancing cycle is complete [steps (1) to (4)], moments are *not* carried back to the hinged end. This is shown in Fig. 4.7.2. In this example, the process ends and a final exact solution is obtained.

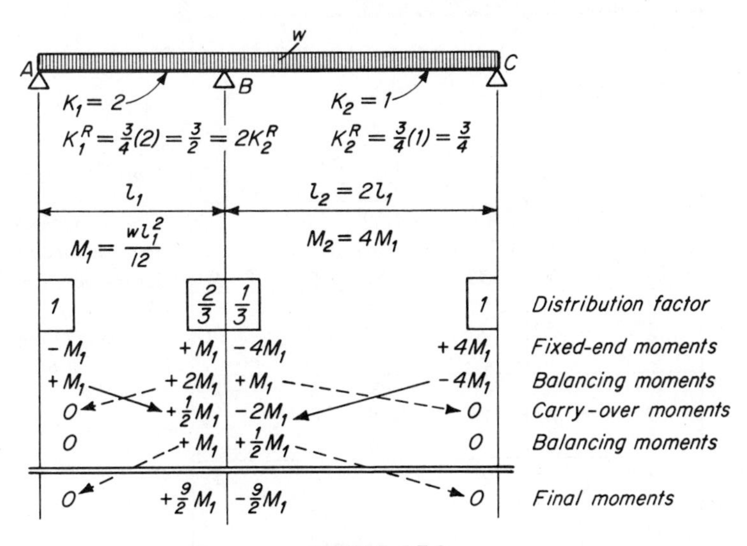

FIGURE 4.7.2

Example 4.7.1. Solve the problem of Fig. 4.7.3 by moment distribution.

In this case, the fixed ends A, C can *always* balance the moments, so that *no* balancing moments are ever added at A or C; i.e., the distribution factors at A and at C are equal to zero. The balancing moments at B are carried over to A and C once. (The moment at A in this example happens to be zero.)

Example 4.7.2. Determine the end moments in the spans of the beam of Fig. 4.7.4, if the fixed moments equal 100 kN·m (75 k·ft) in span 1 and 400 kN·m (300 k·ft) in span 2.

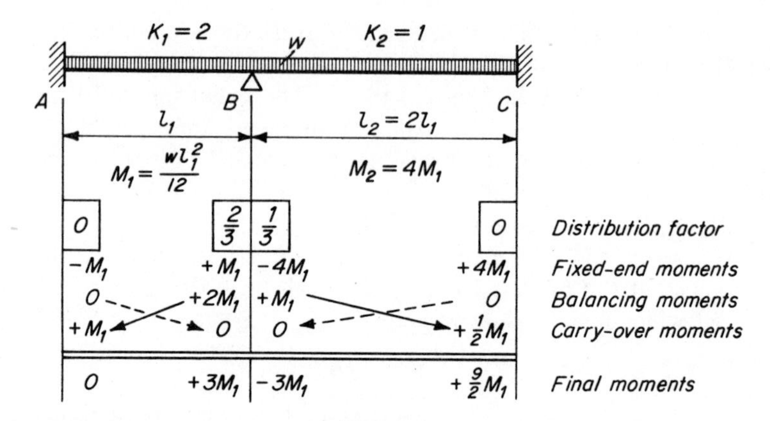

FIGURE 4.7.3

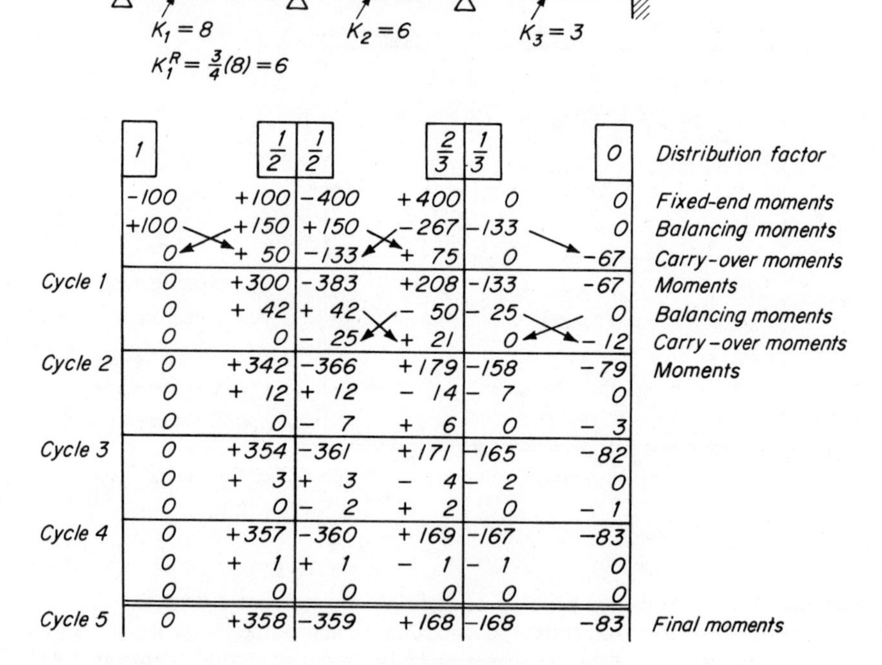

FIGURE 4.7.4

In Fig. 4.7.4 the actual moments are computed after each cycle of balancing and carrying-over. The process is stopped when the corrections are less than 1 kN·m (1 k·ft).

4.8 BEAMS ON ELASTIC SUPPORTS

An elastic spring deflects in proportion to the load applied to it. Calling P the load and δ the displacement (Fig. 4.8.1), the ratio:

$$\frac{P}{\delta} = K \quad \text{kN/m (lb/in.)} \tag{4.8.1}$$

is called the *spring modulus*, or *spring constant*, of the spring. From (4.8.1):

$$P = K\delta, \tag{4.8.2a}$$

$$\delta = \frac{P}{K}. \tag{4.8.2b}$$

A tensile or compressive bar of cross-sectional area A and length l acted upon by an axial load P shortens by:

$$\delta = \frac{Pl}{EA} = \frac{P}{EA/l}; \tag{4.8.3}$$

its spring modulus, by comparison with (4.8.2b), is:

$$K = \frac{EA}{l}. \tag{4.8.4}$$

Elastic beam deflections are also proportional to the applied load and, hence, beams may also be considered as springs. For example, the tip deflection of a cantilever under a concentrated tip load P equals [see Table 4.3.1(e)]:

$$\delta_P = \frac{1}{3}\frac{Pl^3}{EI} = \frac{P}{3EI/l^3},$$

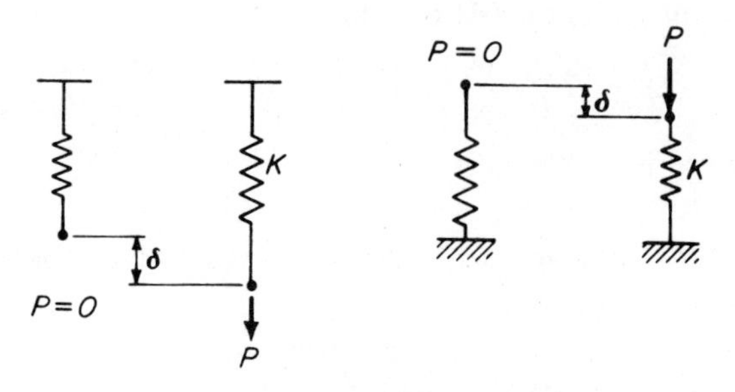

FIGURE 4.8.1

and its spring constant equals:

$$K_P = \frac{3EI}{l^3}.$$

The beam spring constant depends on the type of load. From the tip deflection of a cantilever under a uniform load [Table 4.3.1(d)]:

$$\delta_w = \frac{wl^4}{8EI} = \frac{wl}{8EI/l^3},$$

the corresponding K is seen to equal:

$$K_w = \frac{8EI}{l^3}.$$

Figure 4.8.2 gives spring constants at the point D for cantilevered, simply supported, and fixed-end beams.

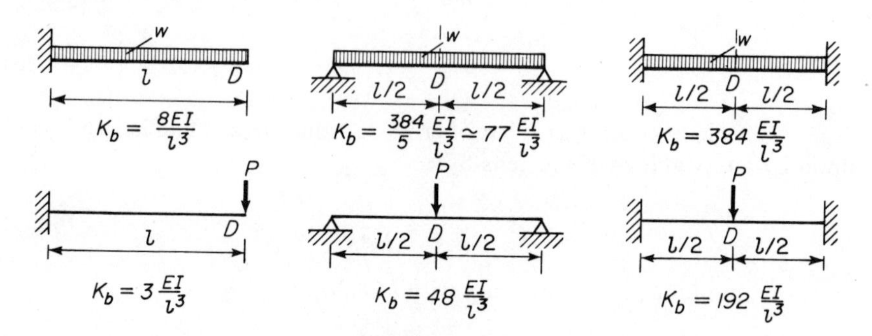

FIGURE 4.8.2

A spring constant may be defined for any load position and at any beam section. For example, the deflection at a section $x = zl$ $(0 \leq z \leq 1)$ of a simply supported beam under a uniform load w may be shown to be given by:

$$\delta(z) = \frac{16}{5}(z^4 - 2z^3 + z)\delta_{\max} \quad \left(z = \frac{x}{l}\right), \tag{4.8.5}$$

where $\delta_{\max} = \frac{5}{384}(wl^4/EI) \approx \frac{1}{77}(wl^4/EI)$ is the midspan deflection. Hence, the spring constant $K(z)$ at a section z is:

$$K(z) = 77\frac{EI}{l^3}\frac{5}{16(z^4 - 2z^3 + z)} \equiv 77\frac{EI}{l^3}f_s(z), \tag{4.8.6}$$

where $f_s(z)$ is given in Table 4.8.1.

Table 4.8.1 SPRING CONSTANTS OF UNIFORMLY LOADED SIMPLY SUPPORTED BEAM

$$f_s(z) = \frac{K(z)}{K(\frac{1}{2})}$$

z	0/1	0.1/0.9	0.2/0.8	0.3/0.7	0.4/0.6	0.5
$f(z)$	∞	3.19	1.68	1.23	1.05	1.00

Similarly, the deflections of a cantilever beam under a uniform load may be shown to be given by:

$$\delta(z) = \frac{1}{3}(z^4 - 4z^3 + 6z^2)\delta_{max} \qquad \left(z = \frac{x}{l}\right), \tag{4.8.7}$$

where x is measured from the fixed end and $\delta_{max} = \frac{1}{8}(wl^4/EI)$ is the tip deflection. Hence, the spring constant at a section z is:

$$K(z) = \frac{8EI}{3} \frac{3}{z^4 - 4z^3 + 6z^2} = \frac{8EI}{l^3} f_c(z), \tag{4.8.8}$$

where $f_c(z)$ is given in Table 4.8.2.

Table 4.8.2 SPRING CONSTANTS OF UNIFORMLY LOADED CANTILEVER BEAM

$$f_c(z) = \frac{K(z)}{K(1)}$$

z	0	0.1	0.2	0.3	0.4	0.5	0.6	0.7	0.8	0.9	1.0
$f(z)$	∞	53.5	14.30	6.88	4.13	2.82	2.10	1.65	1.36	1.16	1.00

Beams are often supported on elastic rather than rigid supports. An elastic support may be provided by the soil, if it deflects in proportion to the load, or by an actual elastic support, as in the case of a beam supported by another beam. As shown by the following examples, the reaction of the elastic support, and hence the bending moment diagram for beams on elastic supports, may often be obtained by considering the support as a spring of spring constant K_s and the beam as another spring of spring constant K_b.

Example 4.8.1. Consider a beam simply supported on two rigid supports A, C (Fig. 4.8.3) and elastically supported at its midspan B on a spring of constant K_s.

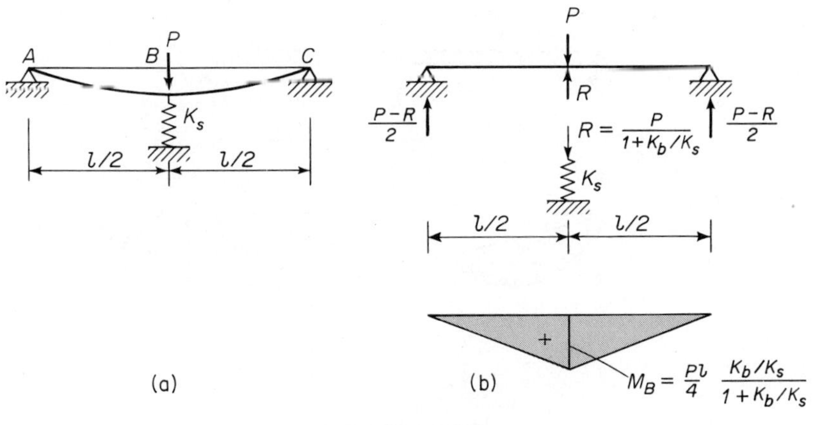

FIGURE 4.8.3

Under a concentrated load P at midspan, the beam deflects by δ and the support reacts with an upward reaction:

$$R = K_s\delta. \tag{a}$$

The beam is acted upon by a load $P - R$ and, from Fig. 4.8.2, its spring constant is $K_b = 48EI/l^3$; hence:

$$P - R = K_b\delta. \tag{b}$$

Equating the values of δ given by (a) and (b) we obtain:

$$\frac{R}{K_s} = \frac{P - R}{K_b} \quad \therefore \quad R = \frac{P}{1 + K_b/K_s}, \tag{4.8.9}$$

and by (a):

$$\delta = \frac{P}{K_s + K_b}. \tag{4.8.10}$$

As K_s approaches infinity, the support becomes rigid, δ approaches zero, and R approaches P, since the entire load is supported by B. Table 4.8.3 gives R/P, and:

$$\frac{M_B}{M_B]_{K_s=0}} = \frac{(P - R)l/4}{Pl/4} = 1 - \frac{R}{P} \quad \text{versus} \quad \frac{K_s}{K_b}.$$

Table 4.8.3

K_s/K_b	1/3	2/5	1/2	2/3	1.0	2	∞
R/P	0.25	0.29	0.33	0.40	0.50	0.67	1.00
$M_B/M_B]_{K_s=0}$	0.75	0.71	0.67	0.60	0.50	0.33	0.00

Example 4.8.2. Solve the problem of Example 4.8.1 for a uniform load w (Fig. 4.8.4).

The elastic support deflection $\delta = R/K_s$ equals the difference between the midspan beam deflection due to $W = wl$ with a spring constant $K_b = \frac{384}{5}(EI/l^3)$

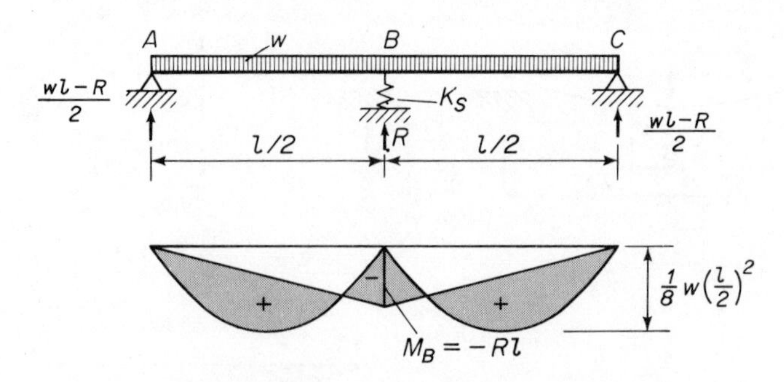

FIGURE 4.8.4

and that due to R with a spring constant $48EI/l^3 = \frac{5}{8}K_b$; hence:

$$\delta = \frac{R}{K_s} = \frac{wl}{K_b} - \frac{R}{\frac{5}{8}K_b} = \frac{1}{K_b}(wl - \tfrac{8}{5}R), \tag{a}$$

from which:

$$R = \left(\frac{5}{8}\right)\frac{wl}{1 + \frac{5}{8}(K_b/K_s)}, \tag{4.8.11}$$

and, by (a):

$$\delta = \frac{wl}{\frac{8}{5}K_s + K_b}. \tag{4.8.12}$$

As K_s approaches infinity, δ approaches zero, and R approaches $\frac{5}{8}wl$, twice the reaction at the fixed end of the beam of Table 4.1.2 (13) for a span $l/2$.

Example 4.8.3. When a moment M_B is applied at the tip of a cantilever [Fig. 4.8.5(a)], the moment M_A at the fixed end is equal to M_B; i.e., *the cantilever carries M_B unchanged from one end of the beam to the other*. We wish to determine the influence on the beam's moment-carrying mechanism of an elastic support at B [Fig. 4.8.5(b)].

From the tip deflection of the cantilever $\delta_0 = \dfrac{1}{2}\dfrac{Ml^2}{EI} = \dfrac{M/l}{2EI/l^3}$, we find the spring constant due to M to be

$$K_{b,M} = \frac{2EI}{l^3}, \tag{4.8.13}$$

by means of which:

$$\delta = \frac{R}{K_s} = \frac{M_B/l}{2EI/l^3} - \frac{R}{3EI/l^3} \quad \therefore \quad R = \frac{\frac{3}{2}M_B/l}{1 + (3EI/l^3)/K_s}, \tag{4.8.14}$$

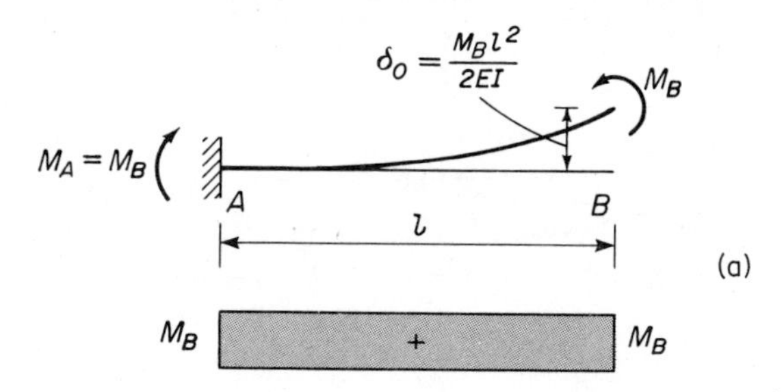

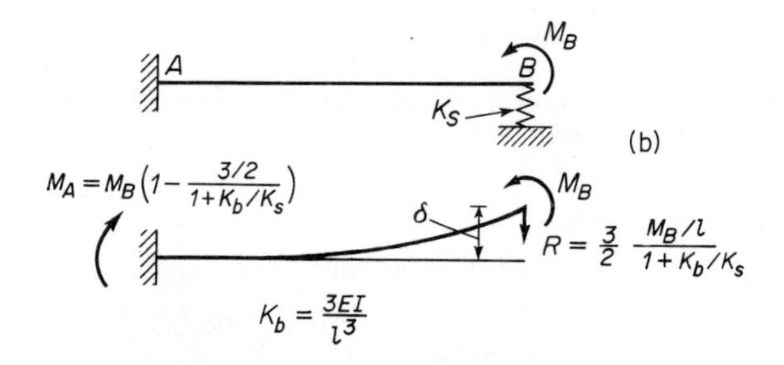

FIGURE 4.8.5

and the moment M_A becomes:

$$M_A = M_B - Rl = M_B\left(1 - \frac{3/2}{1 + (3EI/l^3)K_s}\right). \tag{4.8.15}$$

For K_s approaching infinity, $M_A = -\frac{1}{2}M_B$, as shown by (4.6.2); for $K_s = 6EI/l^3$, $M_A = 0$; for K_s smaller than $6EI/l^3$, M_A is positive; for $K_s = 0$, $M_A = M_B$. It is thus seen that an elastic support with a large K_s dampens the moment M_B and changes its sign in transmitting it to the support A. This result is basic in understanding the damping of boundary moments in thin shells (see Chapter 17).

Example 4.8.4. A uniformly loaded beam is supported on three elastic supports of spring constant K_s (Fig. 4.8.6). Determine how the load $W = wl$ is distributed between the three supports.

From statics:

$$2R_1 + R_2 = W \quad \therefore \quad R_1 = \frac{W - R_2}{2}, \tag{a}$$

and, by (4.8.2b):

$$\delta_1 = \frac{R_1}{K_s} \qquad \delta_2 = \frac{R_2}{K_s}. \tag{b}$$

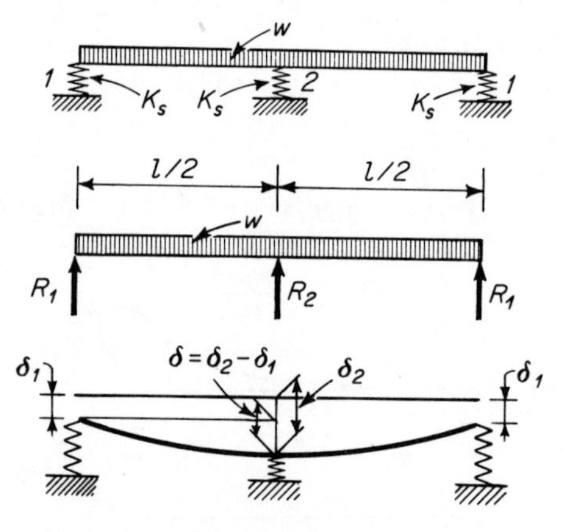

FIGURE 4.8.6

The elastic bending displacement of the beam at 2:

$$\delta = \delta_2 - \delta_1 = \frac{R_2 - R_1}{K_s}, \tag{c}$$

is due to the load $W = wl$ with spring constant $K_b = \frac{384}{5}(EI/l^3)$ and to the reaction R_2 with spring constant $48EI/l^3 = \frac{5}{8}K_b$:

$$\delta = \frac{W}{K_b} - \frac{R_2}{\frac{5}{8}K_b}. \tag{d}$$

Equating (c) and (d) and using (a), we obtain:

$$R_2 = \frac{5}{8}W\frac{1 + \frac{1}{2}(K_b/K_s)}{1 + \frac{15}{16}(K_b/K_s)}, \qquad R_1 = \frac{3}{16}W\frac{1 + \frac{5}{3}(K_b/K_s)}{1 + \frac{15}{16}(K_b/K_s)}. \tag{4.8.16}$$

For K_s approaching infinity, $R_2 = \frac{5}{8}W$, $R_1 = \frac{3}{16}W$, as in Table 4.1.2 (13). Table 4.8.4 gives R_1/W and R_2/W versus K_b/K_s and shows how a stiff beam is the best

Table 4.8.4

K_b/K_s	0	0.5	1.0	2.0	5.0	10.0	∞
R_1/W	0.19	0.23	0.26	0.28	0.31	0.32	0.33
R_2/W	0.62	0.54	0.48	0.43	0.38	0.36	0.33

guarantee of an even distribution of the load to an elastic foundation: if the beam could be made infinitely stiff, δ_1 would be equal to δ_2 and, by (b), the three reactions would be equal.

4.9 SHEAR BEAMS

All the beam deflections evaluated in the preceding sections were obtained without taking into account the contribution of the shear stresses to the deflections. To show that shear deflections are usually small compared to bending deflections, consider the cantilever of Fig. 4.9.1. Indicating by $f_s = \alpha P/A$ the shear stress *at the neutral axis* of the beam cross section, by $G = E/2(1 + \nu) \approx E/2$ the shear modulus, and by γ the shear strain, the shear deflection δ_s equals $\gamma l = (f_s/G)l$, or:

$$\delta_s = \alpha \frac{P}{GA} l \approx 2\alpha \frac{Pl}{EA}. \tag{4.9.1}$$

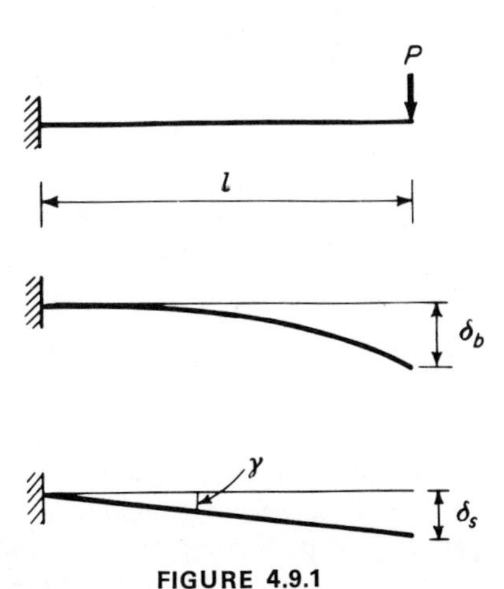

FIGURE 4.9.1

The bending deflection under the same load, from Table 4.3.1 (e), equals:

$$\delta_b = \frac{1}{3}\frac{Pl^3}{EI} = \frac{1}{3}\frac{Pl^3}{EAr^2}, \tag{4.9.2}$$

where r is the radius of gyration of the cross section. Hence, the ratio of bending to shear deflection equals:

$$\frac{\delta_b}{\delta_s} = \frac{1}{3}\frac{Pl^3}{EAr^2}\frac{EA}{2\alpha Pl} = \frac{1}{6\alpha}\left(\frac{l}{r}\right)^2. \tag{4.9.3}$$

For a rectangular cross section, $\alpha = \frac{3}{2}$, $r^2 = h^2/12$, $\delta_b/\delta_s = \frac{4}{3}(l/h)^2$, and for $10 \leq l/h \leq 20$:

$$133 \leq \frac{\delta_b}{\delta_s} \leq 533.$$

For I-beams $\alpha = 1$, and the web area A_w must be used in (4.9.1). Assuming, for example, $A_w = \frac{1}{5}A$:

$$\delta_s = \frac{2Pl}{EA_w} = 10\frac{Pl}{EA}. \tag{a}$$

With a flange area equal to $\frac{2}{5}A$ and a web thickness $t = \frac{1}{5}A/h$, the moment of inertia of the section equals, approximately:

$$I = 2\left(\frac{2}{5}A\right)\left(\frac{h}{2}\right)^2 + \frac{1}{12}(th^3) = \frac{1}{5}Ah^2 + \frac{1}{12}\left(\frac{1}{5}A\right)h^2 = \frac{13}{60}Ah^2,$$

and

$$r^2 = \tfrac{13}{60}h^2, \tag{b}$$

by means of which:

$$\delta_b = \frac{1}{3}\frac{Pl^3}{EA(\frac{13}{60}h^2)} \approx \frac{3}{2}\frac{Pl^3}{EAh^2}. \tag{c}$$

By (a) and (c):

$$\frac{\delta_b}{\delta_s} = \frac{3}{20}\left(\frac{l}{h}\right)^2,$$

and for $10 \leq l/h \leq 20$:

$$15 \leq \frac{\delta_b}{\delta_s} \leq 60. \tag{d}$$

PROBLEMS

4.1 The floor system of an office building consists of a 100 mm (4 in.) concrete slab spanning 1.75 m (6 ft) between 450 mm (18 in.) deep steel beams. The beams in turn span 9 m (30 ft) between 600 mm (24 in.) deep steel girders 9 m (30 ft) long. The floor is covered with vinyl tile, and a plaster ceiling hangs from it. Design the beams and the girders as ideal I-beams with flanges 300 mm (12 in.) wide. (Assume beams and girders to be simply supported.)

4.2 Design the 5 m (16 ft) long, simply supported wood joists spaced 400 mm (16 in.) o.c., for the floor of a private house. Use standard 50 mm (2 in.) wide lumber and consider a dead load of 1 kN/m² (20 psf).

4.3 A column with a load of 2 000 kN (450 k) is picked up at the center of a simply supported girder spanning 14 m (45 ft). The depth of the girder is limited to 1 m (40 in.) and the thickness of its flange plates to 50 mm (2 in.). Determine the width b of the girder flanges in high-strength steel, with $F_y = 345$ MPa (50,000 psi). Ignore the dead load of the girder first; then evaluate the increase in bending stress due to the dead load.

4.4 A steel spandrel beam carries a 100 mm (4 in.) brick wall and 1.75 m (6 ft) of floor loaded with a total load of 8 kN/m² (160 psf). Design the beam for a span of 10 m (32 ft). Note that the weight of the wall is usually assumed to impose on the beam a triangular load, with a maximum at midspan, due to a height of wall equal to half the beam span. Assume the beam to be fixed against rotation at its ends, and of A36 steel.

4.5 Determine the width b of concrete joists 90 cm (3 ft) o.c., fixed at both ends and spanning 10 m (32 ft), required to carry a 75 mm (3 in.) concrete slab, 50 mm (2 in.) of terrazzo, and a 25 mm (1 in.) plaster ceiling for a public space. The depth of the joist is limited to 432 mm (17 in.), and the concrete strength is $f_c' = 20$ MPa (2,900 psi). Note that the width may be governed by space requirements for the placement of bars at the center and may be increased at the ends if required for shear.

4.6 A concrete canopy cantilevers 3.5 m (12 ft) from a building and consists of two boundary beams 2.5 m (8 ft) o.c., with a 100 mm (4 in.) slab between them. Determine b and A_s for the beams if $f_c' = 30$ MPa (4,350 psi) and d is 380 mm (15 in.). Ignore the twist in the beams and assume a L.L. = 1.4 kN/m².

4.7 Determine the maximum allowable span of a 300 mm × 600 mm (12 in. × 24 in.) fixed-end concrete beam required to carry only its own weight for $f_c' = 40$ MPa (5,800 psi).

4.8 What is the concrete slab thickness h required to carry a superimposed load of 5 kN/m² (100 psf) on a simply supported span of 4.5 m (15 ft), if $f_c' =$ 20 MPa (2,900 psi)?

4.9 Compute the deflection of a steel window mullion under wind load. The spacing between mullions is 3 m (10 ft) and the height of the mullion is 9 m (30 ft). The simply supported mullion has the shape of an I-beam with a web 25 mm × 200 mm (1 in. × 8 in.) deep and two flanges 40 mm × 75 mm ($1\frac{1}{2}$ in. × 3 in.). The assumed wind pressure is 1.5 kN/m² (30 psf) and the steel is A36.

4.10 Compute the deflections of the beams of Problems 4.2 and 4.6.

4.11 An aluminum flagpole 18 m (60 ft) high, with a tubular cross section of outside diameter 400 mm (16 in.) and wall thickness 12.5 mm ($\frac{1}{2}$ in.), is acted upon by wind. Determine the maximum stress and the deflection at the top of the pole if the uniform wind load on the pole is 0.7 kN/m² (15 lb/sq ft) of projected area and the tip load due to the force on the flag is 1.5 kN (300 lb).

4.12 Compute the deflection of a 200 mm (8 in.) concrete slab spanning 5.5 m (18 ft), to be used as a simpy supported pedestrian bridge in a public building. How much would you camber up the slab at midspan to eliminate the deflection due to the dead load of the simply supported slab? ($f_c' = 30$ MPa, 4,350 psi).

4.13 The deflection of a 6 m (20 ft) long simply supported concrete beam carrying a 200 mm (8 in.) brick wall, 3.5 m (12 ft) high, is limited to 12.5 mm ($\frac{1}{2}$ in.). If the beam is 300 mm (12 in.) wide, what is its required depth? Do not ignore the dead load of the beam. ($f_c' = 20$ MPa, 2,900 psi)

4.14 What are the maximum moment and maximum shear in a 9 m (30 ft) long, fixed-ended concrete beam 300 mm (12 in.) wide and 750 mm (30 in.) deep, due to a differential support settlement of 25 mm (1 in.)? Is the beam capable of supporting these moments and shears if $f_c' = 30$ MPa (4,350 psi)?

4.15 What is the deflection of the slab of Problem 4.12 if the ends of the slab are fixed? What would be the camber to wipe out the dead-load deflection?

4.16 What is the carrying capacity of 50 mm × 300 mm (2 in. × 12 in.) simply supported Douglas fir joists spaced 400 mm (16 in.) o.c. and spanning 5 m (16 ft)? What is their deflection under full load?

4.17 Draw the bending-moment diagram for a three-span simply supported continuous beam of constant I and equal spans L for: (a) a uniform load w on the center and the right-hand span, (b) a uniform load w on all the spans. Express the moments as fractions of $wL^2/8$.

4.18 A two-span simply supported A36 steel beam carries a uniform live load of 15 kN/m (1 k/ft) and a uniform dead load of 22 kN/m (1.5 k/ft). If each span is 7.5 m (25 ft) long, determine the maximum design moment and the maximum design shear, and choose a suitable wide flange section for the beam.

4.19 The structure of a tool shed added to an existing building consists of frames 3.5 m (12 ft) o.c. The frames consist of 7 m (24 ft) long beams pinned to the existing wall and fixed to columns 3.5 m (12 ft) high. The bases of the columns are hinged. The shed roof consists of steel decking weighing 0.6 kN/m² (12 psf) and carries a 2 kN/m² (40 psf) live load. If the moment of inertia of the beam is 1.5 times that of the column, determine the final design moments and shears of the frames, assuming at first the dead load of the beams to be 5% of the live load they carry.

4.20 A three-span, reinforced concrete bridge consists of 450 mm × 1 200 mm (18 in. × 48 in.) simply supported beams 1.75 m (6 ft) o.c. and a 150 mm (6 in.) slab. In addition to a uniform load of 2.5 kN/m² (50 psf), the bridge must carry a concentrated load of 50 kN (12,000 lb) placed at the center of any span. Compute the maximum and minimum design moments and shear if the side spans are 12 m (40 ft) long and the center span is 15 m (50 ft) long.

4.21 Determine by moment distribution the moments and shears resulting from a 25 mm (1 in.) settlement of one of the interior supports in the beams of Problem 4.20.

4.22 A continuous, three-span simply supported concrete girder of constant cross section has side spans of 6 m (20 ft) and a central span of 9 m (30 ft). It carries 6 m (20 ft) of contributory area loaded with a 10 kN/m^2 (200 psf) live load and a dead load of 15 kN/m^2 (300 psf). Determine by moment distribution the maximum positive and negative moments in the girder by assuming the live load to act: (a) on the central girder only, (b) on the entire girder. What is the maximum shear in the girder?

4.23 A two-span beam of total length l has a rigid right support and is supported by elastic supports of modulus K_s at midspan and at the left support. Determine the moments and reactions for: (a) a uniform load w, (b) a concentrated load P at midspan.

4.24 Determine the beam stiffness I/l in the two-span beam of Fig. 4.8.3. if $P =$ 110 kN (25 k), $l = 12$ m (40 ft), $K_s = 175$ kN/m (1,000 lb/in.), and $R = P/2$.

4.25 What is the ratio of the shear to the bending deflection in a cantilevered I-beam with a tip load P, $A_w = \frac{1}{5}A$, and a span-depth ratio $l/h = 2$?

4.26 A fixed-end beam in a multistory building is 10.5 m (35 ft) long between a core and a perimeter column. The column is outside the glass line and is subjected to temperature variations of +16°C (60°F) and −38°C (−36°F) from the interior ambient temperature. The beam, a W460 × 143 (W18 × 96) of A36 steel, carries a uniform load of 32 kN/m (2.2 k/ft). What temperature moments are induced in a beam on the twentieth floor, 72 m (240 ft) above grade, and what are the total reactions on the column and on the core?

5. Beam-Columns and Buckling

5.1 BENDING AND COMPRESSION OR TENSION [8.2]

The resultant of a load P and a moment $M = Pe$, where e is the *eccentricity* of the load P, is a force parallel to P displaced by e from the line of action of P (Fig. 5.1.1).

When P is applied at the centroid of the section of a column or beam of area A, it produces a direct stress (compression or tension):

$$f_a = \frac{P}{A}. \tag{a}$$

The moment M produces extreme fiber bending stresses:

$$f_b = \pm\frac{Mc}{I}. \tag{b}$$

The resultant of P and M produces fiber stresses (Fig. 5.1.2):

$$f_{1,2} = f_a \pm f_b = \frac{P}{A} \pm \frac{Mc}{I}. \tag{5.1.1}$$

Indicating by:

$$r^2 = \frac{I}{A} \tag{5.1.2}$$

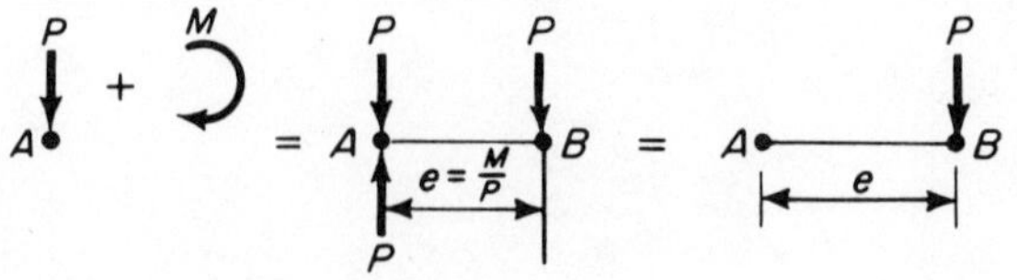

FIGURE 5.1.1

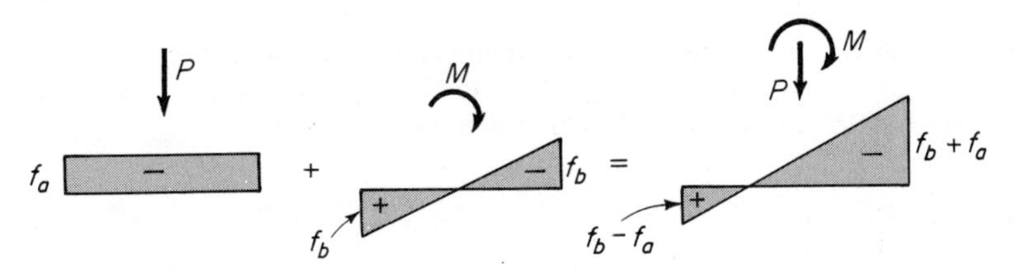

FIGURE 5.1.2

the square of the *radius of gyration* of the section, the combined stresses $f_{1,2}$ become:

$$f_{1,2} = \frac{P}{A}\left(1 \pm \frac{ec}{r^2}\right). \tag{5.1.3}$$

The fiber stresses $f_{1,2}$ are of the same sign if:

$$e \leq \frac{r^2}{c}. \tag{5.1.4}$$

For a rectangular cross section, (5.1.4) gives:

$$r^2 = \frac{1}{12}\frac{bh^3}{bh} = \frac{1}{12}h^2, \qquad c = \frac{h}{2}, \qquad \frac{r^2}{c} = \frac{h}{6} \qquad \therefore \quad e < \frac{h}{6};$$

i.e., the load must fall within the middle third of the section, [Fig. 5.1.3(a)]. For an ideal I-beam of area $2A$:

$$r^2 = \frac{Ah^2/2}{2A} = \frac{h^2}{4}, \qquad c = \frac{h}{2}, \qquad e \leq \frac{h}{2};$$

i.e., the load P must pass the edge of the section [Fig. 5.1.3(b)] before $f_{1,2}$ have opposite sign.

When a material has an allowable direct stress F_a and an allowable bending stress F_b, the allowable *combined* stresses are often determined by the formula:

$$\frac{f_a}{F_a} + \frac{f_{bx}}{F_{bx}} + \frac{f_{by}}{F_{by}} \leq 1.0 \leftarrow \tag{5.1.5}$$

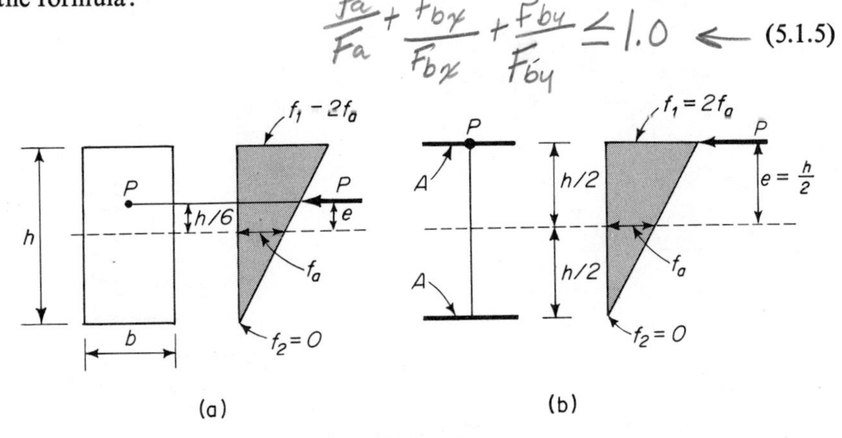

FIGURE 5.1.3

Example 5.1.1. Determine the maximum allowable eccentricity (a) along the web, and (b) at right angles to the web of a load capable of producing a direct stress $f_a = 103.4$ MPa (15 ksi) on a W920 × 446 (W36 × 300) column, if $F_a = F_b =$ 138 MPa (20 ksi).

Case (a) With $f_1 = F = 138$ MPa, $f_a = P/A = 103.4$ MPa, $c_x = 460$ mm, $r_x =$ 386 mm, $A = 56\,970$ mm², (5.1.3) gives:

$$138 = 103.4\left[1 + e\frac{460}{(386)^2}\right] \quad \therefore \quad e = 108 \text{ mm},$$

$$P = f_a A = \frac{103.4}{1\,000} \times 56\,970 = 5\,890 \text{ kN},$$

$$M = Pe = 5\,890\left(\frac{108}{1\,000}\right) = 636.2 \text{ kN}\cdot\text{m (473 k}\cdot\text{ft)}.$$

Case (b) With $r_y = 97$ mm, $c_y = 223$ mm, (5.1.3) gives:

$$138 = 103.4\left[1 + e\frac{223}{(97)^2}\right] \quad \therefore \quad e = 14 \text{ mm},$$

$$P = 5\,890 \text{ kN}, \quad M = 5\,890\left(\frac{14}{1\,000}\right) = 82.5 \text{ kN}\cdot\text{m (61.6 k}\cdot\text{ft)}.$$

Example 5.1.2. Determine the maximum height H of a square smoke stack made out of brick, with outer side a, and a concentric square conduit of side b, capable of withstanding a wind pressure p without developing tensile stresses (Fig. 5.1.4).

Calling ρ the unit weight of brick and A the cross-sectional area of the stack, the total weight of the stack is $W = \rho AH$, and the maximum direct stress due to W is:

$$f_a = \frac{W}{A} = \rho H.$$

The moment at the foot of the stack and the corresponding f_b due to the wind are:

$$M = \frac{1}{2}(pa)H^2, \qquad f_b = \pm\frac{\frac{1}{2}(pa)H^2(a/2)}{\frac{1}{12}(a^4 - b^4)} = \pm\frac{3pH^2}{a^2(1 - b^4/a^4)}.$$

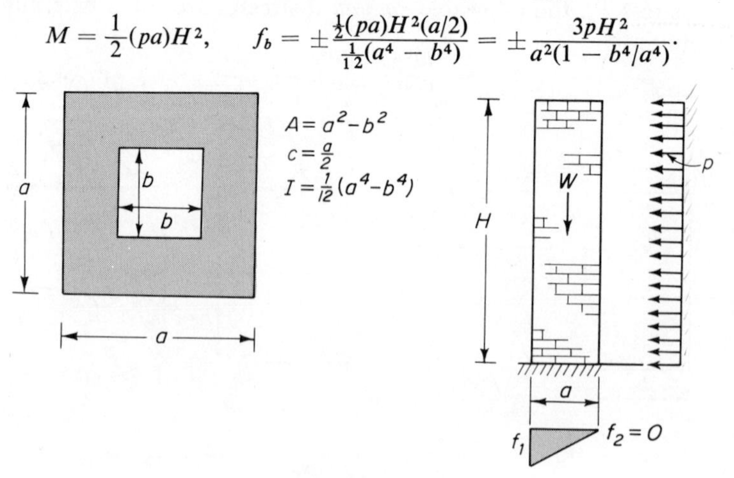

FIGURE 5.1.4

The maximum tensile stress is zero if $f_a = f_b$; hence:

$$pH = \frac{3pH^2}{a^2(1 - b^4/a^4)} \quad \therefore \quad pH = \frac{1}{3}\rho a^2\left(1 - \frac{b^4}{a^4}\right). \tag{5.1.6}$$

With $\rho = 18.85$ kN/m³ (120 lb/cu ft), $a = 1.20$ m (4 ft), $b = 0.60$ m (2 ft):

$$pH = \tfrac{1}{3}[18.85 \times 1.44(1 - 1/2^4)] = 8.48 \text{ kN/m (600 lb/ft)},$$

and $H = 10$ m (30 ft) for $p = 1$ kN/m² (20 psf). The corresponding maximum compression at the foot of the smoke stack is:

$$f_a + f_b = 2f_a = 2\rho H = 2 \times 18.85 \times 10 = 377 \text{ kN/m}^2 = 0.377 \text{ MPa (50 psi)},$$

well below an allowable stress of 0.83 MPa (120 psi).

Example 5.1.3. Design in steel the compressed member AC of the marquee of Example 3.2.4, which carries a distributed load of $4.8 \times 3 = 14.4$ kN/m (1 k/ft).

If the member is simply supported at both ends, its middle section is subjected to a compression $C = 37.4$ kN (8.66 k) (see Example 6.2.1). and a bending moment $M = \frac{1}{8} \times 14.4 \times 3^2 = 16.2$ kN·m (12.5 k·ft). From Appendix A, it is found that a Standard S180 × 22.8 (S7 × 15.3) beam has $A = 2\,890$ mm² and $S = 172\,000$ mm³; hence:

$$f_a = \frac{37.4}{2\,890} = 0.012\,9 \text{ kN/mm}^2 = 12.9 \text{ MPa (1.95 ksi)},$$

$$f_b = \frac{16.2 \times 1\,000}{172\,000} = 0.094\,2 \text{ kN/mm}^2 = 94.2 \text{ MPa (14.4 ksi)}.$$

For an A36 steel, $F_a = F_b = 151$ MPa and (5.1.5) gives:

$$\frac{12.9}{151} + \frac{94.2}{151} = 0.71 < 1.$$

Example 5.1.4. The rectangular prestressed concrete beam of Fig. 5.1.5 has curved tendons with a sag δ pulled to a tension T. Evaluate the live load the beam can carry without developing tensile stresses.

It is shown in Section 9.2 that a cable with a small sag ratio δ/l, carrying a uniform load w, develops a tension T which is practically constant along the cable and equal to the thrust H [see (9.2.7)]:

$$T \sim H - \frac{wl^2}{8\delta}.$$

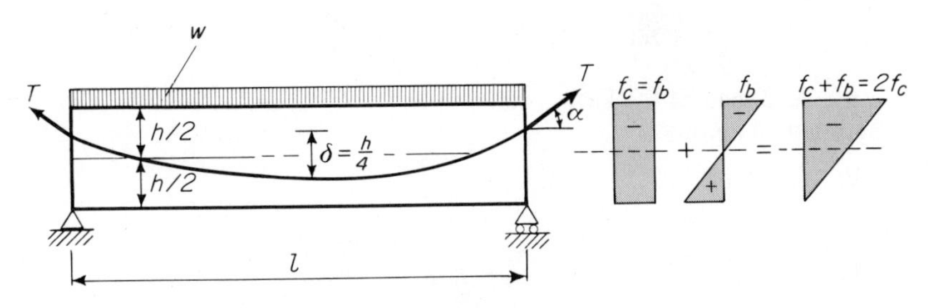

FIGURE 5.1.5

Hence, the load w_t carried by the tendons is given by:

$$w_t = \frac{8T\delta}{l^2}. \tag{a}$$

The prestressing compression $C = H \approx T$, applied at the centroid of the rectangular section of area bh, sets the concrete in compression with a stress:

$$f_c = \frac{T}{A} = \frac{T}{bh}. \tag{b}$$

Calling w_b the additional load carried by the beam in bending, and noticing that, when no tensile stresses are developed, the concrete beam behaves like a beam of homogeneous material, the bending extreme fiber stresses become:

$$f_b = \pm \frac{w_b l^2/8}{bh^2/6} = \pm \frac{3}{4}\left(\frac{w_b}{b}\right)\left(\frac{l}{h}\right)^2. \tag{c}$$

Equating the maximum tensile stress f_b to the compressive stress f_c and solving for w_b, we obtain:

$$\frac{T}{bh} = \frac{3}{4}\left(\frac{w_b}{b}\right)\left(\frac{l}{h}\right)^2 \quad \therefore \quad w_b = \frac{4}{3}\frac{Th}{l^2}. \tag{d}$$

The total distributed load carried by the beam is, thus:

$$w = w_t + w_b = 8\frac{\delta}{l^2}T + \frac{4}{3}\frac{h}{l^2}T. \tag{5.1.7}$$

With $\delta \approx h/4$, the total load is:

$$W = wl = \frac{10}{3}\left(\frac{h}{l}\right)T, \tag{5.1.8}$$

while the maximum compressive stress is given by:

$$f_1 = 2f_c = \frac{2T}{bh}. \tag{5.1.9}$$

For example, by (5.1.9), a beam of section 300 mm $\times$ 600 mm (12 in. $\times$ 24 in.) with an allowable stress F_c = 14 MPa (2 ksi) may develop a tension in the tendons:

$$T = \frac{1}{2}F_c bh = \frac{1}{2} \times \frac{14}{1\,000} \times 300 \times 600 = 1\,260 \text{ kN (288 k)}.$$

With $h/l = \frac{1}{20}$, i.e., for a length $l = 20 \times 0.6 = 12$ m (40 ft), the beam can carry, without developing tensile stresses, a total load given by (5.1.8):

$$W = \frac{10}{3} \times \frac{1}{20} \times 1\,260 = 210 \text{ kN} \quad \text{or} \quad w = \frac{210}{12} = 17.5 \text{ kN/m (1.2 k/ft)}.$$

Since the dead load of the beam is $0.3 \times 0.6 \times 22.78 = 4.1$ kN/m, the allowable live load is 13.4 kN/m.*

By (a) and (d) for $\delta = h/4$:

$$\frac{w_t}{w_b} = \frac{8\delta}{\frac{4}{3}h} = \frac{8h/4}{\frac{4}{3}h} = \frac{3}{2} \quad \therefore \quad w_t = \frac{3}{2}w_b,$$

*Top reinforcement is required in the beam to absorb the excess of the negative moment $T\delta$ over the dead load moment $wl^2/8$ so as to avoid tension in the concrete in the absence of live load.

so that $\frac{3}{2}/(1 + \frac{3}{2})$ or 60% of the load is carried by cable action of the tendons and 40% by bending action of the beam.

The same beam made out of reinforced concrete with 1% tensile steel reinforcement could carry about 10 kN/m, including its own dead load, or 5.9 kN/m of live load. The increase in live load due to prestressing is thus about 130%.

5.2 BEAM-COLUMNS

A beam-column is a structural member carrying both lateral and longitudinal loads, i.e., loads at right angles to its axis and compressive or tensile loads.

The stresses in a beam-column may be approximately evaluated as the sum of the bending and direct stresses, as was done in the examples of Section 5.1, but in so doing one tacitly assumes that the longitudinal load does not contribute substantially to the bending of the element. In computing the bending moment correctly, one must, instead, take into account the lever arm of the longitudinal load due to the bending deflections caused by the transverse load.

The cantilever of Fig. 5.2.1 is acted upon by a transverse load W and a compressive load P. We wish to determine the maximum fiber stresses in the beam, which occur at the section B.

If we ignore the tip deflection δ, the stresses at B are given by (5.1.1) with $M_W = Wl$. But under the action of W, the cantilever tip deflects by $\delta = \frac{1}{3}(Wl^3/EI)$ [see Table 4.3.1 (e)] and the load P adds to the moment $M_W = Wl$ a moment:

$$M_P = P\delta = \frac{P}{3}\frac{Wl^3}{EI}. \tag{a}$$

The ratio of M_P to M_W may be evaluated by noticing that, since the fiber stress f_b due to M_W equals $(Wl)c/I$, multiplication and division of (a) by Wc gives:

$$M_P = \frac{1}{3}\left(\frac{P}{W}\right)\left(\frac{(Wl)c/I}{E}\right)\frac{Wl^2}{c} = \frac{1}{3}\left(\frac{f_b}{E}\right)\left(\frac{l}{c}\right)\left(\frac{P}{W}\right)Wl = \frac{1}{3}\left(\frac{f_b}{E}\right)\left(\frac{l}{c}\right)\left(\frac{P}{W}\right)M_W. \tag{b}$$

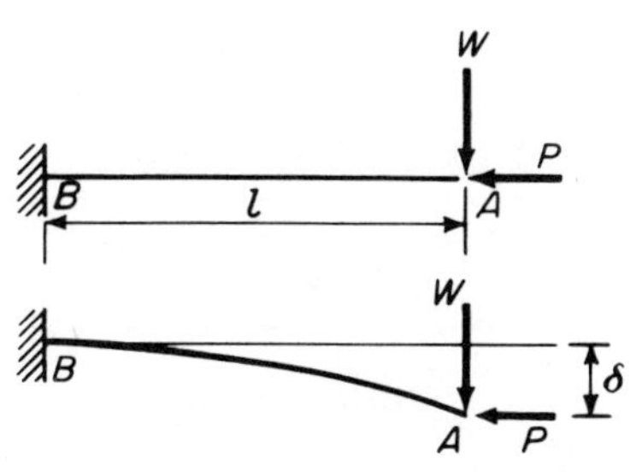

FIGURE 5.2.1

By (b), the total moment at B becomes:

$$M_B = M_W + M_P = \left[1 + \frac{1}{3}\left(\frac{f_b}{E}\right)\left(\frac{l}{c}\right)\left(\frac{P}{W}\right)\right]M_W. \tag{5.2.1}$$

The ratio f_b/E (1/1 200 for steel, 1/2 400 for concrete) is of the order of 1/2 000; the ratio l/c is of the order of 20. Hence, the factor $\frac{1}{3}(f_b/E)(l/c)$ is of the order of 1/300 and:

$$M_B \approx \left(1 + \frac{1}{300}\frac{P}{W}\right)M_W. \tag{5.2.2}$$

Thus, the increase in M_W due to P is unimportant unless P is much larger than W. On the other hand, when P is much larger than W, the moment M_P is comparable to M_W. This added large M_P increases the maximum beam deflection from δ to, say, $\delta + \delta'$ and produces an additional increase $P\delta'$ in M_P, which increases δ, etc. It is thus seen that (5.2.2) is not applicable for $P \gg W$, since in this case a chain reaction may start, capable of producing the collapse of the beam. (This is analogous to the buckling phenomenon discussed in Section 5.3.) Usually, the deflection δ' due to the moment $P\delta$ is negligible in comparison with δ, and (5.2.2) holds.

For the simply supported beam of Fig. 5.2.2, one obtains, similarly, for the midspan moment due to P:

$$M_P = P\delta = P\frac{5}{384}\frac{wl^4}{EI} = \frac{5}{48}\left(\frac{(wl^2/8)c/I}{E}\right)\left(\frac{l}{c}\right)Pl$$

$$= \frac{5}{6}\left(\frac{f_b}{E}\right)\left(\frac{l}{c}\right)\left(\frac{P}{wl}\right)\left(\frac{wl^2}{8}\right) = \frac{5}{6}\left(\frac{f_b}{E}\right)\left(\frac{l}{c}\right)\left(\frac{P}{W}\right)M_W,$$

where $W = wl$ is the total lateral load on the beam. Hence, the maximum moment becomes:

$$M_C = \left[1 + \frac{5}{6}\left(\frac{f_b}{E}\right)\left(\frac{l}{c}\right)\left(\frac{P}{W}\right)\right]M_W. \tag{5.2.3}$$

Indicating by $\bar{f}_b$ the stresses due to M_C and by f_b the stresses due to M_W we obtain:

$$\frac{\bar{f}_b}{f_b} = \frac{M_C}{M_W} = 1 + \frac{5}{6}\left(\frac{f_b}{E}\right)\left(\frac{l}{c}\right)\left(\frac{P}{W}\right). \tag{5.2.3a}$$

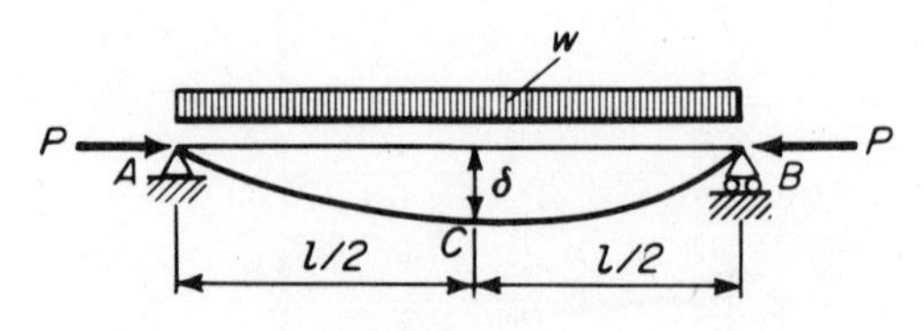

FIGURE 5.2.2

Example 5.2.1. The uniformly loaded, simply supported W310 × 60 (W12 × 40) beam of Fig. 5.2.3 is set on its supports when the ambient temperature is 5°C. Determine its maximum fiber stresses when the temperature is 20°C, if the supports are prevented from moving apart.

The coefficient of thermal expansion of steel is 1×10^{-5} mm/mm/°C. Hence, for a temperature difference of 15°C, the thermal strain is:

$$\epsilon = 15 \times 1 \times 10^{-5} = 15 \times 10^{-5} \text{ mm/mm},$$

and the corresponding compressive stress is:

$$f_c = E\epsilon = 200\,000 \times 15 \times 10^{-5} = 30 \text{ MPa}. \tag{c}$$

The beam area is 7 610 mm², and the total compressive force is:

$$P = \frac{30}{1\,000} \times 7\,610 = 228 \text{ kN}.$$

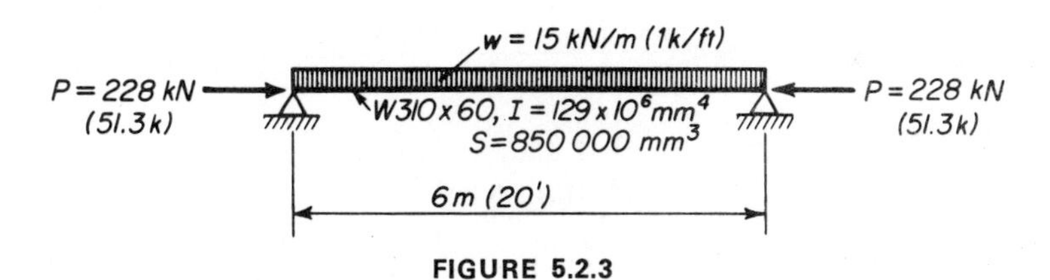

FIGURE 5.2.3

With $S = 850\,000$ mm³ and $w = 15$ kN/m, the maximum bending stress in the uncompressed beam is:

$$f_b = \frac{wl^2/8}{S} = \frac{15 \times 6^2 \times 1\,000}{8 \times 850\,000} = 0.079\,4 \text{ kN/mm}^2 = 79.4 \text{ MPa}. \tag{d}$$

Hence, by (5.2.3a), the maximum bending stress when thermal expansion is prevented, is:

$$\bar{f}_b = \left[1 + \frac{5}{6}\left(\frac{79.4}{200\,000}\right)\left(\frac{6 \times 1\,000}{155}\right)\left(\frac{228}{90}\right)\right]79.4$$

$$= (1 + 0.03)79.4 = 81.78 \text{ MPa},$$

or 3% above f_b. The maximum fiber stress is a compressive stress:

$$f_c + \bar{f}_b = 30.0 + 81.78 \simeq 112.0 \text{ MPa},$$

or 41% higher than the bending stress f_b.

Example 5.2.2. The beam of Fig. 5.2.4(a) is supported on a fixed hinge at A and a roller hinge at B. Its fiber stresses at midspan are, with $S = 1.15 \times 10^6$ mm³ (70.2 in.³):

$$f_b = \frac{15 \times 9^2 \times 1\,000}{8 \times 1.15 \times 10^6} = 0.132 \text{ kN/mm}^2 = 132 \text{ MPa}.$$

What is the maximum fiber stress $\bar{f}_b$ if the horizontal displacement of the right support is prevented?

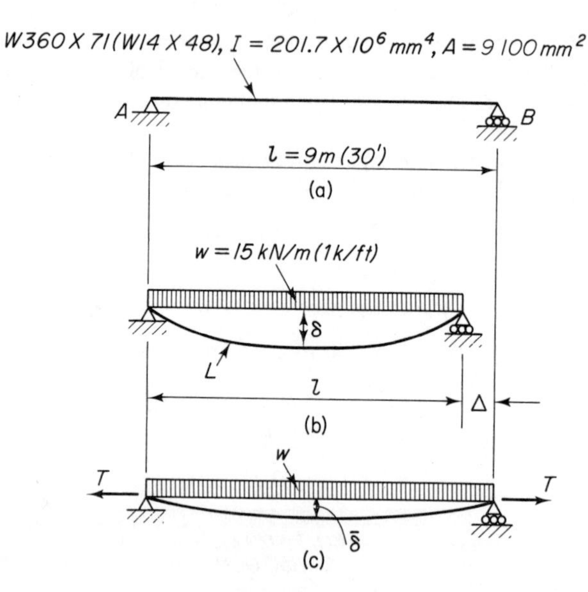

FIGURE 5.2.4

Under the distributed load w, the beam deflects at midspan by:

$$\delta = \frac{5}{384}\frac{wl^4}{EI} = \frac{5}{384}\frac{15 \times 9^4 \times 1\,000^4}{200\,000 \times 201.7 \times 10^6} = 31.8 \text{ mm}.$$

Since the middle axis of the beam is the neutral axis and does not elongate, the deflected beam length is 9 m. Hence, assuming the shape of the deflected beam axis under the small sag-span ratio $\delta/l = 31.8/9\,000 = 0.003\,5$ to be a parabola, we may estimate the horizontal displacement of B as the difference Δ between L and l in Fig. 5.2.4(b). As shown in Section 9.1 [see (9.1.7)], this displacement Δ is given approximately by:

$$\Delta = \frac{8}{3}\left(\frac{\delta}{l}\right)^2 l = \frac{8}{3}\left(\frac{\delta}{l}\right)\delta = 2.67 \times 0.003\,5 \times 31.8 = 0.3 \text{ mm}.$$

To wipe out this horizontal displacement, the beam must be pulled by a force T capable of stretching it by Δ, so that, by (3.2.3) and (3.2.4):

$$T = EA\left(\frac{\Delta}{l}\right) = \frac{8}{3}EA\left(\frac{\delta}{l}\right)^2 = \frac{8}{3} \times \frac{200\,000}{1\,000} \times 9\,100\left(\frac{31.8}{9 \times 1\,000}\right)^2 = 60.6 \text{ kN}.$$

The tensile stress due to T is:

$$f_t = \frac{60.6}{9\,100} = 0.006\,66 \text{ kN/mm}^2 = 6.66 \text{ MPa}.$$

The reduction in the bending stresses due to P is given by the expression in brackets in (5.2.3), with $P = -T$, since the moment $T\delta$ due to T has a sign opposite to that due to W:

$$1 - \frac{5}{6}\left(\frac{f_b}{E}\right)\left(\frac{l}{c}\right)\left(\frac{T}{W}\right) = 1 - \frac{5}{6}\left(\frac{132.1}{200\,000}\right)\left(\frac{9 \times 1\,000}{180}\right)\left(\frac{60.6}{135}\right)$$
$$= 1 - 0.012\,35 = 0.987\,65.$$

The tension T reduces the moment by only 1.24%, and hence carries by *cable action* only this small fraction of the total load. This mechanism becomes more important for large sags: when a beam has large plastic deflections because it is over-stressed in bending, some of its load may still be carried by cable action if the supports are prevented from moving.

The maximum tensile fiber stress in the beam is:

$$\bar{f}_b + f_t = 0.987\,65 \times 132.1 + 6.66 = 137.13 \text{ MPa (19.9 ksi)},$$

and is 4% larger than the bending stress due to the lateral load.

5.3 BUCKLING [5.2, 7.1]

The increase induced by a thrust P in the maximum bending moment due to a lateral load [see (5.2.1), (5.2.3)] may be written in a form which relates it to the structural properties of the beam rather than to the ratio P/W.

For the case of a cantilever with a transverse tip load W (Fig. 5.2.1), the total moment is given approximately by:

$$\begin{aligned} M &= M_W + M_P = Wl + P\delta = Wl + P\frac{1}{3}\frac{Wl^3}{EI} \\ &= \left[1 + \frac{P}{3EI/l^2}\right]M_W. \end{aligned} \tag{5.3.1}$$

For the case of a uniformly loaded, simply supported beam, the approximate maximum moment is given by:

$$\begin{aligned} M &= M_W + M_P = \frac{wl^2}{8} + P\delta = \frac{wl^2}{8} + P\frac{5}{384}\frac{wl^4}{EI} \\ &= \left[1 + \frac{P}{(\frac{48}{5})(EI/l^2)}\right]M_W. \end{aligned} \tag{5.3.2}$$

Equations (5.3.1) and (5.3.2) show that the magnifying factor due to the thrust P contains the ratio of P to a multiple of EI/l^2. This quantity has the dimensions of a load:

$$\left[\frac{EI}{l^2}\right] = \left(\frac{\text{kN}}{\text{mm}^2}\right)\left(\frac{\text{mm}^4}{\text{mm}^2}\right) = \text{kN}$$

and is defined by the material constant E, the cross-sectional bending property I, and the span l of the beam; the coefficient in front of this quantity depends on the boundary conditions of the beam. The physical significance of the load EI/l^2 will now be shown.

It was stated in Section 5.2 that buckling occurs if and only if, when the additional deflection δ' due to $M_P = P\delta$ increases the value of M_P to $P(\delta + \delta')$; the increased M_P, in turn, increases the deflection by δ'' so that M_P becomes $P(\delta + \delta' + \delta'')$, etc., this "chain reaction" phenomenon continues until the beam breaks. Experiments show that, *however small the value of the lateral load*, the chain reaction does or does not occur in a given beam, depending on whether the thrust P is greater or smaller than a given load called the *critical load* P_{cr} for that beam. For $P > P_{cr}$, the beam continues deflecting, i.e., is unstable and breaks; for $P < P_{cr}$, the beam remains in equilibrium, i.e., is stable. Hence, $P = P_{cr}$ is the lowest value of P for which the beam becomes unstable, or the highest for which the beam will still remain in equilibrium under thrust, however small the value of the lateral load.

Let us determine P_{cr} for the *rigid* column of Fig. 5.3.1, which is hinged at A and supported by a spring of spring constant k at B. For $P \leq P_{cr}$, under the action of W and P the column rotates by γ and the point B displaces,

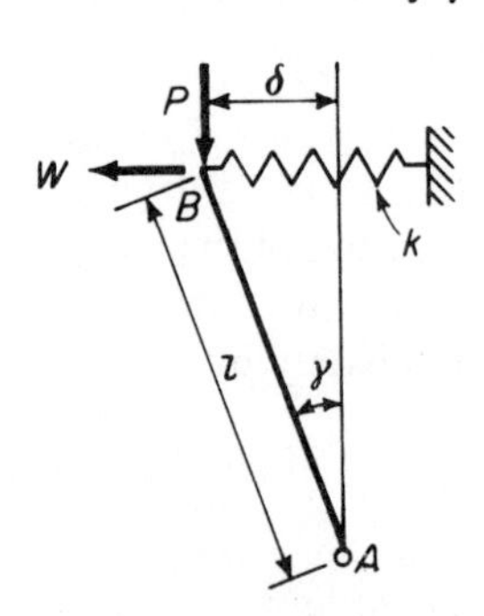

FIGURE 5.3.1

elongating the spring by δ. If the column *remains in equilibrium* under the action of W, P, and the spring reaction $k\delta$, the moment of these forces about A must be zero:

$$(W - k\delta)l \cos \gamma + P\delta = 0,$$

from which:

$$P = \frac{(k\delta - W)l \cos \gamma}{\delta}. \tag{a}$$

Equation (a) shows that there is a value of the load P capable of maintaining the column in equilibrium in a deflected position *even if* $W = 0$:

$$P = kl \cos \gamma, \tag{b}$$

and that the value of this load for which γ is almost zero (and hence $\cos \gamma < 1$), i.e., the value of P at which the column will *start* rotating, equals:

$$P_{cr} = kl. \tag{5.3.3}$$

The value of P_{cr} is seen to depend exclusively on the column length and the spring constant. For $P < P_{cr}$ the column's only position of equilibrium, without a lateral load, is a vertical position.

Writing (5.3.3) as:

$$P_{cr}\delta = k\delta l, \tag{5.3.4}$$

it is seen that the critical load may also be found by equating the moment $P\delta$ of P about A to the moment due to the elastic spring support, $(k\delta)l$.

In order to obtain the critical load for a cantilevered column one must, similarly, equate the moment $P\delta$ of the thrust to the bending moment M_A at the root of the cantilever (Fig. 5.3.2). Since the column deflection is unknown, both moments are unknown; but we may obtain approximate values of P_{cr} by assuming, for consistency's sake, that both the deflection δ and the moment M_A are due to the same lateral load. For example, assuming that the lateral load is a concentrated tip load W:

$$P\delta = P\frac{1}{3}\frac{Wl^3}{EI}, \qquad M_A = Wl,$$

from which, equating $P\delta$ to M_A:

$$P_{cr} = 3\frac{EI}{l^2}. \tag{c}$$

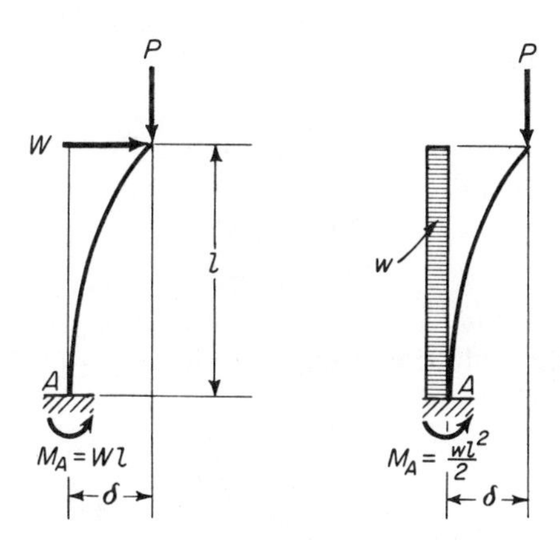

FIGURE 5.3.2

Assuming the lateral load to be uniformly distributed:

$$P\delta = P\frac{1}{8}\frac{wl^4}{EI}, \qquad M_A - \frac{wl^2}{2},$$

from which:

$$P_{cr} = 4\frac{EI}{l^2}. \tag{d}$$

Since P_{cr} is independent of the lateral load and is the *minimum* load capable of maintaining the column in equilibrium, (c) might be a better approximation than (d).

For a simply supported beam (Fig. 5.3.3), a lateral load W at midspan gives:

$$P\delta = P\frac{1}{48}\frac{wl^4}{EI}, \qquad M_C = \frac{Wl}{4} \qquad \therefore \quad P_{cr} = 12\frac{EI}{l^2}, \tag{e}$$

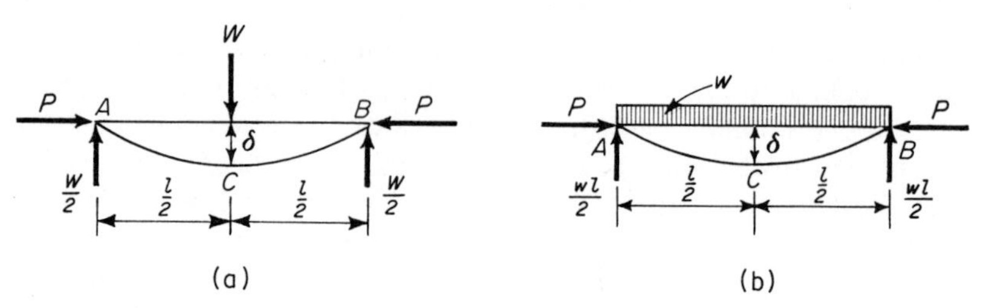

FIGURE 5.3.3

while a uniform load gives:

$$P\delta = P\frac{5}{384}\frac{wl^4}{EI}, \qquad M_C = \frac{wl^2}{8} \qquad \therefore \quad P_{cr} = 9.6\frac{EI}{l^2}, \tag{f}$$

and (f) may be expected to be a better approximation than (e).

The exact value of P_{cr} is obtained by equating at each section x of the beam (Fig. 5.3.4) the moment Py due to P to the bending moment $M(x)$ due to the lateral deflection y. Since, by (4.1.4a), $M(x) = -EIC = -EI\, d^2y/dx^2$:

$$Py = -EI\frac{d^2y}{dx^2},$$

and the *buckling equation:*

$$\frac{d^2y}{dx^2} = -\frac{P}{EI}y \tag{5.3.5}$$

states that *at every section* x the buckled deflection y is such that its second derivative equals the deflection itself multiplied by $-(P/EI)$.

One such deflection is:

$$y = \delta \sin\sqrt{\frac{P}{EI}}\,x, \tag{g}$$

since:

$$\frac{d^2y}{dx^2} = -\frac{P}{EI}\delta \sin\sqrt{\frac{P}{EI}}\,x = -\frac{P}{EI}y.$$

For a simply supported beam of length l, the deflection must be zero at $x = 0$ and $x = l$. The deflection (g) is always zero at $x = 0$, but is zero at $x = l$ if and only if:

$$\delta \sin\sqrt{\frac{P}{EI}}\,l = 0,$$

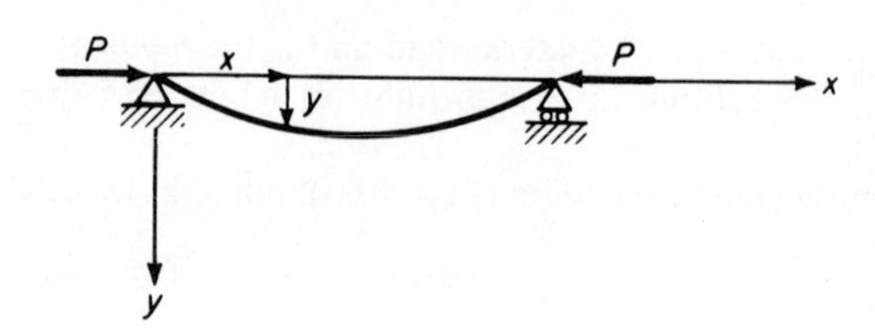

FIGURE 5.3.4

i.e., if either $\delta = 0$, in which case $y \equiv 0$ and the beam is not buckled, or:

$$\sqrt{\frac{P}{EI}}\, l = n\pi \quad \therefore \quad P = n^2\pi^2\frac{EI}{l^2} \quad (n = 1, 2, 3, \ldots). \tag{h}$$

The lowest value of P corresponds to $n = 1$, i.e., to a buckled shape with one-half sine wave and is the exact value of P_{cr}:

$$P_{cr} = \pi^2\frac{EI}{l^2}.^* \tag{5.3.6}$$

Since $\pi^2 = 9.87$, it is seen that the approximation (f) was quite satisfactory.

It must be noticed that the I appearing in (5.3.6), or any other buckling formula, is the least value of I corresponding to a bending deflection permitted by the boundary conditions. For example, if a W (wide flange) beam can bend in a direction parallel to the flanges, $I = I_y$, i.e., $I_{min.}$.

The buckling loads of beams with different types of support conditions are easily found by determining the span between inflection points of the simply supported beam of identical behavior, i.e., the so-called *reduced length* l_r. For example, from Fig. 5.3.5 it is seen that a cantilevered beam of span l behaves like a simply supported beam of span $2l$, so that its critical load is:

$$P_{cr} = \pi^2\frac{EI}{(2l)^2} = \frac{\pi^2}{4}\frac{EI}{l^2} = 2.47\frac{EI}{l^2}. \tag{5.3.7}$$

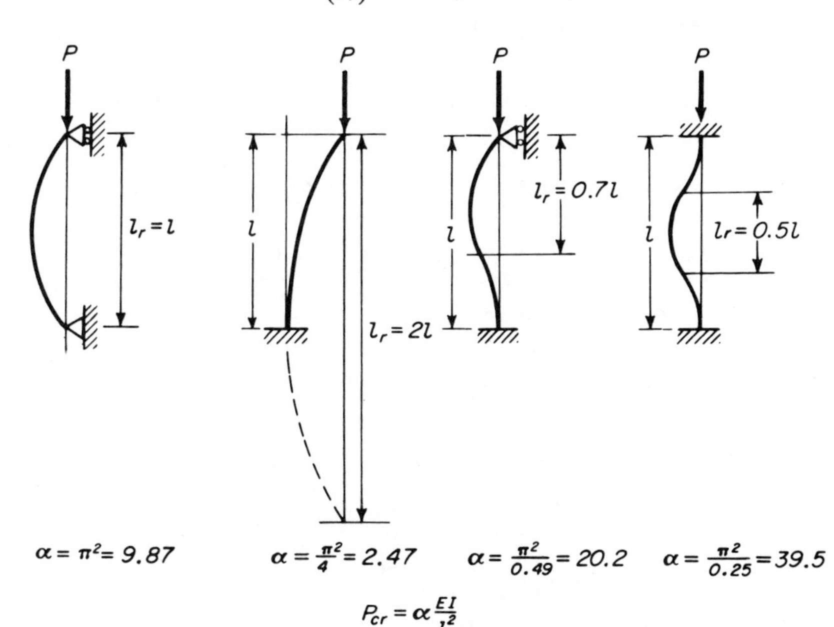

FIGURE 5.3.5

*The higher values of P, corresponding to $n = 2, 3, 4, \ldots$, are of little physical significance unless the beam is so braced as to buckle into 2, 3, 4, . . . half sine waves.

It is thus seen that the magnification factors appearing within the brackets of (5.3.1) and (5.3.2) contain the ratios of P to P_{cr} and may be written as:

$$\beta = 1 + \frac{P}{P_{cr}}.$$

Since β gives correct results only for $P \ll P_{cr}$, and since for small values of a quantity α the expression $1 + \alpha \approx 1/(1 - \alpha)$, it is customary to let $P/P_{cr} = \alpha$ and to write the magnification factor β as:

$$\beta = \frac{1}{1-\alpha} \quad \text{with} \quad \alpha = \frac{P}{P_{cr}}. \tag{5.3.8}$$

The factor β may be used to determine moments, stresses, deflections, or slopes in beam columns with $P \ll P_{cr}$, as soon as P_{cr} is known.

Example 5.3.1. Evaluate P_{cr} and the corresponding $f_{cr} = P_{cr}/A$ for an A36 W360 × 79 (W14 × 53) of length 6 m (20 ft) when it is (a) simply supported, (b) fixed at both ends, and (c) cantilevered.

With $E = 200\,000$ MPa, $I_y = 24 \times 10^6$ mm^4, $A = 10\,060$ mm^2, and $l = 6\,000$ mm,

Case (a)

$$P_{cr,a} = \frac{\pi^2 \times 200\,000(24 \times 10^6)}{1\,000(6\,000)^2} = 1\,315 \text{ kN}, \qquad f_{cr} = 130.7 \text{ MPa} < F_c.$$

Case (b) $\quad P_{cr,b} = 4P_{cr,a} = 5\,260$ kN, $\quad f_{cr} = 522.8$ MPa $> F_c$.

Case (c) $\quad P_{cr,c} = \frac{1}{4}P_{cr,a} = 329$ kN, $\quad f_{cr} = 32.7$ MPa $< F_c$.

In cases (a) and (c) buckling and not strength governs the beam design.

Example 5.3.2. Determine P_{cr} and f_{cr} for an aluminum column of rectangular cross section 50 mm × 75 mm (2 in. × 3 in.), 3.5 m (12 ft) long, fixed at one end and simply supported at the other:

$$P_{cr} = \frac{\pi^2 \times 69 \times 10^3 \times (\frac{1}{12})(75 \times 50^3)}{1\,000(0.7 \times 3.5 \times 1\,000)^2} = 88.6 \text{ kN}, \quad f_{cr} = 23.6 \text{ MPa} < F_c.$$

Example 5.3.3. Compute the maximum deflection of the beam of Fig. 5.2.3.

$$P_{cr} = \frac{\pi^2 \times 200\,000(12.9 \times 10^6)}{1\,000(6\,000)^2} = 7\,073 \text{ kN};$$

$$P = 280 \text{ kN}; \qquad \alpha = \frac{P}{P_{cr}} = 0.04,$$

$$\beta = \frac{1}{1-\alpha} = 1.04,$$

$$\delta = 1.04 \frac{5}{384} \frac{15 \times 6\,000^4}{200\,000(129 \times 10^6)} = 10.2 \text{ mm (0.401 in.)}.$$

Example 5.3.4. Determine P_{cr} and f_{cr} for a concrete, simply supported column 100 mm × 200 mm (4 in. × 8 in.), 4.5 m (15 ft) long.

$$P_{cr} = \frac{\pi^2 \times 20\,000 \times \frac{1}{12} \times 200 \times 100^3}{1\,000(4.5 \times 1\,000)^2} = 162.5 \text{ kN}, \qquad f_{cr} = 8.1 \text{ MPa} < F_c.$$

The values of P_{cr} obtained in this section, and used in the next for analysis and design, were derived under the assumption that the stress P_{cr}/A due to the critical value of the thrust P remains within the elastic range. In practice this happens only for very slender columns, and most columns buckle in the plastic range.

It is also assumed in the derivation that P remains parallel to the column axis after the column buckles, as is usually the case for thrusts due to vertical loads. It is easy to see, instead, that a column cannot buckle under any thrust P which remains tangent to the column axis while the column buckles. In fact, as shown by Fig. 5.3.6, if the column were to buckle to the right, the counterclockwise moment $P\delta$ would tend to straighten it, while a deflection to the left would create a clockwise moment $P\delta$ which would also tend to straighten the column. In other words, the moment $P\delta$ is a stabilizing moment, and the "chain reaction" displacement cannot take place. Thus, one cannot buckle a concrete column by prestressing it through a tendon acting along the column axis, since the tension in the tendon, and hence the equal and opposite compression in the column, would rotate together with the rotation of the ends of the buckled column and remain tangent to the buckled column axis.

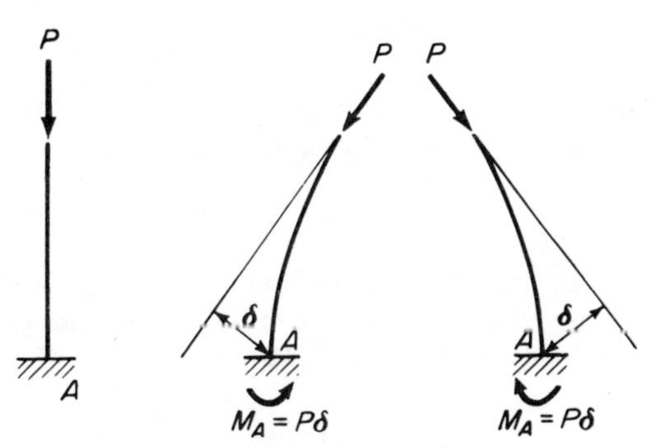

FIGURE 5.3.6

5.4 DESIGN LIMITATIONS DUE TO BUCKLING

The design of compressed columns is limited by buckling considerations. If we set $I = Ar^2$ in (5.3.6) [see (5.1.2)], and divide the equation through by A, the critical stress f_{cr} becomes:

$$f_{cr} = \frac{\pi^2 E}{(l/r)^2}. \tag{5.4.1}$$

The ratio l/r of the reduced, or unbraced, length to the minimum radius of gyration r is called the *slenderness ratio* of the column. Equation (5.4.1) shows that for slender columns with small r and large l, i.e., for large values of l/r, the critical stress may be so small as to become smaller than the allowable stress F_c. In any case, most codes do not permit slenderness ratios larger than 200 for compressed columns and give empirical curves limiting F_c in terms of l/r. For example, Fig. 5.4.1 gives F_c in terms of l/r for steel

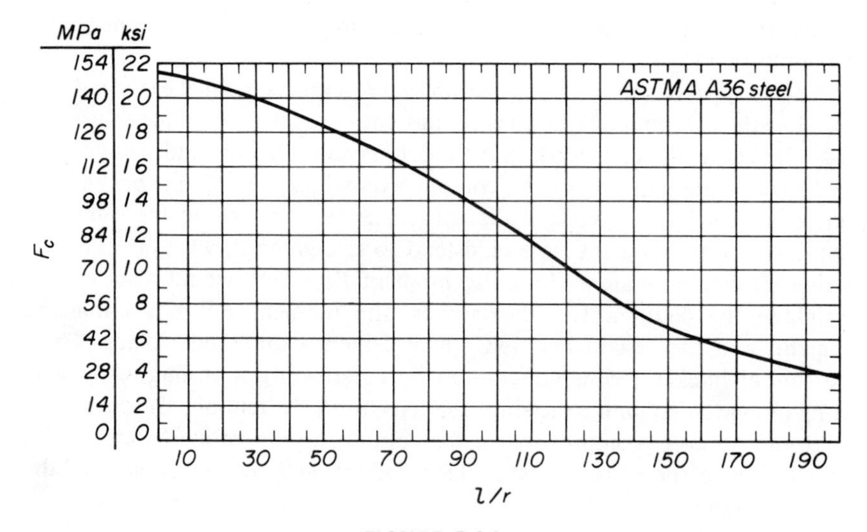

FIGURE 5.4.1

columns, according to the AISC 1971 specifications. When buckling is the critical factor in column design, a factor of safety of 2.5 is customarily used against f_{cr}.

Example 5.4.1. Determine the maximum length allowed by code for a W310 × 60 (W12 × 40) simply supported column, with $l/r = 120$ and the corresponding allowable compressive load on it. With $A = 7\,610$ mm, $r_{min} = 49.3$ mm, and $(l/r) = 120$:

$$l_{max} = 120 \times \frac{49.3}{1\,000} = 5.9 \text{ m (19 ft).}$$

From the graph of Fig. 5.4.1:

$$F_c = 71 \text{ MPa}, \qquad P = 76.10 \times \frac{71}{1\,000} = 540 \text{ kN (121 k).}$$

The maximum length of the same column under the same load, fixed at both ends, would be $2 \times (5.9 \text{ m}) = 11.8$ m. The maximum length of the same wide flange beam used as a cantilever is $\frac{1}{2}(5.9 \text{ m}) = 2.95$ m.

Example 5.4.2. Determine the wall thickness t of a simply supported, compressed circular column of radius R made out of aluminum, given its length l and the compressive load P.

By Table 4.1.1 (10):

$$I = \frac{\pi}{8} D^3 t, \qquad A = \pi D t, \qquad r^2 = \frac{I}{A} = \frac{D^2}{8}, \qquad r = \frac{D}{2\sqrt{2}} = \frac{1}{\sqrt{2}} R,$$

$$f_{cr} = \frac{\pi^2 E}{(l/r)^2} = \frac{\pi^2 E}{(\sqrt{2}\, l/R)^2}, \qquad f_c = \frac{1}{2.5} f_{cr} = \pi^2 \frac{ER^2}{5l^2},$$

$$P = f_c A = \frac{\pi^3}{2.5}\left(\frac{Rt}{l^2}\right) ER^2, \qquad t = \frac{2.5}{\pi^3} \frac{Pl^2}{ER^2} = 0.08 \frac{Pl^2}{ER^3}.$$

An aluminum column 6 m (20 ft) long, 300 mm (1 ft) in diameter with a load of 2 200 kN (500 k) requires a thickness:

$$t = 0.08 \frac{2\,200 \times (6\,000)^2 (1\,000)}{69 \times 10^3 \times 150^3} = 27.2 \text{ mm (1.07 in.)}.$$

The corresponding stress is:

$$f_c = \frac{P}{\pi(2R)t} = \frac{2\,200}{3.14 \times 2 \times 150 \times 27.2} = 0.085\,9 \text{ kN/mm}^2 = 85.9 \text{ MPa (12.5 ksi)}$$

and is below the allowable stress in compression.

Example 5.4.3. A slender cantilevered column may buckle under its own dead load of w kN/m (lb/ft). The critical distributed load $P_{cr} = wl_{cr}$ may be proved to be 3.17 times the value of the same load concentrated at the top of the column. Determine the critical length l_{cr} of such a column.

The dead load per unit length w equals the weight per unit volume ρ times the cross-sectional area A times one. Hence:

$$wl_{cr} = \rho A l_{cr} = 3.17 \frac{\pi^2}{4} \frac{E(r^2 A)}{l_{cr}^2} \quad \therefore \quad l_{cr}^3 = 3.17 \frac{\pi^2}{4} r^2 \frac{E}{\rho}. \tag{5.4.2}$$

The critical height for a W920 × 446 (W36 × 300) with $r_{min} = 97.3$ mm is:

$$l_{cr}^3 = 3.17 \times 2.47 \times 97.3^2 \frac{200\,000}{1\,000 \times 0.000\,077} = 1\,925 \times 10^{11} \text{ mm},^3$$

$$l_{cr} = 57\,740 \text{ mm} = 57.74 \text{ m (189 ft)}.$$

The buckling loads considered in the preceding examples are those capable of bending a beam or column *as a whole*, but a beam may buckle in a variety of ways. For example, the compressed flange of an I-beam may buckle *out* of its own horizontal plane, due to its small thickness (Fig. 5.4.2). This type of instability is called *lateral buckling* and may be seen to involve a twist of the beam's cross section; therefore, it involves both the bending and torsional stiffness of the beam. For example, the load w which will produce lateral buckling in a simply supported beam

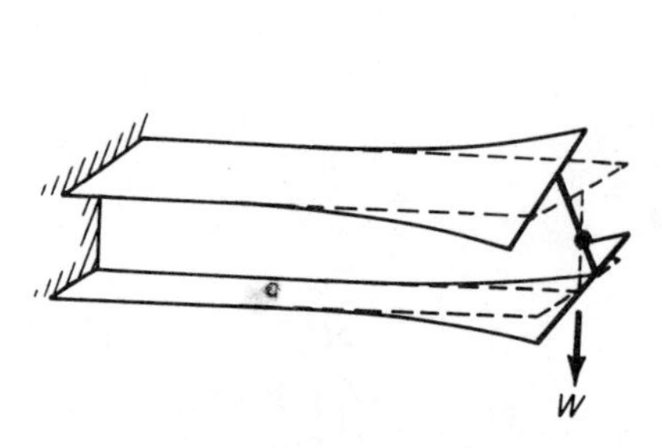

FIGURE 5.4.2

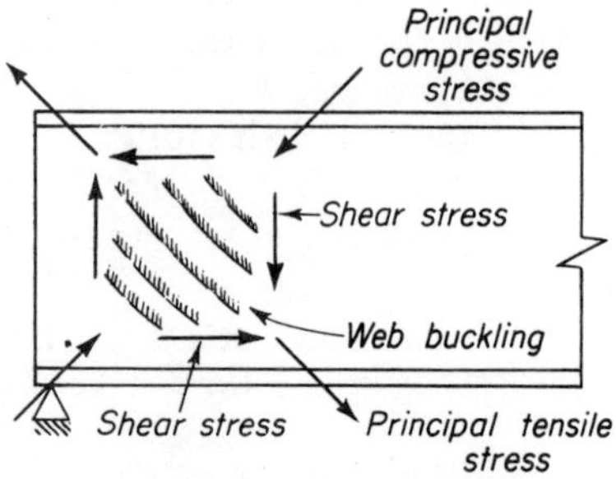

Principal compressive stress
Shear stress
Web buckling
Shear stress
Principal tensile stress

FIGURE 5.4.3

is estimated conservatively by the equation:

$$(wl)_{cr} = \frac{28}{l^2}\sqrt{EI_{min}GK}, \tag{5.4.3}$$

where the *torsional rigidity* GK is the product of the shear modulus $G = E/2(1 + \nu)$ (ν = Poisson's ratio) and the *torsional constant* K of the cross section, with dimensions of mm⁴ (in.⁴), which is given in steel handbooks.*

Example 5.4.4. Determine the critical span for a laterally unbraced simply supported W920 × 446 (W36 × 300) barely capable of just carrying its own weight.

With $w = 4.38$ kN/m, $I_y = 540.8 \times 10^6$ mm⁴, $K = 26.7 \times 10^6$ mm⁴, $E =$ 200 000 MPa, $G = 200\,000/2(1 + 0.3) = 76\,920$ MPa, (5.4.3) gives:

$$l_{cr}^3 = \frac{28}{4.38}\sqrt{200\,000 \times 540.8 \times 10^6 \times 76\,920 \times 26.7 \times 10^6}$$

$$= 9.528 \times 10^{13}\ \text{mm}^3$$

$$l_{cr} = 45\,670\ \text{mm} = 45.67\ \text{m}\ (150\ \text{ft}).$$

It is seen by comparison with the result of Example 4.1.3 that the critical span, 45.67 m is less than one-half the limit span in bending of 116.55 m.

Since any thin compressed part of a structural element may buckle, a variety of other buckling modes may develop in a beam. For example, the upper part of the web of an I beam may buckle in compression near the center of the span, where bending stresses are maximum, or the web may buckle diagonally near the beam support, where the shear stresses and hence the equivalent diagonal compression and tension are maximum (Fig. 5.4.3).

All these possibilities are taken into account by design codes in determining the values of the allowable stresses for compressed elements.

**Steel Design Handbook*, American Institute of Steel Construction, New York, 1980.

PROBLEMS

5.1 The concrete core of a 120 m (400 ft) high office building, in the shape of an I, has a depth of 9 m (30 ft) and a width of 4.5 m (15 ft). The walls are 300 mm (1 ft) thick and carry a load at the base of 1 200 kN/m (80 k/ft), of which 30% is live load. The wind load on the building to be resisted by the core in the strong direction is 13 kN/m (900 lb/ft) of height. If the allowable stress in the concrete wall under dead and live load is 7 MPa (1 ksi), compute the combined stress due to vertical and wind loads on the core, and determine whether the wall thickness is adequate, remembering that the allowable stress under combined gravity and wind loads is 33% higher than under gravity loads only.

5.2 In the core of Problem 5.1 the width of the footing under the core is 6 m (20 ft) and the length is 12 m (40 ft). Determine the thickness of the footing required to avoid tension under the footing.

5.3 A concrete balcony railing is prefabricated as a wall. If the railing is 1 m (3 ft) high, what is the minimum thickness required to avoid tension at its base? (See Chapter 1 for loading.)

5.4 The 2.5 m (8 ft) high cantilevered basement wall of a house carries a load of 150 kN/m (10 k/ft) at the top. The water table extends to a point 1.5 m (5 ft) below the top of the wall. Compute the thickness of the wall if (a) no tension is allowed in the concrete, (b) a minimum compressive stress of 0.2 MPa (30 psi) is specified to prevent cracking. Assume the bottom of the wall to be 2.5 m (8 ft) below grade.

5.5 A W360 × 223 (W14 × 150) column is subjected to a direct force of 2 000 kN (450 k) and moments about the strong axis of 200 kN·m (150 k·ft) and about the weak axis of 65 kN·m (50 k·ft). If $F_a = 117$ MPa (17 ksi), $F_b = 151$ MPa (22 ksi), determine whether the column can safely carry the maximum combined stress.

5.6 Determine P_{cr} and f_{cr} for the following members: (a) a 50 mm × 100 mm (2 in. × 4 in.) Douglas fir stud 2.75 m (9 ft) high, hinged at its top and its bottom; (b) a 200 mm × 300 mm (8 in. × 12 in.) concrete column 4.25 m (14 ft) high, fixed at both its top and its bottom, of 20 MPa concrete; (c) an aluminum column 3.5 m (12 ft) high, in the shape of a rectangular tube with outside dimensions 150 mm × 300 mm (6 in. × 12 in.) and 6 mm ($\frac{1}{4}$ in.) wall thickness, hinged at both ends; (d) a W200 × 46 (W8 × 31) steel column 4 m (13 ft) high, fixed at the bottom and hinged at the top.

5.7 Determine the minimum thickness of a cantilevered concrete wall with $f'_c = 20$ MPa (2,900 psi), 6 m (20 ft) high, carrying a vertical load of 175 kN/m (12 k/ft) with a safety factor against buckling of 3.

5.8 Determine the maximum height of a 200 mm (8 in.) thick, cantilevered concrete wall with $f'_c = 30$ MPa (4,350 psi) which would just buckle under its own weight.

5.9 The flagpole of Problem 4.11 is stayed near its top by three guy wires inclined 60° to the horizontal. The initial tension in each wire is negligible. Determine the maximum stress in the flagpole, assuming it to be hinged at its top and its bottom, and considering only one guy wire to react in tension when the wind blows in the plane of the wire and the pole. Compare the maximum pole deflection δ with the δ_0 of Problem 4.11, if the stress in the guy wire is 207 MPa (30,000 psi).

5.10 What is the critical simply supported span of a laterally unbraced W460 $\times$ 143 (W18 $\times$ 96) beam carrying a superimposed load of 18 kN/m (1,200 lb/ft)?

5.11 The laterally unbraced, simply supported beams of a trellis are made of 20 MPa (2,900 psi) concrete, span 12 m (40 ft), and have a depth $d = 600$ mm (2 ft). Determine the width b required to guarantee a safety factor of 3 against lateral buckling, if the live load on the beams is 1.5 kN/m (100 lb/ft). The torsional constant for a rectangular deep section is $k = \frac{1}{3}db^3$.

6. Trusses

6.1 TRUSSES [6.3]

Trusses are combinations of purely tensile and compressive elements capable of spanning economically distances from 9 m (30 ft) to 300 m (1,000 ft).

A plane truss consists of straight bars, usually of constant cross section. Figure 6.1.1 indicates some of the most commonly used types. The elements of a truss may be made out of wood, concrete, steel, or other metals. In relatively small trusses the compressive elements may be made out of wood; in most trusses the tensile elements are made out of steel and are either standard beams, angles, or tubes. Low capacity trusses spanning up to 30 m (100 ft) are called *steel joists* (Fig. 6.1.2) and may be chosen out of fabricators' catalogs since they are standard structures.

Larger or special trusses are designed under the simplifying assumptions given in the following paragraphs, which allow cross section and sizes to be readily estimated.

Trusses are called *statically determinate* when they can be analyzed by the laws of static equilibrium; they are called *statically indeterminate* when these laws are insufficient to determine the stresses in the bars and, hence, require that the elasticity of the material be taken into account. We shall limit our treatment to statically determinate plane trusses of the types illustrated in Fig. 6.1.1. (Space frames are considered in Chapter 18.)

The basic assumptions of elementary truss analysis are those required to guarantee that each bar element will be under simple tension or simple compression:

1. The bars are pin-connected, i.e., are hinged at the ends.
2. The loads are applied only at the joints, or *panel points*, of the truss.

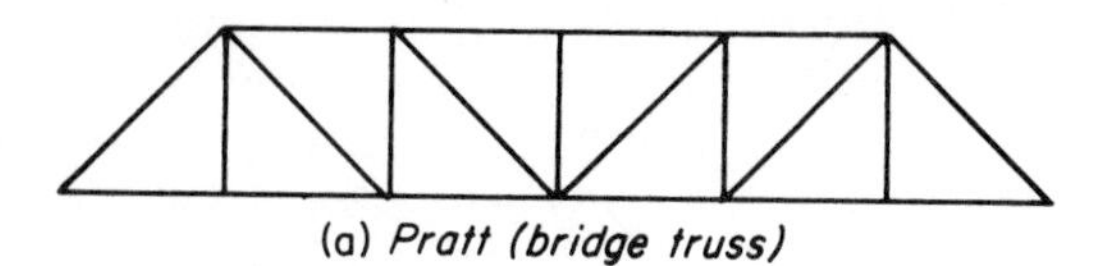

(a) *Pratt (bridge truss)*

(b) *Warren*

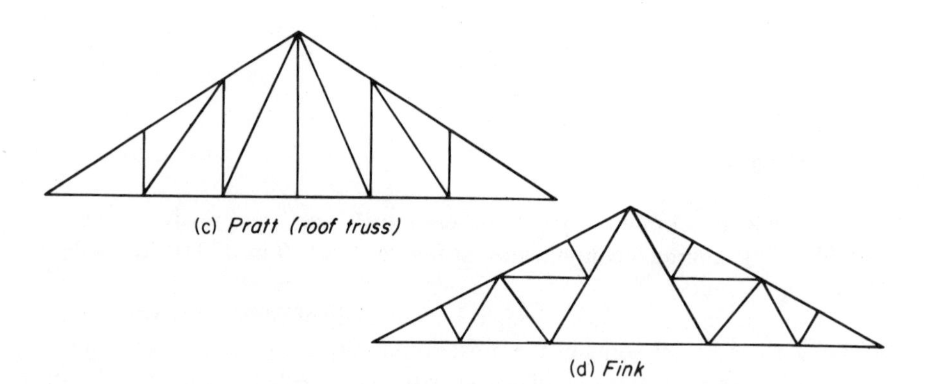

(c) *Pratt (roof truss)*

(d) *Fink*

FIGURE 6.1.1

FIGURE 6.1.2

Since in practice the bars are riveted, bolted, or welded directly to each other or to gusset plates, the bar ends are not free to rotate and so-called *secondary bending moments* are developed at the bar ends. (These are evaluated in Section 6.4 for an elementary case, in order to show the relative importance of bending to direct stresses in trusses.)

Some of the loads acting on a truss may actually be applied at its joints but others, like the dead load of the bars, are not. Hence, additional bending moments are developed in the bars due to these loads. (These bending moments are also mentioned in Section 6.4.)

The design of the gusset plates and the bar connections is a technical engineering task that is not considered here.

6.2 PLANE TRUSS ANALYSIS [6.2]

The stresses in the bars of a statically determinate plane truss can be evaluated by stating that each part of the truss and, in particular, each joint is in equilibrium.

The equilibrium of any part of a truss is guaranteed by satisfying the three equations of plane equilibrium:

$$\sum F_x = 0, \qquad \sum F_y = 0, \qquad \sum M_0 = 0. \tag{6.2.1}$$

Hence, if a section isolating a part of a truss cuts n bars, the three equations (6.2.1) will allow the evaluation of the stresses in three of these bars, provided the stresses in the remaining $n - 3$ be already known.

A joint is in equilibrium if the first two equations (6.2.1) are satisfied. Hence, the equilibrium of a joint allows the evaluation of the stresses in two bars meeting at that joint, provided the stresses in all the remaining bars meeting at that joint be known.

The following examples illustrate the use of the *section method* and the *joint method* of stress determination.

Example 6.2.1. The marquee of Example 3.2.4 is supported by the simplest possible truss, a triangular bracket consisting of a tensile rod AB and a compressive bar AC. The load supported at the joint A is a vertical force of 21.6 kN (5 k). Indicating by T and C the forces in these two bars [Fig. 6.2.1(a)], the first two equations of (6.2.1) give:

$$T \cos 30° - C = 0, \qquad 21.6 - T \sin 30° = 0,$$

from which:

$$T = 43.2 \text{ kN } (10 \text{ k}), \qquad C = 37.4 \text{ kN } (8.66 \text{ k}).$$

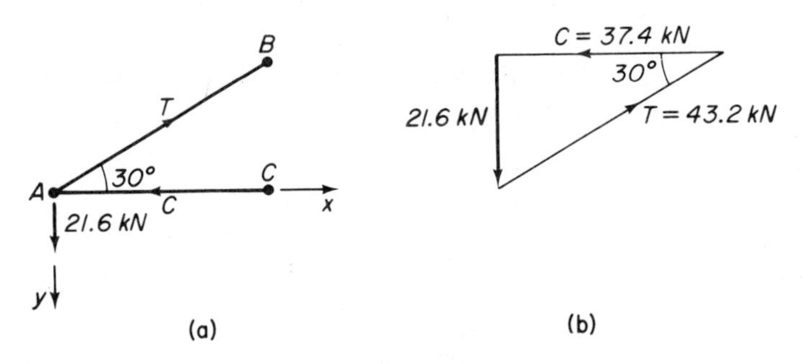

FIGURE 6.2.1

The same result may be obtained graphically by splitting the 21.6 kN load into components in the horizontal and the 30° directions, as shown in Fig. 6.2.1(b), i.e., using a closed force polygon.

The compressive steel bar AC requires a cross section of only 271 mm² (0.43 sq in.) for an allowable stress of 138 MPa (20 ksi) in simple compression but actually requires a much larger section because it is acted upon by distributed loads and, moreover, may buckle in the vertical direction. Its complete analysis was obtained in Example 5.1.3.

Example 6.2.2. The Warren truss with verticals of Fig. 6.2.2 is loaded with equal loads at the evenly spaced, lower panel points and is simply supported at the ends. Since the forces in the bars of the truss are proportional to the loads, we shall indicate the intensity of the panel point loads arbitrarily as 100 units of force. Thus, the reactions are 250 units.

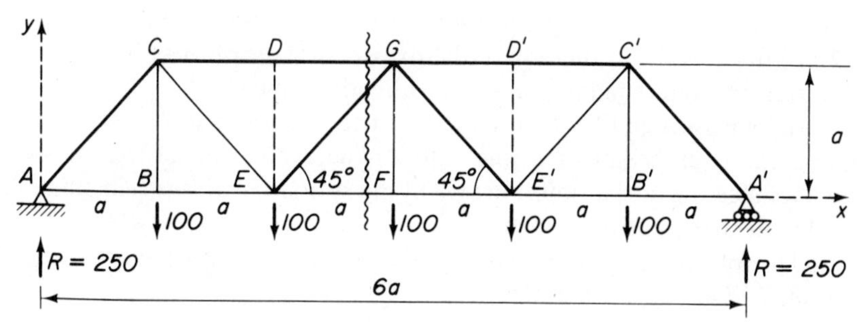

FIGURE 6.2.2

The evaluation of the forces in the bars, considered *positive when tensile* and *negative when compressive*, may be carried out by the joint method, starting at the left support. In this analysis, the action of bar AC on joint A is indicated by F_{CA} and is positive when tensile. The action of bar AC on joint C is indicated by F_{AC}; hence, $F_{CA} = F_{AC}$. The x-axis is positive to the right, and the y-axis positive upward. In writing equilibrium equations, all *unknown* forces are assumed positive, and all *known* forces are entered with a plus or a minus sign, depending on whether they act along the plus or minus axes.

Joint A:

$$\sum F_y = 250 + F_{AC} \cos 45° = 0, \qquad F_{AC} = -354.$$

$$\sum F_x = -354 \cos 45° + F_{AB} = 0, \qquad F_{AB} = +250.$$

Joint B:

$$\sum F_y = -100 + F_{BC} = 0, \qquad F_{BC} = +100.$$

$$\sum F_x = -250 + F_{BE} = 0, \qquad F_{BE} = +250.$$

Joint C:

$$\sum F_y = 354 \cos 45° - 100 - F_{CE} \cos 45° = 0, \qquad F_{CE} = +212.$$

$$\sum F_x = 354 \cos 45° + 212 \cos 45° + F_{CD} = 0, \qquad F_{CD} = -400.$$

Joint D:

$$\sum F_y = F_{DE} = 0.$$

$$\sum F_x = 400 + F_{DG} = 0, \qquad F_{DG} = -400.$$

Joint E:

$$\sum F_y = -100 + 212 \cos 45° + 0 + F_{EG} \cos 45° = 0, \qquad F_{EG} = -71.$$

$$\sum F_x = -250 - 212 \cos 45° - 71 \cos 45° + F_{EF} = 0, \qquad F_{EF} = +450.$$

Joint F:

$$\sum F_y = -100 + F_{FG} = 0, \qquad F_{FG} = +100.$$

By symmetry, all forces in the bars in the right half of the truss are equal to those in the corresponding bars in the left half of the truss.

The section method may be used advantageously when the force in a specific bar is to be evaluated. For example, cutting the truss of Fig. 6.2.2 with a vertical section to the left of GF and taking moments of all the forces acting on the left part of the truss about G, we have F_{EF} as the only unknown force appearing in the equation:

$$250 \times 3a - 100 \times 2a - 100 \times a - F_{EF} \times a = 0,$$

from which $F_{EF} = +450$.

The same analysis may be performed *graphically* as follows (Fig. 6.2.3):

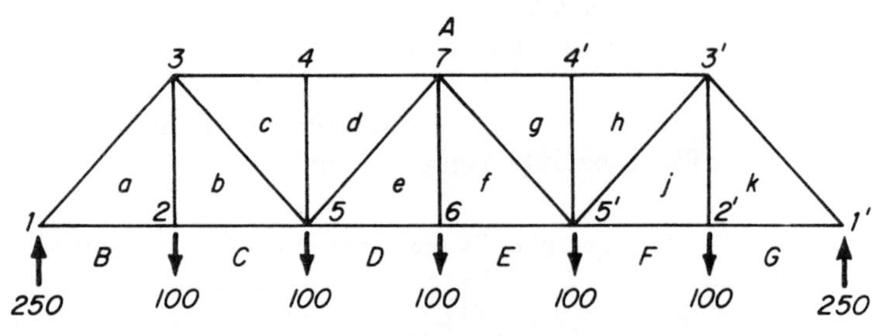

FIGURE 6.2.3

1. Label each interior mesh by means of a lower-case letter.

2. Label each exterior area between two successive loads or reactions by means of an upper-case (capital) letter.

3. Define each load by the two letters labeling the adjoining areas. For example, the center load is the load DE; the left reaction is the load AB.

4. Define each bar by the two letters labeling the adjoining meshes or the adjoining mesh and the adjoining load area. (This definition of bars is known as *Bow's notation*.)

5. Draw in a chosen scale the closed polygon of the loads and the reactions for half the truss $ABCDE$, taking advantage of symmetry (Fig. 6.2.4).

6. Starting at joint 1 and moving clockwise about 1, draw the closed polygon of the equilibrated forces acting at 1, $BAaB$, by drawing aB parallel to the bar aB starting at B, and Aa parallel to the bar Aa starting at A.

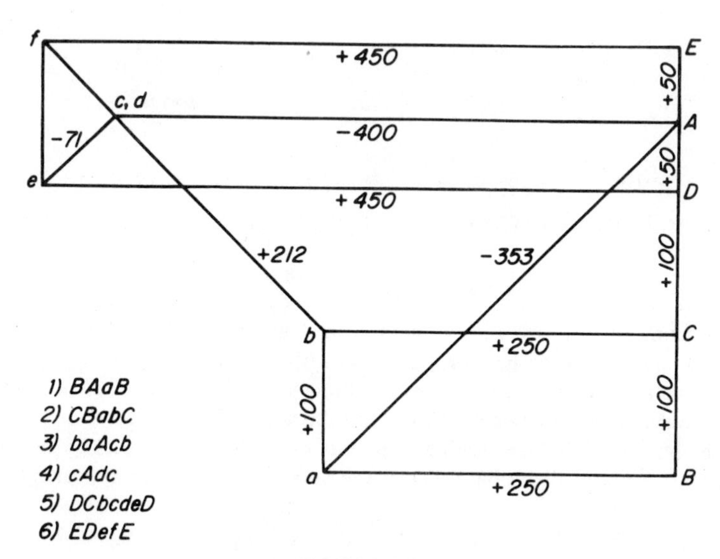

FIGURE 6.2.4

7. Draw the closed polygons for the remaining joints: 2 (*CBabC*), 3 (*baAcb*), 4 (*cAdc*), 5 (*DCbcdeD*), 6 (*EDefE*).

8. The forces in the bars are in the same scale as that used for the applied loads.

9. Notice that the joints must be taken in such a sequence that not more than two bar forces be unknown at a joint.

Example 6.2.3. Determine graphically the forces in the bars of the truss of Fig. 6.2.5, and check the force in the bar *Ac* by the section method, taking moments about the point 4. The solution is shown in Fig. 6.2.6. (Note that only half the diagram need be drawn because of the symmetry of *both* the structure and the loads.)

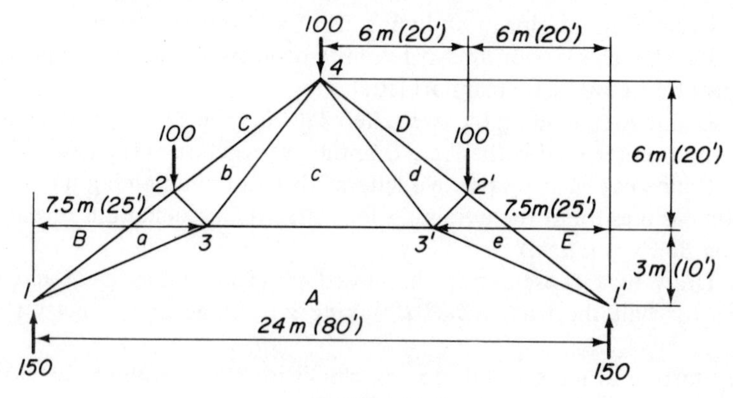

FIGURE 6.2.5

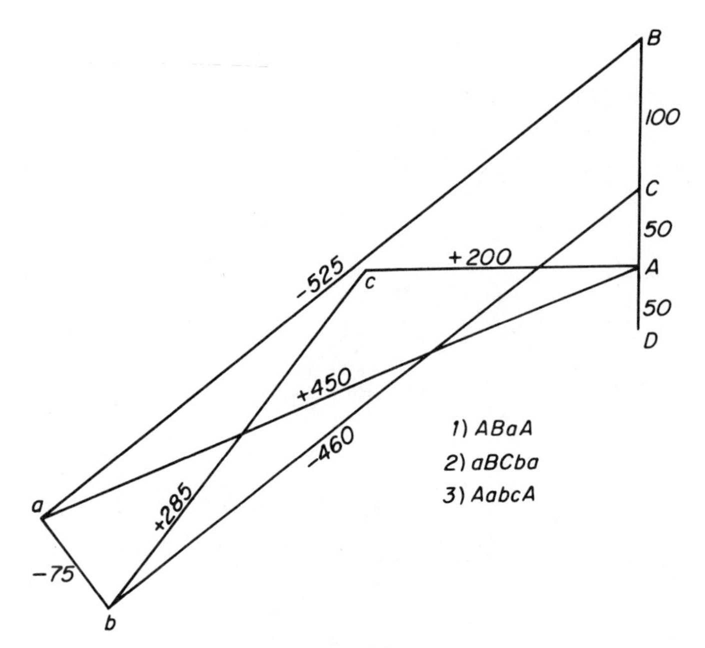

FIGURE 6.2.6

The check of the force in Ac by the moment method gives:

$$150 \times 12 - 100 \times 6 - F_{Ac} \times 6 = 0, \qquad F_{Ac} = +200.$$

If trusses of the type shown in Fig. 6.2.5 are spaced 9 m (30 ft) o.c. and support a roof load of 4.8 kN/m² (100 psf), including the dead load of the roof and a snow load of 2.4 kN/m² (50 psf), the forces indicated by 100 in the diagram represent a load of 4.8 kN/m² (0.1 ksf) over an area of $7.5 \times 9 = 67.5$ m² (750 sq ft), i.e., a load of $4.8 \times 67.5 = 324$ kN (75 k). Hence, the forces in the bars are obtained from those in the diagram by using a multiplication factor 3.24 (0.75). The compressive force in the bar aB is $-525 \times 3.24 = -1\,700$ kN. The tensile force in the bar cA equals $+200 \times 3.24 = 648$ kN. Thus, the compressed bar aB would require a steel area of 12 320 mm² (ignoring buckling) and the tensile bar cA an area of 4 700 mm² for an allowable stress of 138 MPa.

Example 6.2.4. Determine graphically the forces in the bars of the truss of Fig. 6.2.7, where the horizontal forces represent wind loads.

The horizontal reaction at B equals the sum of the horizontal forces, i.e., 200. The equal and opposite vertical reactions are obtained by means of a moment equation about F:

$$R_V \times 24 - 50 \times 9 - 100 \times 4.5 = 0 \quad \therefore \quad R_V = 37.5.$$

The solution is shown in Fig. 6.2.8. The force in bar ab is zero.

If trusses like those in Fig. 6.2.7 are spaced 9 m (30 ft) o.c. and if the wind exerts a pressure of 1.5 kN/m² (30 psf) of vertical roof projection, the force indicated

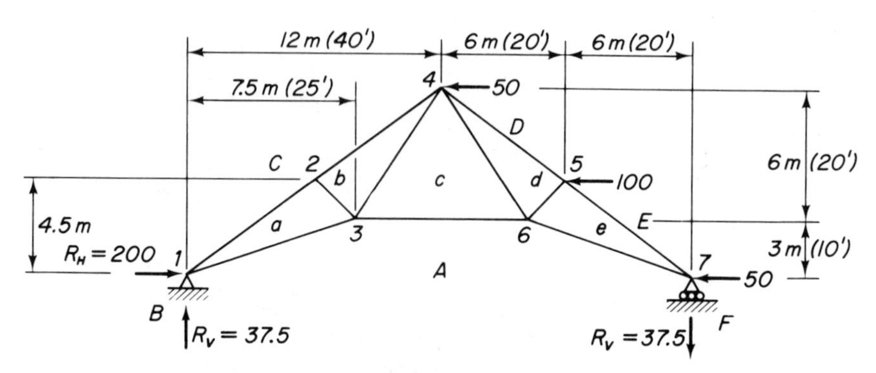

FIGURE 6.2.7

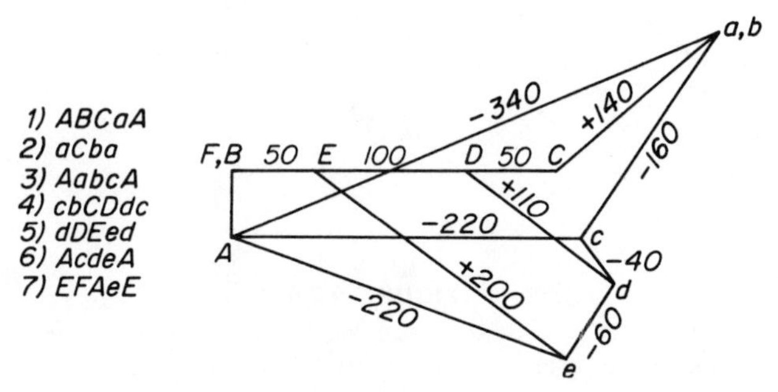

FIGURE 6.2.8

by 100 in the diagram represents a pressure of 1.5 kN/m² (30 psf) over a vertical area of $9 \times 4.5 = 40.5$ m² (450 sq ft), and hence represents a force of $40.5 \times 1.5 = 60.75$ kN (13.5 k). The wind forces in the bars are then obtained by multiplying the forces in the diagram by the factor 0.607 5 (0.135). Thus, the force in the bar *Aa* equals $-340 \times 0.607\,5 = -206$ kN, and the force in the bar *Ee* equals $+200 \times 0.607\,5 = +122$ kN.

Example 6.2.5. We wish to determine the maximum span of a steel truss made of square panels of side a with tensile diagonals (Fig. 6.2.9) and with bars of identical cross section A, capable of barely supporting its own weight.

Assuming that the upper compressive chord of the truss will be prevented from buckling, failure will occur in the most highly stressed bars, i.e., the center bars of the upper and lower chords. Assuming the load of the truss to be uniformly distributed and to be w kN/m (lb/ft), the force H in these bars may be evaluated by taking moments about M, after isolating the portion of the truss to the right of the cut:

$$\frac{wl}{2}\left(\frac{l}{2}\right) - \frac{wl}{2}\left(\frac{l}{4}\right) - Ha = 0, \qquad H = \frac{wl^2}{8a}.^* \tag{a}$$

*It is seen from this result that this bar force is computed by dividing the simply supported beam moment, $wl^2/8$, by the truss depth a.

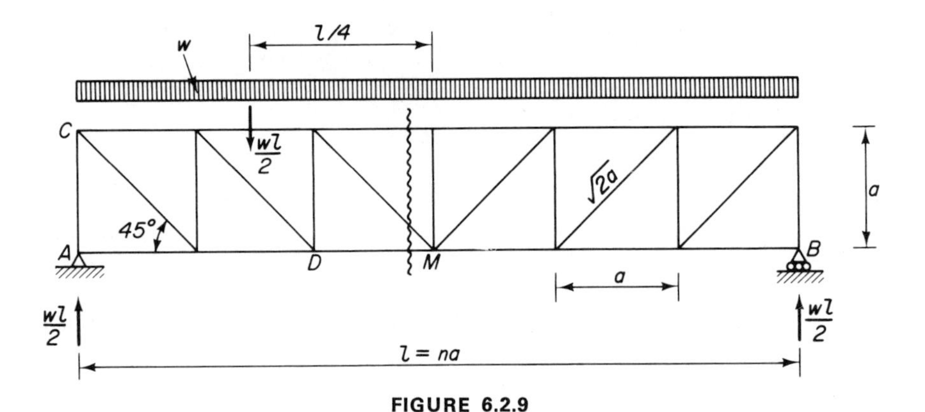

FIGURE 6.2.9

Hence, the highest tension or compression in the bars of the truss equals:

$$f = \frac{H}{A} = \frac{wl^2}{8aA}. \tag{b}$$

The load w kN/m (lb/ft) is the weight of the truss bars per unit length of truss. Each panel has two horizontal bars of length a, one vertical bar also of length a, and one diagonal bar of length $\sqrt{2}\,a$, or altogether $(2 + 1 + 1.41)a = 4.41a$ length of bars and a volume of $4.41aA$ of steel. If we indicate by ρ the unit weight of steel, the weight of one panel is $4.41\,\rho aA$, and the weight per unit length of truss is $w = 4.41\rho A$. Hence, the stress (b) becomes:

$$f = \frac{4.41\rho A l^2}{8aA} = 0.55\rho\frac{l^2}{a}. \tag{c}$$

Indicating by n the number of panels in the truss we obtain:

$$l = na \quad \text{or} \quad n = \frac{l}{a}, \tag{d}$$

and, substituting in (c):

$$\frac{l^2}{a} = nl = n^2 a = \frac{f}{0.55\rho}. \tag{e}$$

With an ultimate steel strength of 414 MPa (60,000 psi) and $\rho = 0.000\,077$ N/mm³ (0.29 lb/cu in.):

$$nl_u = n^2 a = \frac{414}{0.55 \times 0.000\,077} = 9\,775\,700 \text{ mm} = 9\,775 \text{ m (32,070 ft)}. \tag{f}$$

The next to the highest force in the bars of the truss occurs in the diagonal nearest to the support and, by equilibrium of the joint C, is found to equal:

$$F_{AC} = \sqrt{2}\,\frac{wl}{2}. \tag{g}$$

Equating F_{AC} to $F_{DM} = H$ of (a), i.e., equating the maximum force due to shear to the maximum force due to bending:

$$\sqrt{2}\,\frac{wl}{2} = \frac{wl^2}{8a} \quad \therefore \quad \frac{l}{a} = n = 4\sqrt{2} \approx 6.$$

It is thus seen that for $n < 6$ the higher force occurs in the first diagonal and by (f) the maximum span for $n = 6$ equals 1 630 m (but corresponds to a truss 272 m deep), while the ultimate span is inversely proportional to the span-depth ratio n. Table 6.2.1 gives l_u and a versus n.

Table 6.2.1

$n = l/a$	40	30	20	10
a, m (ft)	6 (19.5)	11 (34.7)	24 (78.5)	97 (314.0)
l_u, m (ft)	240 (785)	330 (1,040)	480 (1,570)	970 (3,140)

6.3 TRUSS DEFLECTIONS

The deflections of a truss are due to the lengthening of its tensile bars and shortening of its compressive bars. The assumption of pin-connected members allows the graphical evaluation of these deflections by means of a diagram, called *Williot's diagram*, which will be illustrated by the simple example of the bracket of Fig. 6.3.1(a).

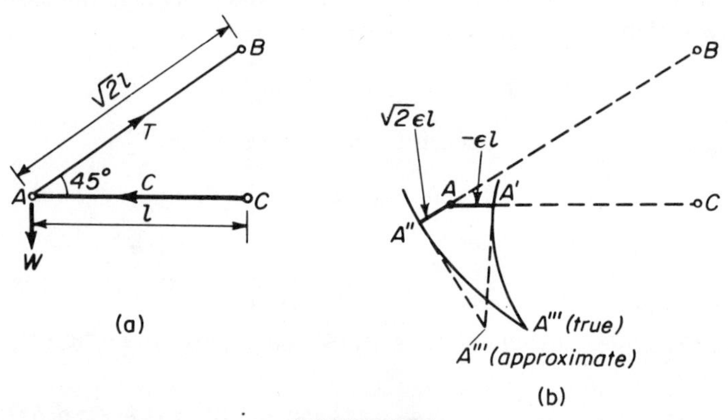

FIGURE 6.3.1

Under the load W, the steel bars AB and AC develop bar forces T and C, respectively. The areas of the bars will be chosen so as to have in each bar the same allowable stress f and, hence, the same strain $\epsilon = f/E$. The elongations of the two bars are then:

$$\Delta l_{AC} = -\epsilon l, \qquad \Delta l_{AB} = \sqrt{2}\, \epsilon l,$$

where a minus sign indicates a shortening. After deformation, the ends A of the two bars are still connected. Since both bars are free to rotate about their support hinges C and B, their ends A' and A'' [Fig. 6.3.1(b)] will swing along circles of radii $l - \epsilon l = l(1 - \epsilon)$ and $\sqrt{2}\, l + \sqrt{2}\, \epsilon l = \sqrt{2}\, l(1 + \epsilon)$, respectively, until they coincide at the intersection A''' of these two circles.

Since ϵl is very small compared to l ($\epsilon = f/E$ is of the order of $165/200\,000 = 1/1\,200$), the arcs of circle $A'A'''$ and $A''A'''$ may be approximated by their tangents at A' and A'', i.e., by the perpendiculars to AC and AB. Thus, the displacement of A may be obtained (Fig. 6.3.2) by drawing a segment $AA' = \epsilon l$ parallel to AC to the right of A, a segment $AA'' = \sqrt{2}\, \epsilon l$

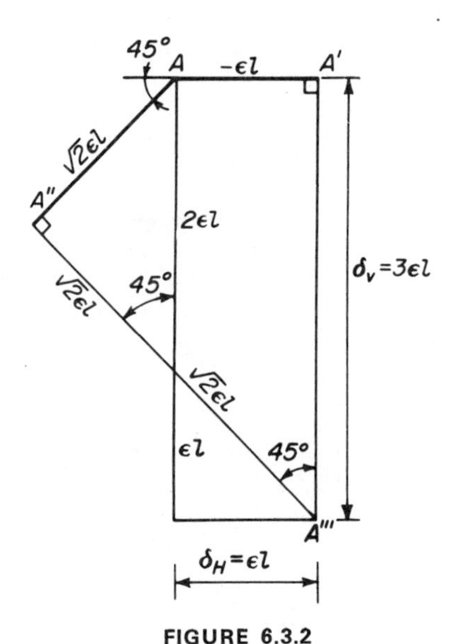

FIGURE 6.3.2

parallel to AB to the left of A, and by determining the intersection A''' of the perpendiculars to AA' and AA''. The total displacement AA''' has a vertical component δ_v and a horizontal component δ_h. For $l = 3$ m (10 ft), $f = 165$ MPa (24 ksi), $E = 200\,000$ MPa (29,000 ksi):

$$\epsilon l = (165/200\,000) \times (3 \times 1\,000) = 2.48 \text{ mm},$$

$$\sqrt{2}\, \epsilon l = 3.5 \text{ mm},$$

$$\delta_v = 7.44 \text{ mm (0.29 in).}, \qquad \delta_h = 2.48 \text{ mm (0.098 in.).}$$

Similar Williot diagrams may be obtained for any statically determinate truss by combining the diagrams for each triangular truss mesh.

Example 6.3.1. Determine the displacement of the point A of the truss of Fig. 6.3.3(a) under the assumption that each bar is under the same stress f and, hence, the same strain ϵ. The diagram of Fig. 6.3.3(b), giving the displacement AA' of A in the scale $\epsilon a = 1$, is obtained as follows:

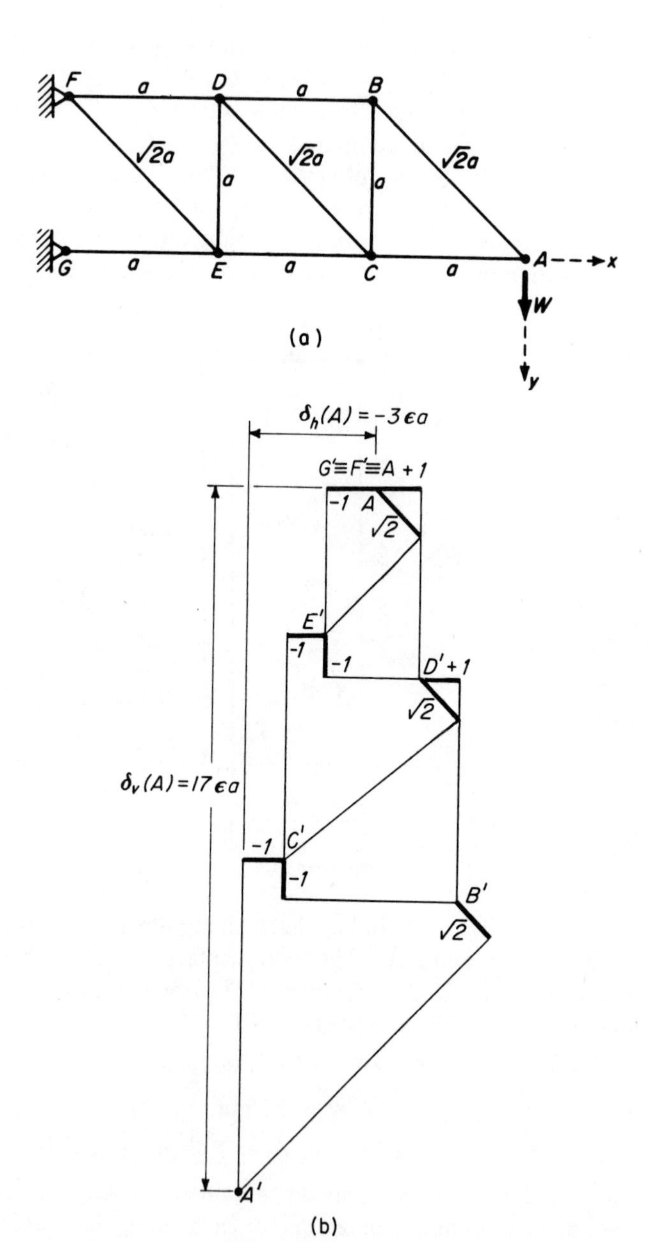
F
a
D
a
B
√2a
a
√2a
a
√2a
G
a
E
a
C
a
A
x
W
(a)
y
δh(A) = −3εa
G′≡F′≡A + 1
−1
A
√2
E′
−1
−1
D′ + 1
√2
δv(A) = 17εa
−1
C′
−1
B′
√2
A′
(b)

FIGURE 6.3.3

1. E' is obtained by combining the shortening -1 of GE starting at G' with the lengthening $\sqrt{2}$ of FE starting at F'.

2. D' is located by combining the shortening -1 of ED starting at E' with the lengthening 1 of FD starting at F'.

3. C' is located by combining the shortening -1 of EC starting at E' with the lengthening $\sqrt{2}$ of DC starting at D'.

4. B' is located by combining the shortening -1 of CB starting at C' with the lengthening 1 of DB starting at D'.

5. A' is located by combining the shortening -1 of CA starting at C' with the lengthening $\sqrt{2}$ of BA starting at B'.

The displacement of A is $F'A'$ and has a $\delta_h = -3(\epsilon a)$, and a $\delta_v = 17(\epsilon a)$. For $a = 3$ m (10 ft) and $\epsilon = 165/200\,000$, $\delta_v = 17 \times (165/200\,000) \times 3\,000 =$ 42 mm (1.65 in.), with a deflection-to-span ratio of $42/(3 \times 3\,000) = 1/214$.

6.4 SECONDARY MOMENTS IN TRUSSES [7.4]

In order to show the relative importance of the bending stresses due to secondary moments in trusses, these stresses will be determined for the simple bracket of Fig. 6.3.1.

In view of the smallness of the rotations β of the bar AB and γ of the bar AC, β and γ are obtained by dividing the displacement components perpendicular to the bars by the bar length. From Figs. 6.3.1 (b) and 6.3.2:

$$\beta = \frac{A''A'''}{AB} = \frac{2\sqrt{2}\,\epsilon l}{\sqrt{2}\,l} = 2\epsilon, \qquad \gamma = \frac{A'A'''}{AC} = \frac{3\epsilon l}{l} = 3\epsilon. \tag{a}$$

From Fig. 6.4.1 the change in the 45° angle BAC is $\beta - \gamma = -\epsilon$.

Let us assume that the bars AB and AC are rigidly connected at A so that the change in angle $-\epsilon$ is prevented. This means that equal and opposite

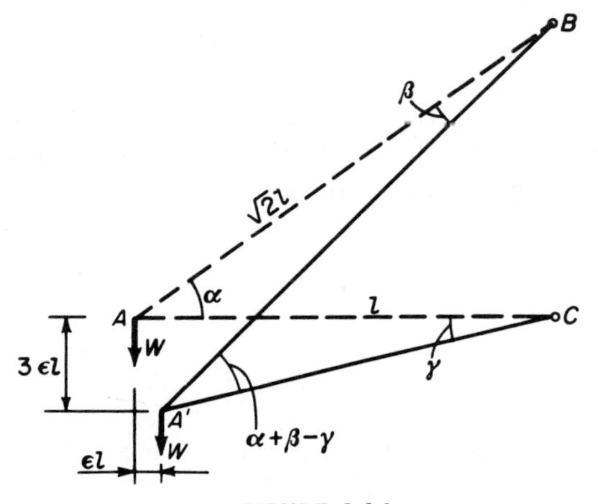

FIGURE 6.4.1

moments M are developed at the ends A of these bars capable of opening up the angle between the bar ends by $+\epsilon$. But, from Table 4.3.1 (g) the rotation of the end A of a simply supported beam due to a moment M_A equals $\phi_A = \frac{1}{3}(M_A l/EI)$; hence, calling I_1 and I_2 the moments of inertia of AB and AC:

$$\phi_1 + \phi_2 = \frac{1}{3}\frac{Ml}{EI_1} + \frac{1}{3}\frac{M\sqrt{2}\,l}{EI_2} = -(\beta - \gamma) = \epsilon = \frac{f}{E},$$

from which:

$$M = \frac{3f}{l/I_1 + \sqrt{2}\,l/I_2} = \frac{3fI_1}{l}\frac{1}{1 + \sqrt{2}\,(I_1/I_2)} \tag{b}$$

Assuming $I_1 = I_2 = I$ and the depths h of AB and AC identical, the bending stresses due to M are:

$$f_b = \frac{Mh/2}{I} = \frac{3}{2}\frac{1}{1 + \sqrt{2}}\left(\frac{h}{l}\right)f, \tag{c}$$

from which the ratio of the bending stress f_b to the direct stress f becomes:

$$\frac{f_b}{f} = \frac{3}{2(1 + \sqrt{2})}\frac{h}{l} = 0.62\frac{h}{l}. \tag{d}$$

For $h/l = 0.1, f_b/f = 0.062 = 6.2\%$.

If the ends B and C are also assumed fixed, the moments at B and C by (4.6.2) are equal to $-\frac{1}{2}M_A$ and:

$$\phi_1 + \phi_2 = \left(\frac{1}{3}\frac{Ml}{EI} + \frac{1}{6}\frac{-\frac{1}{2}Ml}{EI}\right) + \left(\frac{1}{3}\frac{M\sqrt{2}\,l}{EI} + \frac{1}{6}\frac{-\frac{1}{2}M\sqrt{2}\,l}{EI}\right)$$

$$= \frac{1}{4}\frac{Ml}{EI} + \frac{1}{4}\frac{M\sqrt{2}\,l}{EI} = \frac{f}{E}.$$

Hence:

$$M = \frac{4fI}{l}\frac{1}{1 + \sqrt{2}}, \qquad f_b = \frac{2}{1 + \sqrt{2}}\left(\frac{h}{l}\right)f = 0.83\left(\frac{h}{l}\right)f.$$

For $h/l = 0.1, f_b/f = 0.083 = 8.3\%$.

When the bending stresses due to the dead load of the bars are added to those due to the rigid joint at A, the total secondary bending stress may be of the order of 10% to 15% of the direct stress.

PROBLEMS

6.1 A bridge-type Pratt truss (Fig. 6.1.1) spanning 36 m (120 ft) is 3 m (10 ft) deep and consists of 12 equal panels. The truss is one of a series 9 m (30 ft) o.c. and carries a total roof load of 3.5 kN/m^2 (70 psf). Determine the bar forces, if the loads enter only at panel points through roof beams.

6.2 Of the total load on the truss of Problem 6.1, 1.5 kN/m² (30 psf) represents the snow load. Compute the bar forces if the snow load is applied over only half the span.

6.3 The steel joist illustrated in Fig. 6.1.2 is an upside-down Warren truss (Fig. 6.1.1). A series of such trusses is to be used in an office floor spanning 12 m (36 ft). The trusses are 1 m (3 ft) deep, consist of six panels spaced 1 m (3 ft) o.c., and carry a 75 mm (3 in.) concrete slab, a 25 mm (1 in.) plaster ceiling, and a 50 mm (2 in.) lightweight concrete fill. Determine the bar forces in the joists.

6.4 Pratt roof trusses (Fig. 6.1.1), with eight panels 1.75 m (6 ft) wide and a height of 2.5 m (8 ft), are spaced 3.5 m (12 ft) o.c. Determine the bar forces in each truss for a wind load consisting of a pressure on the windward side of 1 kN/m² (20 psf) and a suction on the leeward side of 0.5 kN/m² (10 psf). The truss is supported by a hinge on the windward side and a roller on the leeward side.

6.5 Determine the bar forces in the cantilevered truss of Fig. 6.3.3(a) with loads of 22 kN (5 k) at the joints E, C, and A, for $a = 4.5$ m (15 ft).

6.6 Determine the bar forces in the truss of Problem 6.5 when the depth of the truss equals $a/2$.

6.7 A tower crane has a cantilevered Pratt bridge truss (Fig. 6.1.1) on which the hoist moves along the bottom chord. The truss consists of six panels with a support at the second panel point from the left and a support at the left end. Each panel is 3 m (10 ft) long and the truss is 1.75 m (6 ft) deep. Compute the *maximum* bar force in each member if the 90 kN (20 k) hoist load can be placed at any panel point to the right of the support.

6.8 The corner supports of a square thin shell roof, 15 m (50 ft) on a side, consist of triangular vertical trusses with a vertical tensile chord 9 m (30 ft) high and an inclined compressive chord making an angle of 30° to the vertical. The trusses are divided by horizontal bars into three panels 3 m (10 ft) high. The truss diagonals are compressive bars. The trusses support at their tops vertical reactions of 400 kN (90 k) and horizontal reactions of 310 kN (70 k). Determine the bar forces.

6.9 The truss of Fig. 6.2.7 carries vertical loads of 135 kN (30 k) at panel points 2, 4, and 5. Determine its bar forces.

6.10 Determine the deflection of point A at the tip of the truss of Problem 6.5. Elongations are to be evaluated on the basis of $F_t = F_c = 138$ MPa (20 ksi), $E = 200\,000$ MPa (29,000 ksi).

6.11 Determine the deflection at point A of the truss of Problem 6.6. Elongations are to be computed on the basis of $F_t = F_c = 138$ MPa (20 ksi), $E = 200\,000$ MPa (29,000 ksi).

6.12 Determine the deflection at the center point F of the lower chord of the aluminum truss in Fig. 6.2.2, for $a = 4.5$ m (15 ft). The centerline does not rotate nor move horizontally. Note that in constructing the Williot diagram it must be assumed that one of the bars does not rotate, as was implicitly done in

Example 6.3.1 by assuming that the line *FG* remains vertical. In the case of symmetrical loading this can be accomplished by starting the construction with the central vertical bar of the truss. Assume $F_t = F_c = 69$ MPa (10 ksi), $E = 69\,000$ MPa (10,000 ksi).

6.13 Determine the deflection of the panel points on the upper chord of the truss of Problem 6.3 for $F_t = F_c = 138$ MPa (20 ksi), $E = 200\,000$ MPa (29,000 ksi). Note that in this case the bar assumed not to rotate is the center horizontal bar.

6.14 Compute the deflection at the center of the truss of Problem 6.1, assuming the truss to be an ideal I-beam with a moment of inertia determined as follows: the constant flange area is evaluated on the basis of the maximum top or bottom chord force for $F_t = F_c = 138$ MPa (20 ksi) and is assumed to be concentrated at the top and bottom chords, with the beam depth equal to the truss depth (see Table 4.1.1 (6) for small t).

6.15 Using the approximation described in Problem 6.14, compute the deflection at the center of the truss of Problem 6.2.

6.16 Determine the areas of the two top wide flange bars of the truss of Problem 6.8, assuming the trusses to be made of steel with $F_t = F_c = 103$ MPa (15 ksi), $E = 200\,000$ MPa (29,000 ksi), and taking into account buckling limitations. Evaluate the change in the 30° angle between these two bars and the secondary moments that would appear if this change were prevented by continuity, assuming the far ends of the bars to be hinged. Determine the increase in stress due to the secondary moments.

6.17 The bracket of Fig. 6.2.1 is made of steel. Under an increase in temperature of 28°C (50°F) the bar lengths *AB*, *AC* increase, but the support length *BC* remains unchanged. Determine the moment at *A* in the two bars if the 30° angle is maintained by continuity and the two bars have identical *I*'s, assuming $l = 3$ m (10 ft) and $\alpha = 1 \times 10^{-5}$ mm/mm/°C (0.000 006 in./in./°F).

7. Frames

7.1 THE SIMPLE FRAME UNDER VERTICAL LOADS [8.2]

The simple frame, consisting of a horizontal beam rigidly connected to two vertical columns, is one of the basic components of many structures. Depending on whether the columns are hinged or fixed at their base, the frame is called *hinged* or *fixed* [Figs. 7.1.1(a) and (b)].

The structural behavior of the simple frame is characterized by the rigid connection between beam and columns, which makes the rotations of the ends of the beam equal to those of the tops of the columns. When the frame structure and the load on it are symmetrical about its midspan vertical axis, the tops of the columns do not move laterally. The frame then behaves essentially as a three-span continuous beam, since in such a beam the middle supports also do not move and the rotations of the ends of the middle span are equal to those of the end spans. The only difference between the frame and the continuous beam is that in the frame the angle between two successive members is and remains 90°, while in the continuous beam it is and remains

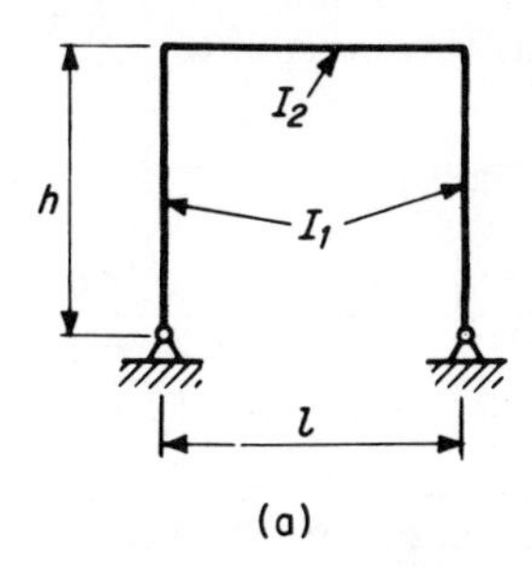

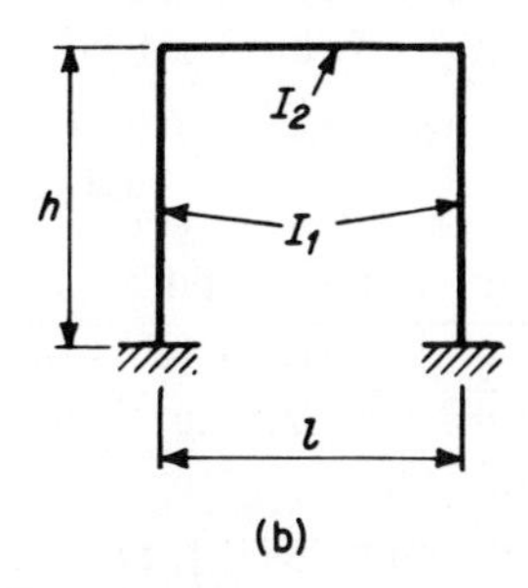

FIGURE 7.1.1

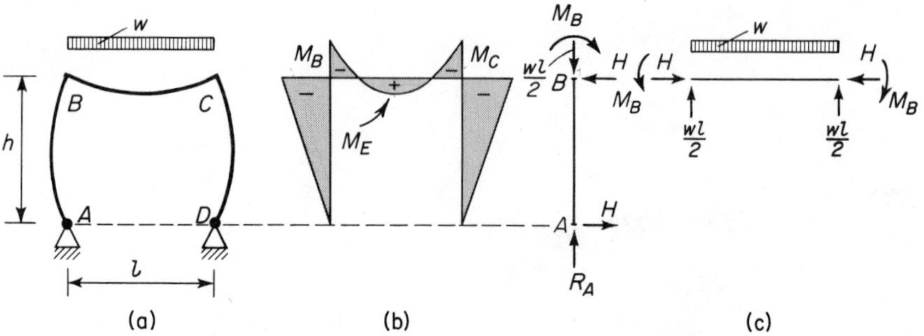

$$M_B = M_C = -\frac{wl^2}{12}\frac{3}{3+2\alpha};\quad M_E = \frac{wl^2}{24}\frac{3+6\alpha}{3+2\alpha};\quad H = \frac{M_B}{h}\quad R_A = R_D = \frac{wl}{2}$$

FIGURE 7.1.2

180°. This difference in angle is responsible for the frame *thrust*, which does not exist in the continuous beam.

The bending moment diagram in the *hinged frame* under a uniform vertical load w (Fig. 7.1.2) is obtained directly from that of the continuous beam of Example 4.6.4 (a), by substituting the height h of the columns for l_1, the length l of the beam for l_2, and by letting:

$$\left(\frac{l_1}{l_2}\right)\left(\frac{I_2}{I_1}\right) = \left(\frac{h}{l}\right)\left(\frac{I_2}{I_1}\right) = \alpha, \tag{7.1.1}$$

so that by equations (a) of Section 4.6:

$$\begin{aligned} M_B &= -\frac{wl^2}{12}\frac{3}{3+2\alpha} \equiv m_B wl^2, \\ M_E &= \frac{wl^2}{8} + M_B = \frac{wl^2}{24}\frac{3+6\alpha}{3+2\alpha} \equiv m_E wl^2. \end{aligned} \tag{7.1.2}$$

It is seen from (7.1.1) and (7.1.2) that as α approaches zero, i.e., as the rigidity of the columns increases, the moments M_B and M_E approach the values of those in a fixed beam, while as α approaches infinity, i.e., the columns decrease in rigidity, M_B approaches zero, $(3 + 6\alpha)/(3 + 2\alpha)$ approaches 3, and M_E becomes $wl^2/8$. Table 7.1.1 gives m_B and m_E as functions of α.

Table 7.1.1 HINGED FRAME

α	0	0.5	1.0	1.5	2.0	2.5	3.0	3.5	4.0	∞
m_B	−0.083	−0.062	−0.050	−0.042	−0.036	−0.031	−0.028	−0.025	−0.023	0
m_E	0.042	0.062	0.075	0.083	0.089	0.094	0.097	0.100	0.103	0.125

The column AB [Fig. 7.1.2(c)] is acted upon by the beam reaction $R_B = wl/2$ and the moment M_B. For vertical equilibrium, the hinge A reacts with an upward reaction $R_A = R_B$. For rotational equilibrium, the moment about B of all forces on the column must equal zero. Hence, $Hh - M_B = 0$, from which:

$$H = \frac{M_B}{h} = m_B\left(\frac{l}{h}\right)(wl). \qquad (7.1.3)$$

(In the continuous beam M_B/h was the *vertical* reaction due to M_B.) Figure 7.1.2(c) shows how both the columns and the beam are under bending, shear, *and* compression, but (7.1.3) indicates that the thrust is usually a small fraction, $m_B(l/h)$, of the weight $W = wl$ on the frame.

Example 7.1.1. A series of hinged steel frames 6 m (20 ft) on center with $h =$ 9 m (30 ft) and $l = 18$ m (60 ft) carry a total roof load of 9.6 kN/m² (200 psf). Determine the dimensions of their beams and columns for $F_b = 165$ MPa (24 ksi).

The uniform load on each beam is $w = 9.6 \times 6 = 57.6$ kN/m, and $wl^2 = 57.6 \times 18^2 = 18\,660$ kN·m. Choosing $\alpha = 1$ and neglecting the effect of the thrust on the beam, we obtain:

$$m_E = 0.075, \quad M_E = 0.075 \times 18\,660 = 1\,400 \text{ kN·m } (1{,}080 \text{ ft·k}),$$

$$S = \frac{1\,400 \times 1\,000^2}{165} = 8.485 \times 10^6 \text{ mm}^3,$$

and we choose a W920 × 238 (W36 × 160) beam with:

$$S_2 = 8.883 \times 10^6 \text{ mm}^3, \qquad I_2 = 4\,060 \times 10^6 \text{ mm}^4.$$

For the column:

$$m_B = -0.050, \qquad M_B = -930 \text{ kN·m } (-720 \text{ ft·k}),$$

$$\frac{wl}{2} = \frac{57.6 \times 18}{2} = 518 \text{ kN } (120 \text{ k}),$$

$$\alpha = \left(\frac{h}{l}\right)\left(\frac{I_2}{I_1}\right) = 1, \qquad I_1 = \frac{h}{l} I_2 = \frac{9}{18} \times 4\,060 \times 10^6 = 2\,030 \times 10^6 \text{ mm}^4.$$

We choose a W840 × 210 (W33 × 141) with $I_1 = 3\,099 \times 10^6$ mm⁴, $S_1 = 7\,342 \times 10^6$ mm³, $A_1 = 268\,400$ mm², and find a stress:

$$f = \frac{930 \times 1\,000}{7\,342 \times 10^6} + \frac{518 \times 1\,000}{268\,400} = 126.7 + 19.3 = 146 \text{ MPa } (18.77 \text{ ksi}).$$

Assuming that the columns are restrained from buckling out of the plane of the frame, the AISC specifications give (see Fig. 5.4.1):

$$h = 9 \text{ m}, \qquad r = 340.4 \text{ mm},$$

$$\frac{h}{r} \simeq 27, \quad F_b = 138.9 \text{ MPa},$$

and from (5.1.5):

$$\frac{f_a}{F_a} + \frac{f_b}{F_b} = \frac{19.3}{138.9} + \frac{126.7}{165} = 0.91 < 1.00,$$

showing that the combined stress is below the allowable value.

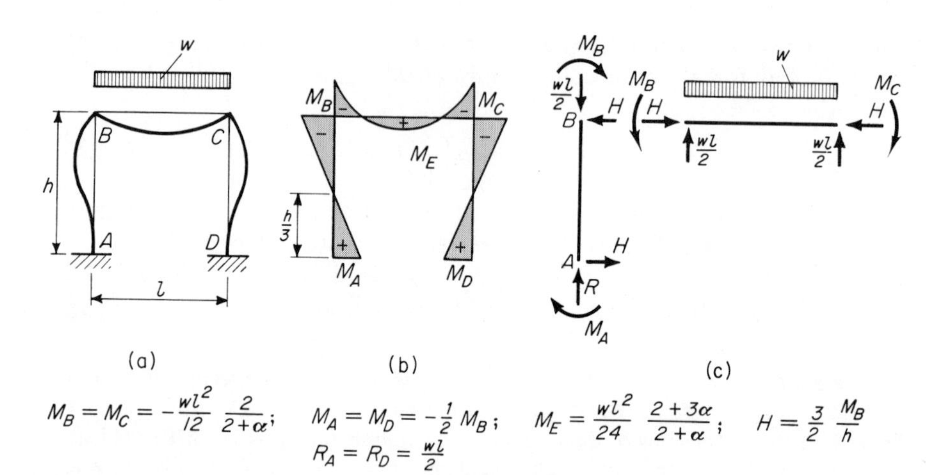

FIGURE 7.1.3

From the continuous beam of Example 4.6.4 (b) one derives the bending moment diagram for the *fixed frame* of Fig. 7.1.3. Table 7.1.2 gives the coefficients for the moments M_B and M_E:

$$M_B = -\frac{wl^2}{12}\frac{2}{2+\alpha} \equiv m_B wl^2, \qquad M_E = \frac{wl^2}{24}\frac{2+3\alpha}{2+\alpha} \equiv m_E wl^2. \tag{7.1.4}$$

Since $M_A = -\frac{1}{2}M_B$, the column has an inflection point at $h/3$. The fixed frame behaves very much like a hinged frame with columns $\frac{2}{3}h$ high, except for the small lateral displacement at the inflection point.

Table 7.1.2 FIXED FRAME

α	0	0.5	1.0	1.5	2.0	2.5	3.0	3.5	4.0	∞
m_B	−0.083	−0.068	−0.056	−0.048	−0.042	−0.038	−0.033	−0.030	−0.028	0
m_E	0.042	0.058	0.069	0.077	0.083	0.088	0.092	0.094	0.097	0.125

Since the moments M_B and M_A are both clockwise, the thrust H is given by:

$$Hh = M_B + |M_A| = \frac{3}{2}M_B \quad \therefore \quad H = \frac{3}{2}m_B\left(\frac{l}{h}\right)(wl). \tag{7.1.5}$$

The thrust in the fixed frame is larger than in the hinged frame both because of the factor $\frac{3}{2}$ in (7.1.5) and because m_B is larger in the fixed frame. Since the inflection point in the column B of the fixed frame is at $h/3$ from A and it is at A in the hinged frame, for any other condition of imperfect fixity of the columns the inflection point falls within the lower third of the column.

Example 7.1.2. A 3.5 m (12 ft) wide reinforced concrete overpass is supported by two fixed frames. The overpass has a span of 12 m (40 ft), a height of 6 m (20 ft), and carries a total load, including the weight of the structure, of 20 kN/m² (400 psf). The frames are 600 mm (2 ft) thick and the concrete strength is $f'_c = 30$ MPa (4,350 psi). Determine the depth d of their beams and the width d' of the columns.

The uniform load on the beam of each frame is $w = 20 \times 1.75 = 35$ kN/m, and $wl^2 = 35 \times 12^2 = 5\,040$ kN·m. Choosing $\alpha = 1$ and again neglecting the effect of the thrust on the beam:

$$m_E = 0.069, \qquad M_E = 0.069 \times 5\,040 = 348 \text{ kN·m (3,180 k·in.)}.$$

The resisting moment of the concrete section of the beam (see Section 4.2) is given by:

$$M_R = M_E = \tfrac{1}{6} f_c b d^2 = \tfrac{1}{6}(0.45 \times f'_c) b d^2$$

$$= \frac{1}{6}\left(0.45 \times \frac{30}{1\,000}\right) \times 600 \times d^2 = 1.35 d^2 \text{ kN·mm}.$$

The beam depth required for bending is, therefore:

$$d = \sqrt{\frac{348 \times 10^3}{1.35}} = 508 \text{ mm} \approx 550 \text{ mm (1.80 ft)}.$$

For the column:

$$m_B = -0.056, \qquad M_B = 0.056 \times 5\,040 = 282 \text{ kN·m}.$$

The required width in bending is:

$$d' = \sqrt{\frac{282 \times 10^3}{1.35}} = 457 \text{ mm} \approx 500 \text{ mm (1.64 ft)}.$$

The shear at the ends of the beam is:

$$V = 35 \times 6 = 210 \text{ kN}, \qquad v_c = \frac{210 \times 1\,000}{600 \times 550} = 0.60 \text{ MPa} > 0.50 \text{ MPa}.$$

The concrete section required to resist this shearing force for an allowable stress $v_c = 0.50$ MPa, given in Table 2.4.2, is:

$$A = \frac{V}{v_c} = \frac{210 \times 1\,000}{0.50} = 420\,000 \text{ mm}^2$$

Since the area provided for bending is only $600 \times 550 = 330\,000$ mm², to satisfy shear requirements the beam depth may be increased to $(420\,000/330\,000) \times 550 \simeq 700$ mm (2.30 ft), or shear reinforcement must be used.

The horizontal shear at the base of the frame is:

$$H = \frac{3}{2}\frac{M_B}{h} = \frac{3}{2}\left(\frac{282}{6}\right) = 70.5 \text{ kN (16.1 k)}.$$

The column is acted upon by the moment $M_B = 282$ kN·m, the shear $H = 70.5$ kN, and the axial force $V = 210$ kN. At an allowable stress of 7 MPa, the area required for axial force is:

$$A' = \frac{210 \times 1\,000}{7} = 30\,000 \text{ mm}^2.$$

Therefore, we will approximately satisfy the combined axial-bending stress condition by increasing the column area by A', i.e., the column depth by $A'/b = 30\,000/600 = 50$ mm. The final column depth is then $d = 500 + 50 = 550$ mm (1.80 ft). The concrete area required to resist the horizontal shearing force at the base of the frame is:

$$A = \frac{70.5 \times 1\,000}{0.50} = 141\,000 \text{ mm}^2,$$

which is less than the area required for bending and compression.

Finally, we can check our assumption for α:

$$\alpha = \frac{6}{12}\left(\frac{700}{550}\right)^3 = 1.03,$$

which, by Table 7.1.2, does not significantly alter the moments.

7.2 THE MULTI-BAY FRAME UNDER VERTICAL LOADS [8.3]

The approximate analysis of a *hinged multi-bay frame* under a vertical uniform load w [Fig. 7.2.1(a)] may be obtained by noticing that the tops of the intermediate columns *for all practical purposes* do not rotate, and do not move laterally, so that (1) the intermediate spans may be considered fixed, and that (2) the end spans may be considered fixed at one end and rigidly connected to the exterior columns at the other [Fig. 7.2.1(b)].

The moments in the end bays of the frame may be obtained (as was done for continuous beams) by stating that the rotation at B is identical for the column and the beams, and that the rotation at C is zero.

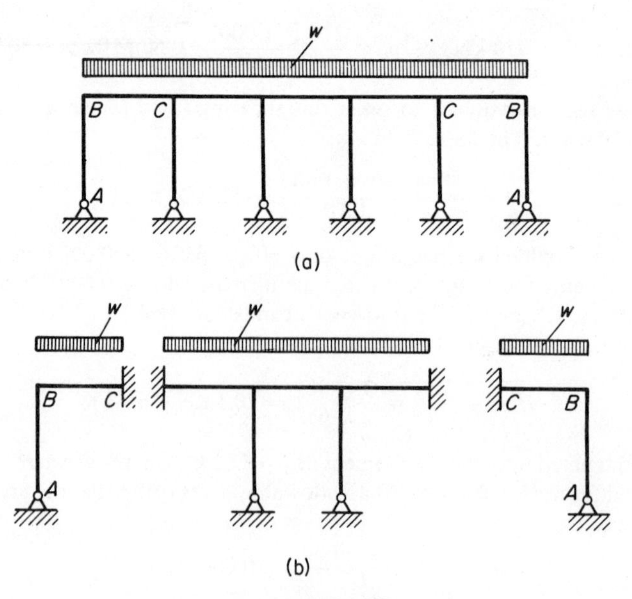

FIGURE 7.2.1

It is thus found that for a hinged frame:

$$M_B = \frac{wl^2}{12}\frac{3}{3+4\alpha}, \qquad M_C = -\frac{wl^2}{8}\frac{2+4\alpha}{3+4\alpha}. \qquad (7.2.1)$$

As α approaches zero, M_B and M_C approach the fixed beam moments $-wl^2/12$; as α approaches infinity, i.e., the column becomes infinitely flexible, M_B approaches zero and M_C approaches $-wl^2/8$, i.e., the moments of a beam fixed at one end and simply supported at the other [see Table 4.1.2 (13)].

Similarly, for a *fixed frame* (Fig. 7.2.2), the end bays may be shown to develop moments:

$$M_B = -\frac{wl^2}{12}\frac{1}{1+\alpha}, \qquad M_A = -\frac{1}{2}M_B, \qquad M_C = -\frac{wl^2}{8}\frac{2+3\alpha}{3+3\alpha}. \qquad (7.2.2)$$

For α approaching zero or infinity, M_B and M_C approach the same values as those in (7.2.1).

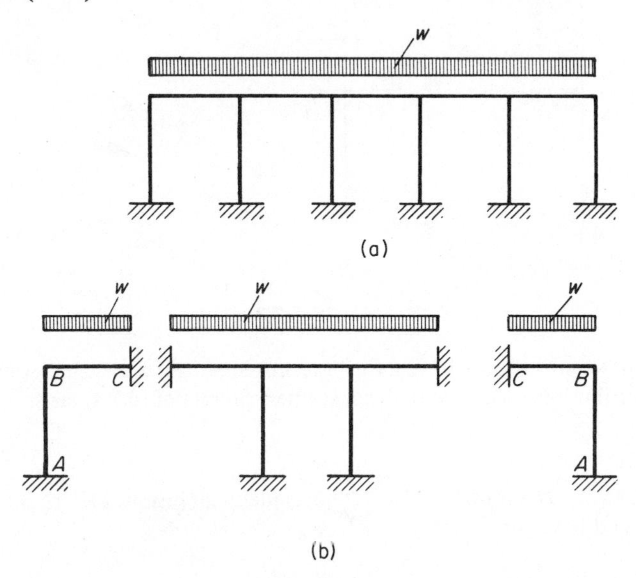

FIGURE 7.2.2

Example 7.2.1. The frame of Fig. 7.2.3(a) has identical beams, and $\alpha = 1$ for all bays. Determine the support moments in the end bays and in the internal bays as a percentage of the fixed-end moments.

By (7.2.1):

$$M_B = \frac{3}{3+4\times 1}100 = 43\%, \qquad [45\%]$$

$$M_C = \frac{3}{2}\frac{2+4\times 1}{3+4\times 1}100 = 129\%, \quad [120\%]$$

$$M_D = 1\times 100 = 100\%. \qquad [95\%]$$

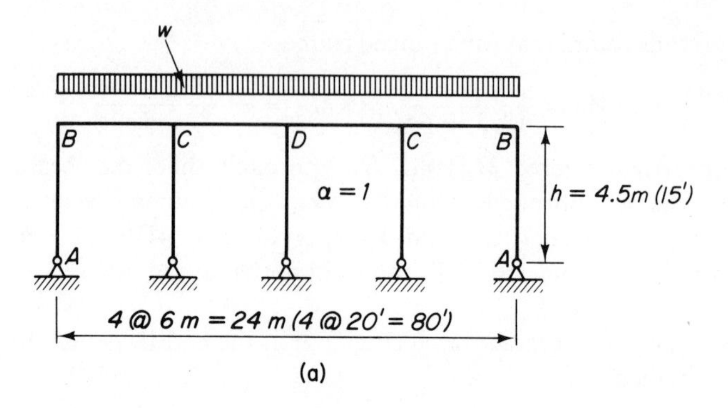

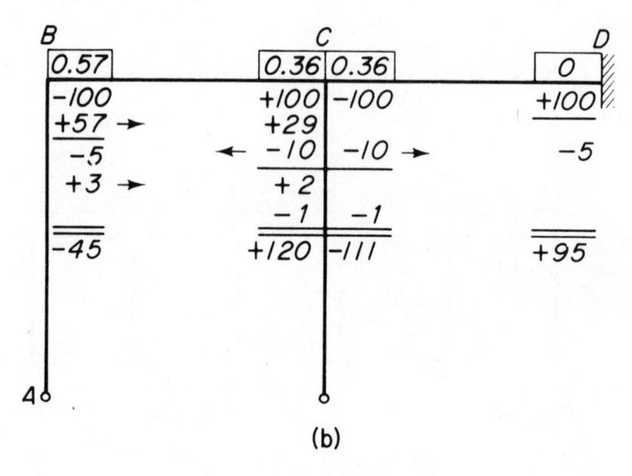

FIGURE 7.2.3

The moments shown in square brackets are the exact values obtained by moment distribution, using the reduced stiffness factors of (4.7.4), as shown in Fig. 7.2.3(b).

Example 7.2.2. The frame of Fig. 7.2.4 has identical beams. Determine the value of α in the end bays for which $M_D = -M_B$.

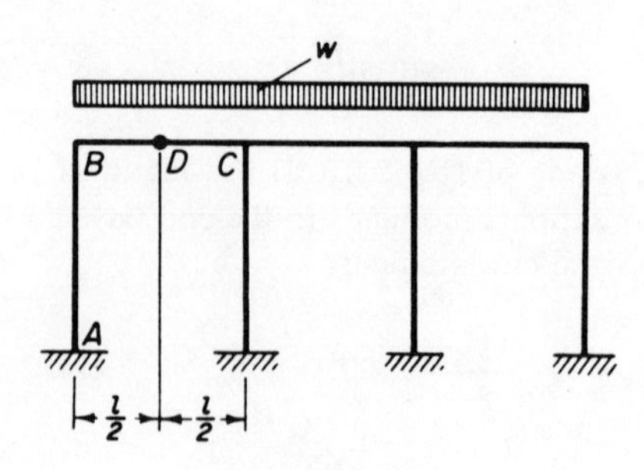

FIGURE 7.2.4

By (7.2.2):

$$M_D = \frac{1}{8}wl^2 + \frac{1}{2}(M_B + M_C)$$

$$= \frac{1}{8}wl^2 - \frac{1}{2}\left(\frac{wl^2}{12}\frac{1}{1+\alpha} + \frac{wl^2}{8}\frac{2+3\alpha}{3+3\alpha}\right) = \frac{wl^2}{8}\frac{2+3\alpha}{6(1+\alpha)}.$$

The set condition requires that:

$$\frac{wl^2}{8}\frac{2+3\alpha}{6+6\alpha} = \frac{wl^2}{12}\frac{1}{1+\alpha} \quad \therefore \quad \alpha = \frac{2}{3}.$$

For this value of α:

$$M_C = -\frac{wl^2}{8}\frac{2+3(\frac{2}{3})}{3+3(\frac{2}{3})} = -\frac{wl^2}{10},$$

$$M_B = -M_D = -\frac{wl^2}{20}.$$

7.3 THE ONE-STORY FRAME WITH SIDESWAY [8.2]

Under the action of unsymmetrical vertical loads or of horizontal loads, a symmetrical simple frame moves horizontally. This horizontal displacement at its top is called *sidesway*.

To consider a simple example of sidesway, let the beam of the simple *hinged frame* of Fig. 7.3.1(a) be loaded uniformly over its left half. Such a load can be considered as the sum of the two loads in Figs. 7.3.1(b) and (c). The symmetrical loading condition (b) does not induce sidesway; the anti-symmetrical condition (c) does. Let us evaluate the sidesway in frame (c).

Horizontal equilibrium in frame (a) requires that the thrust H at A be equal and opposite to the thrust H at D. But, because of the symmetry of frame (b), the thrust H' at A is equal and opposite to that at D; while, because of the antisymmetry of frame (c), the thrust H'' at A is equal and in the same

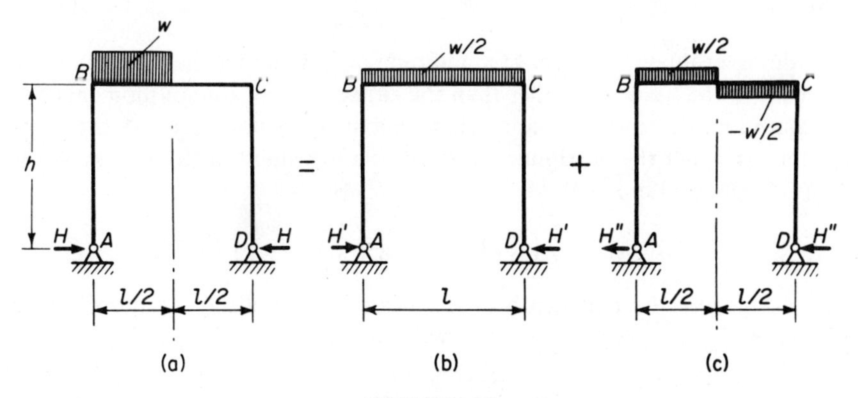

FIGURE 7.3.1

direction as that at D. Hence, by equilibrium of frame (c) in the horizontal direction, $H'' = 0$. Since $H'' = 0$, the columns of frame (c) have zero moments and stay straight, and the beam of frame (c) is simply supported. Moreover, its antisymmetrical deformation (Fig. 7.3.2) introduces an inflection point at midspan, so that the beam acts as two simply supported beams of span $l/2$ under a load $w/2$ acting down on the left beam and a load $w/2$

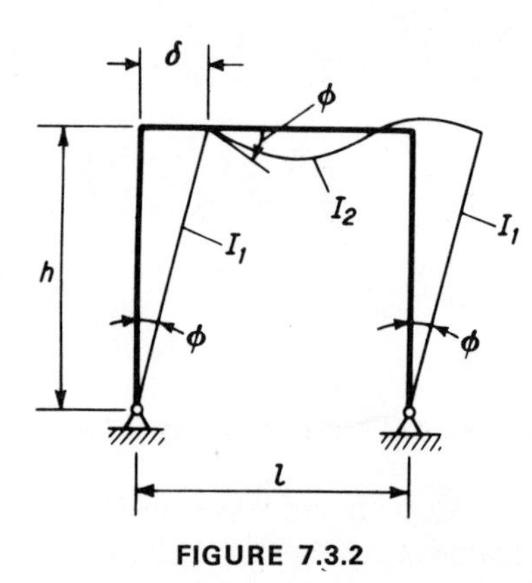

FIGURE 7.3.2

acting up on the right beam. The maximum positive and negative moments in the beam are:

$$M''_{\max} = \pm \frac{1}{8}\frac{w}{2}\left(\frac{l}{2}\right)^2 = \pm\frac{1}{64}wl^2. \tag{7.3.1}$$

The end rotations of the beam are, by Table 4.3.1 (a):

$$\phi = \frac{1}{24EI_2}\left(\frac{w}{2}\right)\left(\frac{l}{2}\right)^3 = \frac{1}{384}\frac{wl^3}{EI_2},$$

and the sidesway δ is given by:

$$\delta = \phi h = \frac{h}{384}\frac{wl^3}{EI_2}. \tag{7.3.2}$$

In order to evaluate the order of magnitude of δ, we notice that the maximum moment in the beam is smaller than the sum of the maximum moments due to conditions (b) and (c), since these maxima do not occur at the same section, and that the maximum moment due to condition (b) is at most the simply supported moment $M'_{\max} = \frac{1}{8}(w/2)l^2$. Hence:

$$M_{\max} < \frac{1}{8}\left(\frac{w}{2}\right)l^2 + \frac{1}{8}\frac{w}{2}\left(\frac{l}{2}\right)^2 = \frac{5}{64}wl^2. \tag{a}$$

Indicating by c the half depth of the beam, and multiplying and dividing (7.3.2) by $\frac{5}{64}c$, the sidesway may be written as:

$$\left.\frac{\delta}{h}\right]_{\text{hinged}} = \frac{1}{30}\frac{(\frac{5}{64}wl^2)c}{EI_2}\left(\frac{l}{c}\right) = \frac{1}{30}\left(\frac{f}{E}\right)\left(\frac{l}{c}\right). \tag{7.3.3}$$

For a steel frame with $l/c = 30$:

$$\frac{\delta}{h} = \frac{f}{E} = \frac{165}{200\,000} \simeq \frac{1}{1\,200}.$$

For a *fixed frame* under vertical loads (Fig. 7.3.3), it may be shown that the sidesway is given by:

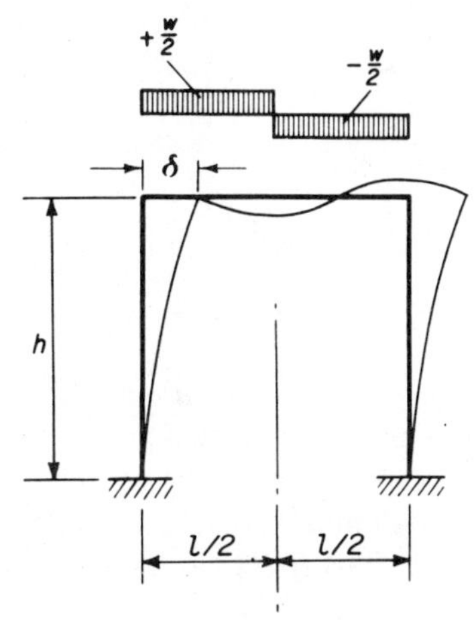

FIGURE 7.3.3

$$\left.\frac{\delta}{h}\right]_{\text{fixed}} = \left(\frac{1}{2}\,\frac{\alpha}{\alpha + \frac{1}{6}}\right)\left.\frac{\delta}{h}\right]_{\text{hinged}} \tag{7.3.4}$$

and hence is always smaller than one-half the hinged-frame sidesway under the same loads.

The sidesway of a frame under horizontal loads is evaluated by similar methods. For example, the *hinged frame* of Fig. 7.3.4 deflects antisymmetrically with thrusts $P/2$, so that:

$$M_B = -M_C = \frac{Ph}{2}; \tag{7.3.5}$$

its sidesway may be shown to be given by:

$$\left.\frac{\delta}{h}\right]_{\text{hinged}} = \frac{2 + 1/\alpha}{12}\,\frac{Ph^2}{EI_1}. \tag{7.3.6}$$

The total wind force $P = pbh$ on the side of a simply supported frame, where b is the spacing of the frames, may be concentrated half at the top and half at the bottom of the column, and its moments may be computed by (7.3.5), with an error of less than 10% for $\alpha \leq 1$ and less than 15% for $\alpha \leq 3$. In view of the antisymmetrical deformation of the frame, the wind force $P/2$ at the top of the frame may be considered as a pressure $P/4$ on the windward side and a suction $P/4$ on the leeward side of the frame without altering the value of the moments [Fig. 7.3.4(c)].

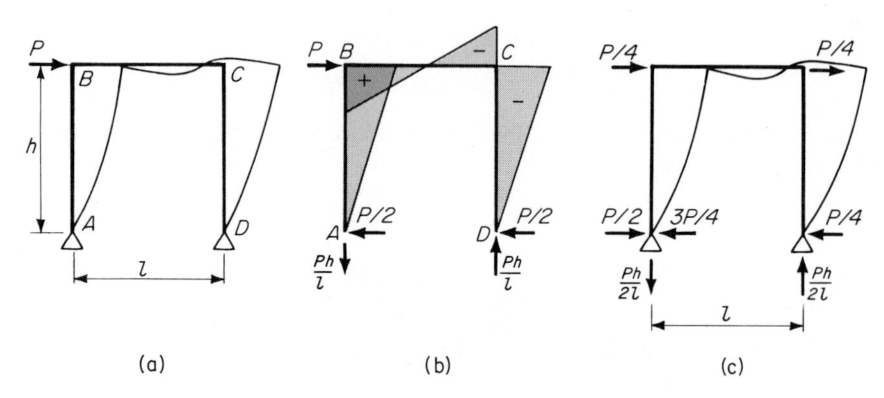

FIGURE 7.3.4

Similarly, the wind force P *at the top of a hinged multi-bay frame* with a constant beam cross section may be evaluated as if each bay acted as a simple frame, developing thrusts proportional to the column stiffness. For identical columns, this simplifying assumption of the so-called *portal method* gives for a frame with n columns a thrust $P/(n-1)$ at the internal columns and $P/2(n-1)$ at the external columns. The corresponding moments at the top of the internal and external columns are, respectively:

$$M_i = \frac{Ph}{n-1}, \qquad M_e = \frac{Ph}{2(n-1)}. \tag{7.3.7}$$

Example 7.3.1. Design the steel frame of Fig. 7.3.5(a). The nonuniform load may be divided into a uniform load of 30 kN/m (2 k/ft) and an antisymmetric load of 15 kN/m (1 k/ft).

The maximum and minimum moments due to the antisymmetric load are, by (7.3.1) with $w/2 \equiv w' = 15$ kN/m:

$$M = \pm\frac{wl^2}{64} = \pm\frac{15 \times (12)^2}{64} = \pm 33.75 \text{ kN}\cdot\text{m (25 k}\cdot\text{ft)}.$$

The moment M_E due to the uniform load, assuming $\alpha = 1.5$, is, by Table 7.1.1:

$$M_E = 0.083 \times 30 \times (12)^2 = 358 \text{ kN}\cdot\text{m (265 k}\cdot\text{ft)}.$$

For A36 steel, with $F_b = 165$ MPa (22 ksi):

$$S = \frac{358 \times 1\,000^2}{165} = 2.17 \times 10^6 \text{ mm}^3,$$

choose a W530 × 101 (W21 × 68) with $S_2 = 2.295 \times 10^6$ mm³, $I_2 = 615.68 \times 10^6$ mm⁴. Similarly:

$$M_B = -0.042 \times 30 \times (12)^2 = -181 \text{ kN}\cdot\text{m}.$$

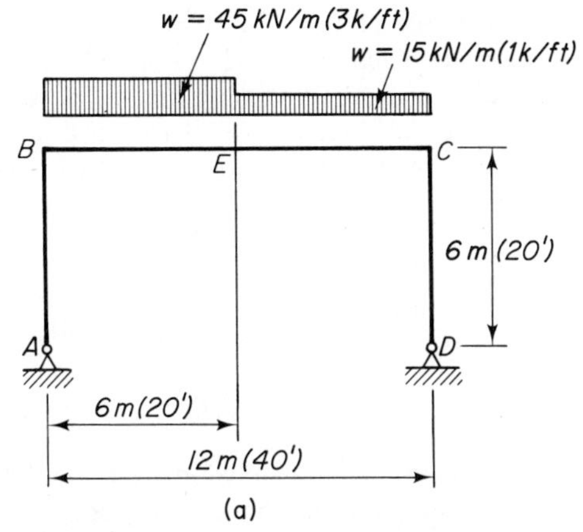

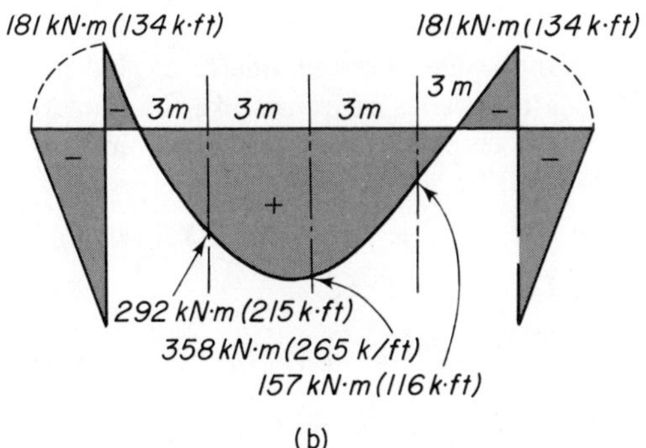

FIGURE 7.3.5

$$\frac{wl}{2} = \frac{30 \times 12}{2} = 180 \text{ kN}, \qquad \frac{w'l}{4} = \frac{15 \times 12}{4} = 45 \text{ kN},$$

$$R_A = 180 + 45 = 225 \text{ kN}, \qquad R_D = 180 - 45 = 135 \text{ kN},$$

$$S_1 = \frac{181 \times 1\,000^2}{165} = 1.095 \times 10^6 \text{ mm}^3.$$

Choosing a W360 × 91 (W14 × 61) with $I = 266.24 \times 10^6$ mm⁴, $S = 1.511 \times 10^6$ mm³, $A = 11\,550$ mm², $r = 151.9$ mm:

$$\frac{P}{A} = \frac{225}{11\,550} = 0.019\,5 \text{ kN/mm}^2 = 19.5 \text{ MPa (2.8 ksi)},$$

$$\frac{M}{S} = \frac{181 \times 1\,000^2}{1.511 \times 10^6} = 119.8 \text{ MPa (17.4 ksi)}.$$

Assuming lateral buckling is prevented, $h/r = (6 \times 1\,000)/151.9 \approx 40$, $F_b = 132.3$ MPa (from Fig. 5.4.1) and, by (5.1.5),

$$\frac{19.5}{132.3} + \frac{119.8}{165} = 0.87 < 1.$$

The combined stress is within that allowable and $\alpha = (6/12)(615.63/266.24) = 1.16$ is reasonably close to the assumed value of 1.5.

Example 7.3.2. The frame of Fig. 7.3.5 represents one of a series spaced 6 m (20 ft) on centers and acted upon by a wind pressure of 1 kN/m² (20 psf) on the left side. Determine the additional wind moments and check whether the wind stresses are within 33% of the vertical load stresses determined in Example 7.3.1, so that by code the frame members do not have to be increased in size. The load P at the top of the frame is:

$$P = 1 \times 6 \times 3 = 18 \text{ kN (4 k)}.$$

By (7.3.5):

$$M_B = \frac{Ph}{2} = \frac{18 \times 6}{2} = 54 \text{ kN·m} < 0.33 \times 181 = 59.7 \text{ kN·m},$$

$$R_A = \frac{Ph}{l} = \frac{18 \times 6}{12} = 9 \text{ kN} < 0.33 \times 225 = 74.3 \text{ kN}.$$

Both the moment and the reaction due to wind are within 33% of those due to the vertical load, and the members do not have to be changed.

Example 7.3.3. Determine the moments due to a wind pressure of 1.5 kN/m² (30 psf) on the left side of a series of frames like the frame in Fig. 7.2.3, spaced 6 m (20 ft) on centers, and check whether they are within 33% of the vertical load moments in Example 7.2.1, for $wl^2/12 = 100$ kN·m.

$$P = 1.5 \times 6 \times 2.25 = 20.25 \text{ kN (4.5 k)}.$$

The moments are, by (7.3.7):

$$M_B = \frac{Ph}{2(n-1)} = \frac{20.25 \times 4.5}{2 \times 4} = 11.4 \text{ kN·m} < 0.33 \times 43 = 14.2 \text{ kN·m},$$

$$M_C = M_D = 11.4 \times 2 = 22.8 \text{ kN·m} < 0.33 \times 129 = 42.6 \text{ kN·m}.$$

Since all the beam moments are within 33% of those due to the vertical loads, no change in the members is required.

The moments in the *fixed frame* of Fig. 7.3.6 may be shown to be given by:

$$M_B = -M_C = \frac{Ph}{2}\frac{3\alpha}{1+6\alpha}, \qquad M_A = -M_D = -\frac{Ph}{2}\frac{1+3\alpha}{1+6\alpha}. \quad (7.3.8)$$

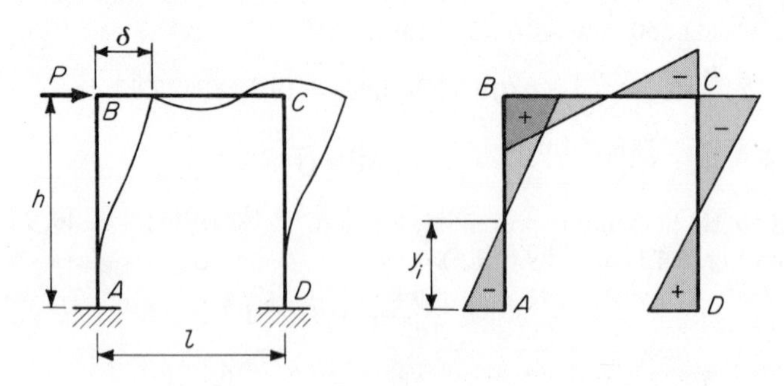

FIGURE 7.3.6

Its sidesway is given by:

$$\frac{\delta}{h}\bigg]_{\text{fixed}} = \frac{2+3\alpha}{1+6\alpha}\frac{Ph^2}{12EI_1}. \tag{7.3.9}$$

and, as α increases, becomes progressively smaller than the hinged-frame sidesway.

The ratio k of the absolute value of M_B to that of M_A is:

$$k = \left|\frac{M_B}{M_A}\right| = \frac{3\alpha}{1+3\alpha},$$

and the height of the inflection point of the column (Fig. 7.3.7) is:

$$y_i = \frac{1}{1+k}h = \frac{1+3\alpha}{1+6\alpha}h. \tag{7.3.10}$$

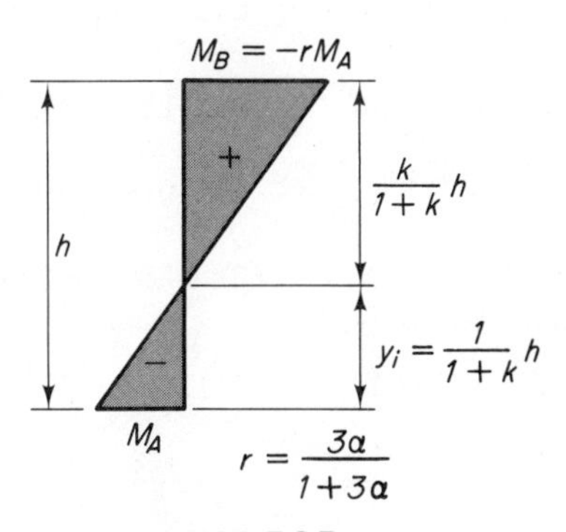

FIGURE 7.3.7

As shown in Table 7.3.1, $y_i = h$ for $\alpha = 0$ and approaches $h/2$ as α increases. Hence, the inflection point is always in the upper half of the column and approaches the column midpoint as the beam stiffness becomes large in comparison with the column stiffness.

Table 7.3.1

α	0	1/4	1/2	1	2	4	8	∞
y_i/h	1	0.7	0.62	0.57	0.54	0.52	0.51	0.50

Since $\alpha > 1$ for most frames and for $\alpha > 1$ the inflection point is practically at $h/2$, while the inflection point in the beam is always at $l/2$, fixed frames under a horizontal force P at their top behave for all practical purposes as frames with hinges at $h/2$ in the columns and at $l/2$ in the beam (Fig. 7.3.8). In this case, the fixed frame is statically determinate, and $M_B = -M_A = Ph/4$.

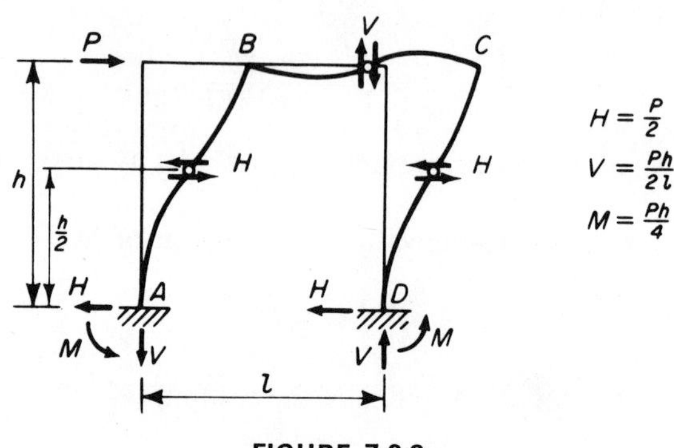

FIGURE 7.3.8

Example 7.3.4. Design the steel frame of Fig. 7.3.9.

The moments due to the uniform load, assuming $\alpha = 1.5$, are obtained from Table 7.1.2:

$$M_E = +0.007\,7 \times 30 \times 12^2 = 333 \text{ kN·m},$$

$$S = \frac{333 \times 1\,000^2}{165} = 2.018 \times 10^6 \text{ mm}^3.$$

Therefore, choose W530 × 92 (W21 × 62), with $S = 2.081 \times 10^6$ mm³, $I = 553.28 \times 10^6$ mm⁴. Similarly:

$$M_B = -0.048 \times 30 \times 12^2 = -207 \text{ kN·m},$$

$$S = \frac{207 \times 1\,000^2}{165} = 1.255 \times 10^6 \text{ mm}^3,$$

$$R_B = \frac{wl}{2} = \frac{30 \times 12}{2} = 180 \text{ kN}.$$

Therefore, choose W310 × 107 (W12 × 72) with $I = 248.35 \times 10^6$ mm⁴, $S = 1\,596 \times 10^6$ mm³, $A = 13\,680$ mm², $r = 134.9$ mm, so that:

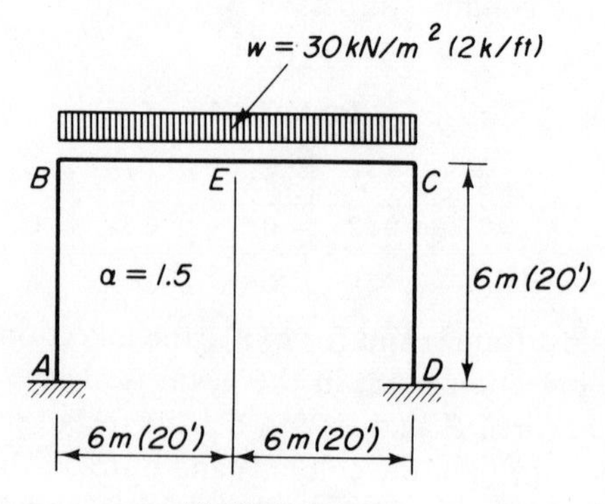

FIGURE 7.3.9

$$\frac{P}{A} = \frac{180 \times 1\,000}{13\,680} = 13.2 \text{ MPa},$$

$$\frac{M}{S} = \frac{207 \times 1\,000^2}{1.596 \times 10^6} = 129.7 \text{ MPa}.$$

Assuming lateral buckling is prevented, $h/r = (6 \times 1\,000)/134.9 = 45$, $F_a = 129.5$ MPa (from Fig. 5.4.1) and:

$$\frac{13.2}{129.5} + \frac{129.7}{165} = 0.89 < 1.$$

The combined stress is within that allowable and $\alpha = (6/12)(553.28/248.35) = 1.11$ is reasonably close to the assumed value of 1.5.

Example 7.3.5. The frame of Fig. 7.3.9 is one of a series spaced 6 m (20 ft) on centers and acted upon by a wind pressure of 1.5 kN/m² (30 psf) on the left side. Determine the additional wind moments and reactions and check whether they are within 33 % of those due to the vertical loads.

By the portal method the moments in the columns of a *fixed* multi-bay frame under a horizontal force P are:

$$M_i = \frac{Ph}{2(n-1)}, \qquad M_e = \frac{Ph}{4(n-1)}. \tag{7.3.11}$$

Therefore, with $P = 6 \times 1.5 \times 3 = 27$ kN, and since $\alpha > 1$:

$$M_B = \frac{Ph}{4} = \frac{27 \times 6}{4} = 40.5 \text{ kN·m} < 0.33 \times 207 = 68.3 \text{ kN·m},$$

$$R_A = \frac{Ph}{2l} = \frac{27 \times 6}{2 \times 12} = 6.75 \text{ kN} < 0.33 \times 180 = 59.4 \text{ kN}.$$

Both the moment and the reaction due to wind are less than 33 % of those due to the vertical loads; therefore, no change is required in the members.

Example 7.3.6. Determine the moments in the reinforced concrete frame of Fig. 7.3.10, one of a series 9 m (30 ft) on centers, acted upon by a wind pressure of 1 kN/m² (20 psf).

The load P acting at the top of the frame is:

$$P = 1 \times 2.25 \times 9 = 20.25 \text{ kN (4.5 k)}.$$

The thrust in each interior column is $P/3 = 6.75$ kN, and half that value in each exterior column. The resulting moments assuming an inflection point at midheight are, respectively:

$$M_C = M_E = 6.75 \times 2.25 = 15.19 \text{ kN·m (11.2 k·ft)},$$

$$M_A = M_G = (6.75/2) \times 2.25 = 7.59 \text{ kN·m (5.6 k·ft)}.$$

The moment diagram is shown for the right half of the frame.

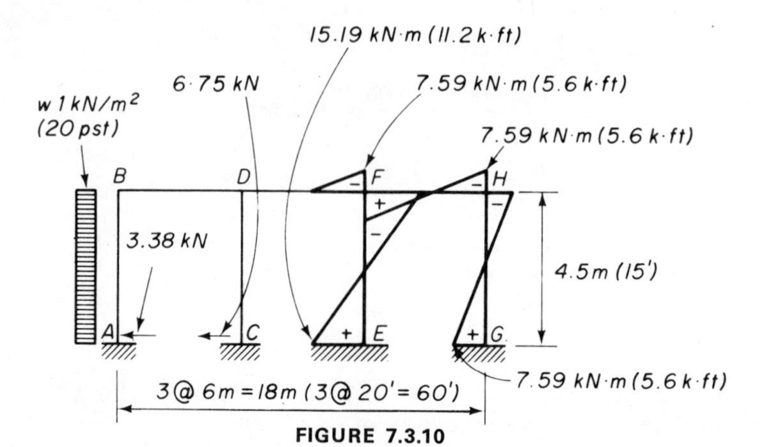

FIGURE 7.3.10

Example 7.3.7. Determine the wind moments in the concrete frame of Fig. 7.3.11, if this frame is one of a series 7.5 m (25 ft) on centers, acted upon by a wind pressure of 1.5 kN/m² (30 psf). The columns and beam cross sections are given in the figure.

We first note that for all practical purposes the stiffness of the beam FH is identical to that of beams BD and DF, since:

$$\left(\frac{I}{l}\right)_{DF} = \frac{300 \times 600^3/12}{6\,000} = 900\,000 \text{ mm}^3,$$

$$\left(\frac{I}{l}\right)_{FH} = \frac{300 \times 750^3/12}{12\,000} = 879\,000 \text{ mm}^3.$$

The total wind force acting at the top of the frame is:

$$P = 1.5 \times 7.5 \times \frac{7.5}{2} = 42.2 \text{ kN } (9.4 \text{ k}).$$

The wind moments in the columns are, by (7.3.11):

$$M_E = \frac{Ph}{2(n-1)} = \frac{42.2 \times 7.5}{2 \times 3} = 52.8 \text{ kN}\cdot\text{m } (39.2 \text{ k}\cdot\text{ft}),$$

$$M_G = \frac{Ph}{4(n-1)} = 26.4 \text{ kN}\cdot\text{m } (19.5 \text{ k}\cdot\text{ft}).$$

The final moments are shown for the right portion of the frame.

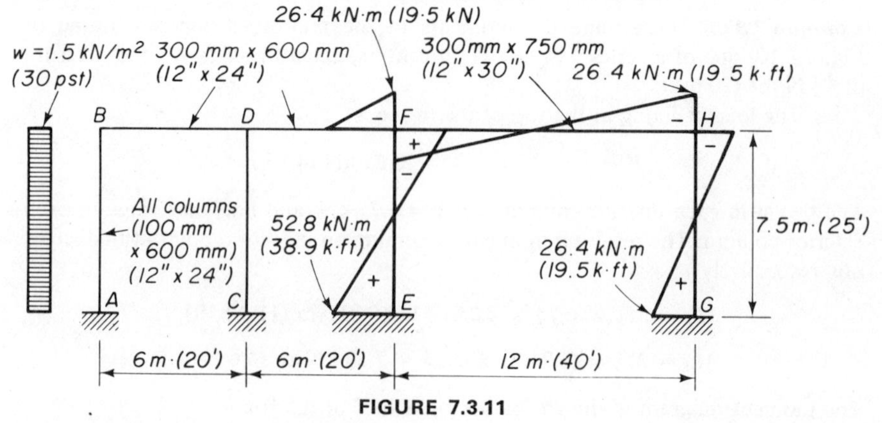

FIGURE 7.3.11

7.4 THE PREVENTION OF SIDESWAY [8.2]

The stresses set up in a frame by lateral forces are usually larger than those set up by vertical forces of equal intensity. For example, a total distributed load $W = wl$ on the beam of a hinged frame with $\alpha = 1$ sets up a maximum moment 0.075 Wl (Table 7.1.1), while the same load W acting laterally on the column and considered acting half at the top and half at the bottom of the column sets up, by (7.3.5), a maximum moment $(W/4)(h) = 0.25\ Wh$. When h is approximately equal to l, the horizontal force moment is about three times the moment due to the vertical force.

Frames may be economically stiffened against lateral forces by the introduction of diagonals [Fig. 7.4.1(a)]. Consider, for example, a hinged frame with $\alpha = 1$ and a sidesway given by (7.3.6):

$$\delta = \frac{2 + 1/\alpha}{12} \frac{Ph^3}{EI_1} = \frac{1}{4} \frac{Ph^3}{EI_1}, \tag{7.4.1}$$

[Fig. 7.4.1(b)]. Due to the sidesway δ, the diagonal distance d [Fig. 7.4.1(d)] increases by:

$$\delta' = \delta \sin \beta = \delta \frac{l}{d},$$

and the unit elongation or strain of AC becomes:

$$\epsilon = \frac{\delta'}{d} = \frac{\delta l}{d^2}, \tag{7.4.2}$$

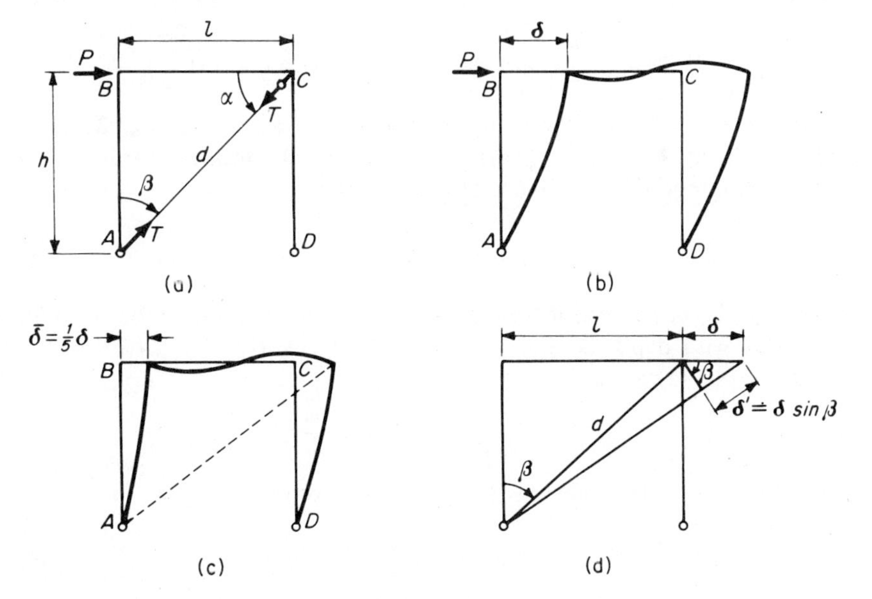

FIGURE 7.4.1

or, by (7.4.1):

$$\epsilon = \frac{1}{4}\frac{Phl}{EI_1}\left(\frac{h}{d}\right)^2. \tag{7.4.3}$$

Since, by (7.3.5), $M_B = Ph/2$, indicating by c the half-width of the column, by f its allowable bending stress, and by $\epsilon_c = f/E$, the allowable bending strain in the column, ϵ may be written as:

$$\epsilon = \frac{1}{2}\frac{M_B c}{EI_1}\left(\frac{l}{c}\right)\left(\frac{h}{d}\right)^2 = \frac{1}{2}\left(\frac{f}{E}\right)\left(\frac{l}{c}\right)\left(\frac{h}{d}\right)^2 = \frac{1}{2}\epsilon_c\left(\frac{l}{c}\right)\left(\frac{h}{d}\right)^2,$$

from which:

$$\frac{\epsilon}{\epsilon_c} = \frac{1}{2}\left(\frac{l}{c}\right)\left(\frac{h}{d}\right)^2 = \frac{1}{2}\left(\frac{h}{c}\right)\left(\frac{l}{h}\right)\frac{h^2}{h^2+l^2} = \frac{1}{2}\left(\frac{h}{c}\right)\frac{l/h}{1+(l/h)^2}. \tag{7.4.4}$$

For $h/c = 20$, $l/h = 1$, $\epsilon = 5\epsilon_c$, i.e., the strain in the diagonal direction is five times the allowable column strain.

If we now introduce in the frame a tensile hinged diagonal connecting A to C, which must develop a strain equal to ϵ_c, the distance AC will increase by $\epsilon_c d$ instead of $5\,\epsilon_c d$. The frame deformation is reduced by a factor of 5 and, consequently, the moments and the bending stresses are also reduced by the same factor.

A rough evaluation of the saving in material due to the diagonal may be obtained by noticing that the horizontal component of the tension T in the diagonal bar, $T\cos\alpha = Tl/\sqrt{l^2+h^2} = T/\sqrt{1+(h/l)^2}$ [Fig. 7.4.1(a)], must reduce the horizontal force P to $P/5$ in order to reduce δ to $\delta/5$:

$$P - T\cos\alpha = \frac{P}{5} \quad \therefore \quad T = \frac{4}{5}\frac{P}{\cos\alpha} = \frac{4}{5}\sqrt{1+\left(\frac{h}{l}\right)^2}\,P.$$

For $h = l$, $T = \sqrt{2}\,\frac{4}{5}P$, the required diagonal bar area is $A_d = \sqrt{2}\,\frac{4}{5}(P/f)$ and its volume $A_d d = \sqrt{2}\,\frac{4}{5}(P/f)\sqrt{2}\,l = \frac{8}{5}(Pl/f)$. On the other hand, the cross-section modulus required by the moment $Ph/2$ is:

$$S = \frac{Ph}{2f}. \tag{a}$$

Assuming the cross sections of the beam and the column to be ideal I's of area A with depth $2c$ and a flange area $A_F = A/2$, $I = 2A_F c^2 = Ac^2$, $S = I/c = Ac$, (a) gives:

$$Ac = \frac{Ph}{2f} \quad \therefore \quad A = \frac{P}{f}\frac{h}{2c}. \tag{b}$$

The volume of the frame bars without diagonal for $h/2c = 10$ is:

$$V = A(2h+l) = 3\frac{h}{2c}\frac{Pl}{f} = 30\frac{Pl}{f}. \tag{c}$$

With P reduced to $P/5$, the volume of the frame bars plus the diagonal is:

$$V' = 3\left(\frac{h}{2c}\right)\frac{(P/5)l}{f} + \frac{8}{5}\frac{Pl}{f} = \frac{38}{5}\frac{Pl}{f} = 7.6\frac{Pl}{f}.$$

Thus, the use of a diagonal bar results in a stiffening of the frame by a factor of 5, with a saving of about 75% in the amount of material used.

This result is of wide significance: triangulated frames are always stiffer and lighter than rectangular frames.

7.5 THE APPROXIMATE ANALYSIS OF MULTISTORY FRAMES [8.3]

A. *Vertical Loads.* The *interior* bays of multistory frames may be analyzed for preliminary design by assuming that hinges form in the beams at a distance $0.2l$ from the ends [Fig. 7.5.1(a)]. This assumption implies that the maximum positive and negative moments [Fig. 7.5.1(c)] equal:

$$M_{\max} = \frac{w(1-0.4)^2l^2}{8} = 0.045wl^2,$$
$$M_{\min} = -w \times 0.3l \times 0.2l - \tfrac{1}{2}w(0.2l)^2 = -0.08wl^2, \tag{7.5.1}$$

which should be compared with $0.042wl^2$ and $-0.084wl^2$ for a fixed beam. The internal columns may be considered momentless, provided the stress be maintained low, say at 70% to 80% of allowable.

For *exterior* bays the hinge points may be assumed at $0.1l$ from the exterior column and at $0.2l$ from the interior column [Fig. 7.5.1(b)], with moments [Fig. 7.5.1(d)]:

$$M_{\max} = \frac{w(0.7l)^2}{8} = 0.061wl^2,$$

$$M_{\min,e} = -w \times 0.35l \times 0.1l - \frac{w(0.1l)^2}{2} = -0.040wl^2, \tag{7.5.2}$$

$$M_{\min,i} = -w \times 0.35l \times 0.2l - \frac{w(0.2l)^2}{2} = -0.09wl^2,$$

which should be compared with $0.07wl^2$ and $-0.125wl^2$ for a beam fixed at one end and simply supported at the other.

B. *Horizontal Loads.* Indicating the spacing of the frames in a building by b, the floor height by h, and the wind pressure by p, the wind force per floor on each frame equals:

$$W = pbh, \tag{7.5.3}$$

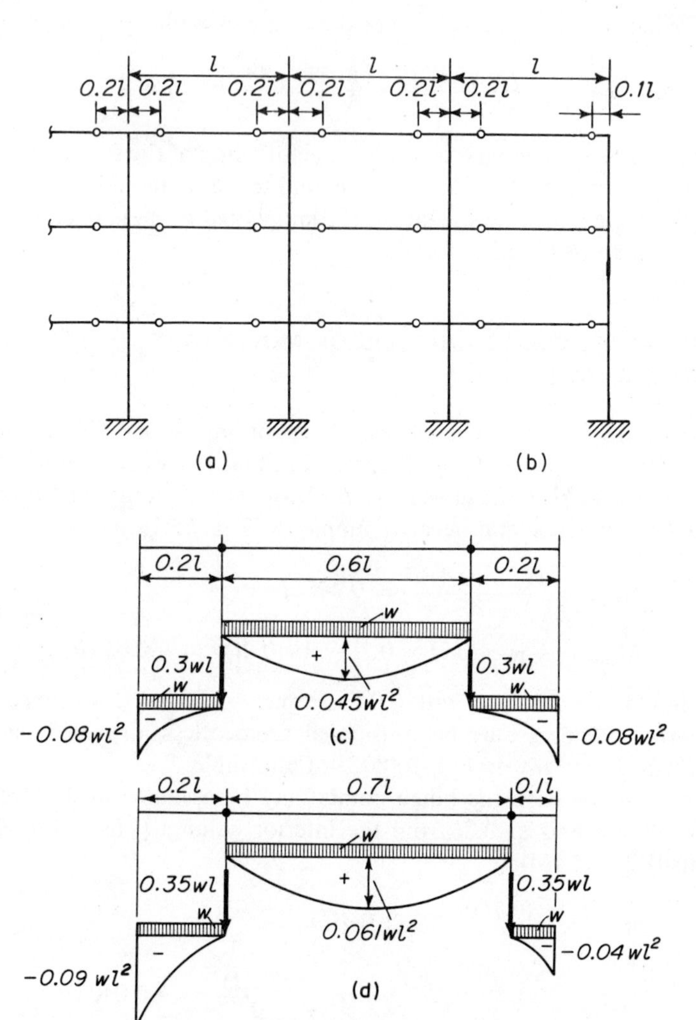

FIGURE 7.5.1

while the top-floor force is $W/2$ (Fig. 7.5.2). The frame analysis by the so-called *cantilever method* (which is applicable to buildings with height-to-width ratios of between 1 and 5) is obtained by assuming (as was done in Fig. 7.3.8 for a simple frame) that (a) inflection points, i.e., hinges appear at midspan of each beam and at midheight of each column, and (b) the column direct stresses vary linearly, i.e., in proportion to their distance from the frame centroidal axis (Fig. 7.5.2). (For columns of equal area, condition (b) states that the column axial forces vary linearly also.) Under these assump-

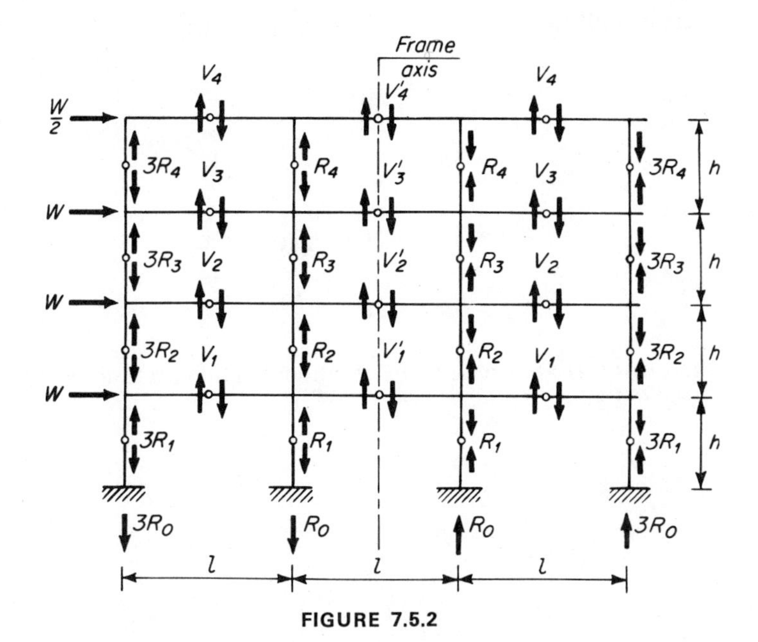

FIGURE 7.5.2

tions, the frame is statically determinate and the direct forces R, the shears V, and the moments M are determined by equilibrium considerations, as will be shown for the frame of Fig. 7.5.2 that has columns of equal area.

By rotational equilibrium, taking moments about the frame axis at the level of the column hinges, we obtain the forces R in the columns:

Level 4:

$$\frac{W}{2}\frac{h}{2} = 2\left(R_4\frac{l}{2} + 3R_4\frac{3l}{2}\right) \quad \therefore \quad R_4 = \frac{1}{40}\left(\frac{h}{l}\right)W,$$

Level 3:

$$\frac{W}{2}\frac{3h}{2} + W\frac{h}{2} = 2\left(R_3\frac{l}{2} + 3R_3\frac{3l}{2}\right) \quad \therefore \quad R_3 = \frac{5}{40}\left(\frac{h}{l}\right)W,$$

Level 2:

$$\frac{W}{2}\frac{5h}{2} + W\frac{3h}{2} + W\frac{h}{2} = 2\left(R_2\frac{l}{2} + 3R_2\frac{3l}{2}\right) \quad \therefore \quad R_2 = \frac{13}{40}\left(\frac{h}{l}\right)W,$$

Level 1:

$$\frac{W}{2}\frac{7h}{2} + W\frac{5h}{2} + W\frac{3h}{2} + W\frac{h}{2} = 2\left(R_1\frac{l}{2} + 3R_1\frac{3l}{2}\right) \quad \therefore \quad R_1 = \frac{25}{40}\left(\frac{h}{l}\right)W$$

By vertical equilibrium at level 0:

$$R_0 = R_1.$$

By vertical equilibrium (Fig. 7.5.3), the shears V equal:

$$V_4 = 3R_4 = \frac{3}{40}\left(\frac{h}{l}\right)W,$$

$$V_3 = 3R_3 - 3R_4 = \frac{12}{40}\left(\frac{h}{l}\right)W,$$

$$V_2 = 3R_2 - 3R_3 = \frac{24}{40}\left(\frac{h}{l}\right)W,$$

$$V_1 = 3R_1 - 3R_2 = \frac{36}{40}\left(\frac{h}{l}\right)W.$$

The thrusts H are obtained by rotational equilibrium about the beam hinges (Fig. 7.5.3):

$$H_4\frac{h}{2} - 3R_4\frac{l}{2} = 0 \quad \therefore \quad H_4 = 3\left(\frac{l}{h}\right)R_4 = \frac{3}{40}W,$$

$$H_3\frac{h}{2} + H_4\frac{h}{2} + 3R_4\frac{l}{2} - 3R_3\frac{l}{2} = 0$$

$$\therefore \quad H_3 = 3\left(\frac{l}{h}\right)(R_3 - R_4) - H_4 = \frac{9}{40}W,$$

$$H_2\frac{h}{2} + H_3\frac{h}{2} + 3R_3\frac{l}{2} - 3R_2\frac{l}{2} = 0$$

$$\therefore \quad H_2 = 3\left(\frac{l}{h}\right)(R_2 - R_3) - H_3 = \frac{15}{40}W,$$

$$H_1\frac{h}{2} + H_2\frac{h}{2} + 3R_2\frac{l}{2} - 3R_1\frac{l}{2} = 0$$

$$\therefore \quad H_1 = 3\left(\frac{l}{h}\right)(R_1 - R_2) - H_2 = \frac{21}{40}W,$$

and by horizontal equilibrium (Fig. 7.5.3):

$$H_0 - H_1 - \frac{W}{2} = 0 \quad \therefore \quad H_0 = H_1 + \frac{W}{2} = \frac{41}{40}W.$$

The thrusts H' are obtained by horizontal equilibrium from Fig. 7.5.3:

$$H'_4 + H_4 - \frac{W}{2} = 0 \qquad \therefore \quad H'_4 = \frac{W}{2} - H_4 = \frac{17}{40}W,$$

$$H'_3 + H_3 - H_4 - W = 0 \quad \therefore \quad H'_3 = W - H_3 + H_4 = \frac{34}{40}W,$$

$$H'_2 + H_2 - H_3 - W = 0 \quad \therefore \quad H'_2 = W - H_2 + H_3 = \frac{34}{40}W,$$

$$H'_1 + H_1 - H_2 - W = 0 \quad \therefore \quad H'_1 = W - H_1 + H_2 = \frac{34}{40}W.$$

The moments M, M', and M'' at the beam ends, at the column bottom, and at the column top, respectively, are obtained by rotational equilibrium of the corresponding elements (Fig. 7.5.3):

$$M_4 = V_4\frac{l}{2} = \frac{3}{80}Wh, \qquad M''_4 = H_4\frac{h}{2} = \frac{3}{80}Wh,$$

$$M_3 = V_3\frac{l}{2} = \frac{12}{80}Wh, \quad M'_3 = H_4\frac{h}{2} = \frac{3}{80}Wh, \quad M''_3 = H_3\frac{h}{2} = \frac{9}{80}Wh,$$

$$M_2 = V_2\frac{l}{2} = \frac{24}{80}Wh, \quad M'_2 = H_3\frac{h}{2} = \frac{9}{80}Wh, \quad M''_2 = H_2\frac{h}{2} = \frac{15}{80}Wh,$$

$$M_1 = V_1\frac{l}{2} = \frac{36}{80}Wh, \quad M'_1 = H_2\frac{h}{2} = \frac{15}{80}Wh, \quad M''_1 = H_1\frac{h}{2} = \frac{21}{80}Wh,$$

$$M'_0 = H_1\frac{h}{2} = \frac{21}{80}Wh.$$

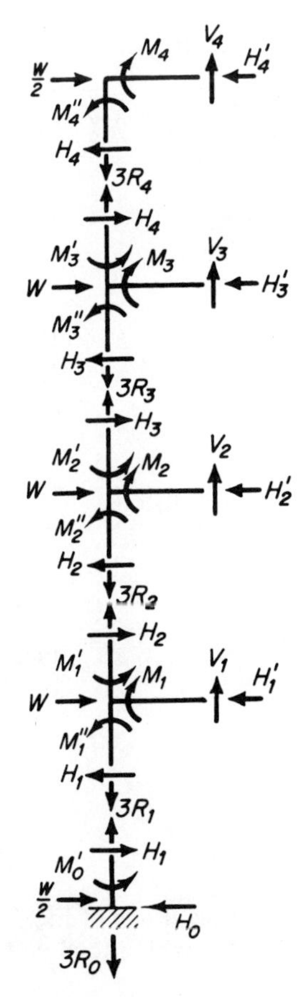

FIGURE 7.5.3

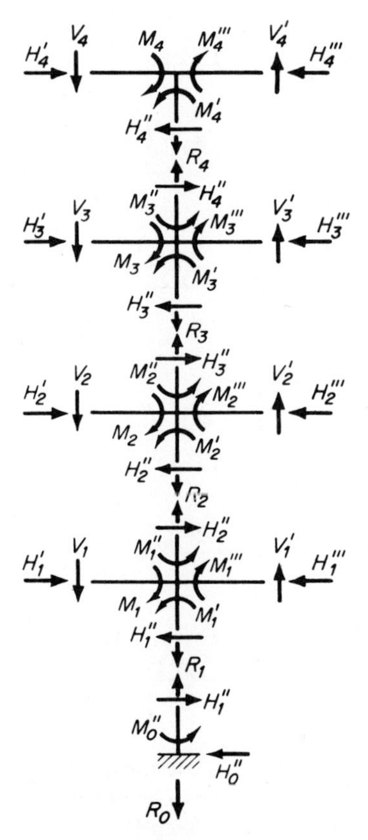

FIGURE 7.5.4

Similarly, from Fig. 7.5.4, the column forces, the shears, and the moments in the interior columns become:

$$V_4 - V_4' + R_4 = 0 \qquad \therefore \quad V_4' = \frac{4}{40}\left(\frac{h}{l}\right)W,$$

$$V_3 - V_3' + R_3 - R_4 = 0 \qquad \therefore \quad V_3' = \frac{16}{40}\left(\frac{h}{l}\right)W,$$

$$V_2 - V_2' + R_2 - R_3 = 0 \qquad \therefore \quad V_2' = \frac{32}{40}\left(\frac{h}{l}\right)W,$$

$$V_1 - V_1' + R_1 - R_2 = 0 \qquad \therefore \quad V_1' = \frac{48}{40}\left(\frac{h}{l}\right)W,$$

$$H_4''\frac{h}{2} - V_4\frac{l}{2} - V_4'\frac{l}{2} = 0 \qquad \therefore \quad H_4'' = \frac{7}{40}W,$$

$$H_3''\frac{h}{2} + H_4''\frac{h}{2} - V_3\frac{l}{2} - V_3'\frac{l}{2} = 0 \quad \therefore \quad H_3'' = \frac{21}{40}W,$$

$$H_2''\frac{h}{2} + H_3''\frac{h}{2} - V_2\frac{l}{2} - V_2'\frac{l}{2} = 0 \quad \therefore \quad H_2'' = \frac{35}{40}W,$$

$$H_1''\frac{h}{2} + H_2''\frac{h}{2} - V_1\frac{l}{2} - V_1'\frac{l}{2} = 0 \quad \therefore \quad H_1'' = \frac{49}{50}W; \quad H_0'' = H_1'',$$

$$M_4 = V_4\frac{l}{2} = \frac{3}{80}Wh, \qquad M_4' = H_4''\frac{h}{2} = \frac{7}{80}Wh,$$

$$M_4''' = V_4'\frac{l}{2} = \frac{4}{80}Wh,$$

$$M_3 = V_3\frac{l}{2} = \frac{12}{80}Wh, \quad M_3'' = H_4''\frac{h}{2} = \frac{7}{80}Wh, \quad M_3' = H_3''\frac{h}{2} = \frac{21}{80}Wh,$$

$$M_3''' = V_3'\frac{l}{2} = \frac{16}{80}Wh,$$

$$M_2 = V_2\frac{l}{2} = \frac{24}{80}Wh, \quad M_2'' = H_3''\frac{h}{2} = \frac{21}{80}Wh, \quad M_2' = H_2''\frac{h}{2} = \frac{35}{80}Wh,$$

$$M_2''' = V_2'\frac{l}{2} = \frac{32}{80}Wh,$$

$$M_1 = V_1\frac{l}{2} = \frac{36}{80}Wh, \quad M_1'' = H_2''\frac{h}{2} = \frac{35}{80}Wh, \quad M_1' = H_1''\frac{h}{2} = \frac{49}{80}Wh,$$

$$M_1''' = V_1'\frac{l}{2} = \frac{48}{80}Wh,$$

$$M_0'' = H_1''\frac{h}{2} = \frac{49}{80}Wh.$$

Example 7.5.1. Determine the vertical load moments in the frame of Fig. 7.5.5, assuming the frame to be one of a set 6 m (20 ft) on centers with a total uniform load of 10 kN/m² (200 psf).

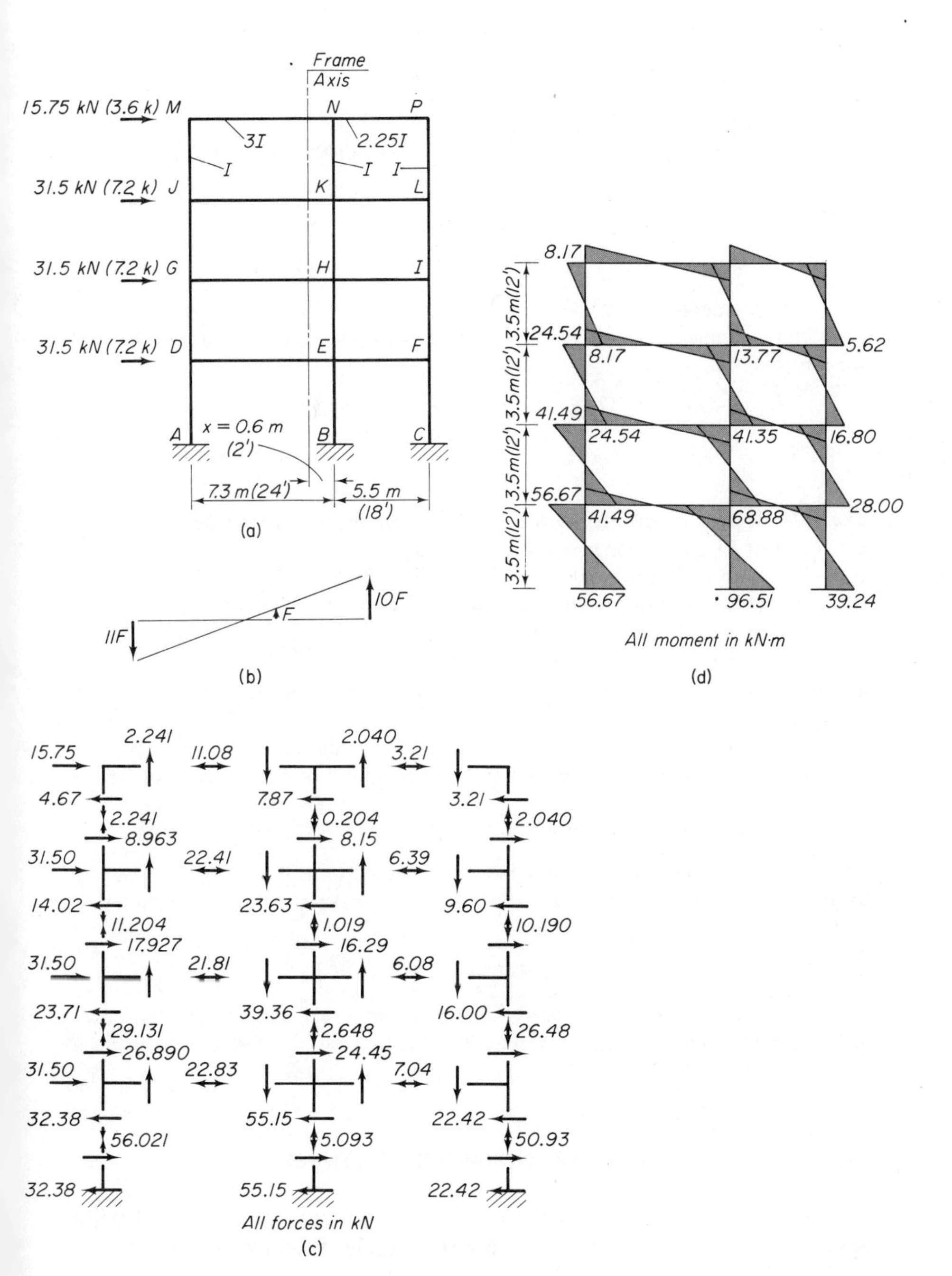

Frame
Axis
15.75 kN (3.6 k) M
N
P
3I
2.25I
I
I
I
31.5 kN (7.2 k) J
K
L
31.5 kN (7.2 k) G
H
I
31.5 kN (7.2 k) D
E
F
A
x = 0.6 m
(2')
B
C
7.3 m(24')
5.5 m
(18')
(a)
11F
F
10F
(b)
8.17
3.5 m(12')
24.54
8.17
13.77
5.62
3.5 m(12')
41.49
24.54
41.35
16.80
3.5 m(12')
56.67
41.49
68.88
28.00
3.5 m(12')
56.67
96.51
39.24
All moment in kN·m
(d)
15.75
2.241
11.08
2.040
3.21
4.67
7.87
3.21
2.241
0.204
2.040
8.963
8.15
31.50
22.41
6.39
14.02
23.63
9.60
11.204
1.019
10.190
17.927
16.29
31.50
21.81
6.08
23.71
39.36
16.00
29.131
2.648
26.48
26.890
24.45
31.50
22.83
7.04
32.38
55.15
22.42
56.021
5.093
50.93
32.38
55.15
22.42
All forces in kN
(c)

FIGURE 7.5.5

By (7.5.2) we obtain, for example:

$$wl_{MN}^2 = 10 \times 6 \times (7.3)^2 = 3\,197 \text{ kN}\cdot\text{m},$$
$$wl_{NP}^2 = 10 \times 6 \times (5.5)^2 = 1\,815 \text{ kN}\cdot\text{m},$$
$$M_{MN} = M_{MJ} = -0.04 \times 3\,197 = -128 \text{ kN}\cdot\text{m } (-92 \text{ ft}\cdot\text{k}),$$
$$M_{PN} = M_{PL} = -0.04 \times 1\,815 = -73 \text{ kN}\cdot\text{m } (-52 \text{ ft}\cdot\text{k}),$$
$$M_{NM} = -0.09 \times 3\,197 = -288 \text{ kN}\cdot\text{m } (-207 \text{ ft}\cdot\text{k}),$$
$$M_{NP} = -0.09 \times 1\,815 = -163 \text{ kN}\cdot\text{m } (-117 \text{ ft}\cdot\text{k}).$$

The moments in the interior columns may be assumed to be the difference between the moments of adjacent beams at the top floor, and one-half the difference for a typical floor:

$$M_{NK} = M_{NM} - M_{NP} = -288 + 163 = -125 \text{ kN}\cdot\text{m } (-92 \text{ ft}\cdot\text{k}),$$
$$M_{KH} = \tfrac{1}{2}(M_{KJ} - M_{KL}) = \tfrac{1}{2}(-288 + 163) \simeq -63 \text{ kN}\cdot\text{m } (-46 \text{ ft}\cdot\text{k}).$$

Example 7.5.2. Determine by the cantilever method the wind load moments in the frame of Fig. 7.5.5 for $p = 1.5$ kN/m² (30 psf).

The frame axis is located at the center of gravity of all the columns; therefore, taking moments about B and assuming the area of all columns to be equal, we have:

$$3x = 1 \times 7.3 - 1 \times 5.5 \quad \therefore \quad x = 0.6 \text{ m}.$$

The forces in kN (kips) resulting from the wind load are shown on the left of Fig. 7.5.5 (a).

The distribution of column reactions based on the cantilever assumption is shown in Fig. 7.5.5 (b), and the moment of the column reactions about the frame axis is:

$$F \times 0.6 + 10F \times 6.1 + 11F \times 6.7 = 135.3F.$$

The column reactions for the top story are obtained by equating this moment to the moment of the wind forces taken about a horizontal plane through the assumed hinges of the top floor [Fig. 7.5.5(c)]:

$$135.3F = 15.75 \times 1.75, \quad F = 0.2040 \text{ kN}, \quad 10F = 2.04 \text{ kN}, \quad 11F = 2.241 \text{ kN}.$$

The same procedure is followed for the other levels, with the results shown in Fig. 7.5.5 (c). The moments are obtained directly from the shears and are given for the columns in Fig. 7.5.5 (d).

7.6 LATERAL STIFFNESS OF FRAMED STRUCTURES

The lateral deflection of the top of a frame under the action of wind loads, called the *wind drift*, may be approximately evaluated by the assumpti ons of the cantilever method.

If the floor beams were infinitely rigid, the lateral deflection δ_i' of each column would be the sum of the deflections of two cantilever beams of length $h/2$ under the action of the wind shears F_i [Fig. 7.6.1(a)]:

$$\delta_i' = 2\frac{1}{3}\frac{F_i(h/2)^3}{EI_1} = \frac{1}{12}\frac{F_ih^3}{EI_1}. \tag{a}$$

The vertical shears V_i [Fig. 7.6.1(b)] may be approximated by:

$$V_il = (F_i + F_{i+1})\frac{h}{2} \quad \therefore \quad V_i = \frac{h}{2l}(F_i + F_{i+1}) \approx \frac{h}{l}F_i,$$

when the number of stories is high and F_{i+1} does not differ substantially from F_i; hence, the antisymmetrical moments M_i at the beam ends are given by:

$$M_i = V_i\frac{l}{2} = F_i\frac{h}{2}.$$

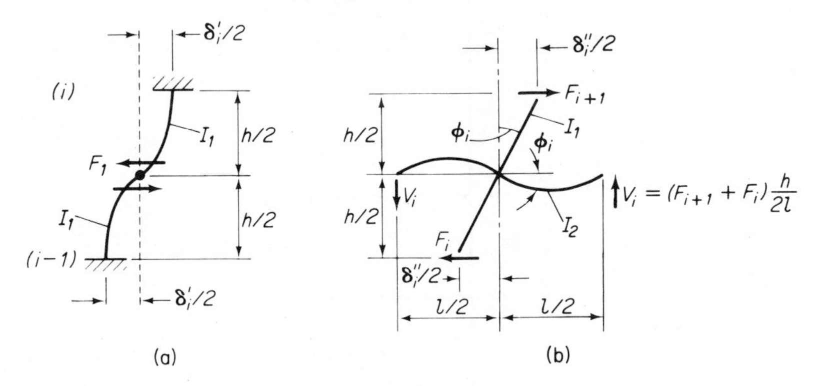

FIGURE 7.6.1

Under the moments M_i the rotation ϕ_i at the beam ends is given by Table 4.3.1 (g):

$$\phi_i = \frac{1}{3}\frac{M_il/2}{EI_2} = \frac{1}{12}\frac{F_ihl}{EI_2}.$$

If the columns were infinitely rigid, the rotation ϕ_i would produce a relative displacement of the hinges [Fig. 7.6.1(b)]:

$$\delta_i'' = \phi_ih = \frac{1}{12}\frac{F_ih^2l}{EI_2} = \left(\frac{l}{h}\frac{I_1}{I_2}\right)\frac{1}{12}\frac{F_ih^3}{EI_1} \equiv \beta\delta_i' \tag{b}$$

where:

$$\beta = \frac{l}{h}\frac{I_1}{I_2} = \frac{1}{\alpha}. \tag{7.6.1}$$

The total displacement of the ith floor relative to the $(i - 1)$st floor is thus [Fig. 7.6.1(b)]:

$$\delta_i = \delta_i' + \delta_i'' = \delta_i'\,(1 + \beta). \tag{7.6.2}$$

For the frame of Fig. 7.6.2 with wind loads $W = pbh$, where b is the frame spacing, and $m + 1$ identical columns and m identical beams, we obtain, approximately:

$$\delta'_4 = \frac{1}{12}\frac{W}{2m}\frac{h^3}{EI_1}, \quad \delta'_3 = \frac{1}{12}\frac{3W}{2m}\frac{h^3}{EI_1}, \quad \delta'_2 = \frac{1}{12}\frac{5W}{2m}\frac{h^3}{EI_1}, \quad \delta'_1 = \frac{1}{12}\frac{7W}{2m}\frac{h^3}{EI_1},$$

and, by (7.6.1) and (7.6.2):

$$\delta_4 = \frac{1}{12}\frac{W}{2m}\frac{h^3}{EI_1}(1 + \beta), \quad \delta_3 = \frac{1}{12}\frac{3W}{2m}\frac{h^3}{EI_1}(1 + \beta),$$

$$\delta_2 = \frac{1}{12}\frac{5W}{2m}\frac{h^3}{EI_1}(1 + \beta), \quad \delta_1 = \frac{1}{12}\frac{7W}{2m}\frac{h^3}{EI_1}(1 + \beta).$$

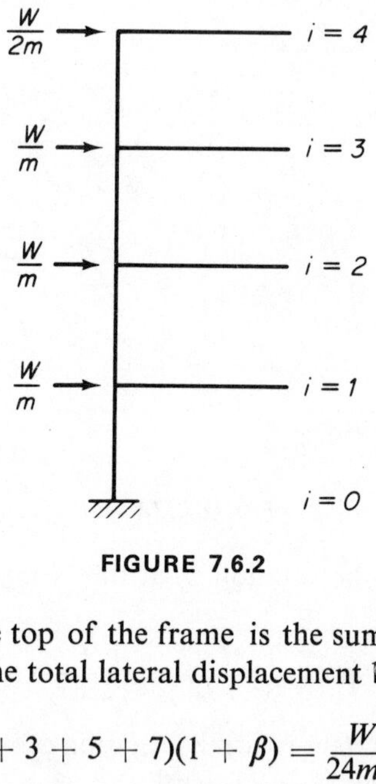

FIGURE 7.6.2

The displacement at the top of the frame is the sum of all the relative displacements δ_i, so that the total lateral displacement becomes:

$$\delta = \frac{Wh^3}{24mEI_1}(1 + 3 + 5 + 7)(1 + \beta) = \frac{Wh^3}{24mEI_1}16(1 + \beta). \quad (7.6.3)$$

For a building with n stories, noticing that the sum of the first n odd integers equals n^2, we obtain:

$$\delta = \frac{Wh^3(1 + \beta)}{24mEI_1}n^2. \quad (7.6.4)$$

Example 7.6.1. Evaluate the lateral deflection at the top of a 13-story, 4-bay concrete building with a floor height h of 3.5 m (12 ft), if its frames are 6 m (20 ft) on centers, the wind pressure is 1.5 kN/m^2 (30 psf), all the concrete columns have an average cross section of 600 × 300 mm (2 ft × 1 ft), and the ratio $\beta = (l/h)(I_1/I_2) = 1$.

With:

$$W = pbh = 1.5 \times 6 \times 3.5 = 31.5 \text{ kN}, \quad n = 13, m = 4,$$

$$I = \tfrac{1}{12} \times 300 \times 600^3 = 5\,400 \times 10^6 \text{ mm}^4.$$

(7.6.4) gives:

$$\delta = \frac{31.5 \times 1\,000(3\,500)^3 13^2(1 + 1)}{24 \times 4 \times 21\,700 \times 5\,400 \times 10^6}$$

$$= 40.6 \text{ mm (1.59 in.)}.$$

The ratio of building height to wind drift is:

$$\frac{H}{\delta} = \frac{13 \times 3\,500}{40.6} = 1\,121.$$

Since this ratio should range between 500 and 1 000, it is acceptable on the conservative side.

A more accurate evaluation of the wind drift may be obtained by locating the column inflection point according to (7.3.10), where $\alpha = 1/\beta$, and by assuming that the I_1 of the columns varies linearly from a value $I_{1,0}$ at the bottom floor to a value $I_{1,n}$ at the top floor and letting:

$$\alpha = \frac{I_{1,n}}{I_{1,0}}, \qquad \beta_0 = \frac{l}{h}\frac{I_{1,0}}{I_2}. \tag{7.6.5}$$

The resulting wind drift may be shown to be given by:

$$\delta = k(\alpha, \beta_0)\frac{Wh^3}{mEI_{1,0}}n^2. \tag{7.6.6}$$

Table 7.6.1 gives the coefficient k as a function of α and β_0.

Example 7.6.2. For the frame of Example 7.6.1 assume the columns of the top floor to be 300 × 300 mm and $\beta_0 = 1$. Determine the wind drift, referring to Table 7.6.1.

$$I_{1,n} = \tfrac{1}{12}(300 \times 300^3) = 675 \times 10^6 \text{ mm}^4,$$

$$I_{1,0} = \tfrac{1}{12}(300 \times 600^3) = 5\,400 \times 10^6 \text{ mm}^4,$$

$$\alpha = \frac{I_{1,n}}{I_{1,0}} = \frac{675}{5\,400} = 0.125.$$

Interpolating linearly from Table 7.6.1 for $\beta_0 = 1$, we obtain:

$$k = 0.0876 + 0.25(0.080\,8 - 0.087\,6) = 0.085\,9,$$

$$\delta = 0.085\,9\,\frac{31.5 \times 1\,000(3.5 \times 1\,000)^3}{4(21\,700)(5\,400 \times 10^6)}(13^2) = 41.8 \text{ mm (1.65 in.)},$$

$$\frac{H}{\delta} = \frac{13 \times 3\,500}{41.8} = 1\,088.$$

The ratio of building height to wind drift is still conservatively acceptable.

Table 7.6.1

$k(\alpha, \beta_0)$

α \ β_0	0.00	0.25	0.50	0.75	1.00	2.00	3.00	4.00	5.00	10.00
0.0	0.083 3	0.088 4	0.093 2	0.097 7	0.102 1	0.117 5	0.130 6	0.141 7	0.151 4	0.186 2
0.1	0.068 9	0.073 9	0.078 7	0.083 2	0.087 5	0.102 8	0.115 5	0.126 2	0.135 5	0.168 0
0.2	0.062 2	0.067 3	0.072 0	0.076 5	0.080 8	0.095 8	0.108 2	0.118 6	0.127 5	0.158 1
0.3	0.057 6	0.062 6	0.067 4	0.071 8	0.076 1	0.090 8	0.102 9	0.113 0	0.121 6	0.150 5
0.4	0.054 0	0.059 0	0.063 8	0.068 2	0.072 4	0.086 9	0.098 8	0.108 6	0.116 8	0.144 3
0.5	0.051 1	0.056 1	0.060 8	0.065 2	0.069 4	0.083 7	0.095 3	0.104 8	0.112 8	0.138 9
0.6	0.048 7	0.053 7	0.058 4	0.062 7	0.066 9	0.081 0	0.092 3	0.101 6	0.109 3	0.134 2
0.7	0.046 5	0.051 6	0.056 2	0.060 6	0.064 7	0.078 6	0.089 7	0.098 7	0.106 1	0.130 1
0.8	0.044 7	0.049 7	0.054 4	0.058 7	0.062 7	0.076 5	0.087 3	0.096 1	0.103 3	0.126 3
0.9	0.043 1	0.048 2	0.052 7	0.057 0	0.061 0	0.074 6	0.085 2	0.093 8	0.100 8	0.122 9
1.0	0.041 6	0.041 8	0.042 4	0.043 2	0.044 2	0.049 4	0.055 5	0.061 6	0.067 4	0.090 4

Framed buildings are often stiffened laterally by so-called *shear walls.* These walls are solid, thin walls enclosing the core of the building. Their stiffness is so much greater than the lateral stiffness of the frame that, usually, they resist between 50% and 90% of the wind load. Vertical trusses provide similar lateral stiffness, as shown in Section 7.4.

Example 7.6.3. The building of Example 7.6.1 consists of five parallel frames spaced 6 m (20 ft) on centers. Determine the deflection of the top of the building if two shear walls 12 m (40 ft) deep and 300 mm (1 ft) thick carry the wind load together with the frames.

The wind load per meter of height of the building is $1.5 \times 4 \times 6 = 36$ kN/m. Assume each shear wall to carry half of 80% of the load or 14.4 kN/m. If the shear walls are built-in into a rigid foundation, they act as cantilever beams and their tip deflection equals:

$$\delta = \frac{1}{8}\frac{wl^4}{EI} = \frac{1}{8}\frac{(14.4)(13 \times 3.5 \times 1\,000)}{21\,700(\frac{1}{12} \times 300 \times 12\,000^3)} = 6.61 \text{ mm (0.26 in.)}.$$

The actual proportion of the wind load carried by the shear walls is obtained approximately by considering the relative stiffness of the walls and the frames. Each frame of Example 7.6.1 carries a load $w = 9$ kN/m and deflects at the top by $\delta = 40.6$ mm. Their relative stiffness is $w/\delta = 9/40.6 = 0.222$ kN/m/mm. The relative stiffness of the shear walls is $14.4/6.61 = 2.178$ kN/m/mm. Since the building has two walls and five frames, the portion of the wind load carried by the walls is:

$$\frac{2 \times 2.178}{2 \times 2.178 + 5 \times 0.222} = 0.797 \simeq 80\%.$$

If the shear walls' moment of inertia varies linearly from I at the bottom to zero at the top:

$$\delta = \frac{1}{6}\frac{wl^4}{EI}.$$

In this case, the shear walls carry only 71% of the wind load.

7.7 VIERENDEEL TRUSSES

Vierendeel trusses (Fig. 7.7.1) are horizontal multi-bay frames well suited to span large distances while leaving unimpeded passage through their bays. They may be used to support an entire building on a few columns or to hang a number of floors of a building from its roof. Vierendeel trusses are often used in bridge design.

The analysis and design of a Vierendeel truss may be carried out approximately by the cantilever method.

The direct forces in the upper and lower chords of the truss are obtained by dividing the simple-beam moment at any section by the truss depth d (Fig. 7.7.1). The direct forces in the verticals are half the difference between

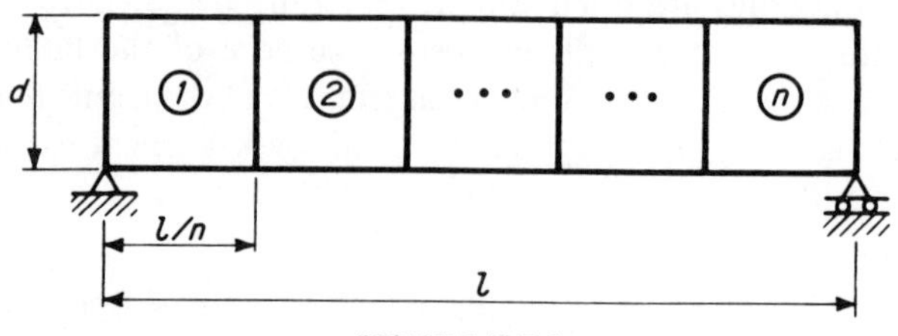

FIGURE 7.7.1

the shears at the midpoint hinges right and left of the verticals. The bending moments in the beams and verticals are obtained by assuming inflection points at the midpoint of each element. This simplified analysis gives reliable results for beams and verticals of approximately the same size.

Example 7.7.1. A Vierendeel steel truss is used to support on corner columns a square building 30 m (100 ft) on a side, 6 stories high, with a core 12 m (40 ft) square [Fig. 7.7.2(a)]. The truss consists of ten panels 3 m × 3.5 m (10 ft × 12 ft) each, with the chords at the level of the first and second floors [Fig. 7.7.2(b)]. The dead load and live load per floor are 9.72 kN/m² (200 psf). Determine the dimensions of the essential truss elements.

The contributory area per meter of truss *per floor* is approximately (30 − 12)/2 = 9 m, so that the load on the truss carrying 6 floors is 6 × 9 × 9.72 = 525 kN/m. The reactions are 525 × 15 = 7 875 kN. The maximum simple-beam moment is $M = wl^2/8 = 525 \times 30^2/8 = 59\,060$ kN·m. The maximum tension and compression in the chords is 59 060/3.5 = 16 875 kN. The shear in panel 1 is $V_1 = 7\,875 - 525 \times 1.5 = 7\,088$ kN and the moment in the chords of panel 1 is (7 088/2)1.5 = 5 315 kN·m. The moment in the left vertical of panel 1 is also 5 315 kN·m; in the right, 9 450 kN·m. The shears in panel 5 equal 7 875 − 13.5 × 525 = 788 kN, and the moments (788/2) × 1.5 = 590 kN·m.

Hence, the support verticals must be designed for a compressive force of 7 875 − 3 544 = 4 331 kN and a moment of 5 315 kN·m and the midspan chords for a direct force of 16 875 kN and a moment of 590 kN·m. Any other vertical carries half the load on one panel, or 525 × 1.5 = 788 kN·m, and moments smaller than 9 450 kN·m.

Considering each element as an ideal I beam, the elements must have an area A and a depth h such that, with an allowable stress f [see (5.1.1)]:

$$f = \frac{P}{A} + \frac{M}{S} = \frac{P}{A} + \frac{M}{Ah/2}. \tag{a}$$

For the chord at midspan, (a) gives, with $h = 600$ mm (24 in.) and $f = 165$ MPa (20 ksi):

$$165 = \frac{16\,875\,000}{A_c} + \frac{590 \times 1\,000^2}{300A_c}, \quad 165A_c = 18\,842\,000$$

$$\therefore \quad A_c = 114\,200 \text{ mm}^2 \text{ (177 sq in.)}.$$

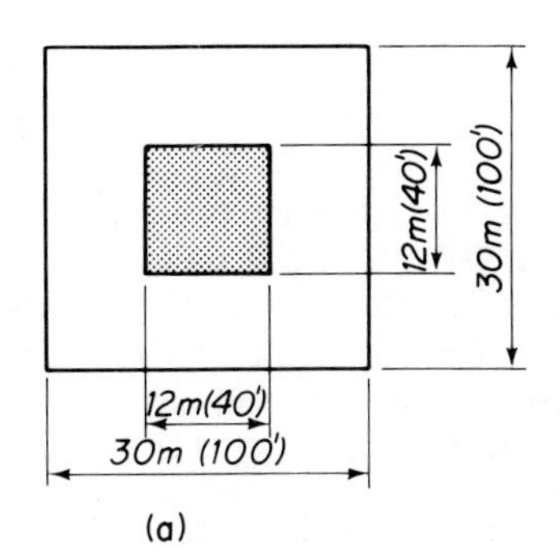

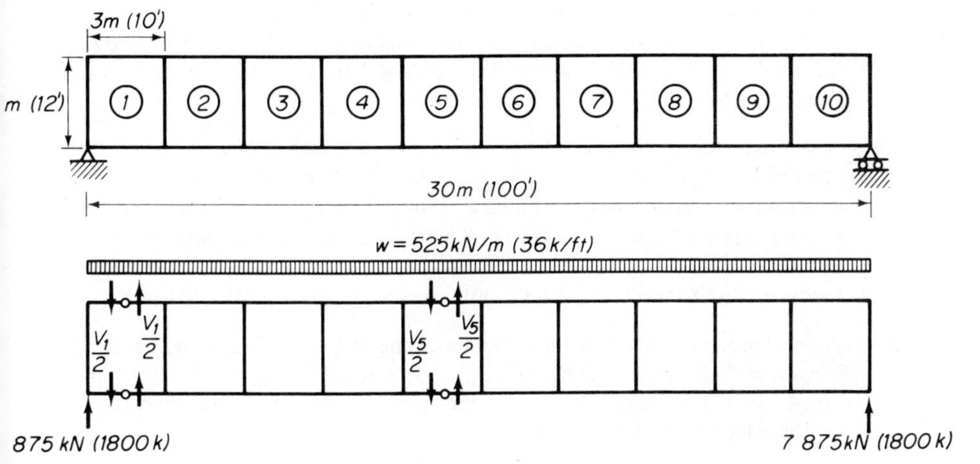

FIGURE 7.7.2

For the support vertical, (a) gives, with $h = 900$ mm:

$$165 = \frac{4\,331\,000}{A_v} + \frac{5\,315 \times 1\,000^2}{450A_v}, \quad 165A_v = 16\,142\,000$$

$$\therefore \quad A_v = 97\,800 \text{ mm}^2 \text{ (152 sq in.).}$$

For the vertical between panels 1 and 2, (a) gives, with $h = 600$ mm:

$$165 = \frac{788\,000}{A_v} + \frac{9\,450 \times 1\,000^2}{300A_v}, \quad 165A_v = 32\,288\,000$$

$$\therefore \quad A_v = 195\,600 \text{ mm}^2 \text{ (303 sq in.).}$$

By comparison with a triangulated truss, a Vierendeel truss is seen to be an inefficient structure, since it carries shears by bending in the top and bottom chords rather than by direct forces in the diagonals. All the same, Vierendeel trusses with many bays and many stories are used as exterior or interior supporting structures in some modern buildings because they afford freedom of openings.

PROBLEMS

7.1 A concrete frame is formed by a 200 mm (8 in.) slab spanning 4.25 m (14 ft) between 200 mm (8 in.) walls 2.75 m (9 ft) high. The frame carries a roof live load of 2 kN/m^2 (40 psf) and is assumed to be hinged at the base. Compute the maximum moments and shears and the reactions of the frame, and determine the reinforcement per unit of width in its members, assuming $f_c' = 30$ MPa (4,350 psi), $f_s = 165$ MPa (24 ksi).

7.2 An auditorium is enclosed by a series of hinged steel frames 7.5 m (25 ft) o.c. spanning 21 m (70 ft). The frames are 10.5 m (35 ft) high and consist of W920 (W36) beams and columns. Choose the members for the frame from among the available W920 (W36) sections, assuming A36 steel and a superimposed load of 4 kN/m^2 (80 psf).

7.3 A two-bay frame consists of two 300 mm × 900 mm (12 in. × 36 in.) concrete beams 12 m (40 ft) long, and three 300 mm × 450 mm (12 in. × 18 in.) concrete columns 4.5 m (15 ft) high and hinged at the base. The frame carries a superimposed load of 26 kN/m (1.8 k/ft). Evaluate the maximum moments and shears in the frame, and determine the corresponding bending reinforcement, for the superimposed load only. Assume $f_s = 165$ MPa (24 ksi).

7.4 A continuous A36 steel beam is rigidly connected to steel columns which are hinged at their bases. The beam spans four 9 m (30 ft) bays, and the columns are 4.5 m (15 ft) high. Determine the member size, if the beam carries a uniform load of 15 kN/m (1 k/ft).

7.5 The frame of Problem 7.1 is acted upon by a lateral wind load equal to 20% of the live load. Determine the moments and stresses due to this load and the adequacy of the members under the combined loading.

7.6 The side walls of the auditorium of Problem 7.2 are subjected to a wind pressure of 1.5 kN/m^2 (30 psf). Determine whether the wind stresses induced in the frame are within 33% of the dead and live load stresses.

7.7 The frame of Problem 7.3 is subjected to a lateral force at the top of 18 kN (4 k). Determine the additional moments and shears induced in the frame and whether the corresponding stresses are within 33% of the stresses of Problem 7.3.

7.8 The beam of Problem 7.4 supports a reciprocating compressor which induces a longitudinal force of 27 kN (6 k). Determine the additional moments in the beam and columns, and revise the member sizes for the combined forces, if necessary.

7.9 A six-bay fixed-end frame supports the back of a grandstand. Each bay is 9 m (30 ft) long and 6 m (20 ft) high. The frame supports five rows of seats with a total width of 3 m (10 ft). Allowing for a 1.25 m (4 ft) aisle per bay, what is the lateral force due to swaying of the spectators in the direction of the frame? Choose the member sizes if the superimposed dead load is 2 kN/m^2 (40 psf) and the frames are of A36 steel and are spaced 6 m (20 ft) o.c.

7.10 A four-bay fixed-end frame is one of a series spaced 7.5 m (25 ft) o.c. Each bay spans 9 m (30 ft); the height of the frame is 4.5 m (15 ft). Determine the moments in all the members if a wind load of 1.5 kN/m^2 (30 psf) acts on the sidewall.

7.11 A 3-story frame with 3 m (10 ft) story height consists of four 6 m (20 ft) bays. It is acted upon by a wind load of 1.25 kN/m^2 (25 psf). If the frames are spaced at 7.5 m (25 ft) o.c., determine the moments, shears, and axial force due to wind.

7.12 Determine the moments, shears, and axial forces in a 6-story one-bay building subjected to horizontal earthquake forces applied at the floor levels and equal to 0.1 of the floor weights. The building consists of frames 6 m (20 ft) o.c. with bays of 7.5 m (25 ft) and story height of 3.75 m (12 ft 6 in.). Total floor dead and live loads are 8.5 kN/m^2 (175 psf).

7.13 Evaluate the deflection at the top of the building of Problem 7.12 if the beams are W460 × 95 (W18 × 64) and the columns W360 × 79 (W14 × 53), on the average.

7.14 Determine the moments due to vertical loads in the frame of Problem 7.12.

7.15 The total deflection of a multistory building due to lateral forces is the sum of the "shear deflections" due to the bending deflection of the frame members (see Section 7.6), and the "cantilever deflections" due to the beam action of the frame (lengthening of the columns on one side, shortening on the other). Determine the cantilever deflection of the frame of Problem 7.12. Note that the moment of inertia of the cantilever beam is computed by assuming an ideal I beam in which the frame columns are the flanges.

7.16 A concrete Vierendeel truss consists of two floors of a building and of the walls between the two floors. The truss is 3.5 m (12 ft) high and spans 28 m (96 ft) with walls 3.5 m (12 ft) o.c. Each floor is 300 mm (12 in.) thick and the walls are also 300 mm (12 in.) thick. The live load on both floors is 5 kN/m^2 (100 psf). Determine the moments and shears in the walls and floors due to the truss action. For the floors, superimpose these moments to the moments in the slab, considered as a fixed-end member spanning between the walls. Determine the haunch which would be required in each bay to resist the moments, assuming $f_c = 30$ MPa (4,350 psi).

7.17 A four-bay steel Vierendeel truss picks up a column at its center vertical. The corresponding load is 260 kN (60 k). The truss depth is 2.75 m (9 ft), and its bays are 3 m (10 ft) wide. Determine the moments, shears, and axial forces in the members.

7.18 Evaluate the percentage of wind load carried by the shear walls of a building consisting of seven one-bay frames and two shear walls spaced 6 m (20 ft) o.c., if the building is 48 m (160 ft) long, 15 m (50 ft) wide, and 108 m (360 ft) high, with floor heights of 3.6 m (12 ft); the shear walls are 300 mm (12 in.) thick and 9 m (30 ft) deep; and the columns have average dimensions of 600 mm × 600 mm (2 ft × 2 ft). Assume a wind load of 1.5 kN/m^2 (30 psf) and a ratio $\beta = 0.8$.

7.19 Determine the maximum moment in the columns of the frames and in the shear walls of the building of Problem 7.18, due to a wind load of 1.5 kN/m^2 (30 psf). Use the portal method for the determination of the moment in the column.

8. Torsion

8.1 TORSION OF CIRCULAR AND HOLLOW BEAMS [5.3]

The sections of a twisted beam rotate and, hence, tend to slide with respect to adjoining sections, thus giving rise to *torsional shear stresses.*

If in a beam of circular cross section the relative rotation of two sections *one unit length apart* is measured by the angle θ (Fig. 8.1.1), the change in angle of a fiber on the beam surface, i.e., the shear strain γ on the beam surface, is given by:

$$AB = R \times \theta = 1 \times \gamma \quad \therefore \quad \gamma = R\theta. \tag{8.1.1}$$

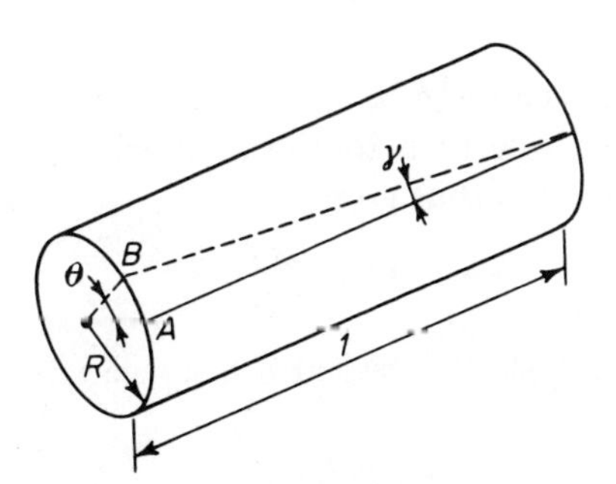

FIGURE 8.1.1

The shear strain equals zero on the beam axis and is proportional to the distance of the fiber from the axis of the beam, so that at a distance r from the axis, γ equals $r\theta$. Hence, the shear stress f_s at a distance r from the axis is given by:

$$f_s = G\gamma = Gr\theta, \tag{8.1.2}$$

where G is the shear modulus, $G = E/2(1 + \nu)$.

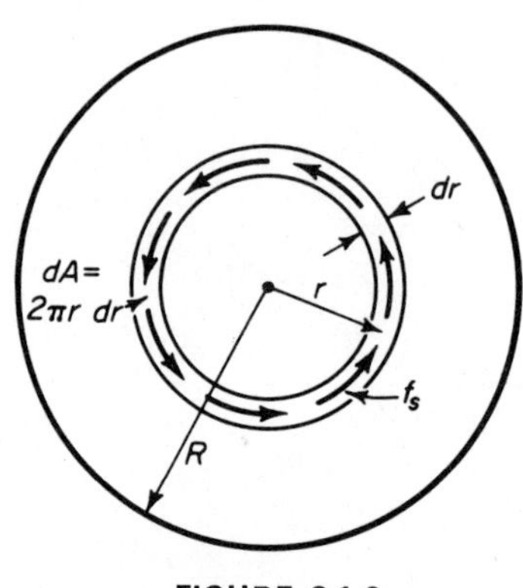

FIGURE 8.1.2

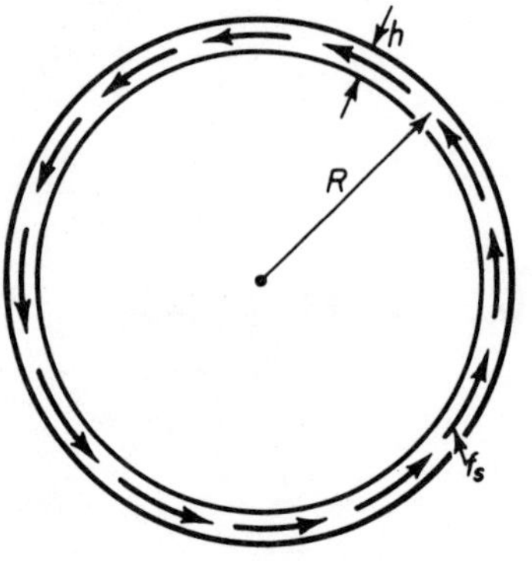

FIGURE 8.1.3

The external *twisting moment*, or *torque* M_t, is equilibrated by the moment of the shear stresses about the center of the section (Fig. 8.1.2):

$$M_t = \int_0^R f_s r\, dA = \int_0^R (Gr\theta) r (2\pi r\, dr) = 2\pi \frac{R^4}{4} G\theta = \pi \frac{R^4}{2} G\theta.$$

In terms of the parameter:

$$I_p = \int_0^R r^2 (2\pi r\, dr) = \frac{\pi R^4}{2} = \frac{\pi D^4}{32} \quad (D = 2R), \tag{8.1.3}$$

called the *polar moment of inertia* of the cross section about the beam axis, the rotation *per unit of beam length* becomes:

$$\theta = \frac{M_t}{GI_p}. \tag{8.1.4}$$

The rotation ϕ of the free end of a beam of length l, with respect to the end built-in against rotation, due to a torque M_t applied at the free end, is proportional to its length and thus given by:

$$\phi = \theta l = \frac{M_t l}{GI_p}. \tag{8.1.5}$$

By (8.1.2) and (8.1.4), the shear stress at a distance r from the beam axis becomes:

$$f_s = \frac{M_t r}{I_p}, \tag{8.1.6}$$

and the maximum shear stress, which occurs at the beam surface, is:

$$f_{s,\max} = \frac{M_t R}{I_p} = \frac{2}{\pi}\frac{M_t}{R^3} = \frac{16}{\pi}\frac{M_t}{D^3}. \tag{8.1.7}$$

Since the fibers of the beam located near its axis are practically unstressed, it is economical to remove them and to resist torsion by means of hollow circular sections. When the tubular section has a thickness h which is small with respect to its radius, the shear stress in it is virtually constant and the torque of the shear stress becomes (Fig. 8.1.3):

$$M_t = (2\pi Rh)f_s R,$$

from which:

$$f_s = \frac{M_t}{2\pi R^2 h}. \tag{8.1.8}$$

Example 8.1.1. A steel post 6 m (20 ft) high carries a sign on a horizontal bracket 2 m (6 ft) long (Fig. 8.1.4). The area of the sign is 1 m × 2 m (3 ft × 6 ft). Determine the torsional stress in a hollow post of circular cross section of radius $R = 300$ mm (1 ft) and thickness $h = 6$ mm ($\frac{1}{4}$ in.) due to a wind pressure of 2.5 kN/m² (50 psf). What is the rotation of the top of the post?

The torque due to the wind is:

$$M_t = 2.5 \times (1 \times 2) \times 1.3 = 6.5 \text{ kN·m. } (57.6 \text{ k·in.})$$

and by (8.1.8) the shear stress equals:

$$f_s = \frac{6.5 \times 1\,000^2}{2\pi(300)^2(6)} = 1.9 \text{ MPa } (277 \text{ psi}).$$

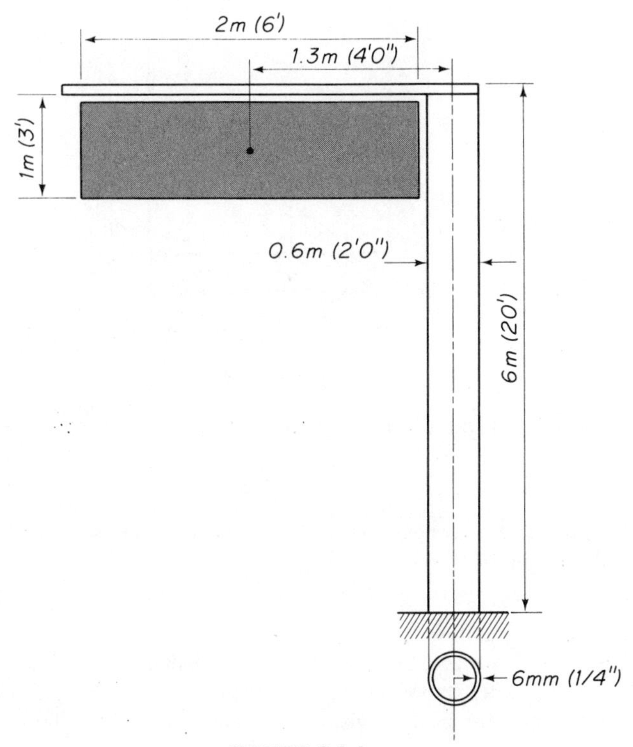

FIGURE 8.1.4

The rotation at the top of the post, with

$$I_p = \int_{R-h/2}^{R+h/2} r^2(2\pi r\, dr) \approx 2\pi Rh \times R^2 = 2\pi R^3 h \tag{a}$$

and $G = 76\,000$ MPa (11,000 ksi) becomes by (8.1.5):

$$\phi = \frac{6.5(6)(1\,000)^3}{76\,000 \times 2\pi(300)^3(6)} = 5.04 \times 10^{-4} \text{ rad}$$

$$= \frac{180}{\pi} \times 5.04 \times 10^{-4} \text{ deg} = 0.029 \text{ deg} = 1.73 \text{ min.}$$

The torsional stress in a box section may be evaluated by the same method used for a hollow circular section, since the stress f_s in it is practically constant. For a rectangular box section of sides a, b, and small thickness h (Fig. 8.1.5), one obtains, by taking moments about the center of the section:

$$M_t = \left(2ah \times \frac{b}{2} + 2bh \times \frac{a}{2}\right) f_s = 2abhf_s = 2Ahf_s,$$

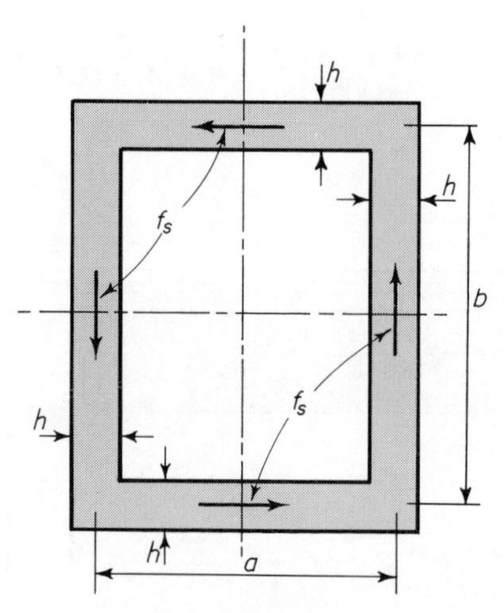

FIGURE 8.1.5

where A is the area of the full cross section, so that:

$$f_s = \frac{M_t}{2Ah} \tag{8.1.9}$$

Equation (8.1.9) may be used for any hollow section of small thickness; (8.1.8) is a particular case of (8.1.9).

Example 8.1.2. A hollow spandrel beam, made out of steel, is twisted by the bending moment of a beam, built-in into it at midspan, which is 9 m (30 ft) long, carries 30 kN/m (2 k/ft), and is fixed at the other end [Fig. 8.1.6 (a)]. Determine the thickness of the spandrel beam walls, if the spandrel section is 300 mm × 600 mm (1 ft × 2 ft).

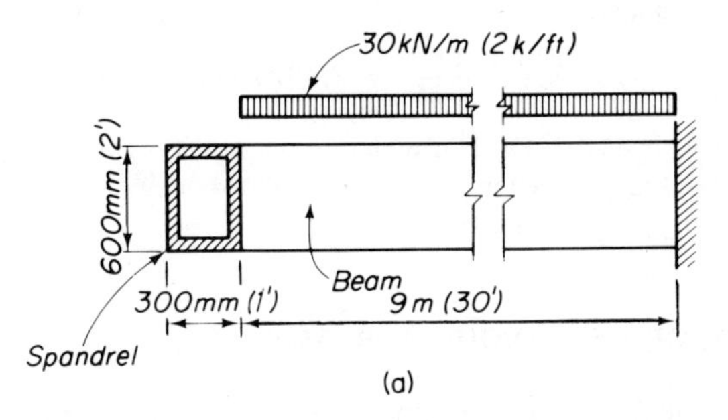

(a)

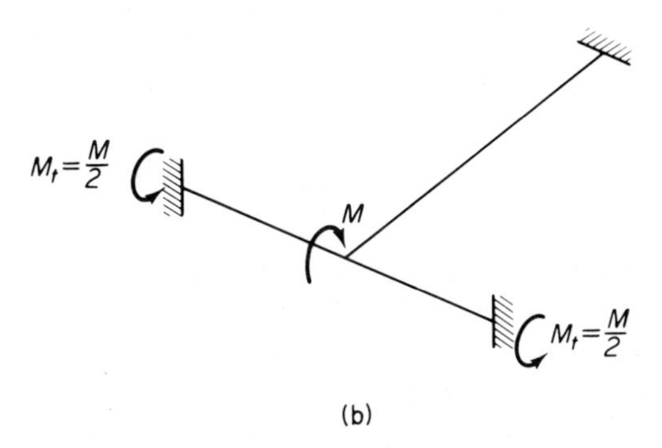

(b)

FIGURE 8.1.6

Due to the elasticity of the spandrel, the end of the beam built-in into it will rotate and its end moment M will be smaller than the fixed end moment $wl^2/12$; we shall assume it equals $wl^2/16$:

$$M = \tfrac{1}{16} \times 30 \times 9^2 = 152 \text{ kN}\cdot\text{m} \ (113 \text{ k}\cdot\text{ft}).$$

The maximum torque M_t in the spandrel occurs at its ends and equals $M/2$, since the two equal end torques equilibrate the applied moment M [Fig. 8.1.6 (b)]:

$$M_t = \tfrac{152}{2} = 76 \text{ kN}\cdot\text{m}.$$

By (8.1.9), the thickness h for an allowable shear stress of 35 MPa (5 ksi) is given by:

$$h = \frac{76 \times 1\,000^2}{2 \times 300 \times 600 \times 35} = 6 \text{ mm (0.25 in.)}.$$

The allowable stress in torsion is kept low because the spandrel beam develops additional shears in bending due to the beam reaction. The total load on the beam is $30 \times 9 = 270$ kN. Ignoring the dead load, its end reactions equal approximately 135 kN.* Under a center load of 135 kN, the end reactions of the spandrel equal 67.5 kN. This 67.5 kN shear is taken by the two *vertical* sides of the box, of area $2 \times 300 \times 6 = 3\,600 \text{ mm}^2$ [see (4.1.6)]. Hence, the additional shear due to the

*The reactions are dependent on the degree of restraint at each support.

vertical reactions equals:

$$f_s = \frac{67.5 \times 1\,000}{2 \times 3\,600} = 9.4 \text{ MPa (1.25 ksi)},$$

and the total shear is 44.4 MPa. For a slenderness ratio of the walls $b/h = 6\,000/6 = 100$ and no buckling stiffeners on the box plates, the ASTM specifications for A36 steel allow an $f_s = 58$ MPa (8.4 ksi).

8.2 TORSION OF RECTANGULAR BEAMS [5.3]

The rotation θ per unit length of beam and the maximum shear stress $f_{s,\max}$ in beams of *rectangular* cross section are related to the torque M_t by the equations:

$$\theta = \frac{M_t}{GC} \quad (C = \beta cb^3), \tag{8.2.1}$$

$$f_{s,\max} = Gb\theta \approx \left(3 + 1.8\frac{b}{c}\right)\frac{M_t}{cb^2}. \tag{8.2.2}$$

where GC is the *torsional rigidity* of the section, b the smaller and c the larger side of the cross section, and β is given in Table 8.2.1 versus c/b. The shear stress $f_{s,\max}$ occurs at the middle of the *large* sides c.

For very thin sections ($b \ll c$), θ and $f_{s,\max}$ become:

$$\theta = \frac{M_t}{G(\frac{1}{3}cb^3)}, \qquad f_{s,\max} = Gb\theta = \frac{M_t}{\frac{1}{3}cb^2}. \tag{8.2.3}$$

These formulas are derived by a nonelementary theory, which also proves that the torsional rigidity of an *open* section built by means of thin rectangular sections is the sum of the rigidities of each section:

$$GC]_{\text{total}} = G\sum C = G\sum \tfrac{1}{3}cb^3. \tag{8.2.4}$$

Equation (8.2.4) may be used, for example, to obtain the torsional rigidity of a wide flange or a channel beam, and hence its rotation:

$$\theta = \frac{M_t}{GC_{\text{total}}}.$$

The maximum torsional shear in W beams may be obtained by multiplying the $G\theta$ by the $b = \bar{b}$ of the flanges, including an average stress concentration factor of 2, to take into account the stress concentration at the re-entrant corners. Thus:

$$f_{s,\max} = 2\bar{b}\frac{M_t}{\sum \frac{1}{3}cb^3} \quad (\bar{b} = b \text{ of flanges}). \tag{8.2.5}$$

Table 8.2.1

c/b	1	1.5	2.0	2.5	3	4	6	10	∞
β	0.141	0.196	0.229	0.246	0.264	0.281	0.299	0.312	0.333

Example 8.2.1. The hollow post of Example 8.1.1 is replaced by a W360 × 211 (W14 × 142). Determine its $f_{s,\max}$ due to torsion.

From the dimensions of the flanges (see Appendix A) we obtain:

$$c = 394 \text{ mm}, \quad b = \bar{b} = 27 \text{ mm}, \quad \tfrac{1}{3}cb^3 = 258.50 \times 10^4 \text{ mm}^4,$$

from those of the web:

$$c = 375 \text{ mm}, \quad b = 17.3 \text{ mm}, \quad \tfrac{1}{3}cb^3 = 64.72 \times 10^4 \text{ mm}^4,$$

and, hence by (8.2.5):

$$C_{\text{total}} = 2 \times 258.5 \times 10^4 + 64.72 \times 10^4 = 581.72 \times 10^4 \text{ mm}^4,$$

$$f_{s,\max} = \frac{2 \times 27 \times 6.5 \times \times 1\,000^2}{581.72 \times 10^4} = 60.34 \text{ MPa (8.75 ksi)}.$$

Example 8.2.2. Determine the torque M_t and the maximum torsional stress in the beams AB and CD of the rectangular balcony of Fig. 8.2.1, due to a total load w (kN/m²) on the balcony, under the assumption that the load spans in the direction parallel to AB.

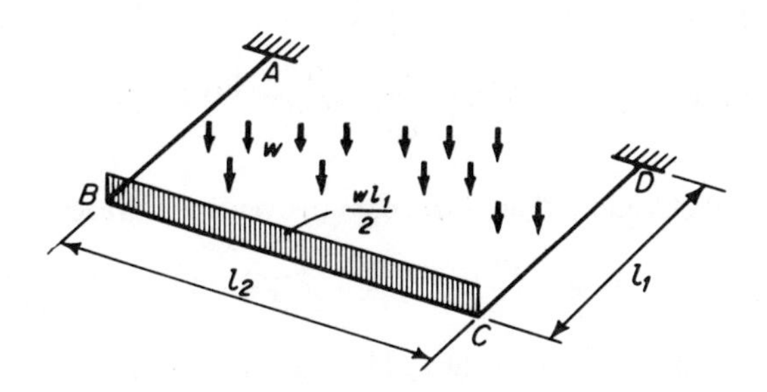

FIGURE 8.2.1

The load per unit length of beam BC is $\bar{w} = wl_1/2$. The ends of beam BC rotate by:

$$\phi_1 = \frac{1}{24}\frac{\bar{w}l_2^3}{EI_2} \quad \text{[see Table 4.3.1(a)]}$$

due to the load, and by:

$$\phi_2 = -\frac{1}{2}\frac{M_t l_2}{EI_2} \quad \text{[see Table 4.3.1(c)]}$$

due to the bending moments M at B and C, which are torsional moments M_t for the beams AB and CD.

Due to the twisting moments M_t, the beams AB and CD rotate at B and C by an angle, given by (8.2.1), $\phi_3 = \phi l_1 = M_t l_1/GC$.

Since the beam BC is rigidly connected to the beams AB and CD:

$$\phi_1 + \phi_2 = \phi_3,$$

or:

$$\frac{1}{24}\frac{\bar{w}l_2^3}{EI_2} - \frac{1}{2}\frac{M_t l_2}{EI_2} = \frac{M_t l_1}{GC},$$

from which:

$$M_t = \frac{\frac{1}{24}\bar{w}l_2^3/EI_2}{(\frac{1}{2}l_2/EI_2) + (l_1/GC)}.$$

If all the beams have rectangular cross section of depth h substantially larger than their width b, and ν is assumed equal to zero:

$$EI_2 = E\frac{bh^3}{12}, \qquad G = \frac{E}{2(1+\nu)} \approx \frac{E}{2},$$

$$GC \approx G\frac{1}{3}hb^3 \approx E\frac{hb^3}{6},$$

and the torque becomes:

$$M_t = \frac{\bar{w}l_2^2/12}{1 + (l_1/l_2)(h/b)^2}.$$

For $l_1 = \frac{1}{2}l_2 = 3$ m (10 ft), $h = 3b = 1$ m (3 ft), $w = 10$ kN/m² (0.2 ksf):

$$M_t = \frac{1}{1 + (\frac{1}{2})(\frac{3}{1})^2} \times \frac{1}{12}\left(10 \times \frac{3}{2}\right)(6)^2 = 8.18 \text{ kN}\cdot\text{m},$$

and by (8.2.2) the maximum torsional stress in AB and CD is:

$$f_{s,\max} = [3 + 1.8(\tfrac{1}{3})]\frac{8.18 \times 1\,000}{1\,000\left(\frac{1\,200}{3}\right)^2} = 0.27 \text{ MPa (39 psi)}.$$

The beams AB and CD are loaded, moreover, by the reactions V of beam BC:

$$V = \frac{1}{2}\left(10 \times \frac{3}{2}\right) \times 6 = 45 \text{ kN}.$$

and hence develop a maximum shear $V = 45$ kN and a maximum moment $M = 45 \times 3 = 135$ kN·m. The vertical shear stress in these beams by (4.1.5) equals:

$$f_s = 1.5\frac{45 \times 1\,000}{\left(\frac{1\,000}{3}\right)(1\,000)} = 0.20 \text{ MPa (29 psi)}$$

and must be added to the torsional shear stress.

8.3 CIRCULAR BALCONY BEAMS

Circular balcony beams are usually built-in at their ends and develop at each section both bending and torsion. Table 8.3.1 gives the bending and twisting moments in circular balcony beams of rectangular cross section with a radius R and half-opening angle ϕ_0 varying from 30° to 90°, uniformly loaded with a load w kN/m (lb/ft) (Fig. 8.3.1).*

*"Circular Beams Transversely Loaded," P. J. Spyropoulos, *Journ. A.C.I.*, October, 1963, in which $E/G = 2.35$ and β has the values shown in Table 8.2.1.

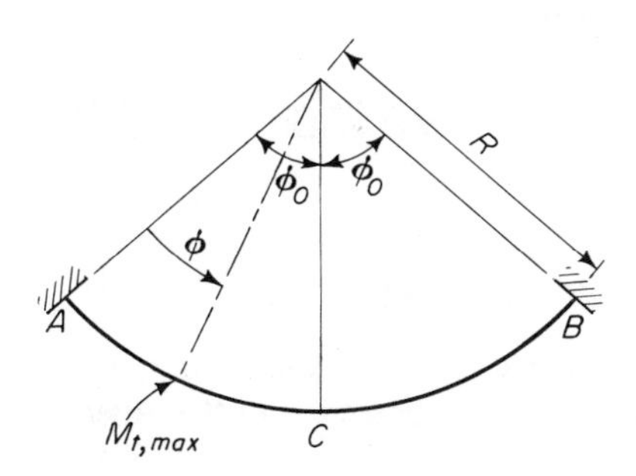

FIGURE 8.3.1

M_C, M_A are the bending moments at C and A, $M_{t,A}$ is the twisting moment at A, $M_{t,\max}$ is the maximum twisting moment, and $\Delta\phi$ is the angle range within which the maximum twisting moment occurs.

The table, combining the results of four tables in which the ratio of depth h to width b varies from 1 to 4, gives the largest value obtained by the coefficients as h/b varies between 1 and 4. Hence, the values in the table are on the safe side. The angle $\Delta\phi$ varies with h/b, and the table indicates its range, i.e., its largest and smallest values.

Table 8.3.1

ϕ_0°	M_C/wR^2	M_A/wR^2	$M_{t,A}/wR^2$	$M_{t,\max}/wR^2$	$\Delta\phi$ for $M_{t,\max}$
30°	0.04	−0.10	−0.005	0.008	17°–14°
45°	0.09	−0.24	−0.057	0.025	24°–21°
60°	0.15	−0.44	−0.073	0.052	30°–27°
75°	0.21	−0.69	−0.160	0.085	34°–33°
90°	0.27	−1.00	−0.300	0.121	38°

Example 8.3.1. Determine the maximum total shear stress in a circular, concrete balcony beam of radius 3 m (10 ft) with an opening of 120°, loaded with a load $w = 10$ kN/m (0.75 k/ft). The beam cross section is 900 mm × 450 mm (3 ft × 1.5 ft).

The maximum twisting moment for $\phi_0 = 120/2 = 60°$ occurs at A and equals:

$$M_t = -0.073 \times 10 \times 3^2 = -6.57 \text{ kN·m} \ (-58.2 \text{ k·in.}).$$

The opening angle is $2\phi_0 = 120/57.3 = 2.1$ radians and the beam length equals:

$$L = R(2\phi_0) = 3 \times 2.1 = 6.3 \text{ m}.$$

Hence, the vertical reaction at A equals:

$$V = 10 \times \frac{6.3}{2} = 31.5 \text{ kN}.$$

With $c = 900$ mm, $b = 450$ mm, (8.22) gives:

$$f_{s,\max} = [3 + 1.8(\tfrac{1}{2})]\frac{6.57 \times 1\,000^2}{900 \times 450^2} = 0.14 \text{ MPa (19 psi)}.$$

The shear stress due to V equals:

$$f_s = 1.5\frac{31.5 \times 1\,000}{900 \times 450} = 0.12 \text{ MPa (17 psi)},$$

and the maximum total shear stress, at the middle of the vertical sides, is 0.26 MPa. Since the allowable shear stress in concrete is of the order of 0.70 MPa, the section is safe in shear.

8.4 TORSION AND BENDING OF UNSYMMETRICAL SECTIONS

It was pointed out in Section 4.1 that the vertical shear in a wide flange beam is absorbed, for all practical purposes, by shear stresses evenly distributed in the web. This does not imply that there is no shear in the flanges. In fact, by cutting the flanges of a bent beam with a vertical plane ABC (Fig. 8.4.1), we see that equilibrium in the horizontal direction of the two portions of flanges to the left of this plane requires horizontal shear

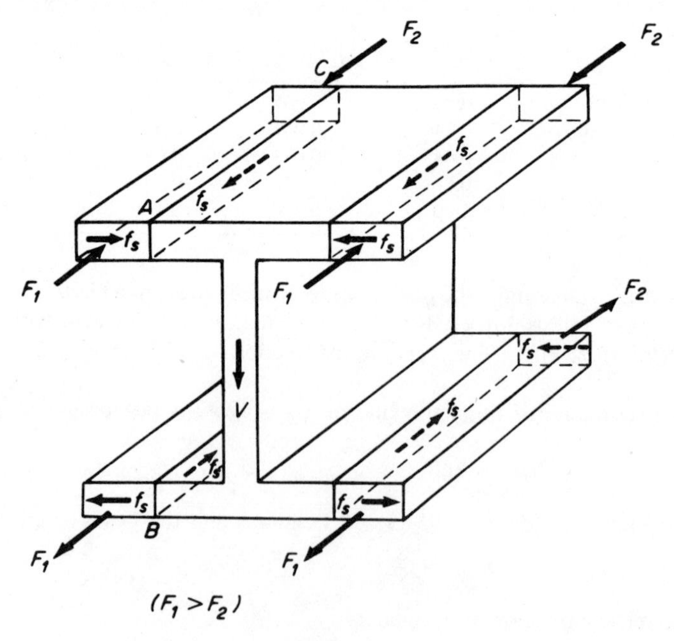

FIGURE 8.4.1

stresses f_s to equilibrate the difference in the normal forces due to bending $F = F_1 - F_2$. Hence, for rotational equilibrium, horizontal shear stresses f_s will also develop on the flange cross sections.

In a beam of symmetrical cross section, the horizontal shears f_s in the cross sections of each flange (Fig. 8.4.1) add up to two equal and opposite forces, and the resultant of the shear stresses in the web is a vertical shear V parallel to the axis of the web.

In a nonsymmetrical section, instead, the horizontal shears do not add up to zero. For example, in a channel beam (Fig. 8.4.2) the vertical shears have a resultant V, the horizontal shears are equivalent to a couple $M = (f_s tb)h$, and the resultant of V and of this couple M is a vertical force V displaced to the right by an amount $e = M/V$. Therefore, in order not to introduce torsion in a channel beam, a vertical load V must be applied not at the centroid O or at the web A, but at the point B. If the load is applied

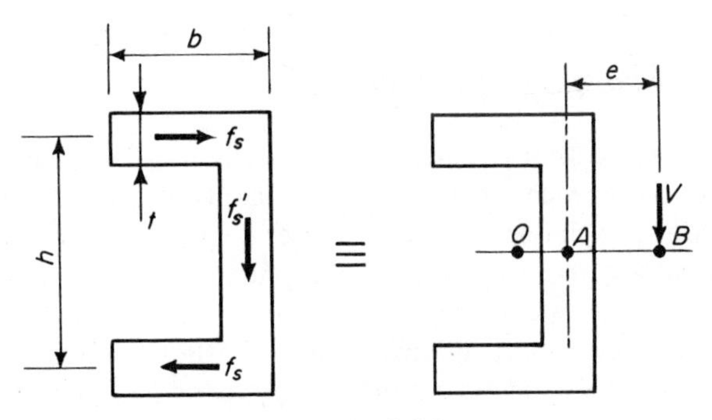

FIGURE 8.4.2

at A, the beam will twist under a twisting moment Ve. The point B at which a load must be applied to avoid twist is called the *center of shear* or *center of twist* of the section.

For example, since an angle develops shears parallel to its sides (Fig. 8.4.3), the vertical resultant of the angle shears is located at B and, to obtain equilibrium without twist, a vertical force must be applied at B.

Example 8.4.1. Determine the maximum *torsional* shear stress in an angle of sides b and thickness t due to a load V applied at its centroid and making a 45° angle with the sides (Fig. 8.4.3).

The centroid O is at horizontal distance $(\sqrt{2}/2)b$ from the center of twist B; hence, the twisting moment M_t equals $(\sqrt{2}/2)Vb$. The maximum shear stress due to M_t, by (8.2.5), with $\bar{b} = t$ equals:

$$f_{s,t} = 2t\frac{(\sqrt{2}/2)Vb}{2(\frac{1}{3}bt^3)} = \frac{3\sqrt{2}}{2}\frac{V}{t^2}.$$

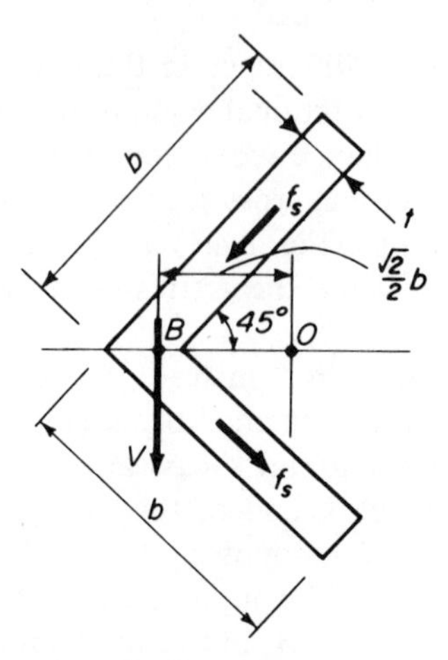

FIGURE 8.4.3

8.5 THE SOAP-BUBBLE ANALOGY

Since the evaluation of torsional stresses is based on a nonelementary theory (due to the French engineer de St. Venant), it is interesting to realize that the qualitative behavior of these stresses may often be visualized on the basis of an analogy between the torsional phenomenon and a phenomenon seemingly unrelated to it, the equilibrium of a soap bubble.

It may be shown mathematically that, if a hole with the shape of the cross section of a twisted bar is cut out of a plate, and a film of soap solution is spread over the hole and slightly blown up by air pressure, the *slope* of the soap bubble is proportional to the *shear stress* in the twisted bar. Most of the shear stress characteristics mentioned in previous sections may be easily grasped by use of this *soap-bubble analogy*.

1. The slope of the soap bubble over a circular hole (Fig. 8.5.1) is zero at the center, maximum at the boundary, and constant all around the boundary. Hence, the maximum shear stress occurs at the boundary of the

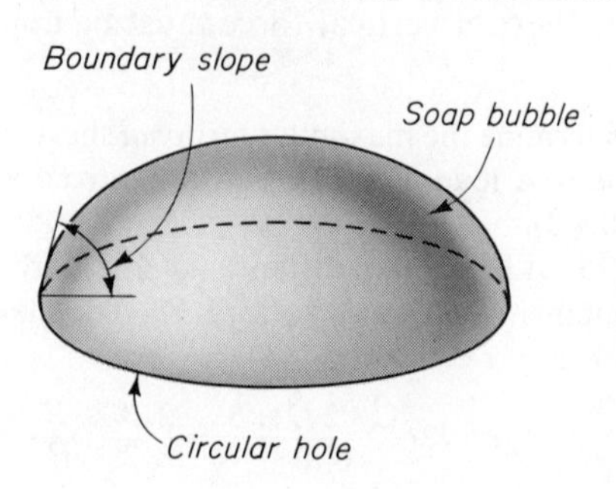

cross section, and the portions of the cross section around the center are minimally stressed.

2. The soap bubble over a rectangular hole (Fig. 8.5.2) has a maximum slope at the middle of the long sides, i.e., at the points *nearest* the center of the section. Hence, the maximum shear stress develops at these points. The bubble has zero slope at the center but also at the corners, since at the

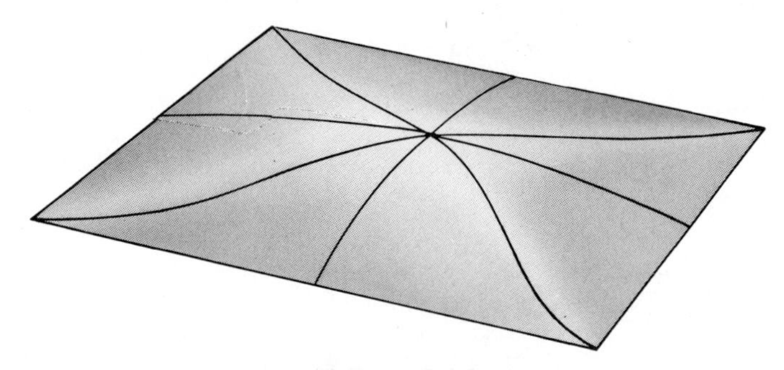

FIGURE 8.5.2

corners the horizontal boundaries of the hole keep the soap film down. Hence, no shear stress is developed at the corners, and the corners could be cut out without noticeably increasing the value of the shear stresses in the section.

3. The bubble over a narrow rectangular hole has the shape of a cylinder, except near the short sides of the hole. Hence, the shear stress on the long sides of a thin cross section is practically constant.

4. The bubble over a hole representing a notched circular cross section (Fig. 8.5.3) has a much greater slope at the inner corner (A) of the notch than at any other point of the boundary; hence, stress concentration takes place in shear at the notch (see also Section 11.3).

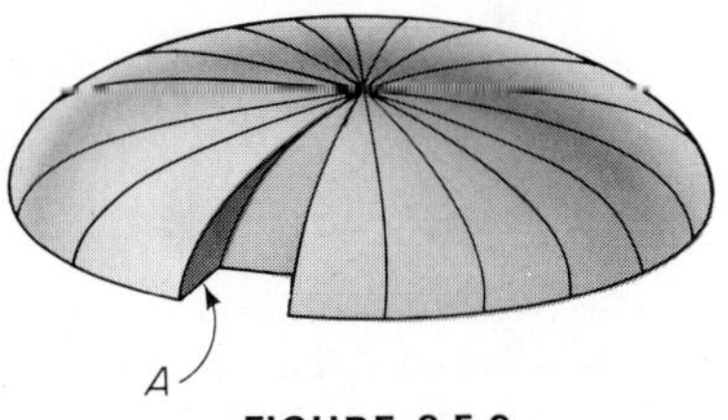

FIGURE 8.5.3

It may also be proved that the twisting resistance of a cross section is proportional to the volume of the soap bubble, but that the analogy for hollow sections requires the inner boundary of the hole to be set at a higher level than the outer boundary. The greater torsional rigidity of thin hollow sections in comparison with that of thin open sections is thus immediately visualized. In the open section of Fig. 8.5.4 (a), the volume of the bubble is small in comparison with the volume of the bubble in Fig. 8.5.4 (b), in which the inner boundary of the closed hollow section is higher than the outer

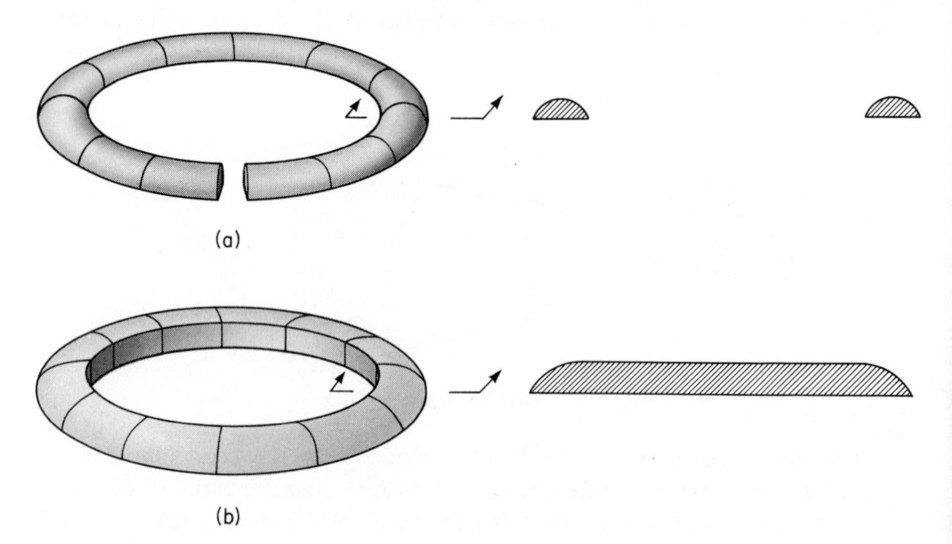

FIGURE 8.5.4

boundary. The ratio of the stiffnesses of the two rings may be shown to be $(3R/h)$, so that (8.1.8) for the open ring becomes:

$$f_s = \frac{3R}{h}\left(\frac{M_t}{2\pi R^2 h}\right) = \frac{3}{2}\frac{M_t}{\pi R h^2}. \tag{8.5.1}$$

PROBLEMS

8.1 A sign 1 m (3 ft) square is attached at the top to one side of an aluminum pole. If the pole is 4.5 m (15 ft) high, with 150 mm (6 in.) outside diameter and with walls 5 mm ($\frac{3}{16}$ in.) thick, what are the stresses due to torsion, shear, and bending at the base of the pole for a 1.5 kN/m² (30 psf) wind force?

8.2 A fixed-end steel beam with a hollow rectangular cross section, $b = 200$ mm (8 in.), $d = 300$ mm (12 in.), spans 6 m (20 ft). A vertical concentrated load of 65 kN (15 k) is supported by the beam at the center of the span on one of the vertical faces of the beam. What is the required wall thickness for an allowable total shear stress of 103 MPa (15 ksi)?

8.3 If the aluminum pole of Problem 8.1 is replaced by a steel W150 × 23.1 (W6 × 15.5), what is the maximum total shear stress at the base of the pole?

8.4 In the balcony beam shown in Fig. 8.2.1, $l_1 = 3.5$ m (12 ft), $l_2 = 7.0$ m (24 ft), $w = 12$ kN/m² (0.25 ksf), $b = 250$ mm (9 in.), $h = 750$ mm (2.5 ft). Determine the total shear stress at A and D.

8.5 A circular balcony beam with an opening angle of 150° and a radius of 4.5 m (15 ft) is loaded with a distributed load $w = 13$ kN/m (0.9 k/ft). Determine the maximum total shear stress in the rectangular beam if $b = 400$ mm (16 in.), $d = 750$ mm (30 in.).

8.6 A 3 m (10 ft) long horizontal beam is cantilevered at right angles from the center of a fixed-end beam which is 6 m (20 ft) long. A load of 1 kN (225 lb) is lifted by means of a pulley at the end of the cantilever. Determine the maximum bending moment, twisting moment, and vertical shear in the 6 m (20 ft) beam, and establish its dimensions if it is made of Douglas fir and is of rectangular cross section.

8.7 A 9 m (30 ft) fixed-end spandrel of reinforced concrete supports one end of two 6 m (20 ft) long fixed-end beams 3 m (10 ft) o.c. The beams carry a 125 mm (5 in.) concrete slab and a load of 5 kN/m² (100 psf). Determine the dimensions of the spandrel. Assume $f'_c = 30$ MPa (4,350 psi).

8.8 A building 72 m (240 ft) high, 30 m (100 ft) wide, and 18 m (60 ft) deep is acted upon by a 1.5 kN/m² (30 psf) wind. The wind action is absorbed by an eccentric square core 6 m (20 ft) on a side. The centroid of the core is 9 m (30 ft) from a short wall and 9 m (30 ft) from a long wall of the building. Determine the thickness required of the core walls at the bottom of the building in order to absorb the bending and shear stresses due to the wind, assuming $f'_c = 30$ MPa (4,350 psi).

8.9 The dynamic force of an earthquake is often considered to be equivalent to 0.05 of the dead weight of the building plus one-fourth of its live load. Determine the maximum bending moment, twisting moment, and shear due to an earthquake in the core of the building of Problem 8.8, and establish whether the wind or the earthquakes govern the design of the core if the dead load per floor (including columns and partitions) is 7.5 kN/m² (150 psf), the live load is 5 kN/m² (100 psf), and the building has 20 stories.

8.10 The railing of a stadium is 1.0 m (3.5 in.) high and is built-in into 6 m (20 ft) long concrete spandrel beams, fixed into columns. Determine the maximum torsional shear produced in the spandrels by the horizontal thrust on the railing (see Chapter 1 for its intensity) if the spandrel cross section is 300 mm × 450 mm (12 in. × 18 in.) and the railing is supported by closely spaced posts.

8.11 Determine the twisting-moment capacity of an A36 steel I-beam with flanges 610 × 25.4 mm (24 in. × 1 in.) and web 920 mm × 25.4 mm (36 in. × 1 in.).

8.12 A boxed A36 steel beam 300 × 600 mm (1 ft × 2 ft) has walls 25 mm (1 in.) thick and thus has an area of 45 000 mm² (72 sq in.). The same amount of material may be used to build an I-beam with flanges 460 × 25 mm (18 in. × 1 in.) and a web 920 mm × 25 mm (36 in. × 1 in.). Compare the maximum bending and twisting moments which these two cross sections can resist.

9. Cables

9.1 CABLE GEOMETRY [6.1]

Cables are tension elements capable of transmitting vertical loads horizontally by developing a horizontal thrust through their sag.

The geometry of a cable is defined by the horizontal distance l between its two supports (assumed for simplicity to be at the same level) and its sag h (Fig. 9.1.1).

The cable length L can be evaluated in terms of l and the *sag-span ratio*:

$$r = \frac{h}{l}. \tag{9.1.1}$$

The values of r are usually small, in most cases of the order of $\frac{1}{10}$. For a single load applied at midspan (Fig. 9.1.1) one finds, by Pythagoras' theorem:

$$L_1 = 2\sqrt{\left(\frac{l}{2}\right)^2 + h^2} = 2\left(\frac{l}{2}\right)\sqrt{1 + 4\frac{h^2}{l^2}}$$

$$= l\sqrt{1 + 4r^2} \tag{9.1.2}$$

Since for small values of x:

$$\sqrt{1 + x} \approx 1 + \tfrac{1}{2}x \quad (x \ll 1), \tag{a}$$

a good approximation of L_1 is given by:

$$L_1 = l(1 + 2r^2) \quad (r \ll 1). \tag{9.1.3}$$

Similarly, for two equal loads applied at the third points (Fig. 9.1.2), the length L_2 becomes:

$$L_2 = 2\sqrt{\left(\frac{l}{3}\right)^2 + h^2} + \frac{l}{3} = \frac{l}{3}(1 + 2\sqrt{1 + 9r^2}),$$

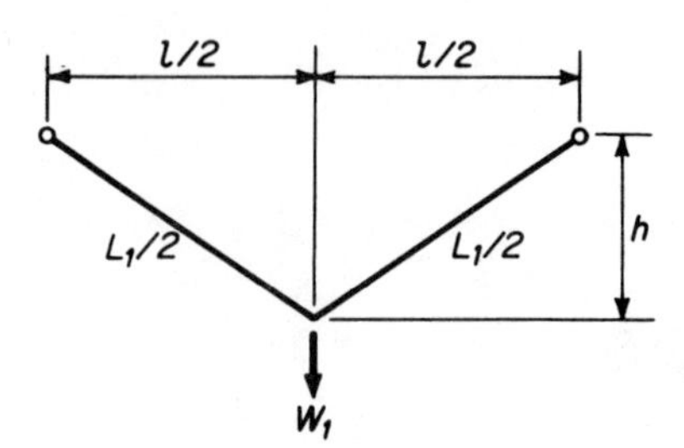

FIGURE 9.1.1

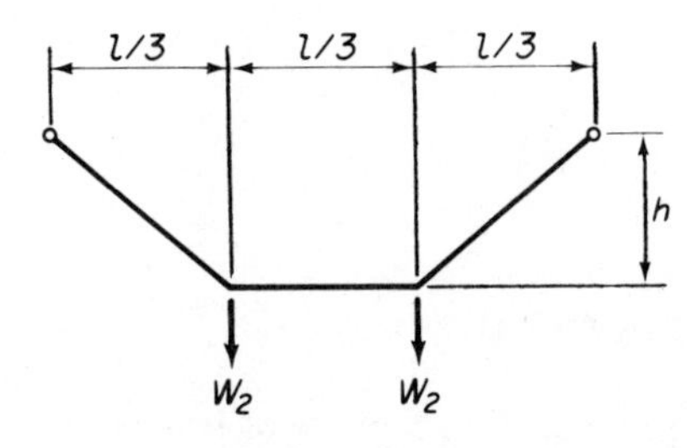

FIGURE 9.1.2

and, by (a), for $r \ll 1$:

$$L_2 = \frac{1}{3}\left[1 + 2\left(1 + \frac{9}{2}r^2\right)\right] = l(1 + 3r^2). \tag{9.1.4}$$

Whatever the number of equally spaced, identical loads, L is greater than L_1 and less than L_2, as shown by Table 9.1.1.

Table 9.1.1

Number of Loads	1	2	3	4	5	6	∞
L/l	$1 + 2r^2$	$1 + 3r^2$	$1 + 2.5r^2$	$1 + 2.8r^2$	$1 + 2.6r^2$	$1 + 2.7r^2$	$1 + 2.67r^2$

The shape assumed by a cable under concentrated loads is called a *funicular* or *string polygon*; it depends on the intensity of the loads and their location. If a single load is located at a distance al ($a < 1$) from the left support (Fig. 9.1.3), the sag h occurs under the load and, letting $b = 1 - a$, the length of the cable equals:

$$\begin{aligned} L &= \sqrt{a^2l^2 + h^2} + \sqrt{b^2l^2 + h^2} \\ &= l\left(a\sqrt{1 + \frac{r^2}{a^2}} + b\sqrt{1 + \frac{r^2}{b^2}}\right) \\ &\approx l\left[a + 1 + \frac{1}{2}\frac{r^2}{a^2}\right) + b\left(1 + \frac{1}{2}\frac{r^2}{b^2}\right)\right] = l\left(1 + \frac{1}{2ab}r^2\right). \end{aligned} \tag{9.1.5}$$

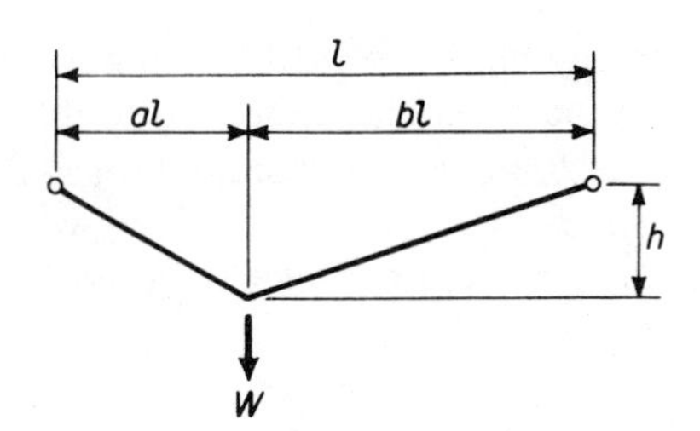

FIGURE 9.1.3

Table 9.1.2 gives the value of $c = 1/2ab$ in terms of a.

Table 9.1.2

$a =$	0.1	0.2	0.3	0.4	0.5
	0.9	0.8	0.7	0.6	0.5
$c = 1/2ab$	5.5	3.1	2.4	2.1	2.0

The funicular polygon for a set of *symmetrical* loads may be readily obtained graphically, as shown in Fig. 9.1.4, where the loads on the left half of the cable are drawn along a vertical line from A to m (*half* the load at mid-span, if any, must be included) and the end points of the loads are connected

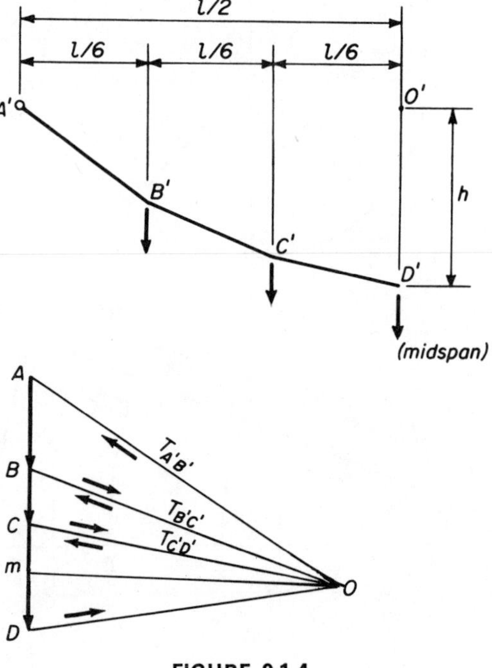

FIGURE 9.1.4

with the point O arbitrarily chosen on the horizontal through m. Drawing $A'B'$ parallel to AO from the left cable support up to the line of the first load, $B'C'$ parallel to BO up to the line of the second load, etc., one obtains the funicular polygon, in which the vertical scale is given by $O'D' = h$. The triangles ABO, BCO, etc., represent graphically the equilibrium of the loaded points B', C', etc., of the cable.

Example 9.1.1. A circular roof carrying a uniform load w (kN/m²) (inclusive of the dead load) consists of slabs supported on cables hanging from an outer compression ring of radius R (m) and a small tension inner ring located h (m) below the outer ring [Fig. 9.1.5(a)]. Its sag ratio $r = h/2R$ is small enough for the load w

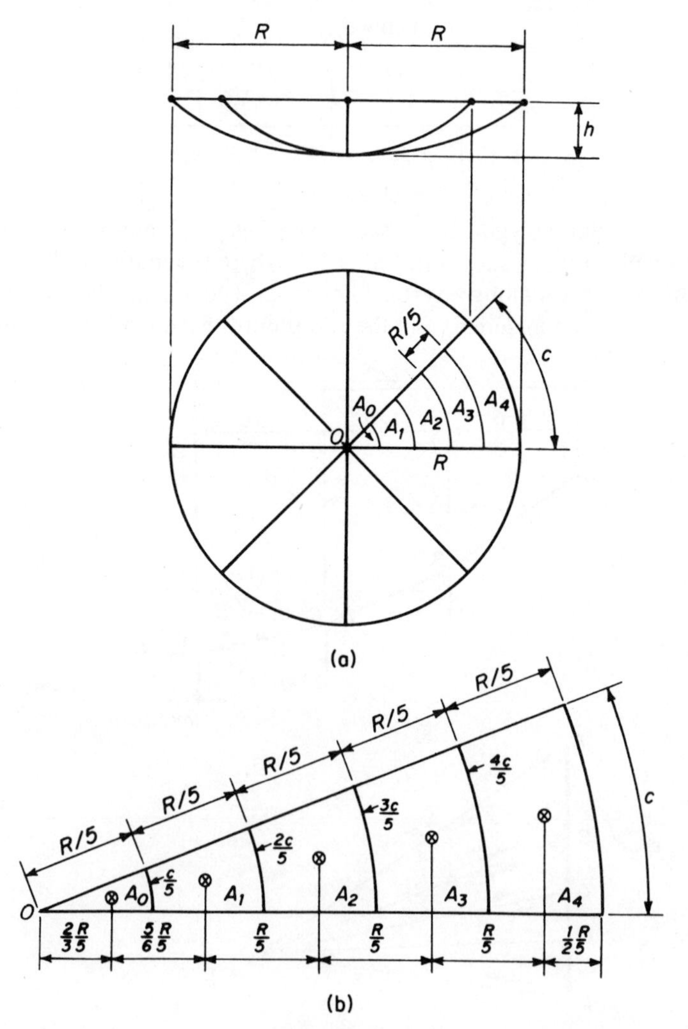

FIGURE 9.1.5

to be considered uniform in horizontal projection and the inner ring is small enough to be considered a point. Determine the funicular shape of the cables, if these are spaced by c (m) along the outer ring.

Subdividing the roof sector supported by one cable into, say, five parts of equal length $R/5$, the areas A_0 of the curvilinear triangle, and A_1 to A_4 of the curvilinear trapezoids [Fig. 9.1.5(b)], are given approximately by:

$$A_0 = \frac{1}{2}\left(\frac{R}{5}\right)\left(\frac{c}{5}\right) = \frac{Rc}{50},$$

$$A_1 = \frac{1}{2}\left(\frac{R}{5}\right)\left(\frac{c}{5} + \frac{2c}{5}\right) = 3\frac{Rc}{50} = 3A_0,$$

$$A_2 = 5A_0, \qquad A_3 = 7A_0, \qquad A_4 = 9A_0.$$

The centroid of A_0 is at $\frac{2}{3}(R/5)$ from O; the centroids of the trapezoids are, for all practical purposes, equidistant from their circular sides.*

The funicular polygon approximating the cable curve can be obtained by applying loads $W_n = wA_n$ at the centroids of the areas A_n. Figure 9.1.6 shows the construction of the funicular polygon with $W = w(Rc/50)$. In order to construct the funicular polygon for a given sag-span ratio, say $h = 2R/8$, first close the polygon with the line BA and draw Ok parallel to BA. Since the specified closing line is AB', draw kO' parallel to AB'. From the new pole O' construct a new force polygon and the required funicular polygon AB'.

As the number of evenly spaced, identical vertical loads on a cable increases indefinitely, the funicular polygon approaches the *funicular curve* for a load uniformly distributed *horizontally*, such as the weight of the roadway of a suspension bridge. This curve is shown in Section 9.2 to be the quadratic parabola (Fig. 9.1.7):

$$y = 4h\left(\frac{x}{l} - \frac{x^2}{l^2}\right). \tag{9.1.6}$$

The length L of the parabolic cable for small r is given by:

$$L = l(1 + \tfrac{8}{3}r^2) \quad (r \ll 1). \tag{9.1.7}$$

The shapes of the cables mentioned so far were obtained under the assumption that the weight of the cable is negligible in comparison with the applied loads. This is often the case in practice. The shape assumed by a

*Calling n (= 0, 1, 2, 3, 4) the order of the areas A_n, the distance of their centroids from the outer side of the trapezoid is:

$$\bar{y}_n = \frac{2}{3}\,\frac{3n+1}{2n+1}\left(\frac{1}{2}\right)\left(\frac{R}{5}\right),$$

so that:

$$\bar{y}_0 = \frac{1}{3}\left(\frac{R}{5}\right), \quad \bar{y}_1 = \frac{8}{9}\left(\frac{1}{2}\right)\left(\frac{R}{5}\right), \quad \bar{y}_2 = \frac{14}{15}\left(\frac{1}{2}\right)\left(\frac{R}{5}\right),$$

$$\bar{y}_3 = \frac{20}{21}\left(\frac{1}{2}\right)\left(\frac{R}{5}\right), \quad \bar{y}_4 = \frac{26}{27}\left(\frac{1}{2}\right)\left(\frac{R}{5}\right).$$

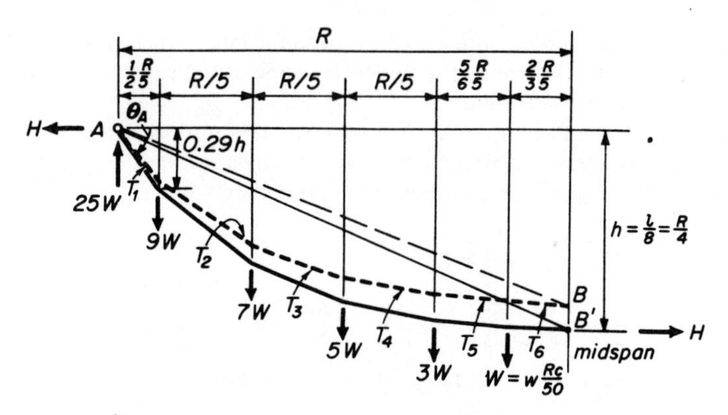

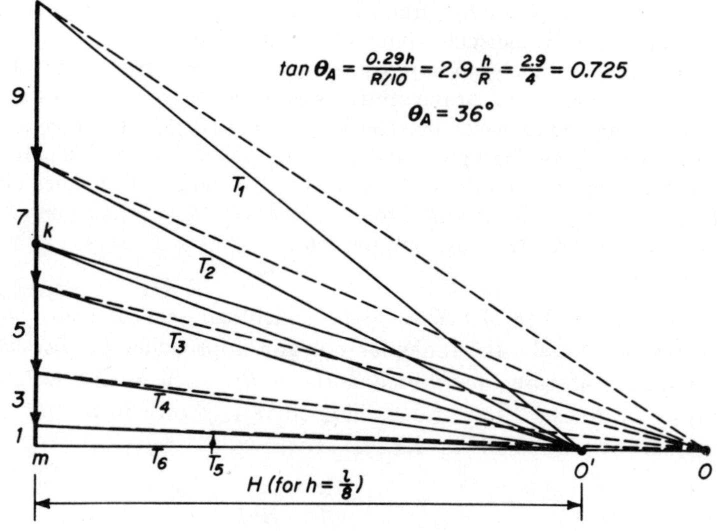

FIGURE 9.1.6

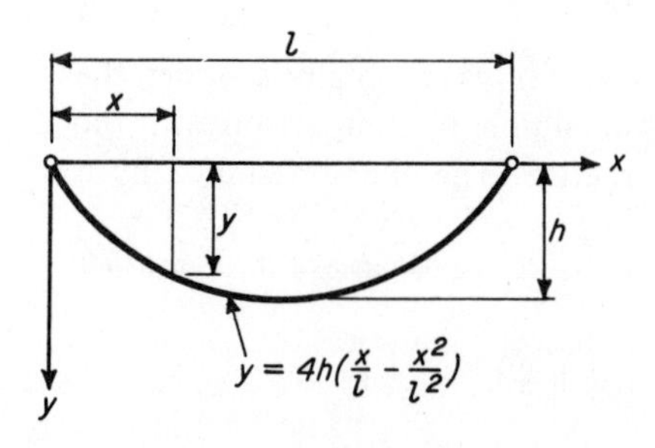

FIGURE 9.1.7

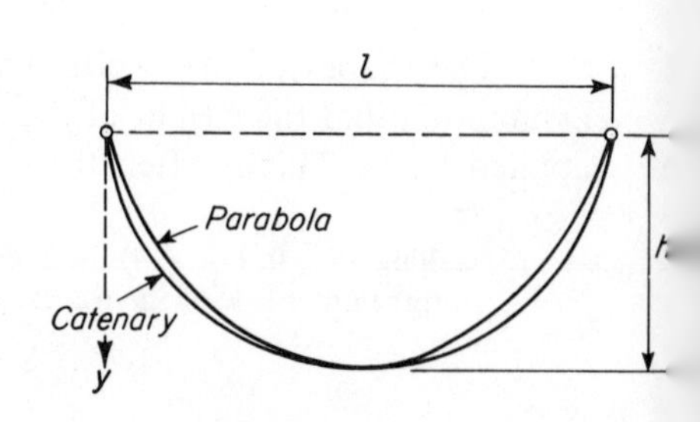

FIGURE 9.1.8

cable of constant cross section under its own weight is a curve called a *catenary:* it is the funicular for a load distributed uniformly *along the length of the cable*, rather than horizontally. As shown in Fig. 9.1.8, the catenary is lower than the parabola of equal sag, since the weight of the cable per unit of *horizontal* projection is greater toward the cable supports, where the cable slope is greater. For $r \leq \frac{3}{10}$, the catenary and the parabola of equal sag are virtually indistinguishable. The catenary is approximately $20r^2$ per cent longer than the equal-sag parabola; for $r = \frac{1}{10}$, the parabola has a length 1.0267 l and the catenary 1.0288 l.

9.2 CABLE STRESSES [6.1]

Cables are so flexible that, for all practical purposes, they cannot develop bending stresses; hence, the tensile force T at any section x of a cable (Fig. 9.2.1) acts in the direction of the tangent to the cable, since a component normal to the cable would produce bending. The tangent force

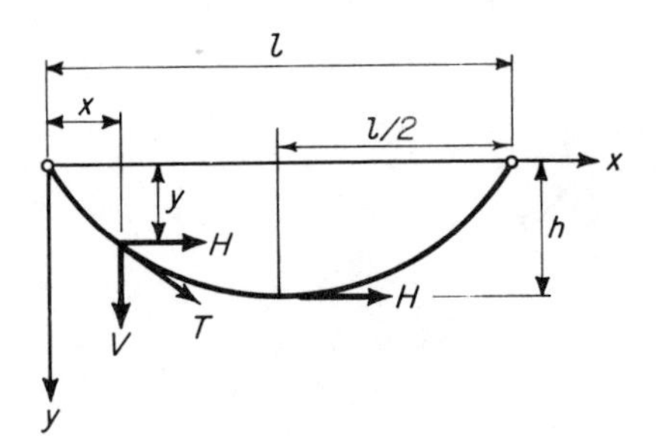

FIGURE 9.2.1

T can be split into a vertical component V and a horizontal component H, called the *thrust*. In a uniformly loaded cable, H is the value of T at the lowest point, where the tangent to the cable is horizontal (Fig. 9.2.1).

The values of T in the sections of a cable carrying concentrated loads are obtained by statics, as shown in Fig. 9.1.4.

The equilibrium in translation of any portion of the cable requires that the sums of all the vertical and of all the horizontal forces on it be zero. Equilibrium in rotation requires that the sum of the moments of the forces about any point be equal to zero. Choosing any point D on the cable as the point about which moments are taken is equivalent to stating that the bending moment is equal to zero at any point of the cable.

For the case of Fig. 9.2.2(a), equilibrium of the entire cable in the x- and y-directions and in rotation about B gives, respectively:

$$H_B - H_A = 0, \qquad R_A + R_B = W, \qquad R_A l - Wbl = 0,$$

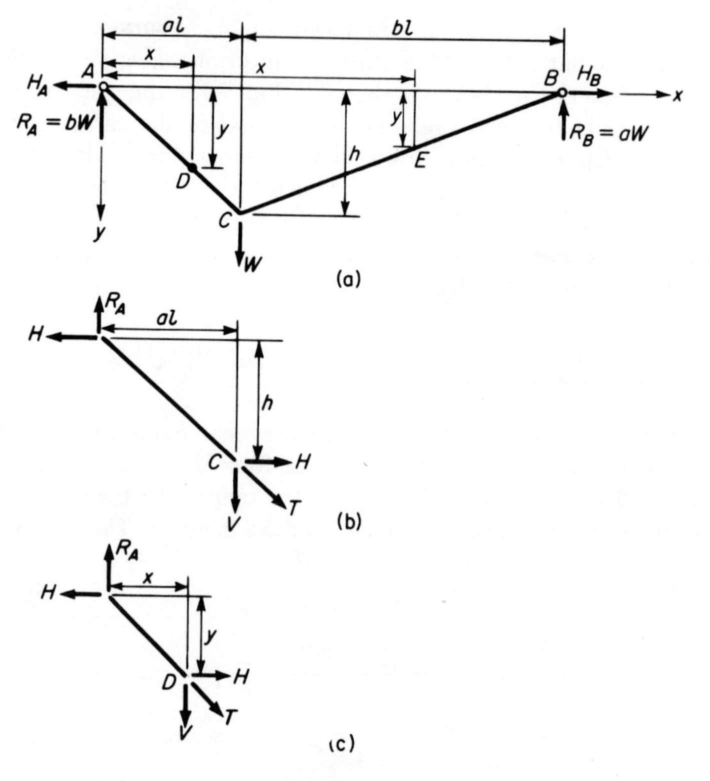

FIGURE 9.2.2

from which:

$$H_A = H_B = H, \qquad R_A = bW, \qquad R_B = aW. \tag{9.2.1a}$$

Similarly, rotational equilibrium of the cable portion between A and C [Fig. 9.2.2(b)] gives, taking moments about C:

$$Hh - R_A al = 0, \qquad H = ab\left(\frac{l}{h}\right)W. \tag{9.2.1b}$$

The thrust is thus obtained by *dividing by the sag h the moment of the vertical forces about C*, $M_V = R_A(al) = (bW)(al)$, i.e., the bending moment at C of a simply supported beam of span l under the same loads. H is seen to be proportional to the load and the span and inversely proportional to the sag. H is greatest when W is located at the midspan point $a = b = \frac{1}{2}$, where the product ab is greatest. Its value is then:

$$H_{max} = \frac{1}{4}\left(\frac{l}{h}\right)W. \tag{9.2.2}$$

For $r = h/l = \frac{1}{10}$, $H_{max} = 2.5W$.

By (9.2.1 a, b), the tension in the cable at any section D to the left of the load and at any section E to the right of the load equals, respectively:

$$T_D = \sqrt{H^2 + R_A^2} = H\sqrt{1 + \left(\frac{R_A}{H}\right)^2} = H\sqrt{1 + \left(\frac{r}{a}\right)^2},$$
$$T_E = \sqrt{H^2 + R_B^2} = H\sqrt{1 + \left(\frac{r}{b}\right)^2}. \tag{9.2.3}$$

When the load is at midspan, $a = b = \frac{1}{2}$ and:

$$T_D = T_E = T = H\sqrt{1 + 4r^2} \approx H(1 + 2r^2). \tag{9.2.4}$$

For $r = \frac{1}{10}$, $T = 1.02H$. In general, for small values of r, T is practically equal to H.

For two loads $W/2$ applied at the third points (Fig. 9.2.3), $R_A = R_B = W/2$, and a moment equation about C gives:

$$H = \left(\frac{W}{2}\right)\left(\frac{l}{3}\right)\left(\frac{1}{h}\right) = \frac{1}{6}\left(\frac{l}{h}\right)W. \tag{9.2.5}$$

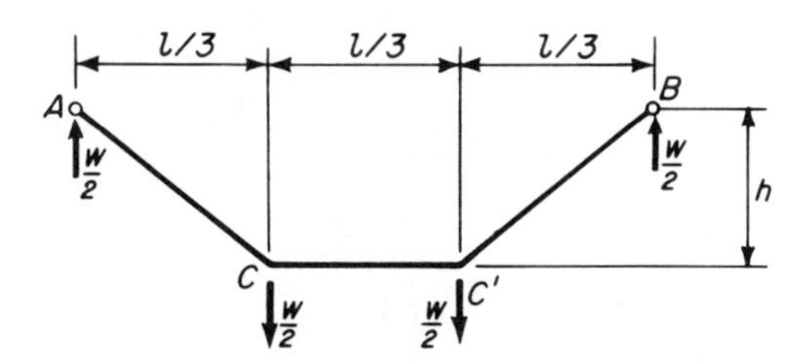

FIGURE 9.2.3

The tension T between C and C' equals H and between A and C is given by:

$$T = H\sqrt{1 + \frac{(W/2)^2}{H^2}} \approx H\left(1 + \frac{9}{2}r^2\right). \tag{9.2.6}$$

For $r = \frac{1}{10}$, T is only $4\frac{1}{2}\%$ larger than H.

For a load w (kN/m) uniformly distributed *horizontally* [Fig. 9.2.4(a)], the vertical reactions are $R_A = R_B = wl/2$, and the thrust is obtained by equating to zero the moments about C of R_A, H, and the load $wl/2$ on the left half of the cable:

$$Hh - \left(\frac{wl}{2}\right)\left(\frac{l}{2}\right) + \left(\frac{wl}{2}\right)\left(\frac{l}{4}\right) = 0,$$

from which:

$$H = \frac{wl^2}{8h}. \tag{9.2.7}$$

Once again, H is equal to the midspan moment $wl^2/8$ of a simply supported beam under the same loads divided by the sag:

$$H = \frac{M_{ss}}{h}. \tag{9.2.8}$$

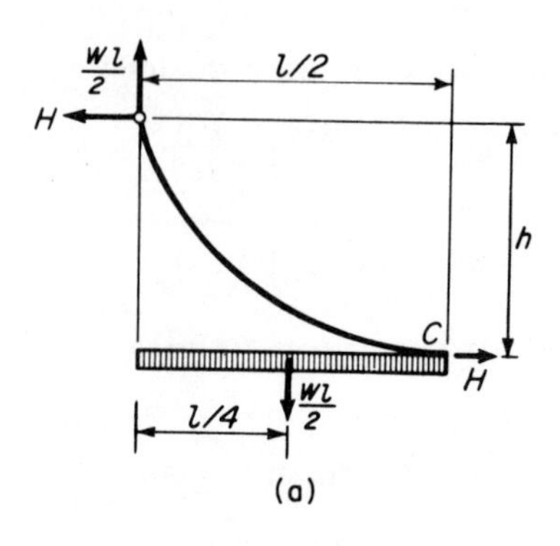

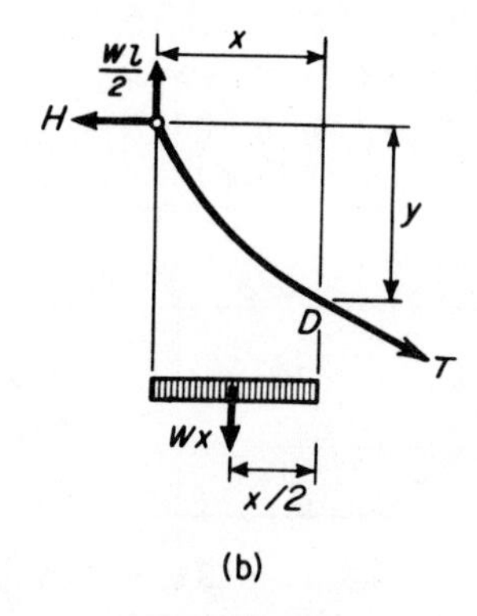

FIGURE 9.2.4

The maximum tension T at the supports of a uniformly loaded cable is given by:

$$T_{\max} = H\sqrt{1 + \left(\frac{wl/2}{H}\right)^2} = H\sqrt{1 + 16r^2} \approx H(1 + 8r^2). \qquad (9.2.9)$$

To determine the cable shape, we take moments about a point D of coordinates x, y [Fig. 9.2.4(b)]:

$$Hy - \frac{wl}{2}x + wx\left(\frac{x}{2}\right) = 0,$$

from which:

$$y = \frac{w}{2H}(lx - x^2). \qquad (a)$$

Substituting in (a) the value of H given by (9.2.7), we obtain:

$$y = 4h\left(\frac{x}{l} - \frac{x^2}{l^2}\right), \qquad (9.2.10)$$

the equation of a quadratic parabola.

When a cable carries a uniform load plus symmetrical, concentrated, vertical loads, its shape consists of a series of parabolic arcs, but its thrust, and hence for small sags its tension T, can always be evaluated by (9.2.8). For example, the thrust in a cable carrying a uniform load w and two concentrated loads $W_1 = wl/2$ at the third points equals:

$$H = \frac{1}{h}\left(\frac{wl^2}{8} + \frac{wl}{2}\frac{l}{3}\right) = \frac{14}{48}\frac{wl^2}{h} \approx \frac{2}{7}\left(\frac{l}{h}\right)wl = \frac{1}{7}\left(\frac{l}{h}\right)W,$$

where $W = 2wl$ is the total load on the cable.*

The tensile stress f_t in the cable is:

$$f_t = \frac{T}{A_r},$$

where A_r is the cable *resisting area*. The minimum cable resisting area is:

$$A_r = \frac{T}{F_t}, \tag{b}$$

where F_t is the allowable stress in tension.

Due to the twisted-strand cable construction, the cable resisting area is approximately two-thirds of its gross area A. The allowable stress F_t in a cable is about one-third of its ultimate strength F_u (see Table 9.2.4).

Example 9.2.1. A passageway between two buildings spans 15 m (50 ft), is 3 m (10 ft) wide, and must carry 5 kN/m² (100 psf) of live load plus its own dead load, estimated at 5 kN/m² (100 psf). We wish to suspend the roadway from two cables with a sag of 3 m (10 ft). Determine the cable diameter for $F_t = 345$ MPa (50 ksi).

With:

$$l = 15 \text{ m} \qquad h = 3 \text{ m}, \qquad r = \frac{h}{l} = \frac{1}{5},$$

$$w = \tfrac{1}{2}(5 + 5)(3) = 15 \text{ kN/m (1 k/ft)},$$

(9.2.7) gives:

$$H = \frac{15 \times (15)^2}{8 \times 3} \approx 141 \text{ kN (31 k)},$$

and, by (9.2.9):

$$T_{\max} = [1 + 8(\tfrac{1}{5})^2](141) \approx 186 \text{ kN (41 k)}.$$

With $F_t = 345$ MPa (50 ksi), (b) gives:

$$A_r = \frac{186(1\,000)}{345} = 539 \text{ mm}^2,$$

and a gross area:

$$A = \tfrac{3}{2}A_r = 809 \text{ mm}^2,$$

with a diameter $d = 32$ mm (1.25 in.).

*For a distributed load with different intensities $w_1, w_2, \ldots$ along different sections of the cable, the evaluation of H requires the previous location of the section where the beam moment becomes maximum. That section is easily identifiable in most cases.

Example 9.2.2. Determine the length l_u of the longest cable barely capable of carrying its own weight, assuming the sag-span ratio to be small.

The weight per unit length of a cable of resisting area A_r is equal to the volume, $A_r \times 1$, of a unit length of cable times the weight ρ per unit volume of the cable steel:

$$w = A_r \rho.$$

Hence, by (9.2.7) and (9.2.9):

$$T_{\max} = (1 + 8r^2)\frac{wl^2}{8h} = = (1 + 8r^2)\frac{wl}{8h/l} = \frac{1 + 8r^2}{8r} A_r \rho l, \tag{a}$$

from which:

$$f_t = \frac{T_{\max}}{A_r} = \frac{1 + 8r^2}{8r} \rho l.$$

Equating f_t to the cable's ultimate strength F_u and solving for l_u, we obtain:

$$l_u = \frac{8r}{1 + 8r^2} \frac{F_u}{\rho}.$$

For $F_u = 1\ 380$ MPa (200,000 psi), $\rho = 77$ kN/m³ (0.29 lb/cu in.), and $r = \frac{1}{10}$:

$$l_u = \frac{\frac{8}{10}}{1 + (\frac{8}{100})} \frac{1\ 380 \times 1\ 000^2}{77 \times 1\ 000} = 13\ 275 \text{ m} = 13.275 \text{ km (8.25 mi)}.$$

For $r = \frac{1}{3}$, $l_u \approx 25.30$ km (15.72 mi).

Example 9.2.3. Given a cable of span l, uniformly loaded in horizontal projection, a small sag makes the cable length L small, i.e., almost equal to its minimum value l, but makes $T_{\max}$ large and hence requires a large gross area A. Instead, a large sag makes L larger, but $T_{\max}$ and hence A small. For what value of r is the cable volume $V = AL$ minimum?

Indicating by W the total load on the cable:

$$W = wl,$$

(9.2.7) and (9.2.9) give:

$$T_{\max} = \frac{1 + 8r^2}{8r} W,$$

and hence a gross area:

$$A = \frac{3}{2} A_r = \frac{3}{2} \frac{T_{\max}}{F_t} = \frac{3}{2} \frac{1 + 8r^2}{8r} \frac{W}{F_t}.$$

The cable length for sags of less than $\frac{3}{10}$ is given by (9.1.7); hence, the cable volume equals:

$$V = AL = \left(\frac{3}{2} \frac{1 + 8r^2}{8r}\right)\left(1 + \frac{8}{3} r^2\right)\frac{Wl}{F_t}. \tag{a}$$

For given W, l, and F_t, the volume is minimum for that value of r which makes the coefficient:

$$C(r) = \frac{(1 + 8r^2)(1 + \frac{8}{3}r^2)}{8r}$$

a minimum, i.e., for $r \approx 0.3$, as can be seen by trial and error. Thus, by (a), the minimum volume equals:

$$V_{\min} = 1.35\frac{Wl}{F_t};$$

by (9.1.7) the optimal length for a span l is:

$$L = 1.24l,$$

and the optimal sag is:

$$h = 0.3l.$$

Example 9.2.4. Determine the cross section of the cables holding a circular roof of the type considered in Example 9.1.1 (Fig. 9.1.6).

With a roof load w_0 (kN/m²), the load per unit length of cable varies linearly from cw_0 (kN/m) at the outer ring of radius R to zero at the center, so that the unit load on the cable is (Fig. 9.2.5):

$$w(x) = \frac{cw_0}{R}x \quad \text{kN/m.}$$

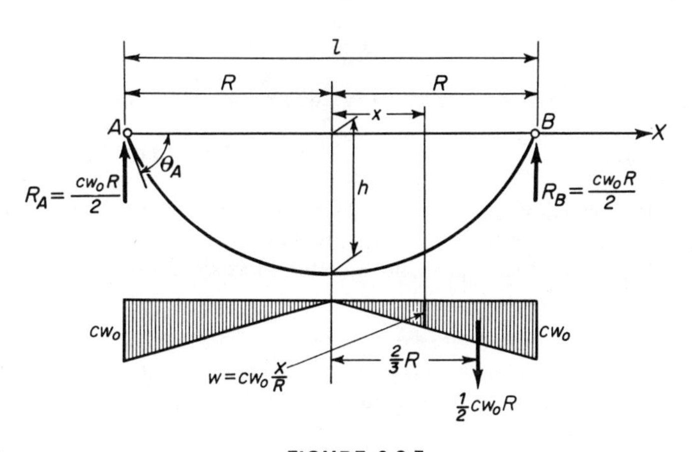

FIGURE 9.2.5

The reactions equal:

$$R_A - R_B = \tfrac{1}{2}cw_0R,$$

and the midspan beam-moment equals:

$$M_{ss} = \tfrac{1}{2}(cw_0R)(R) - \tfrac{1}{2}(cw_0R)(\tfrac{2}{3}R) = \tfrac{1}{6}cw_0R^2.$$

Hence, by (9.2.8), with $l = 2R$ and $r = l/h$:

$$H = \frac{1}{6}\frac{cw_0R^2}{h} = \frac{cw_0R}{12r}.$$

$T_{\max}$ may be obtained by noticing that, if θ_A is the angle between the tangent to the cable and the horizontal at A:

$$H = T_{\max}\cos\theta_A \quad \therefore \quad T_{\max} = \frac{H}{\cos\theta_A}. \tag{a}$$

From Fig. 9.1.6, $\theta_A = 36°$, $\cos\theta_A = 0.81$, and $T_{\max} = (1/0.81)H = 1.23H.$

For $r = \frac{1}{8}$, $R = 45$ m (150 ft), $c = 2\pi R/100 = 2.83$ m (9.4 ft), $h = 11.5$ m (37.5 ft), and $w_0 = 5$ kN/m² (100 psf), we obtain:

$$H = \frac{2.83 \times 5 \times 45}{12(1/8)} = 425 \text{ kN (95.5 k)},$$

$$T_{\max} = 1.23 \times 425 \simeq 525 \text{ kN (118.0 k)},$$

and with $F_t = 345$ MPa (50,000 psi):

$$A_r = \frac{525 \times 1\,000}{345} = 1\,522 \text{ mm}^2 \qquad A = \frac{3}{2} \times 1\,522 = 2\,283 \text{ mm}^2 \text{ (3.54 sq in.)},$$

$$d = 53.9 \text{ mm (2.12 in.)}.$$

It may be shown that for a parabola with a small sag:

$$\cos\theta_A \approx 1 - 8r^2, \qquad \frac{1}{\cos\theta_A} \approx 1 + 8r^2. \tag{9.2.11}$$

Approximating $\cos\theta_A$ in this problem by (9.2.11) we obtain [by (a)]:

$$T_{\max} = \frac{(1 + 8r^2)}{12r} c w_0 R, \tag{9.2.12}$$

and for $r = \frac{1}{8}$, $R = 45$ m, $c = 2\pi R/100 = 2.83$ m, $w_0 = 5$ kN/m²:

$$T_{\max} = \frac{(1 + 8/64)}{12(1/8)} 2.83 \times 5 \times 45 = 477.5 \text{ kN (106 k)},$$

an acceptable approximation for $T = 525$ kN, which was obtained graphically. Thus, (9.2.11) allows an evaluation of $T_{\max}$ without requiring the derivation of the funicular polygon, and this is probably a better approximation of $T_{\max}$, since it is based on a funicular curve.

Example 9.2.5. Determine the reactions and the sag h_D at D for the cable shown in Fig. 9.2.6, whose supports are not on the same level. (The sag h_C is given and equals 4 m).

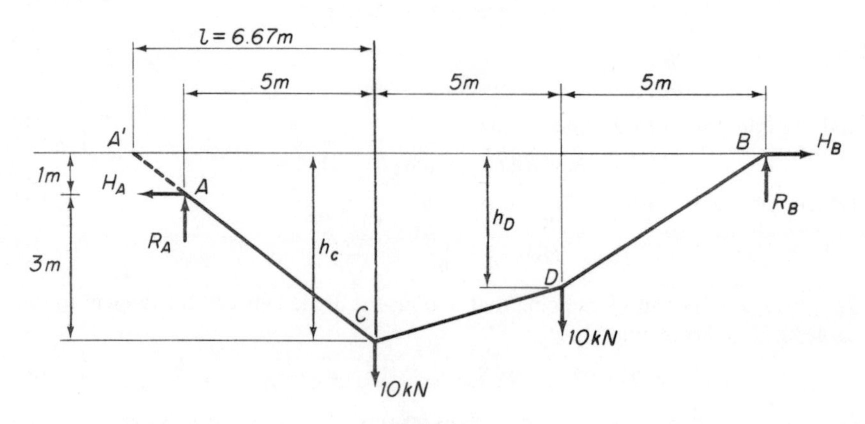

FIGURE 9.2.6

The point A' is at the same level as B on the prolongation of the polygon side AC, so that:

$$l = (4)\tfrac{5}{3} \doteq 6.67 \text{ m}.$$

Equilibrium in rotation about A' gives:

$$10(6.67) + 10(11.67) - R_B(16.67) = 0 \quad \therefore \quad R_B = 11.0 \text{ kN}.$$

Vertical equilibrium requires that:

$$R_A + R_B - 10 - 10 = 0,$$

from which:

$$R_A = 20 - 11 = 9.0 \text{ kN}.$$

Equating to zero the moment about C of the loads to the left of C gives:

$$5R_A - 3H_A = 0.$$

Therefore:

$$H_A = \tfrac{5}{3}(9) = 15.0 \text{ kN},$$

and for horizontal equilibrium:

$$H_B = H_A = 15.0 \text{ kN}.$$

To determine h_D, equate to zero the moment about D of the loads to the right of D, which gives:

$$5R_B - h_D H_B = 0,$$

$$h_D = 5\frac{R_B}{H_B} = 5\left(\frac{11.0}{15.0}\right) = 3.33 \text{ m}.$$

9.3 CABLE DISPLACEMENTS AND CABLE STABILITY [6.1, 2.6]

As a cable stretches elastically under load, its original length L increases by ΔL and its sag h by Δh. The lowering Δh of the lowest point in the cable may be readily evaluated *for a single load at midspan*. Indicating by $\Delta r = \Delta h/l$ the increase in the sag-span ratio r, we have from Table 9.1.1:

$$L = l(1 + 2r^2), \tag{9.3.1a}$$

$$L + \Delta L = l[1 + 2(r + \Delta r)^2] \approx l(1 + 2r^2 + 4r\,\Delta r), \tag{9.3.1b}$$

where Δr is assumed to be small enough for $2(\Delta r)^2$ to be negligible in comparison with the other terms in parentheses. Subtracting (9.3.1a) from (9.3.1b) we obtain:

$$\Delta L = 4r(l\,\Delta r) = 4r\,\Delta h. \tag{9.3.2}$$

The elongation ΔL can also be computed as the product of the strain $\epsilon = f_t/E$ times L:

$$\Delta L = \frac{f_t}{E}L = \frac{f_t}{E}l(1 + 2r^2). \tag{9.3.3}$$

Equating (9.3.2) to (9.3.3) and solving for $\Delta h/l$, we obtain:

$$\frac{\Delta h}{l} = \frac{1 + 2r^2}{4r}\frac{f_t}{E}. \tag{9.34}$$

For $r = 0.1$, (9.3.4) gives:

$$\frac{\Delta h}{l} = 2.55\frac{f_t}{E}. \tag{9.3.4a}$$

For $E = 200\,000$ MPa (29×10^6 ksi) and $f_t = 690$ MPa (100,000 psi), the elastic displacement is $\Delta h = \frac{1}{115}l$; for $f_t = 345$ MPa (50,000 psi), $\Delta h = \frac{1}{230}l$. For $r = 0.3$ and $f_t = 690$ MPa (100,000 psi), $\Delta h = \frac{1}{295}l$.

In a *uniformly loaded* cable, the stress, and hence the strain, vary from point to point along the cable, but an estimate of the displacement at midspan may be obtained by making use of an average strain, determined by means of the average between $T_{\min} = H$ and $T_{\max}$. Thus, by (9.2.9):

$$T_{\max} = H(1 + 8r^2), \qquad T_{\text{av}} = \tfrac{1}{2}(H + T_{\max}) = H(1 + 4r^2),$$

$$f_{t,\text{av}} = \frac{T_{\text{av}}}{A}, \qquad \epsilon_{\text{av}} = \frac{f_{t,\text{av}}}{E}, \tag{9.3.5}$$

$$\Delta L = \epsilon_{\text{av}} L = (1 + 4r^2)\frac{HL}{EA}.$$

But, by (9.1.7):

$$L = l(1 + \tfrac{8}{3}r^2),$$

$$L + \Delta L = l[1 + \tfrac{8}{3}(r + \Delta r)^2] \approx l(1 + \tfrac{8}{3}r^2 + \tfrac{16}{3}r\,\Delta r),$$

$$\therefore \quad \Delta L = \tfrac{16}{3}r(l\,\Delta r) = \tfrac{16}{3}r\,\Delta h. \tag{9.3.6}$$

Equating (9.3.6) to (9.3.5) and solving for $\Delta h/l$, we obtain:

$$\frac{\Delta h}{l} = \frac{3 + 8r^2}{16r}\frac{f_{t,\text{av}}}{E}. \tag{9.3.7}$$

For $r = 0.1$, (9.3.7) gives:

$$\frac{\Delta h}{l} = 1.93\frac{f_{t,\text{av}}}{E}, \tag{9.3.7a}$$

or, with $f_{t,\text{av}} = 690$ MPa (100,000 psi), $\Delta h = \frac{1}{150}l$. Equation (9.3.7a) may be used to determine the elastic displacement at midspan for most cables loaded by nonconcentrated loads.

The small elastic displacements of a cable must not be confused with its large movements due to a *change* in loading. For example, let h_a be the sag of a cable due to a concentrated load at al, and h the sag in the same cable due to a concentrated load at $l/2$. Equating the corresponding cable lengths given by (9.1.5) and letting $r_a = h_a/l$, $r = h/l$, we obtain:

$$l\left(1 + \frac{1}{2ab}r_a^2\right) = l(1 + 2r^2),$$

from which, solving for r_a/r:

$$\frac{r_a}{r} = \frac{h_a}{h} = 2\sqrt{ab}. \tag{9.3.8}$$

Table 9.3.1 gives the values of h_a/h in terms of a.

Table 9.3.1

a	0.1	0.2	0.3	0.4	0.5
h_a/h	0.6	0.8	0.92	0.98	1.0

As a concentrated load moves from the midspan point ($a = 0.5$) to the tenth point ($a = 0.1$), the sag decreases from h to $0.6h$, i.e., by 40%, and the midspan point (Fig. 9.3.1) moves up by:

$$\Delta y = 0.4h + (\tfrac{4}{9})(0.6h) = \tfrac{2}{3}h.$$

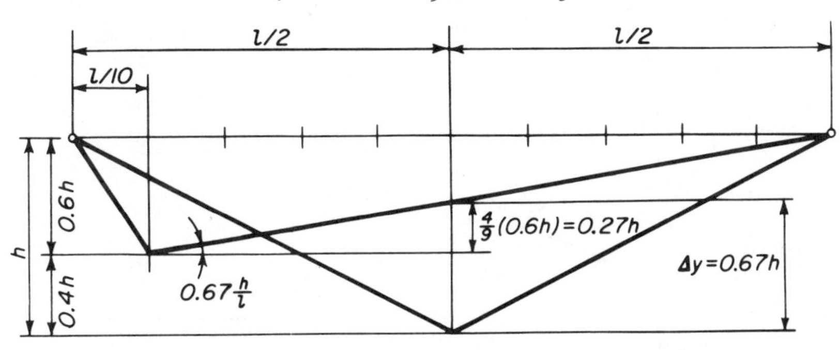

FIGURE 9.3.1

For $h/l = \frac{1}{10}$, the movement at midspan equals $(\frac{2}{3})(\frac{1}{10})l = l/15$ and is 10 to 15 times the elastic displacement given by (9.3.4a).

It is thus seen how, in order to carry changing loads by simple tension, cables must change shape: cables are *unstable* structures. In order to reduce excessive movements due to changing loads, which would be incompatible with practical structural requirements, cables must be *stabilized*: this is usually done by *prestressing* them.

Prestressing may be obtained either by preloading the cable or by stressing it through the action of transverse forces. Stabilization through stiffening trusses is used in suspension bridges. The stiffening effect of the truss is due to its flexural rigidity and to its weight. The effect of the weight alone is illustrated by the example of Fig. 9.3.2.

Indicating by h the sag of a cable stabilized by a uniform load of w(kN/m) [Fig. 9.3.2(a)], it may be shown that the sag of the same cable under a midspan concentrated load [Fig. 9.3.2(b)] equals $1.15h$, and that the cable section at $x = l/10$ has an ordinate equal to $0.23h$. The sag of the cable under the stabilizing load w *and* a midspan concentrated load $W_1 = wl/10$ [Fig. 9.3.2(c)] may be shown to equal $1.04h$. The cable section at $x = l/10$ may be shown to have, in this case, an ordinate equal to $0.35h$.

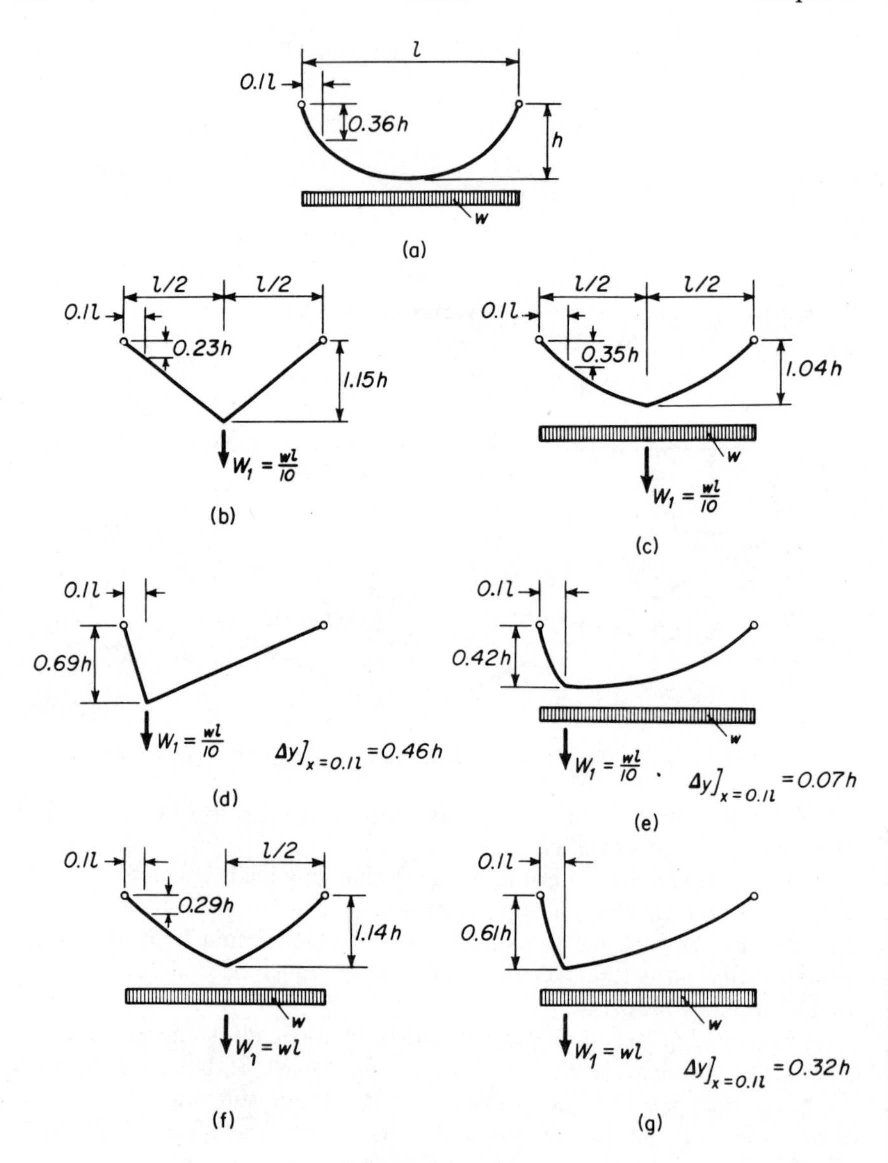

FIGURE 9.3.2

If the concentrated load is now moved to the section $x = l/10$, this section moves down by $0.46h$ (to $0.69h$) in the unstabilized case [Fig. 9.3.2(d)], and by only $0.07h$ (to $0.42h$) in the cable stabilized by the uniform load [Fig. 9.3.2(e)]. Hence, the uniform load reduces the displacement of the concentrated load at $x = l/10$ by 0.46/0.07, or 6.5 times.

On the other hand, for $W_1 = wl$ [Fig. 9.3.2(f), (g)] the uniform load reduces the displacement at $x = l/10$ by a factor of only 1.5, indicating that the ratio of uniform load to concentrated load needed to stabilize the cable is high.

Wind loads are one of the essential causes of cable instability. Relatively small wind velocities may actually induce aerodynamic oscillations capable of seriously damaging or even destroying a cable-supported structure, as in the case of the Tacoma Narrows Bridge which collapsed in 1940. Aerodynamic instability may be avoided, at times, by changing the natural periods of oscillation P_n of the cable. For cables with a small sag the periods P_n are given by:

$$P_n = \frac{2l}{n}\sqrt{\frac{w/g}{T}} \quad (n = 1, 2, 3, \ldots), \tag{9.3.9}$$

where g is the acceleration of gravity 9.81 m/sec² (32.2 ft/sec²). For $n = 1$, (9.3.9) gives the *fundamental period* P_1 of the up and down motion of the cable. For $n = 2$, we obtain the period for the antisymmetrical oscillations of the cable, which produce severe torsion in a suspension bridge when the oscillations of the two cables are out of phase.

Example 9.3.1. A bicycle wheel roof is supported by two sets of cables, connecting an outer compression ring to a center hub. The cables of the two sets are kept apart by vertical compressive spreaders which put the cables in tension (Fig. 9.3.3). Determine the change in period of a lower cable of a bicycle wheel roof due to the spreading of the cables.

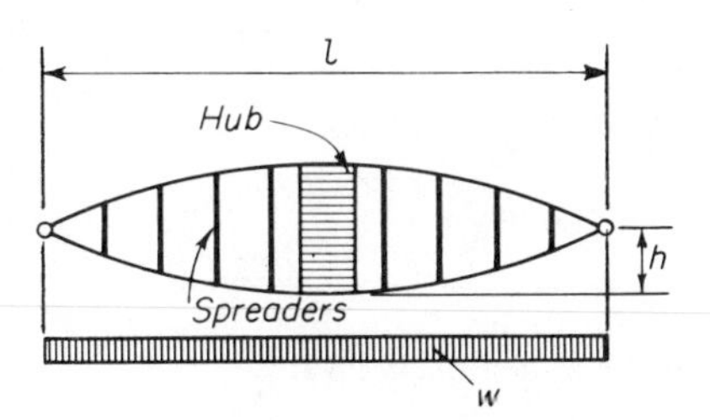

FIGURE 9.3.3

Let the lower cables have the following characteristics: $l = 30$ m (100 ft), $h = 3$ m (10 ft), ($r = 0.1$), $w = 7$ kN/m (500 lb/ft), $F_t = 345$ MPa (50,000 psi) and, hence, by (9.2.9) and (9.2.7):

$$T = H(1 + 8r^2) = \frac{wl^2}{8h}(1 + 8r^2) = 283.5 \text{ kN (67.5 k)},$$

$$A = \frac{283.5 \times 1\,000}{345} = 822 \text{ mm}^2 \text{ (1.35 sq in.)},$$

$$L = (1 + \tfrac{8}{3}r^2)l = 30.8 \text{ m (103 ft)},$$

and, by (9.3.9), a fundamental period ($n = 1$):

$$P_1 = 2 \times 30\sqrt{\frac{7/9.81}{283.5}} = 3.01 \text{ sec.}$$

Let us spread apart the two identical sets of upper and lower cables so as to increase the sag of the lower cables by 100 mm (4 in.). By (9.3.6):

$$\Delta L = \frac{16}{3}\left(\frac{1}{10}\right)100 = 53.3 \text{ mm},$$

and, hence:

$$\epsilon = \frac{\Delta L}{L} = \frac{53.3}{30.8 \times 1\,000} = 0.001\,73 \text{ mm/mm},$$

$$\Delta f_t = E\epsilon = 200\,000 \times 0.001\,73 = 346 \text{ MPa (52.5 ksi)},$$

$$\Delta T = A\,\Delta f_t = 822 \times \frac{346}{1\,000} = 284.4 \text{ kN (71 k)},$$

$$T' = T + \Delta T = 568 \text{ kN (142 k)},$$

$$f_t' = \frac{568 \times 1\,000}{822} = 691 \text{ MPa (103 ksi)},$$

$$P_1' = 2 \times 30\sqrt{\frac{7/9.81}{568}} = 2.17 \text{ sec} = 0.72P_1.$$

The period P_1 is shortened by 28%, with an increase in stress from 345 MPa to 691 MPa.

It is worth noticing that an increase in cable tension of 284.4 kN could have been obtained by a uniform load given approximately by (9.2.7):

$$\frac{wl^2}{8h} = 284.4 \quad \therefore \quad w = \frac{284.4 \times 8 \times 3}{30^2} = 7.58 \text{ kN/m},$$

but the cable period would then have been:

$$P_1 = 2 \times 30\sqrt{\frac{(7 + 7.58)/9.81}{568}} = 3.07 \text{ sec},$$

and the additional tension in the cable would have left the period practically unchanged, because of the added mass on it.

PROBLEMS

9.1 A cable spanning 30 m (100 ft) carries a moving trolley weighing 35 kN (8 k). The cable is 33 m (110 ft) long and the trolley moves to within 3 m (10 ft) of either end of the span. Determine the positions of the trolley [to the nearest 3 m (10 ft)] causing the maximum and minimum tension in the cable and the value of these tensions.

9.2 When two loads of 55 kN (12 k) each are placed at the third points of a cable spanning 18 m (60 ft), the cable sags 3.6 m (12 ft). Vertical cable guys are attached to the points of application of the loads. What is the force in the guy wires if the load on the right is increased by 35 kN (8 k)? Determine the force in the cable before and after the application of the additional load, assuming cable and guy wires to be inextensible.

9.3 In the cable of Problem 9.2 what are the vertical displacements of the cable at the point of application of the loads if no guys are used to balance the added load?

9.4 What is the thrust in a cable of span l and sag h, carrying a uniform load w plus three concentrated loads $W = wl/4$ at the quarter points?

9.5 A series of rigid beams project 15 m (50 ft) from the back of a grandstand to form a roof and are supported as shown in the adjoining sketch. Each beam carries a uniformly distributed load from the roof of 15 kN/m (1 k/ft). Determine the cable tension T_1, T_2, and T_3. See Fig. P9.5.

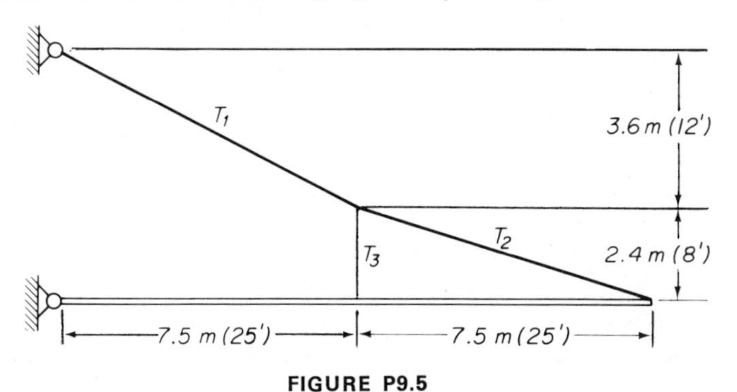

FIGURE P9.5

9.6 In order to increase the rigidity of a cable-supported roof, a form of prestressing may be used. Consider, for instance, a roof spanning 24 m (80 ft), consisting of a series of parallel cables 600 mm (2 ft) o.c. with a sag of 3 m (10 ft), covered with panels of concrete 75 mm (3 in.) thick. The concrete panels are hung on the cables and uniformly loaded with sandbags weighing 5 kN/m² (100 psf). The joints between the panels are then grouted. The sandbags are removed once the grout has set. The concrete panels are thus compressed by the tension in the cables, making the roof monolithic and adding to the bending resistance of the prestressed slabs. If the live load is 1.5 kN/m (30 psf) and shortening of the cables due to the transfer of stress in prestressing the slab is neglected, what is the maximum cable tension and the required cable diameter for $F_t = 550$ MPa (80 ksi)?

9.7 A cable car weighing 20 kN (4,000 lb) crosses a 120 m (400 ft) wide ravine. If one anchorage is 30 m (100 ft) higher than the other, and the length of the cable is 132 m (440 ft), determine the tension in the two portions of the cable when the car is (a) at midspan, (b) at the quarter point nearest to the higher support.

9.8 A cable with a span of 22.5 m (75 ft) and a sag of 4.5 m (15 ft) carries a uniform load of 7.5 kN/m (500 lb/ft). (a) Determine the cable size for $F_t = 550$ MPa (80 ksi). (b) Determine the elastic displacement at the center.

9.9 Determine the fundamental period of the cable of Problem 9.6. Determine the fundamental period when the roof is not prestressed.

9.10 A horizontal suspended roof 6 m (20 ft) above ground has a center span 30 m (100 ft) long and two side spans 15 m (50 ft) long. The roof cables, 6 m (20 ft) o.c., are supported by two internal piers 9 m (30 ft) high. The ends of each cable are anchored to a compressive concrete beam which hangs from the cables and supports a concrete roof slab spanning between beams. The slab is 18 cm (7 in.) thick, the beam is 300 mm (1 ft) wide and 600 mm (2 ft) deep, the roof load is 4 kN/m^2 (80 psf). Determine the compression in the beam, the maximum tension in the cable, and the loads W on the piers.

9.11 Two men each weighing 0.8 kN (160 lb) lift a weight of 0.25 kN (50 lb) by pulling at the two ends of a cable spanning 9 m (30 ft) between two pulleys 9 m (30 ft) above the ground. The weight is attached to the center of the cable, and the two men are 9 m (30 ft) away from the verticals through the pulleys. What is the maximum height to which the men can lift the weight if they can pull on the cable with a force equal, at most, to their weight?

9.12 A cable is stabilized by guy wires attached at its third points and inclined 60° to the horizontal on either side of the cable. If the cable sag is 6 m (20 ft) and its span is 27 m (90 ft), what is the maximum tension in the cable due to a tension of 1 kN (200 lb) in the guy wires?

9.13 A tightrope walker weighs 0.8 kN (160 lb). What is the minimum sag in a 6 mm ($\frac{1}{4}$ in.) wire with $F_t = 620$ MPa (90,000 psi) spanning 30 m (100 ft) necessary to support the man at (a) the center of the span, (b) the quarter point?

9.14 What is the minimum sag at midspan for the rope of Problem 9.13 if three men, weighing 0.8 kN (160 lb) each, stop at the center of the span on a bicycle, assuming a point loading for the bicycle?

9.15 The center pole of a circus tent weighs 0.8 kN/m (50 lb/ft) and is 15 m (50 ft) long. In order to erect it, the pole is hinged at its base, and a rope is attached at a point 3.5 m (12 ft) from its top. The rope then goes over a pulley atop a vertical auxiliary pole 4.5 m (15 ft) high and is pulled by a winch at its foot. If the maximum allowable tension in the rope is 22 kN (5,000 lb), what is the maximum horizontal distance between the foot of the auxiliary pole and the point where the rope is attached to the main pole, for the rope to just begin to lift the main pole without exceeding the rope's capacity?

10. Arches

10.1 FUNICULAR ARCHES [6.4, 8.4]

Let a cable take its funicular shape under a given set of loads [Fig. 10.1.1(a)]. Overturn the cable and, then, let its shape be "frozen" by assuming that the cable can stand compression [Fig. 10.1.1(b)]. The frozen cable becomes a *funicular polygonal arch*, every section of which is under simple compression. [A funicular arch cannot be as slender as a cable of the same material, say steel, since a very slender arch would buckle under compression (Fig. 10.1.2).] The *pressure line* of the polygonal arch, i.e., the line of the resultant force on each section, coincides with the centerline of the arch. The compressive thrust in the arch [Fig. 10.1.1(b)] exerts an *outward* pressure on the arch abutments. It can be computed by (9.2.8) and is inversely proportional to the arch *rise* h. To reduce the cost of the abutments, arch rises are sometimes larger than cable sags.

Arch action differs from cable action in one basic respect: if the loading on a cable changes, the cable changes shape and *remains funicular* (it can do so because of its flexibility); on the other hand, if the loads on a funicular arch change, the arch maintains its shape because of its stiffness and therefore *cannot* become funicular for the *new* loading condition. While a cable is funicular, i.e., momentless, for all loads, an arch is momentless for only *one* loading condition; any other condition introduces bending in the arch. It is easy to compute such arch bending in an elementary case.

Example 10.1.1. The parabolic arch of Fig. 10.1.3 (a) is funicular for a load w (kN/m) uniform in horizontal projection. Determine the maximum bending moment due to a load P concentrated at its crown D for $h = l/2$.

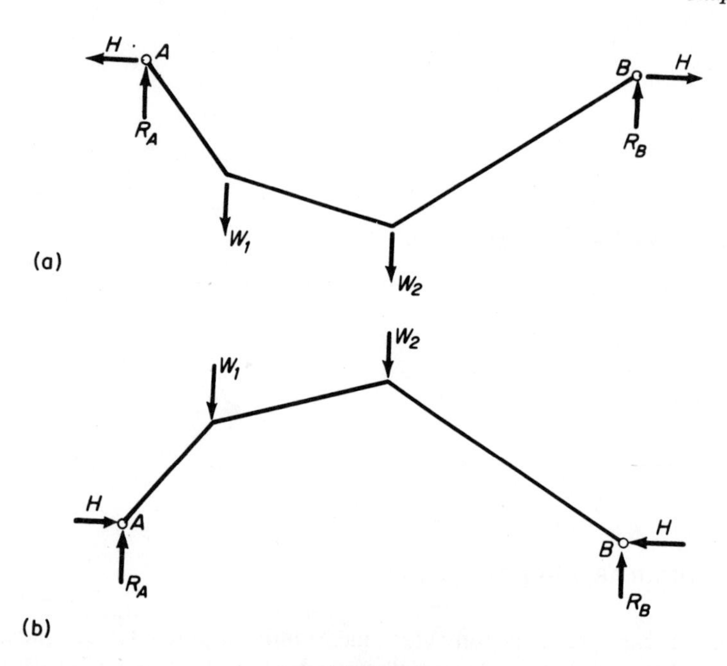

FIGURE 10.1.1

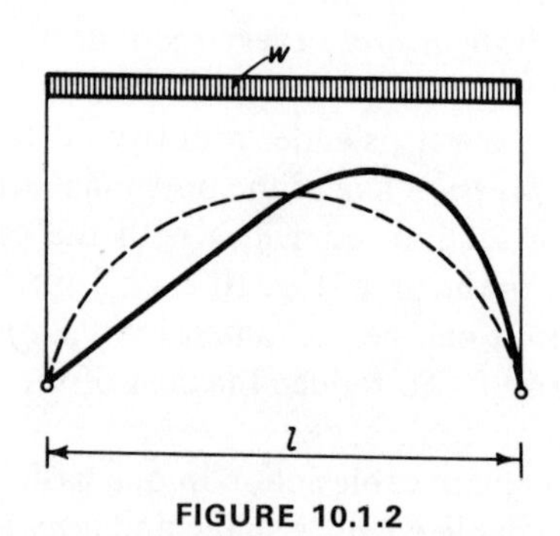

FIGURE 10.1.2

By (9.1.6), the centerline of the arch, with the y-axis positive *upward*, is the parabola:

$$y = 4h\left(\frac{x}{l} - \frac{x^2}{l^2}\right), \tag{a}$$

which can be drawn *by tangents* as shown in Fig. 10.1.4. The thrust in the arch due to the uniform load equals, by (9.2.8):

$$H = \frac{wl^2}{8h} \tag{b}$$

and is the horizontal component of the variable compression C in the arch. At any section of the arch, C is tangent to the centerline of the arch, since the parabolic arch is funicular for a load w uniform in horizontal projection. Hence, the maximum

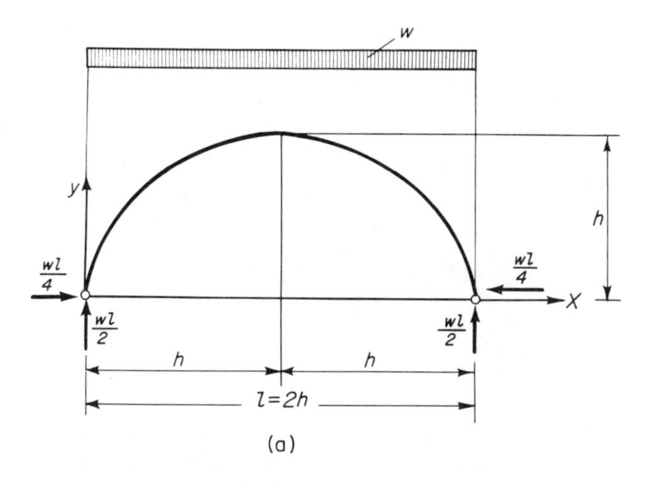

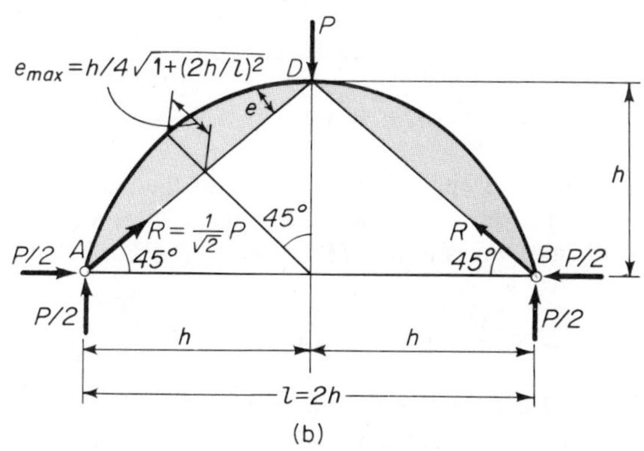

FIGURE 10.1.3

value $C_{\max}$ of C, which occurs at the abutments, is given by:

$$C_{\max} \cos \theta_A = H \quad \therefore \quad C_{\max} = \frac{H}{\cos \theta_A}, \tag{c}$$

where (Fig. 10.1.4):

$$\tan \theta_A = \frac{2h}{l/2} \quad \therefore \quad \cos \theta_A = \frac{1}{\sqrt{1 + \tan^2 \theta_A}} = \frac{1}{\sqrt{1 + 16h^2/l^2}},$$

so that:

$$C_{\max} = H\sqrt{1 + 16h^2/l^2} = \frac{wl^2}{8h}\sqrt{1 + 16h^2/l^2}. \tag{10.1.1}$$

At any other section x:

$$\tan \theta = \frac{dy}{dx} = 4\frac{h}{l}\left(1 - 2\frac{x}{l}\right), \quad \frac{1}{\cos \theta} = \sqrt{1 + 16\frac{h^2}{l^2}\left(1 - 2\frac{x}{l}\right)^2}, \tag{d}$$

and:

$$C(x) = H\sqrt{1 + 16\frac{h^2}{l^2}\left(1 - 2\frac{x}{l}\right)^2}. \tag{10.1.2}$$

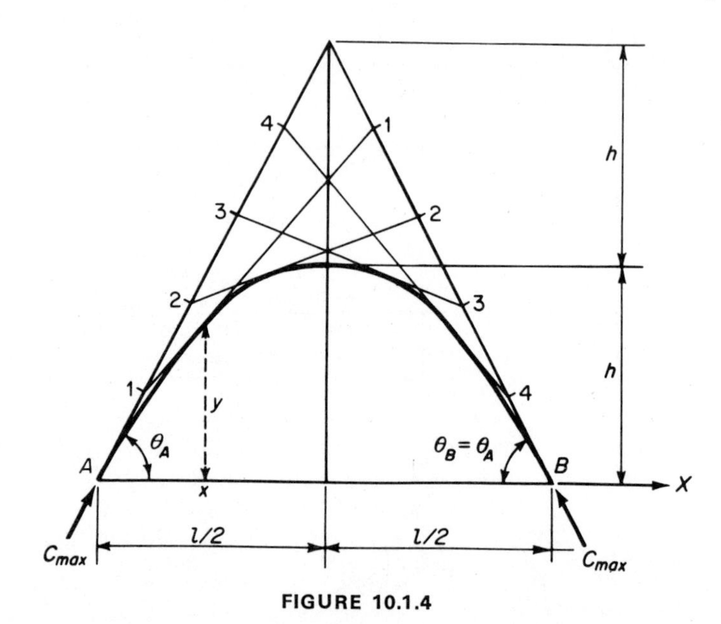

FIGURE 10.1.4

If a concentrated load P is applied at the top D of the parabolic arch [Fig. 10.1.3 (b)], the corresponding vertical reactions V equal $P/2$, and the simply supported beam moment at midspan is $M_{ss} = (P/2)(l/2) = (P/2)h$, so that the thrust equals:

$$H = \frac{M_{ss}}{h} = \left(\frac{1}{h}\right)\left(\frac{Ph}{2}\right) = \frac{P}{2}. \tag{e}$$

The resultant R of V and H is a 45° force of magnitude $P/\sqrt{2}$, and the pressure line for the concentrated load P at D is the triangle ADB. Hence, at every section of the arch, the resultants R exert a bending moment $M = (P/\sqrt{2})e$, where the eccentricity e is the normal distance between the 45° line of action of R and the centroid of the section. The eccentricity e is maximum at $x = h/2$ and equals $(h/4)\sqrt{1 + (2\,h/l)^2}$, so that the maximum bending moment equals:

$$M_{\max} = \frac{\sqrt{2}}{2}\left(\frac{Ph}{4}\right)\sqrt{2} = 0.125Ph. \tag{f}$$

The bending moments are responsible for a compressive stress at the inner fibers and a tensile stress at the outer fibers of the arch. The maximum compressive stresses resulting at any section from P and w may be obtained approximately by (5.1.1) with $P = R + C$ and $M = Re$, since the component of R perpendicular to the centerline of the arch is usually small.

Example 10.1.2. Determine the ultimate span l_u of a reinforced concrete arch of parabolic shape, small rise, and constant cross section A, capable of supporting its own weight.

The maximum funicular compression in the arch is provided by the same formula that gives the maximum tension in a cable, i.e., by (a) of Example 9.2.2:

$$C_{max} = \frac{1 + 8r^2}{8r} A\rho_c l \quad \left(r = \frac{h}{l}\right), \tag{a}$$

where ρ_c is the weight of concrete per unit volume. The compressive stress in the concrete thus becomes:

$$f_c = \frac{C_{max}}{A} = \frac{1 + 8r^2}{8r} \rho_c l.$$

Substituting the ultimate stress F_u for f_c and solving for the corresponding ultimate span l_u, we obtain:

$$l_u = \frac{8r}{1 + 8r^2} \frac{F_u}{\rho_c}. \tag{10.1.3}$$

For $F_u = 40$ MPa (5,800 psi), $\rho_c = 22.8$ kN/m³ (145 lb/cu ft), one obtains $F_u/\rho_c =$ 1 754 m (5,755 ft), which by means of (10.1.3) provides the spans of Table 10.1.1.

Table 10.1.1

	r	1/10	1/8	1/6	1/4	1/3	1/2
l_u	m	1 300	1 560	1 910	2 340	2 475	2 340
	(ft)	4,265	5,115	6,265	7,675	8,120	7,675

The maximum ultimate span is obtained for $r = \frac{1}{3}$ and is about 2.475 km (1.54 mi). With a coefficient of safety of 4, the maximum safe span is 619 m (2,030 ft) long. The longest concrete bridge built to date (the Krk Bridge in Yugoslavia) is 390 m (1,280 ft) long, or one-half the span evaluated above on the basis of compressive stresses due to the weight of the arch only.

Example 10.1.3. Evaluate the compression in the outer ring of the circular roof of Example 9.2.4.

The circle is the funicular shape for a uniform *radial* external pressure p_0 or internal tension t_0, as can be seen by obtaining the limit of the funicular shape of a cable acted upon by a set of evenly spaced, equal radial forces. The compression C in a circular ring of radius R due to t_0 (or p_0) kN/m (lb/ft) is found by considering the equilibrium of half a ring [Fig. 10.1.5 (a)]. The x-components of t_0 (or p_0) on two symmetrically located elements Δs cancel, and the y-components equal:

$$2(t_0 \Delta s) \cos\theta = 2t_0(\Delta s \times \cos\theta) = 2t_0 \Delta R \quad (\text{or } 2p_0 \Delta R),$$

where ΔR is the projection of Δs on the diameter AB. The total force in the y-direction equals $2t_0R$ (or $2p_0R$) and is equilibrated by the two compressive forces C; hence:

$$C = t_0 R \quad (\text{or } p_0 R). \tag{10.1.4}$$

Indicating by c (m) the spacing of the cables on the ring and by w (kN/m²) the weight of the roof, the vertical load on the cables (kN/m) varies linearly from wc

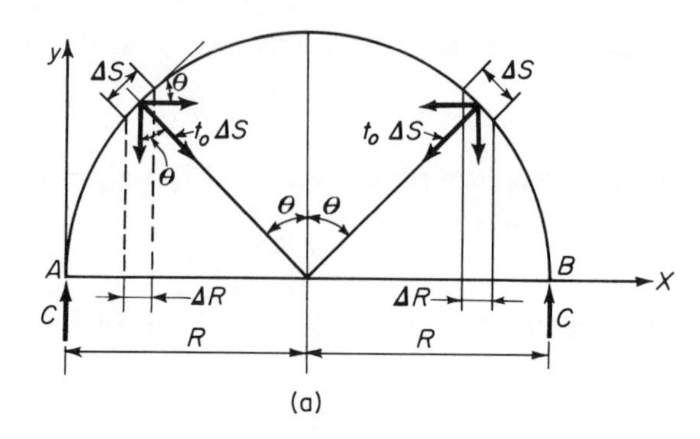

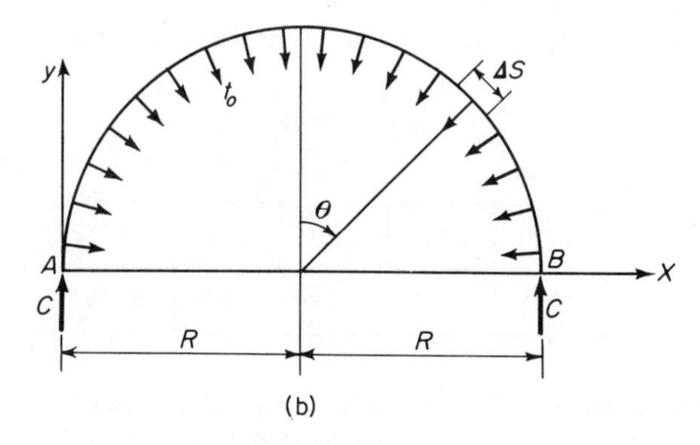

FIGURE 10.1.5

at the ring to zero at the center (Fig. 10.1.6). Hence, the total load on half a cable is $(wc)R/2$ and is applied at a distance $\frac{2}{3}R$ from the center. The simple-beam moment at the cable midspan is (Fig. 10.1.6):

$$M_{ss} = \left(\frac{wcR}{2}\right)R - \left(\frac{wcR}{2}\right)\left(\frac{2R}{3}\right) = \frac{wcR^2}{6},$$

and the thrust equals:

$$H = \frac{M_{ss}}{h} = \frac{wcR^2}{6h}.$$

The thrust t_0 per unit of ring perimeter is:

$$t_0 = \frac{H}{c} = \frac{wR^2}{6h},$$

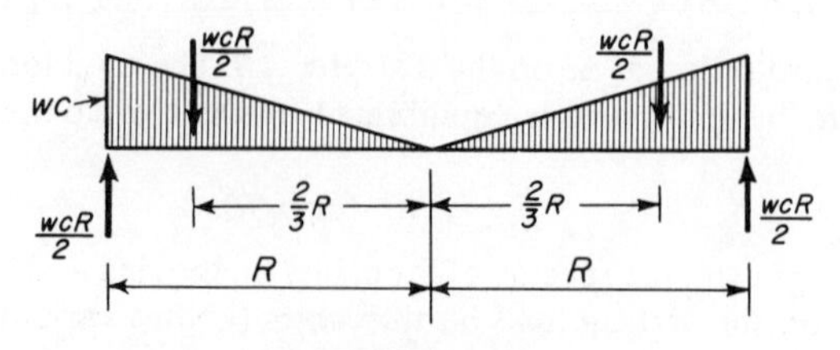

FIGURE 10.1.6

and the compression in the ring, by (10.1.4), becomes:

$$C = t_0 R = \frac{wR^3}{6h}. \tag{10.1.5}$$

For a roof with $R = 45$ m (150 ft), $h = 9$ m (30 ft), and $w = 5$ kN/m² (0.1 ksf), one obtains:

$$C = \frac{5 \times (45)^3}{6 \times 9} = 8\,438 \text{ kN } (1{,}896 \text{ k}).$$

A concrete ring with $F_c = 6.90$ MPa (1 ksi) requires an area of 1 223 000 mm² (1,895 sq in.), or a cross section of 800 × 1 600 mm (32 in. × 63 in.).

10.2 THREE-HINGE ARCHES [8.4]

In the arch of Example 10.1.1, the pressure line due to a concentrated load goes through the crown of the arch, so that the section D is momentless and could be built as a hinge. The arch behaves in this case as a *three-hinge arch* and is statically determinate. Because it is statically determinate, a three-hinge arch is insensitive to differential settlements of its abutments and to changes in its centerline due to thermal expansion.

The stresses in a three-hinge arch are readily determined by statics.

Example 10.2.1. The arch of Fig. 10.2.1 (a) carries a load P at the quarter point. Determine the stresses in the arch.

The vertical (simple-beam) reactions of the arch are $V_1 = \frac{3}{4}P$, $V_2 = \frac{1}{4}P$. The resultant R_2 of V_2 and H must go through the hinge D; hence, the thrust is given by [Fig. 10.2.1(b)]:

$$H \tan \theta_B = \frac{P}{4}, \qquad \tan \theta_B = \frac{h}{l/2} = 2r, \qquad H = \frac{P}{8r}, \tag{a}$$

and the resultant R_2 by:

$$R_2 = \sqrt{H^2 + (P/4)^2} = \frac{P}{4}\sqrt{1 + \left(\frac{1}{2r}\right)^2} = \frac{P}{8r}\sqrt{1 + 4r^2}. \tag{b}$$

The resultant R_1 equals:

$$R_1 = \sqrt{H^2 + \left(\frac{3P}{4}\right)^2} = \frac{P}{8r}\sqrt{1 + 36r^2}. \tag{c}$$

In the left half of the arch, the eccentricity produces compression in its outer fibers and tension in its inner fibers; in the right half of the arch, it produces compression in its inner fibers and tension in its outer fibers.

Example 10.2.2. A scalloped, small rise dome of parabolic cross section and radius R [Fig. 10.2.2(a)] consists of a series of meridional elements of semicircular cross section [Fig. 10.2.2(b)] meeting at the top of the dome. The elements are supported by a ring with a torsional rigidity small enough for the meridional elements to be considered hinged into it. The elements decrease linearly in cross section

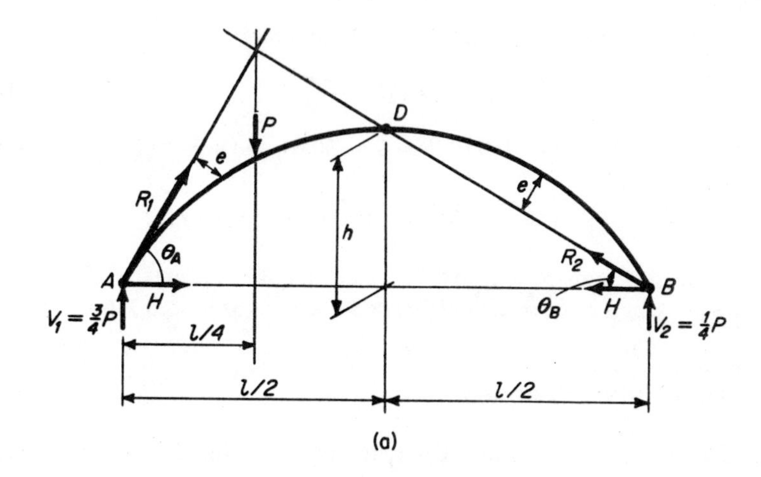

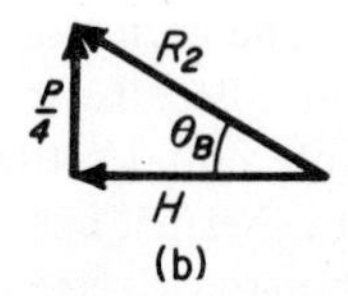

FIGURE 10.2.1

toward the top of the dome, and their thickness t is small enough for the elements to be considered hinged at the top. Hence, two diametrically opposite elements 1, 2 [Fig. 10.2.2(b)] constitute a three-hinge arch. Determine the maximum moment in this arch due to the dead load of the dome, assuming the slope of the dome small enough for $2R$ to be practically equal to the arch length L.

Indicating by $c = 2\pi R/n$ (n = number of elements) the diameter of the cross section of a meridional element *at the ring* [Fig. 10.2.2(b)], its perimeter is:

$$\frac{\pi}{2}c = \frac{\pi}{2}\frac{2\pi R}{n} = \frac{\pi^2}{n}R,$$

and its volume per unit of radial length $(\pi^2/n)Rt$. Indicating by ρ_c the specific weight of concrete, the weight per unit of radial length of element at the ring is:

$$w_0 = \frac{\pi^2}{n}\rho_c Rt \text{ (kN/m, lb/ft).} \tag{a}$$

Since the weight decreases linearly, the weight at a distance x from the ring is:

$$w(x) = w_0\left(1 - \frac{x}{R}\right) = w_0 - w_0\frac{x}{R}. \tag{b}$$

The constant dead load component w_0 does not produce moments in the parabolic arch, so that only the component $-w_0(x/R)$ need be considered (Fig. 10.2.3). The simple-beam moment due to this load is given by:

$$M_{ss}(x) = \left(-\frac{w_0 R}{2}\right)x + \frac{1}{2}\left(w_0\frac{x}{R}\right)(x)\left(\frac{x}{3}\right) = -\frac{w_0 R^2}{2}\left(\frac{x}{R} - \frac{1}{3}\frac{x^3}{R^3}\right) \tag{c}$$

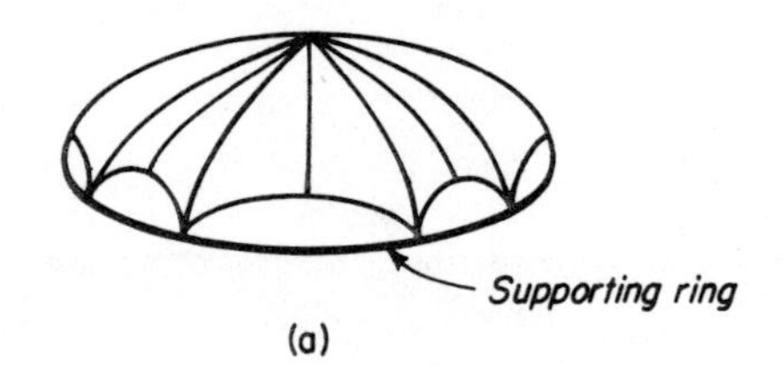

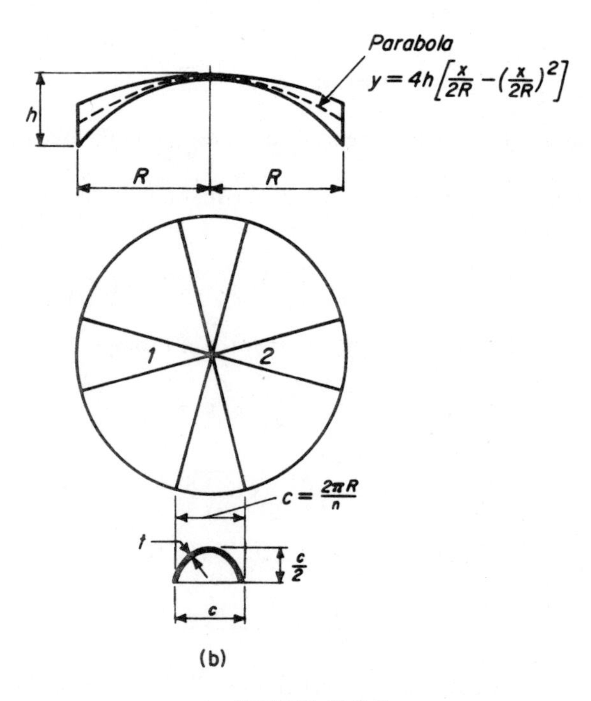

FIGURE 10.2.2

and the thrust H by:

$$H = \frac{M_{ss}]_{\text{at } x=R}}{h} = -\frac{w_0 R^2}{3h}. \tag{d}$$

Hence, by (a) of Section 10.1 with a span $l = 2R$ and letting:

$$\frac{x}{R} = z, \tag{e}$$

the bending moment in the arch becomes:

$$M(x) = M_{ss}(x) - Hy = -\frac{w_0 R^2}{2}\left(\frac{x}{R} - \frac{1}{3}\frac{x^3}{R^3}\right) + \frac{w_0 R^2}{3h} 4h\left[\frac{x}{2R} - \left(\frac{x}{2R}\right)^2\right]$$

$$= \frac{w_0 R^2}{6}(z - 2z^2 + z^3). \tag{f}$$

The maximum value of M occurs at $z = \frac{1}{3}$ and equals:

$$M_{\max} = \tfrac{2}{81} w_0 R^2. \tag{g}$$

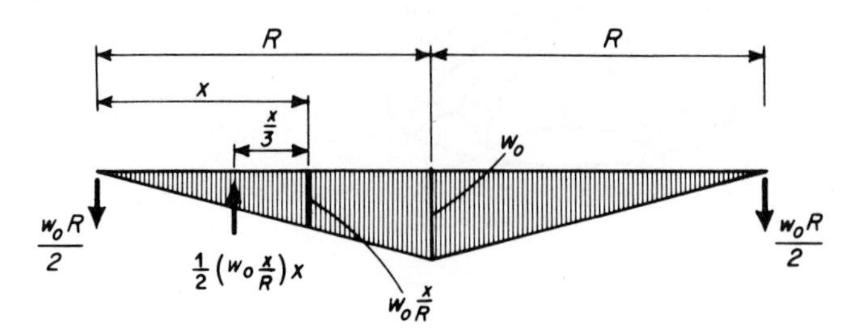

FIGURE 10.2.3

$M_{\max}$ induces tension in the inner fibers and compression in the outer fibers of the arch.

The compression in the arch at $x = R/3$ is the sum of the compression due to w_0, which, by (10.1.2), equals:

$$C_1 = \frac{w_0(2R)^2}{8h}\sqrt{1 + 16\frac{h^2}{l^2}\left(1 - 2\frac{R/3}{2R}\right)^2} = \frac{w_0 R^2}{2h}\sqrt{1 + \left(\frac{8}{3}\frac{h}{l}\right)^2},$$

and of the tension T due to $-w_0(x/R)$, which by (d) of Example 10.1.1 equals:

$$T = \frac{H}{\cos\theta]_{\text{at } x=R/3}} = -\frac{w_0 R^2}{3h}\sqrt{1 + \left(\frac{8}{3}\frac{h}{l}\right)^2}$$

or:

$$C = C_1 + T = \frac{w_0 R^2}{6h}\sqrt{1 + \left(\frac{8}{3}\frac{h}{l}\right)^2}. \tag{h}$$

C may be shown to be parallel to the arch at $x = R/3$. Hence, the fiber stresses at the section $x = R/3$ are given by:

$$f_{o,i} = \frac{C}{A} \pm \frac{M_{\max}}{S_{o,i}}. \tag{i}$$

By Table 4.1.1 (12), the area and the section moduli $S_{o,i}$ of the arch at $x = R/3$, where the diameter of the element is reduced to $\frac{2}{3}c = \frac{4}{3}(\pi R/n)$, are given by:

$$A = \frac{2}{3}\frac{\pi^2}{n}Rt = \frac{6.58}{n}Rt, \qquad S_o = S_2 = 0.83\left(\frac{2}{3}\frac{\pi R}{n}\right)^2 t = \frac{3.64}{n^2}R^2 t,$$

$$S_i = S_1 = 0.47\left(\frac{2}{3}\frac{\pi R}{n}\right)^2 t = \frac{2.06}{n^2}R^2 t,$$

by means of which (i) gives:

$$f_o = -\left(\frac{w_0 R^2}{6h}\sqrt{1 + \left(\frac{8}{3}\frac{h}{2R}\right)^2}\bigg/\frac{6.58}{n}Rt\right) - \left(\frac{2}{81}w_0 R^2\bigg/\frac{3.64}{n^2}R^2 t\right)$$

$$= -\frac{w_0}{t}\left(\frac{n}{39.5}\sqrt{1.78 + \frac{R^2}{h^2}} + \frac{n^2}{148}\right),$$

$$f_i = -\left(\frac{w_0 R^2}{6h}\sqrt{1 + \left(\frac{8}{3}\frac{h}{2R}\right)^2}\bigg/\frac{6.58}{n}Rt\right) + \left(\frac{2}{81}w_0 R^2\bigg/\frac{2.06}{n^2}R^2 t\right)$$

$$= -\frac{w_0}{t}\left(\frac{n}{39.5}\sqrt{1.78 + \frac{R^2}{h^2}} - \frac{n^2}{84}\right).$$

For $n = 12$, $R/h = 4$, we obtain:

$$f_o = -\frac{w_0}{t}\left(\frac{12}{39.5}\sqrt{17.78} + \frac{144}{148}\right) = -(1.27 + 0.97)\frac{w_0}{t} = -2.24\frac{w_0}{t},$$

$$f_i = -\frac{w_0}{t}\left(1.27 - \frac{144}{84}\right) = -(1.27 - 1.72)\frac{w_0}{t} = 0.45\frac{w_0}{t},$$

and by (a):

$$f_o = -2.24\frac{\pi^2}{12}\rho_c\frac{Rt}{t} = -1.83\rho_c R, \qquad f_i = 0.37\rho_c R.$$

For $R = 9$ m (30 ft), $\rho_c = 2\,320$ kg/m³ (145 lb/cu ft) $= 22.8$ kN/m³ (0.084 lb/cu in.), $f_o = -0.376$ MPa (-55 psi), $f_i = 0.076$ MPa (11.0 psi).

10.3 TWO-HINGE ARCHES [8.4]

Two-hinge arches [Fig. 10.3.1(a)] are commonly built of steel, concrete, or wood. They are statically indeterminate, and their thrust is correctly evaluated by taking into account the elasticity of the arch. But it is easy to prove that the thrust H_2 of a two-hinge arch does not usually differ substantially from the thrust H_3 of a geometrically identical three-hinge arch. For example, calling M_c the bending moment at the crown C of a uniformly loaded, two-hinge arch, and taking moments about C [Fig. 10.3.1(b)], one obtains:

$$H_2 h - \frac{wl}{2}\left(\frac{l}{2}\right) + \frac{wl}{2}\left(\frac{l}{4}\right) + M_c = 0,$$

$$\therefore \quad H_2 = \frac{wl^2}{8h}\left(1 - \frac{M_c}{wl^2/8}\right) = H_3\left(1 - \frac{M_c}{M_{ss}}\right), \qquad (10.3.1)$$

where M_{ss} is the midspan moment of the simply supported beam of span l. In any correctly designed arch, M_c is small in comparison with M_{ss}, since

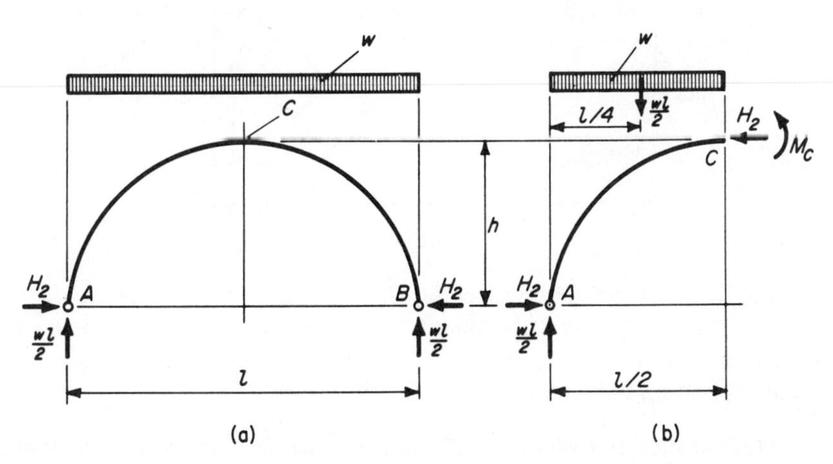

FIGURE 10.3.1

the shape of the arch is chosen as close as possible to the funicular of the main load. Hence, the ratio $\alpha = M_c/M_{ss}$ is much less than 1. Table 10.3.1

Table 10.3.1 VALUES OF $\alpha = M_c/M_{ss}$ FOR PARABOLIC TWO-HINGE ARCHES

$$H_{\bullet} = \frac{wl^2}{8h}(1 - \alpha)$$

h/l \ t_0/l	1/10	1/20	1/30
1/12	0.225	0.056	0.025
1/8	0.099	0.025	0.011
1/4	0.023	0.006	0.003
1/2	0.004	0.001	0.000

gives the factor α for uniformly loaded, parabolic arches, with a rectangular cross section varying as $A_0/\cos\theta$ (Fig. 10.3.2), in terms of h/l and t_0/l, where A_0 and t_0 are the area and the depth of the crown section at C. The table

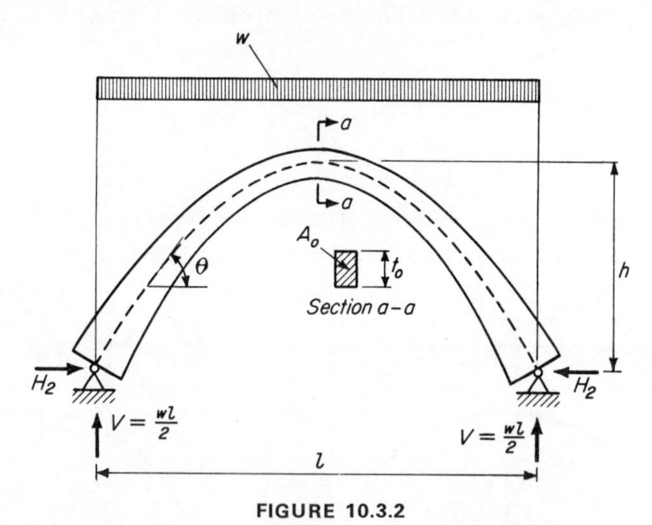

FIGURE 10.3.2

indicates that in most practical cases H_2 may be assumed equal to $H_3 = wl^2/8h$.* Table 10.3.2 gives the thrust H_2 for two-hinge, circular arches of constant cross section under a load w uniform in horizontal projection, and the corresponding three-hinge thrust, H_3.

*The variation of α with t_0/l is due to the consideration of the arch shortening because of compression.

Table 10.3.2 THRUST IN TWO-HINGE AND THREE-HINGE CIRCULAR ARCHES

h/l	0.10	0.15	0.20	0.25	0.30	0.35	0.40	0.45	0.50
H_2/wl	1.23	0.82	0.61	0.48	0.39	0.33	0.28	0.24	0.21
H_3/wl	1.25	0.83	0.63	0.50	0.42	0.36	0.31	0.28	0.25

It is readily seen that two-hinge arches are sensitive to horizontal displacements of the abutments. Consider, for example, a parabolic arch with span l and a small rise h ($r = h/l \ll 1$), defined by the centerline equation:

$$y = 4h\left(\frac{x}{l} - \frac{x^2}{l^2}\right), \tag{a}$$

whose centerline length is approximately given by [see (9.1.7)]:

$$L = l(1 + \tfrac{8}{3}r^2). \tag{b}$$

If one of the supports is displaced horizontally by δ while the length L remains unchanged, and we indicate by $\Delta r = \Delta h/l$ the corresponding reduction in r, we obtain by (b), under the assumption that δ/l is much smaller than 1 and Δr much smaller than r:

$$L = l(1 + \tfrac{8}{3}r^2) = (l + \delta)[1 + \tfrac{8}{3}(r - \Delta r)^2] = (l + \delta)(1 + \tfrac{8}{3}r^2 - \tfrac{16}{3}r\,\Delta r),$$

from which, neglecting terms in $(\delta/l)r\,\Delta r$ and $(\delta/l)r^2$:

$$1 + \frac{8}{3}r^2 = \left(1 + \frac{\delta}{l}\right)\left(1 + \frac{8}{3}r^2 - \frac{16}{3}r\,\Delta r\right) \approx 1 + \frac{8}{3}r^2 - \frac{16}{3}r\,\Delta r + \frac{\delta}{l}$$

$$\therefore \quad \Delta r = \frac{3}{16}\frac{\delta}{rl} = \frac{3}{16}\frac{\delta}{h}. \tag{10.3.2}$$

By (a), the curvature c at the top of the arch equals:

$$c = \frac{d^2y}{dx^2}\bigg]_{x=l/2} = -\frac{8h}{l^2} = -\frac{8}{l}r, \tag{c}$$

from which the change Δc in c due to a change Δr in r (neglecting the change δ in l) is given by:

$$\Delta c = -\frac{8}{l}\Delta r = -\frac{8}{l}\left(\frac{3}{16}\right)\frac{\delta}{h} = -\frac{3}{2}\left(\frac{\delta}{l}\right)\frac{1}{h}. \tag{10.3.3}$$

The bending moment due to Δc is, by (4.1.4 a):

$$\Delta M = -EI\,\Delta c = \frac{3}{2}\left(\frac{\delta}{h}\right)\frac{EI}{l},$$

and, indicating by t the arch depth, the corresponding bending stress f_b becomes:

$$f_b = \frac{\Delta M(t/2)}{I} = \frac{3}{4}\left(\frac{\delta}{l}\right)\left(\frac{t}{h}\right)E. \tag{10.3.4}$$

For example, for $\delta/l = 1/1\,000$ and $t/h = 1/5$, (10.3.4) gives $f_b = (3/20\,000)E$, or $f_b = 3$ MPa (450 psi) for concrete and $f_b = 30$ MPa (4,350 psi) for steel, i.e., one-third of the allowable stress for concrete and one-fifth of that for steel.

10.4 FIXED ARCHES [8.4]

Arches with abutments capable of preventing the rotation of their end sections are called *fixed arches*. They are more rigid than two-hinge arches and hence more sensitive to settlements of the abutments. The thrust H in a uniformly loaded, fixed arch (Fig. 10.4.1) is obtained by taking moments about C:

$$\frac{wl}{2}\left(\frac{l}{2}\right) - \frac{wl}{2}\left(\frac{l}{4}\right) + M_A - M_C - Hh = 0$$

$$\therefore \quad H = \frac{wl^2}{8h}\left(1 - \frac{M_C - M_A}{wl^2/8}\right) = H_3\left(1 - \frac{M_C - M_A}{M_{ss}}\right). \quad (10.4.1)$$

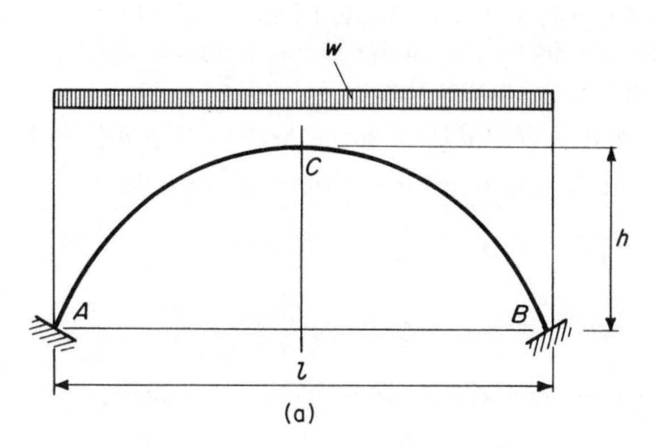

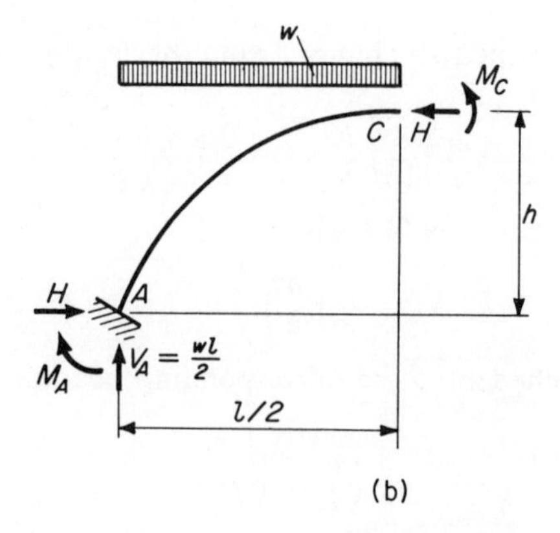

FIGURE 10.4.1

The statically indeterminate moments M_A and M_C are obtained by taking into consideration the elasticity of the arch. Table 10.4.1 gives H and M_A

Table 10.4.1 THRUST AND END MOMENT IN CIRCULAR FIXED ARCHES

h/l	0.10	0.15	0.20	0.25	0.30	0.35	0.40	0.45	0.50
H/wl	1.26	0.83	0.61	0.52	0.44	0.38	0.34	0.30	0.28
M_A/wl^2	0.001	0.002	0.004	0.007	0.010	0.013	0.018	0.022	0.027

for circular arches of constant cross section, uniformly loaded horizontally, in terms of h/l. Once M_A and H are known, the crown moment M_C is given by:

$$M_C = M_{ss} + M_A - Hh. \tag{10.4.2}$$

The thrust H does not differ substantially from the thrust in a three-hinge arch (see Table 10.3.2) but is larger than the thrust in a two-hinge arch, since M_A is positive [see (10.4.1) and (10.3.1)]. The haunch moment M_A is very sensitive to the arch shape and differs substantially in circular and parabolic arches.

Example 10.4.1. Determine the maximum positive and negative moments in a fixed, semicircular arch uniformly loaded in horizontal projection by a load w.

By Table 10.4.1, $H = 0.28wl$, $M_A = 0.027wl^2$. The moment at a section x (Fig. 10.4.2) is:

$$M(x) = \frac{wl}{2}x - \frac{wx^2}{2} + M_A - Hy, \tag{a}$$

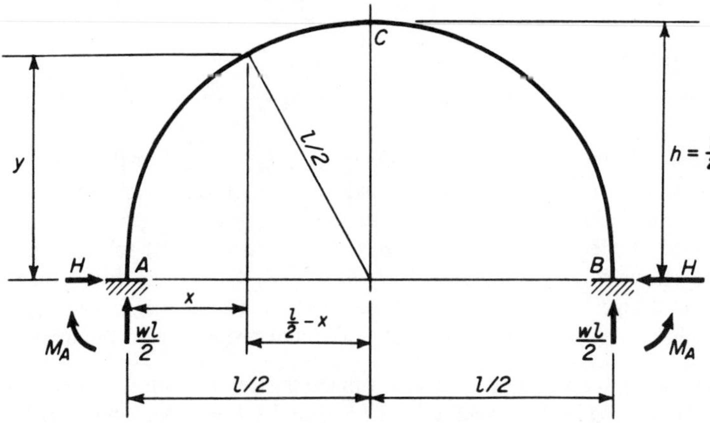

FIGURE 10.4.2

where:

$$y = \sqrt{\left(\frac{l}{2}\right)^2 - \left(\frac{l}{2} - x\right)^2} = \frac{l}{2}\sqrt{1 - \left(1 - \frac{2x}{l}\right)^2}. \tag{b}$$

Hence, letting $x/l = z$:

$$M(x) = wl^2\left[\tfrac{1}{2}z(1 - z) + \frac{M_A}{wl^2} - \frac{H}{wl}\sqrt{z(1 - z)}\right]$$

$$= wl^2[\tfrac{1}{2}z(1 - z) + 0.027 - 0.28\sqrt{z(1 - z)}]. \tag{10.4.3}$$

Table 10.4.2 gives $M(x)$ versus x/l and shows that the maximum M is M_A; the minimum M occurs at about $x/l = 0.1$ and equals $-0.013wl^2$.

Table 10.4.2

x/l	0.0	0.1	0.2	0.3	0.4	0.5
$M(x)/wl^2$	+0.027	−0.012	−0.005	+0.003	+0.010	+0.012

The maximum fiber stresses in this arch occur at A; if the arch cross section is rectangular ($b \times c$), they equal:

$$f = -\frac{wl/2}{bc} \pm \frac{0.027wl^2}{bc^2/6} = -\frac{wl}{bc}\left(0.5 \pm 0.162\frac{l}{c}\right).$$

For $l/c = 20$ the maximum compression equals 74.8 w/b and the minimum compression 54.8 w/b. The corresponding bending stresses in a simply supported beam would equal:

$$f_b = \pm\frac{wl^2/8}{bc^2/6} = \pm\frac{3}{4}\frac{w}{b}\left(\frac{l}{c}\right)^2 = \pm 300\frac{w}{b}.$$

PROBLEMS

10.1 A house in the shape of an A frame is 5.5 m (18 ft) wide at its base and 7.5 m (24 ft) high. The wood frames, spaced 2.5 m (8 ft) o.c., consist of two inclined legs which are hinged where they meet at the top and which may be assumed hinged at the base. The siding consists of 50 mm (2 in.) solid wood planks and plastic roofing weighing 0.1 kN/m² (2 psf). Determine the size of the legs of the frame for a 1.5 kN/m² (30 psf) wind load on one side and a vertical live load of 0.5 kN/m² (10 psf) of horizontal projection on the other side.

10.2 Determine the ultimate span of a funicular steel arch in A36 steel, capable of supporting its own weight for $h/l = \frac{1}{8}$ and $h/l = \frac{1}{4}$, neglecting buckling.

10.3 Determine the minimum thickness of a cylindrical tube of A36 steel with a diameter of 3 m (10 ft), capable of sustaining the uniform, average, hydrostatic, external pressure due to a depth of 30 m (100 ft) of water, ignoring buckling.

10.4 The frame shown in the accompanying sketch, with $a = b = 6$ m (20 ft) and $c = 4.5$ m (15 ft), $d = 1.5$ m (5 ft), is one of a series 3.5 m (12 ft) o.c. It is fabricated of laminated Douglas fir members and carries a roof load of 7.5 cm (3 in.) of solid wood planks, plus roofing and insulation, and a 2 kN/m² (40 psf) snow load. Determine the depth required at the haunch *B* and at the base *A* if the width is 200 mm (8 in.).

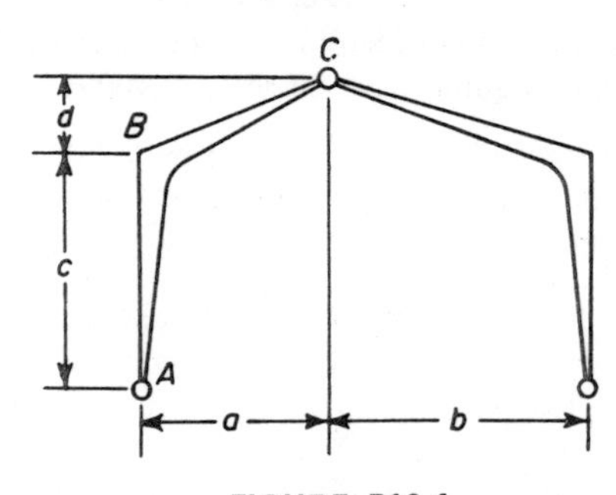

FIGURE P10.4

10.5 A series of frames like that shown in Problem 10.4, with $a = b$, are used to span 30 m (100 ft) over an auditorium. The frames are 9 m (30 ft) o.c., with $c = 6$ m (20 ft), $d = 3$ m (10 ft). The roof dead load is 2.5 kN/m² (50 psf), and the live load is 1.5 kN/m² (30 psf). If the frame is fabricated of standard steel A36 wide flange sections, what are the required member sizes: (a) for live load over the total span, (b) over half the span?

10.6 An unsymmetrical frame of the type shown in Problem 10.4, with $a = 15$ m (50 ft), $b = 3$ m (10 ft), $c = 4.5$ m (15 ft), $d = 3$ m (10 ft), is used as a north-light roof over a factory. The long arm carries a total load of 45 kN/m (3.0 k/ft) and the short arm 22.5 kN/m (1.5 k/ft), both in horizontal projection. Determine the member sizes in A36 steel.

10.7 A spherical dome covering a sports stadium is 120 m (400 ft) in diameter and has a rise of 18 m (60 ft). The dome consists of 30 equally spaced radial steel ribs, hinged at the base, and meeting at a crown hinge. Compute the reactions and moments in the ribs if the total load on the roof of 3.5 kN/m² (70 psf) is assumed to act over the horizontal projection of the dome.

10.8 A series of parallel, two-hinged circular arches of constant cross section 3.5 m (12 ft) o.c. span 30 m (100 ft) over an arena. If their rise is 9 m (30 ft) and the total load on the roof is 3 kN/m² (60 psf), which may be assumed to act on the horizontal projection, determine the area of a laminated fir arch required by compression only. What is the moment at the crown of the arches?

10.9 If the arch of Problem 10.8 is fixed, determine the depth required at its base, if the width of the laminated fir member is 450 mm (18 in.).

10.10 Determine the depth required at the base of the arch of Problem 10.8 if it is fixed, rises 12 m (40 ft) instead of 9 m (30 ft) and its width is 450 mm (18 in.).

10.11 A two-hinge circular arch has a span of 15 m (50 ft) and a rise of 3 m (10 ft). It carries a total load of 12 kN/m (0.8 k/ft). Determine the diameter of an A36 steel tie rod for the arch and the area of 20 MPa (2,900 psi) concrete columns supporting the arch at its ends.

10.12 For the arch of Problem 10.11 determine the area of concrete buttresses tangent to the arch, capable of supporting the arch in simple compression, if no tie rod is used. Assume $f'_c = 20$ MPa (2,900 psi).

10.13 The arch of Problem 10.11 is one of a series of three hinged arches 6 m (20 ft) o.c., acted upon by a lateral wind load of 1.5 kN/m^2 (30 psf) of vertical projection. Determine the vertical reactions and the thrust in the arch.

10.14 A two-hinge parabolic concrete arch of low rise carries a total load of 60 kN/m (4 k/ft), spans 30 m (100 ft), and has a cross section 600 mm (2 ft) deep and 300 mm (1 ft) wide. Determine its minimum rise if the allowable stress in compression 7 MPa (1 ksi).

10.15 A two-hinge parabolic arch spans 60 m (200 ft), has a 9 m (30 ft) rise. The arch is 900 mm (3 ft) deep and 300 mm (1 ft) wide at the crown. Determine the moment at the crown due to a rise of 16°C (30°F) in the temperature of the air, and the amount of additional tensile and compressive grade 60 steel needed to absorb this moment. [Evaluate the change in rise Δh due to a change in arch length ΔL by (9.3.6.).]

10.16 A cylindrical water tank of radius 6 m (20 ft) and depth 6 m (20 ft), made of concrete, is prestressed by tendons wrapped around its outer surface. Determine the tendon cross section per meter (foot) of tank depth at the bottom of the tank, required to ensure that the concrete will not be in tension when the tank is full. (Assume $F_t = 550$ MPa (80,000 psi).)

10.17 Determine the moment at the crown of the arch of Problem 10.15 due to a horizontal displacement of its right support of 150 mm (6 in.). Evaluate the tensile and compressive grade 60 steel needed to absorb this moment.

10.18 An opening 3 m (10 ft) wide and 4.5 m (15 ft) high is to be cut in a 600 mm (2 ft) thick brick wall which is 9 m (30 ft) high. A parabolic brick arch with a rise of 0.6 m (2 ft) is built on top of the opening to support the upper part of the wall. Determine the depth of the arch within one-half brick and the thrust exerted by it on the adjoining part of the wall, assuming the arch to be 600 mm (2 ft) thick.

10.19 A monumental arch of stainless steel has a parabolic shape, spans 30 m (100 ft), and is 60 m (200 ft) high. The tubular cross section at the crown is 600 mm (2 ft) wide by 1.2 m (4 ft) deep; the cross section at the abutments is 1.2 m

(4 ft) wide by 2.4 m (8 ft) deep. Determine the dead-load thrust of the arch, assuming the cross section to vary linearly along the span from crown to abutments and the thickness to be 25 mm (1 in.). (The arch may be considered funicular for its dead load.)

10.20 Determine the thrust and the equation $y = f(x)$ of a funicular arch with a span l and a rise h, carrying a triangular distributed load $p = p_0(2x/l)$ kN/m (lb/ft) on its left half $(0 \leq x \leq l/2)$ and a symmetrical triangular load $p = p_0[2(l - x)/l]$ kN/m (lb/ft) on its right half $(l/2 \leq x \leq l)$.

11. Plane Grid Systems

11.1 RECTANGULAR GRID SYSTEMS [10.1, 10.2]

Plane grid systems have the essential property of transferring loads in two or more directions; they are advantageously used to build large floors and to cover large areas requiring flat roofs. The characteristic features of a two-way transfer of loads are clearly illustrated by the behavior of two beams at right angles to each other, spanning a rectangular area of sides l_1, l_2 (Fig. 11.1.1), and carrying a concentrated load P at their intersection.

It will be assumed that beam 1 (full line) rests on beam 2 (dashed line) and that both beams are simply supported. Under the action of the load P, beam 1 deflects and carries down beam 2, which exerts an upward reaction X on beam 1. Hence, beam 2 carries a load X and beam 1 a load of $P - X$. The midspan deflections δ_1 and δ_2 of the two beams are equal; hence, indicating the moments of inertia of the two beams by I_1 and I_2 and assuming that the beams are made of the same material, by Table 4.3.1(b):

$$\delta_1 = \frac{(P - X)l_1^3}{48EI_1} = \frac{Xl_2^3}{48EI_2} = \delta_2,$$

from which:

$$X = \frac{P}{1 + (l_2/l_1)^3(I_1/I_2)}. \tag{11.1.1}$$

For two beams with the same I the ratio of the load X carried by beam 2 to the total load P becomes:

$$\frac{X}{P} = \frac{1}{1 + (l_2/l_1)^3}, \tag{11.1.1a}$$

and is given versus l_1/l_2 in Table 11.1.1.

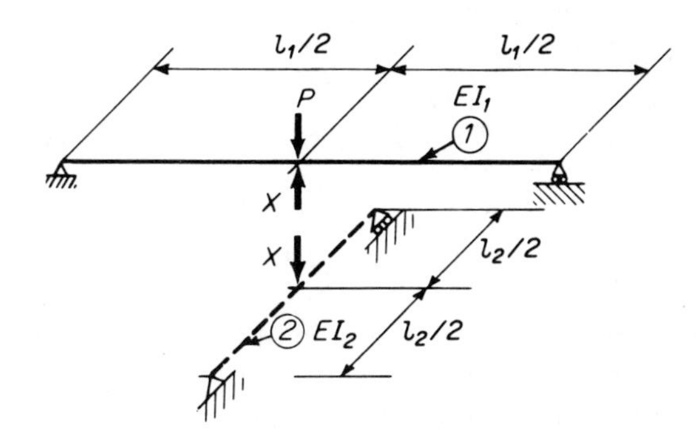

FIGURE 11.1.1

Table 11.1.1

l_1/l_2	1	2	3	4	5
X/P	1/2	8/9	27/28	64/65	125/126

It is seen from this table that as soon as the ratio of sides becomes greater than 2, the shorter beam carries virtually the entire load and the two-way transfer vanishes.

An equal sharing of load between two beams of unequal length may be obtained by using beams with different moments of inertia. When the load is equally shared, X equals $P/2$, so that by (11.1.1):

$$\frac{P}{2} = \frac{P}{1 + (l_2/l_1)^3(I_1/I_2)} \quad \therefore \quad \frac{I_1}{I_2} = \left(\frac{l_1}{l_2}\right)^3, \tag{a}$$

i.e., the moments of inertia of the beams must be in the ratio of the cubes of their spans.

On the other hand, to obtain the most economical design, the maximum stresses should be equal in both beams. Letting:

$$r = \left(\frac{l_2}{l_1}\right)^3 \left(\frac{I_1}{I_2}\right), \tag{b}$$

the loads carried by beams 2 and 1 are given by:

$$X = P\frac{1}{1+r}, \qquad P - X = P\left(1 - \frac{1}{1+r}\right) = P\frac{r}{1+r},$$

and the respective moments at midspan become:

$$M_1 = \frac{(P-X)l_1}{4} = \frac{Pl_1}{4(1+r)}r, \qquad M_2 = \frac{Xl_2}{4} = \frac{Pl_2}{4(1+r)} = \frac{Pl_1}{4(1+r)}\left(\frac{l_2}{l_1}\right).$$

Indicating by h_1 and h_2 the beam depths, their respective maximum bending stresses are given by:

$$f_1 = \frac{M_1(h_1/2)}{I_1} = \frac{Pl_1}{8(1+r)}\frac{rh_1}{I_1}, \qquad f_2 = \frac{Pl_1}{8(1+r)}\left(\frac{l_2}{l_1}\right)\frac{h_2}{I_2}.$$

If f_1 is to be equal to f_2:

$$\frac{rh_1}{I_1} = \left[\left(\frac{l_2}{l_1}\right)^3\left(\frac{I_1}{I_2}\right)\right]\left(\frac{h_1}{I_1}\right) = \left(\frac{l_2}{l_1}\right)\left(\frac{h_2}{I_2}\right) \quad \therefore \quad \frac{h_1}{h_2} = \left(\frac{l_1}{l_2}\right)^2, \tag{c}$$

i.e., the beam depths must be in the ratio of the squares of their spans.

When $l_1 > l_2$ some improvement in load sharing may be obtained at times by fixing (if possible) the ends of the longer beam. In this case, equating deflections, we obtain by (4.4.5) and Table 4.3.1(b):

$$\frac{(P-X)l_1^3}{192EI_1} = \frac{Xl_2^3}{48EI_2} \quad \therefore \quad X = \frac{P}{1 + 4(l_2/l_1)^3(I_1/I_2)}. \tag{11.1.2}$$

Table 11.1.2 gives X/P for $I_1 = I_2$.

Table 11.1.2

l_1/l_2	1	1.6	2	3	4	5
X/P	1/5	1/2	2/3	27/31	16/17	125/129

The table shows that although in this case the load is more evenly shared, this advantage still vanishes rapidly as $l_1 \gg l_2$.

Finally, it may be noted that the sharing of the load depends also on the load distribution on beam 1, as shown by the following example.

Example 11.1.1. Derive the value of the reaction X of beam 2, if beam 1 in Fig. 11.1.1 is uniformly loaded.

Equating deflections, we obtain by Table 4.3.1(a) and (b):

$$\frac{5}{384}\frac{wl_1^4}{EI_1} - \frac{Xl_1^3}{48EI_1} = \frac{Xl_2^3}{48EI_2},$$

from which:

$$X = \frac{5}{8}\frac{wl_1}{1 + (l_2/l_1)^3(I_1/I_2)}. \tag{11.1.3}$$

A comparison of (11.1.3) with (11.1.1) shows that the sharing of a uniform load is even less pronounced than that of a concentrated load and that beam 2 will never carry more than five-eighths of the total load.

The two-way behavior of *rectangular grids*, consisting of "upper" and "lower" beam systems, may be similarly obtained by equating deflections at all the intersecting points, or *nodal points*, of the two systems.

The evaluation of the reactions of grid systems is facilitated in Table 11.1.3, which gives the deflections δ_k at the evenly spaced points k of a simply supported beam due to a unit load applied at one of these points, say the point i. In this table, n is the number of subdivisions in the beam. It must be remembered that the *influence coefficient* δ_k due to a unit load at i is equal to the δ_i due to a unit load at k.

Table 11.1.3*

n	δ at k =	Unit load applied at i =					
		1	2	3	4	5	Factor l^3/EI
2	1	1					1/48
3	1	8					
	2	7	8				1/486
4	1	9					
	2	11	16				1/768
	3	7	11	9			
5	1	32					
	2	45	72				1/3,750
	3	40	68	72			
	4	23	40	45	32		
6	1	25					
	2	38	64				
	3	39	69	81			
	4	31	56	69	64		
	5	17	31	39	38	25	1/3,888

Example 11.1.2. A simply supported rectangular grid consists of two "upper" beams of span l_1 and two "lower" beams of span l_2 intersecting at the third points of their spans (Fig. 11.1.2). The loads P carried at each beam intersection may be considered as the resultant of a uniform load on the contributory area surrounding each nodal point (Fig. 11.1.2). Determine the reactions X of beams 2 on beams 1, which, by symmetry, are all equal.

From Table 11.1.3, the deflection under a load P applied at the third point $k = 1$ of a beam equals $(8/486)(Pl^3/EI)$, while the deflection at the unloaded third point $k = 2$ of the same beam equals $(7/486)(Pl^3/EI)$. The deflection at either third point due to two third-point loads is, thus, $[(7 + 8)/486](Pl^3/EI)$. Hence, equating displacements at the intersection of two beams:

$$\frac{15}{486}\frac{(P - X)l_1^3}{EI_1} = \frac{15}{486}\frac{Xl_2^3}{EI_2},$$

*This table was extracted from "Modern Grid Structures," by F. S. Makowski, *Architectural Science Review*, Sydney, Australia, July, 1960.

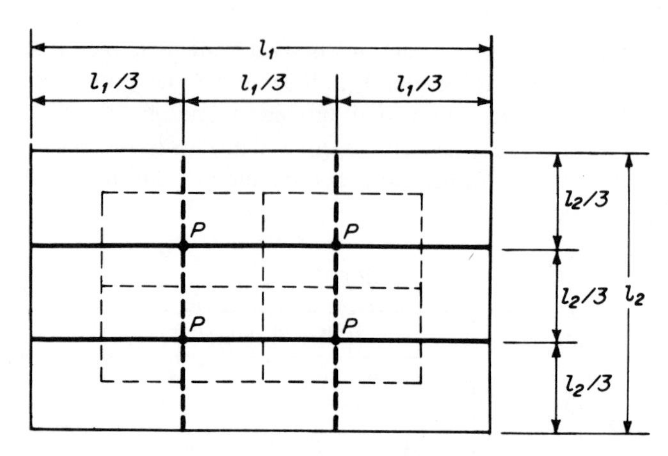

FIGURE 11.1.2

from which:

$$X = \frac{P}{1 + (l_2/l_1)^3(I_1/I_2)}, \tag{a}$$

which is identical to (11.1.1). For $I_1 = I_2$, $l_2 = \frac{1}{2}l_1$, (a) gives $X = \frac{8}{9}P$, $P - X = \frac{1}{9}P$; the maximum moments in the long and the short beams are, respectively:

$$\begin{aligned} M_1 &= (P - X)\left(\frac{l_1}{3}\right) = \frac{1}{27}Pl_1 = 0.037Pl_1, \\ M_2 &= X\left(\frac{l_2}{3}\right) = \left(\frac{8P}{9}\right)\left(\frac{l_1}{6}\right) = \frac{8}{54}Pl_1 = 0.148Pl_1. \end{aligned} \tag{b}$$

For $l_1 = l_2$ and $I_1 = I_2$, the "upper" and "lower" systems of Example 11.1.2 carry half the load each, and their maximum bending moments are equal to half the bending moment M in a one-way beam system. If the allowable stress in the one-way and the two-way systems is to be equal, letting the respective section moduli be S' and S'', we must have:

$$f_b = \frac{M}{S'} = \frac{M/2}{S''} \quad \therefore \quad \frac{S'}{S''} = 2. \tag{c}$$

For ideal I-beams of area A and depth h, $S = Ah/2$ [see Table 4.1.1 (6)]. Indicating by A', A'', and h', h'' the beam areas and depths in the one-way and two-way systems of ideal I-beams, we obtain:

$$\frac{S'}{S''} = \frac{A'h'}{A''h''},$$

and if $A' = 2A''$, so that the amount of material in the two-way system equals that in the one-way system, by (c):

$$\frac{S'}{S''} = \frac{2A''h'}{A''h''} = 2 \quad \therefore \quad h' = h'',$$

with no change in depth of construction for the two-way system. A reduction in the depth of construction of the two-way system, which is often economically interesting, may be obtained through an increase in material by making $2A'' > A'$, since, in this case with $(A'/A'')(h'/h'') = 2$:

$$\frac{h''}{h'} = \frac{A'}{2A''} < 1.$$

Example 11.1.3. The grid of Example 11.1.2 has "upper" beams 1, 3 and "lower" beams 2, 4, with $l_1^3/EI_1 = l_2^3/EI_2$, and carries a single load P at node 1 (Fig. 11.1.3). Determine the reactions at all nodes, considering upward reactions as positive, and evaluate the maximum bending moments in the beams.

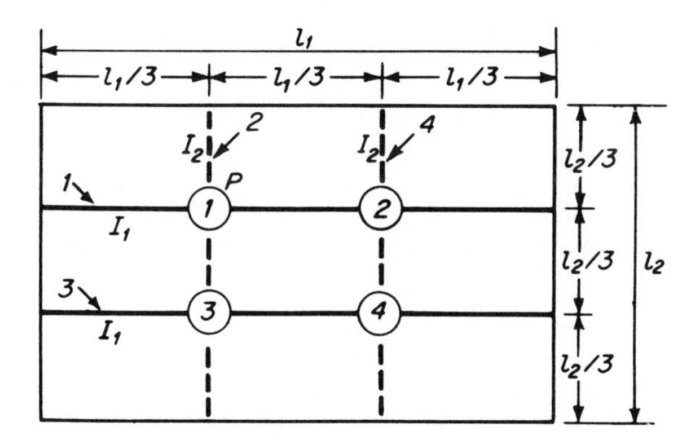

FIGURE 11.1.3

Deleting the common factor $(\frac{1}{486})(l_1^3/EI_1) = (\frac{1}{486})(l_2^3/EI_2)$, we find that the equations stating the equality of the displacements of the two beam systems at their intersections are, by Table 11.1.3:

Node (1): $\quad 8(P - X_1) - 7X_2 = 8X_1 + 7X_3,$

Node (2): $\quad 7(P - X_1) - 8X_2 = 8X_2 + 7X_4,$

Node (3): $\quad -8X_3 - 7X_4 = 7X_1 + 8X_3,$

Node (4): $\quad -7X_3 - 8X_4 = 7X_2 + 8X_4,$

or:

$$\begin{aligned} 16X_1 + 7X_2 + 7X_3 \quad\quad &= 8P, \\ 7X_1 + 16X_2 \quad\quad + 7X_4 &= 7P, \\ 7X_1 \quad\quad + 16X_3 + 7X_4 &= 0, \\ 7X_2 + 7X_3 + 16X_4 &= 0. \end{aligned}$$

The roots of this system are:

$$X_1 = \tfrac{1}{2}P, \qquad X_2 = \tfrac{7}{32}P, \qquad X_3 = -\tfrac{7}{32}P, \qquad X_4 = 0.$$

Since the load P at node 1 deflects beam 1 downward, beam 1 *pushes* down on beam 2 at node 1 and on beam 4 at node 2. Beam 2 *pulls* down the unloaded beam 3 at node 3. Hence, X_2 is positive and X_3 is negative; by symmetry, X_2 and X_3 have the same absolute value. The reaction at node 4, due to the two equal and opposite forces X_2, X_3, is zero. The beams of the two systems must be rigidly connected so as to be able to react both upward and downward.

If beam 1 had to carry the load P alone, its maximum moment would be:

$$M' = \left(\frac{2}{3}P\right)\left(\frac{l_1}{3}\right) = \frac{2}{9}Pl_1.$$

The same beam in the two-way system is loaded as shown in Fig. 11.1.4; its left reaction equals:

$$R_1 = \frac{2}{3}\left(\frac{P}{2}\right) + \frac{1}{3}\left(-\frac{7}{32}P\right) = \frac{25}{96}P,$$

and its maximum moment becomes:

$$M'' = \left(\frac{25}{96}P\right)\left(\frac{l_1}{3}\right) = \frac{25}{288}Pl_1.$$

The ratio of moments is thus:

$$\frac{M''}{M'} = \left(\frac{25}{288}\right)\left(\frac{9}{2}\right) = \frac{25}{64} = 39\%.$$

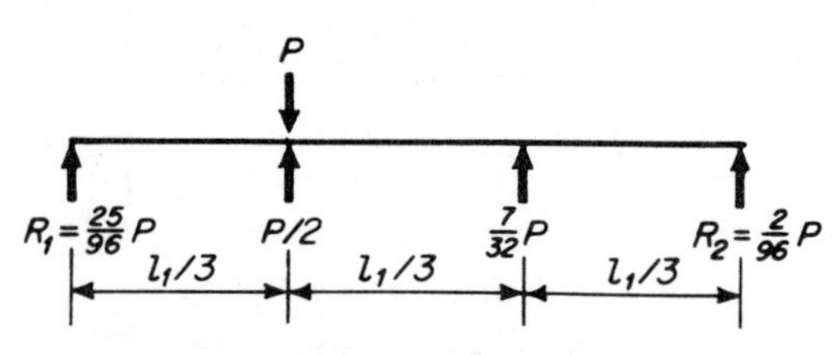

FIGURE 11.1.4

The beams of the upper system in the examples of this section were considered simply supported on the beams of the lower system. In practice, the beams of the two systems are rigidly connected when made out of steel or wood, and monolithically poured when made out of concrete. Therefore, the bending deflections of the beams of one system entail twisting of the beams of the other (Fig. 11.1.5). The torsional stresses developed by the twisting action help transfer an additional part of the load from one system to the other and increase the stiffness of the two-way system (see Section

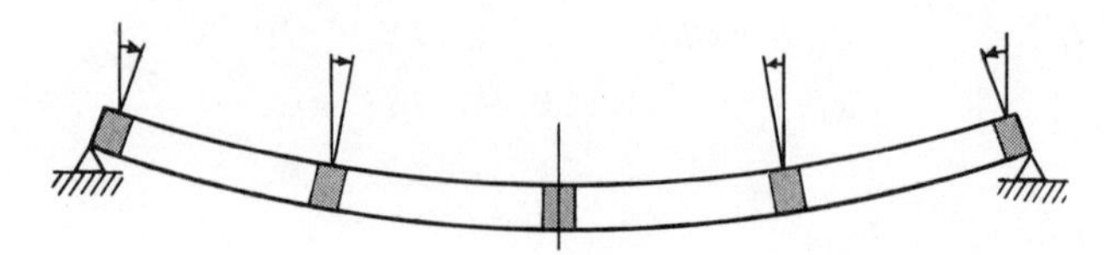

FIGURE 11.1.5

12.1). This increase in stiffness is virtually negligible for I-beams but not entirely unimportant for beams of rectangular cross section.

Because of all the behavior characteristics mentioned above, span-to-depth ratios of between 20 and 40 are common for rectangular grids, as against ratios of between 10 and 20 for one-way systems. (It should be noticed that, unless a reduction in span-to-depth ratios is considered essential, the use of one-way steel systems is always more economical than the use of two-way systems because of the high cost of the connections at the nodes of two-way systems.)

When the number of nodal points in the grid is large, the determination of the reactions between the two beam systems requires the solution of a large number of equations, a task that can be conveniently performed only with the help of an electronic computer. In this case, practical estimates of the load sharing between the two beam systems may be readily obtained by assuming that the beams are so numerous and so closely spaced as to act like a continuous plate (see Example 12.2.3 and Section 18.2).

11.2 SKEW GRIDS AND MULTIPLE GRIDS [10.3]

The reduction of two-way action in rectangular grids with $l_1 \gg l_2$ may be avoided by the use of *skew grids*, since these grids consist of beams of comparable lengths even when $l_1 > l_2$ and, moreover, have the longer beams supported by shorter beams.

For example, in the skew grid of Fig. 11.2.1, where $l_2 < l_1$, the four short beams have a length $\frac{1}{2}\sqrt{l_1^2 + l_2^2} = \frac{1}{2}l_1\sqrt{1 + (l_2/l_1)^2} < l_1$; the diagonal beams have a length $\sqrt{l_1^2 + l_2^2} = l_1\sqrt{1 + (l_2/l_1)^2} > l_1$, but, being supported

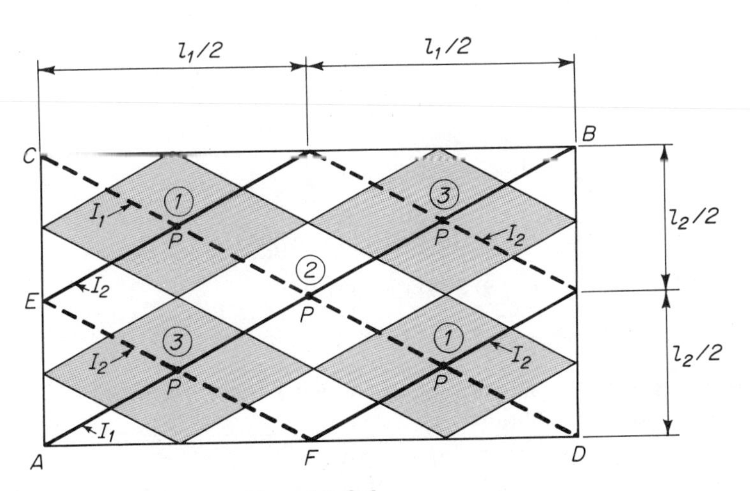

Fig. 11.2.1

on the shorter, stiffer beams, they behave as beams partially fixed at the ends. Moreover, the total load on the grid is carried by five nodes, rather than four, as in Fig. 11.1.3, so that for uniform loading each concentrated load is approximately four-fifths of the concentrated loads shown in Fig. 11.1.3.

Example 11.2.1. Evaluate the maximum bending moments in the grid of Fig. 11.2.1.

Indicate by l the length of the diagonal bars AB and CD, by X_1, X_2, X_3 the reactions at the nodes 1, 2, 3, by r the ratio I_1/I_2 of the moments of inertia of the long and short bars. Noticing that the short bars have a length $l/2$ and making use of Table 11.1.3, we find that the equations stating the equality of the displacements at the nodes become:

Node (1):

$$(P - X_1)\frac{(l/2)^3}{48EI_2} = \frac{l^3}{768EI_1}(9X_1 + 11X_2 + 7X_1),$$

Node (2):

$$\frac{l^3}{768EI_1}[11(P - X_3) + 16(P - X_2) + 11(P - X_3)]$$

$$= \frac{l^3}{768EI_1}(11X_1 + 16X_2 + 11X_1),$$

Node (3):

$$\frac{l^3}{768EI_1}[9(P - X_3) + 11(P - X_2) + 7(P - X_3)] = \frac{(l/2)^3}{48EI_2}X_3,$$

or:

$$\begin{aligned} (16 + 2r)X_1 + 11X_2 &= 2rP, \\ 11X_1 + 16X_2 + 11X_3 &= 19P, \qquad \left(r = \frac{I_1}{I_2}\right) \\ 11X_2 + (16 + 2r)X_3 &= 27P. \end{aligned}$$

The roots of these equations are:

$$X_1 = -\frac{77 + 148r - 64r^2}{(16 + 2r)(14 + 32r)}P,$$

$$X_2 = \frac{7 + 16r}{14 + 32r}P = \frac{1}{2}P,$$

$$X_3 = \frac{301 + 688r}{(16 + 2r)(14 + 32r)}P,$$

which, for $I_1 = \frac{1}{2}I_2$, i.e., $r = \frac{1}{2}$, give:

$$X_1 = -\frac{27}{102}P, \qquad X_2 = \frac{1}{2}P, \qquad X_3 = \frac{129}{102}P.$$

The reactions R at the supports of the diagonal beam AB [Fig. 11.2.2 (a)] are obtained by vertical equilibrium:

$$2R = \left(3 - \frac{129}{102} - \frac{1}{2} - \frac{129}{102}\right)P \quad \therefore \quad R = -\frac{3}{204}P.$$

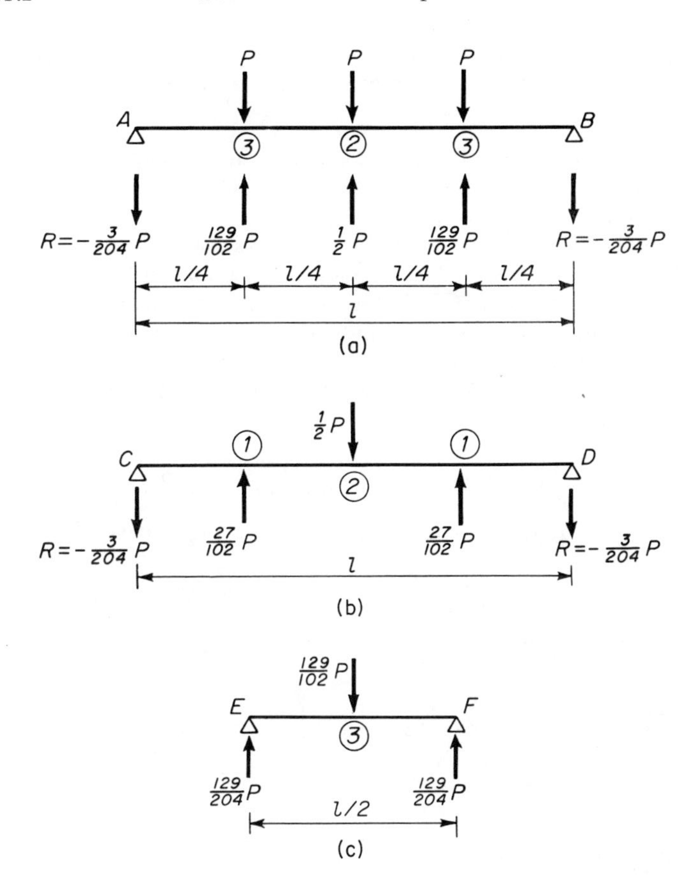

FIGURE 11.2.2

The negative sign of R shows that the reactions act downward on the beam and introduce partial restraints at its ends. The same restraints occur in beam CD [Fig. 11.2.2 (b)]. The maximum moments in the diagonal beam AB and in the short beam EF equal, respectively [Fig. 11.2.2 (a), (c)]:

$$M_1 = \left[\left(-\frac{3}{204}\right)\left(\frac{1}{2}\right) + \left(\frac{27}{102}\right)\left(\frac{1}{4}\right)\right]Pl = \frac{24}{408}Pl = 0.059Pl,$$

$$M_2 = \left(\frac{129}{204}\right)\left(\frac{1}{2}\right)\left(\frac{1}{2}\right)Pl = \frac{129}{816}Pl = 0.158Pl.$$

For $l_2 = \frac{1}{2}l_1$, $l = l_1\sqrt{1 + (\frac{1}{2})^2} = 1.12l_1$, and with P equal to four-fifths of P in Fig. 11.1.3, $M_1 = 0.053Pl_1$, $M_2 = 0.141Pl_1$. The maximum moment M_2 is seen to be 5% smaller than the moment M_2 of (b) in the rectangular grid of Example 11.1.2.

The analysis of skew grids with a high number of nodes by plate analogy is dealt with in Section 18.3.

Because of the more advantageous load distribution provided by skew grids, their span-to-depth ratios commonly vary between 30 and 60.

Example 11.2.2. Determine the loads carried by and the maximum moments in each of the three identical beam systems of the triangular grid of Fig. 11.2.3, which is loaded by a load P at each internal point.

The three beams meeting at 1 are identically loaded and identically supported; hence, each carries, at 1, a load $P/3$. At point 2, *two* beams of length $3a$ carry the load at their third points, and one of length $4a$ carries the load at its quarter point. Noting that by symmetry the short beams carry equal loads at the third points, and that the long beams carry a load $P/3$ at the midpoint and equal loads at the quarter points, by Table 11.1.3, and calling X_2 the reaction of *each* short beam on the long beams, we obtain:

$$2X_2\frac{(8+7)}{486}\frac{(3a)^3}{E(2I)} = (P-2X_2)\frac{(9+7)}{768}\frac{(4a)^3}{EI} + \frac{P}{3}\left(\frac{11}{768}\right)\frac{(4a)^3}{EI},$$

$$2X_2 = 0.94P, \qquad X_2 = 0.47P, \qquad P - 2X_2 = 0.06P.$$

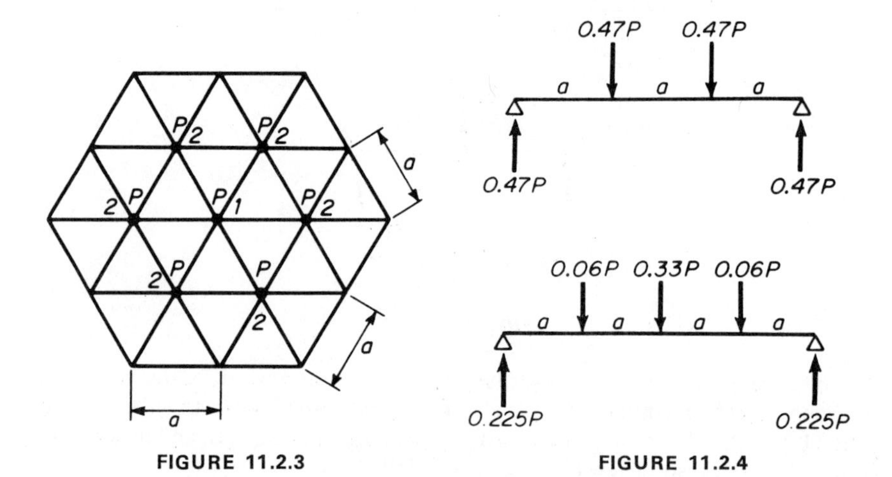

FIGURE 11.2.3

FIGURE 11.2.4

The maximum bending moments in the short and long beams, from Fig. 11.2.4, are:

$$M_2 = 0.47Pa, \qquad M_1 = (0.225 \times 2a - 0.06a)P = 0.39Pa.$$

The analysis of triangular grids with a high number of nodes by plate analogy is dealt with in Section 18.4.

Example 11.2.3. Determine the maximum moments in the beams of the quadruple grid of Fig. 11.2.5. All beams have the same I, and each internal node carries a load P.

At node 1, the load P is carried at midspan by two perpendicular beams of length l, supported by two diagonal beams of length $\sqrt{2}\ l$, which also support four

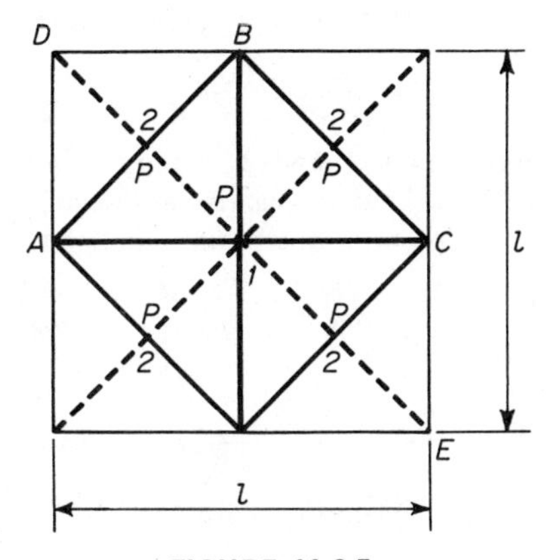

FIGURE 11.2.5

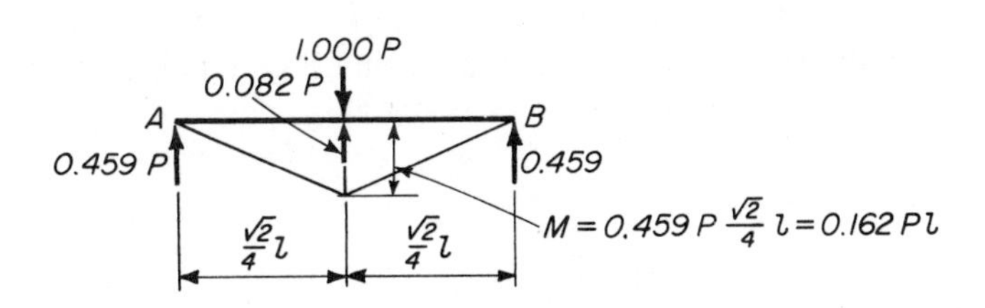

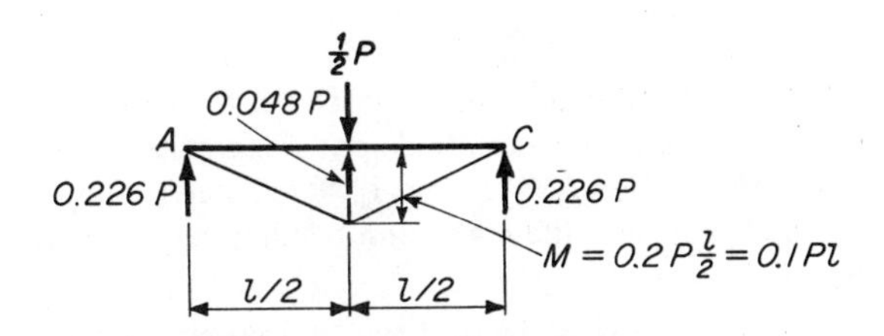

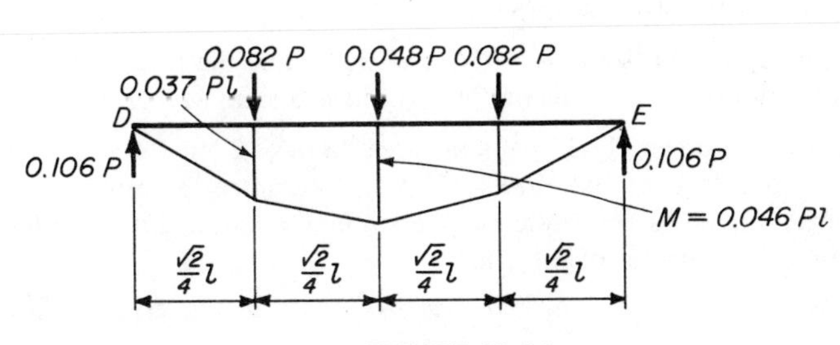

FIGURE 11.2.6

short diagonal beams. Hence, at node 1, by Table 11.1.3:

$$\frac{16}{768}(P - 2X_1)\frac{l^3}{E(2I)} = \frac{16}{768}2X_1\frac{(\sqrt{2}\,l)^3}{E(2I)} + \frac{11}{768}4X_2\frac{(\sqrt{2}\,l)^3}{E(2I)}, \tag{a}$$

where X_1 is the reaction at node 1, and X_2 is the reaction at node 2 of each long diagonal beam. At node 2, the load P is carried at the quarter point of one beam of length $\sqrt{2}\,l$ and at the midpoint of one beam of length $(l/\sqrt{2})$. Hence:

$$\frac{16}{768}(P - X_2)\frac{(l/\sqrt{2})^3}{EI} = \frac{11}{768}X_1\frac{(\sqrt{2}\,l)^3}{EI} + \left(\frac{9+7}{768}\right)X_2\frac{(\sqrt{2}\,l)^3}{EI}. \tag{b}$$

Equations (a) and (b) simplify to:

$$11X_1 + 18X_2 = 2P \quad 122X_1 + 124X_2 = 16P,$$

from which:

$$X_1 = 0.048P, \qquad X_2 = 0.082P.$$

Figure 11.2.6 indicates the loads on the beams and the maximum bending moments in each one of them.

The analysis of quadruple grids with a high number of nodes by plate analogy is dealt with in Section 18.5.

PROBLEMS

11.1 A simply supported rectangular grid consists of three upper beams of length l_1 and three lower beams of length l_2. A load P is carried at each of the equally spaced nodal points. For $I_1 = I_2$ and $l_2 = l_1/3$ determine the loads carried by each of the beams and the maximum bending moment and the reactions in each beam.

11.2 Determine the loads carried by and the maximum moments in each beam of the grid of Problem 11.1, for $l_2 = 2l_1/3$.

11.3 In Example 11.1.2 assume $I_1 = 2I_2$ and determine the ratio l_2/l_1 for which the maximum bending moments M_1, M_2 have a ratio $M_1/M_2 = 2$.

11.4 The grid of Problem 11.2 spans an area 18 m × 27 m (60 ft × 90 ft) and carries a total load of 10 kN/m² (200 psf). Determine the required depth of its 20 MPa (2,900 psi) concrete members if b = 700 mm (27 in.). What is the deflection at the center of the grid?

11.5 What is the ratio I_2/I_1 which makes $M_2 = M_1$ in the grid of Fig. 11.1.2 if $l_1 = 1.5l_2$?

11.6 The roof shown in Fig. 11.2.3, with a = 7.5 m (25 ft), carries a total load of 6.5 kN/m² (130 psf). Determine the depth of 20 MPa (2,900 psi) concrete grid members 450 mm (18 in.) wide. Compute the deflection at the center of the grid.

11.7 Determine the maximum bending moments in the grid of Fig. 11.2.1 for $l_1 = l_2 = l$ and $I_1 = I_2 = I$ due to a load P at each internal node.

11.8 Find the maximum moments in the beams of the grid shown in Fig. 11.2.5 when a load P is applied at node 1 only.

11.9 The roof over a square area l meters (ft) on a side, under a uniform load w, may be covered by grids $l/3$ on a side or $l/4$ on a side. Determine the ratio of the depths of the two grids, assuming the same allowable stress F_b and the same width b in the rectangular cross-section beams of the two grids. Compare the amount of material in the two grids. Apply the results to a grid with $l = 18$ m (60 ft), $b = 200$ mm (8 in.), $w_{LL} = 1.9$ kN/m² (40 psf) and $f'_c =$ 30 MPa (4,350 psi).

11.10 A beam connects the opposite corners of a square area of side l and carries a load P at midspan. A beam, at right angles to the first, supports it at: (a) midspan, (b) a quarter point, (c) a sixth point from one of the corners. The two beams have the same EI. Determine the reaction of the first beam nearest the supporting beam and whether the moment in the diagonal beam changes sign in case (a), (b), or (c). (Use Table 11.1.3.)

11.11 A rectangular grid covers a long rectangular area. Its identical beams are oriented at 45° to the two bounding walls, which are l meters (ft) apart. The beams are supported $l/2$ meters (ft) apart along the support walls. Determine the bending moments in the beams if the grid is uniformly loaded. (Concentrate the uniform load corresponding to the contributory areas at each nodal point of the grid.)

11.12 Solve Problem 11.11 if the beam supports are spaced by $l/3$ along the boundary walls.

11.13 One-way joists 3 m (10 ft) o.c. are supported on two walls 6 m (20 ft) apart. A distribution rib connects four joists at their midspan. If a load P acts at midspan of the first of the four joists and the distribution rib has the same EI as the joists, how is the load P shared by the four joists?

11.14 A circular skylight of radius R is supported by a grid consisting of six beams which connect points spaced by 60° on its boundary. Each beam connects one point with the point 120° around the boundary. Determine the bending moments in each beam if two loads P act on two diametrically opposed points of the grid.

11.15 Solve Problem 11.14 if a single load P acts on a nodal point.

11.16 Solve Problem 11.14 if the support points are spaced by 72° and two loads P act on two adjoining nodal points of the grid, calling l the length of the beams. (The distance between the nodal points and the support points is $l \times \sin 18°/\cos 36° = 0.382l$.) [*Hint:* $\delta_{11}/\delta_{12} = \frac{14}{13}$.]

11.17 In the grid of Fig. 11.1.3 $l_1 = 3l_2$ and each nodal point carries a load P. What should the ratio of the moments of inertia I_1 and I_2 be for the beams to develop equal maximum moments?

11.18 A balcony railing consists of two simply supported horizontal steel pipes 3 m (10 ft) long and one vertical steel pipe 1.2 m (4 ft) high, bracing the horizontal pipes at midspan. The horizontal pipes are 600 mm (2 ft) apart, and the vertical pipe is cantilevered. Determine the maximum stresses in the pipes if their diameter is 76 mm (3 in.), their thickness is 6 mm ($\frac{1}{4}$ in.), and the railing supports a concentrated horizontal force of 1 kN (200 lb) at midspan of the upper pipe.

11.19 Solve Problem 11.18 if the horizontal pipes are braced by vertical pipes at their third points, and 1 kN (200 lb) forces act at the third points of the top pipe.

12. Plates

12.1 PLATE ACTION [10.4]

A flat slab is a plate of constant thickness whose lateral dimensions are large in comparison with its thickness. A slab may have a variety of shapes (rectangular, circular, polygonal, etc.) and may have interior boundaries, so as to become a ring, if circular, a rectangle with centered or uncentered openings, etc.

Flat slabs may be supported along their entire boundary (e.g., on walls) or at isolated points (e.g., on columns); part of their boundary may be unsupported. The conditions along the boundary may be of simple support, of total or of partial fixity. A point supported slab may be rigidly connected to the columns or simply supported by them. It is thus seen that a variety of conditions influence the behavior of a flat slab under the concentrated or distributed loads it is designed to carry.

While bending and transverse shear stresses are developed in a beam on sections perpendicular to the axis of the beam (i.e., in one direction), at a point P of a plate, bending and transverse shear stresses can be developed on sections perpendicular to *any* direction lying in the plane of the plate: *a plate is a two-dimensional structural element* [Fig. 12.1.1 (a)].

Moreover, because of continuity in the plane of the plate, besides the bending stresses f_b and the transverse shears f_s typical of beam action, a plate may develop on any perpendicular section shears f_{hs} parallel to the plate's *neutral plane* [Fig. 12.1.1 (b)]. These shears are also distributed linearly across the plate thickness, and their resultant is a twisting moment on the plate sections [Fig. 12.1.1 (b)]. Thus, plates carry loads by two basic mechanisms: bending and transverse shear (i.e., beam action) and torsion.

The bending mechanism is analogous to the action developed by a

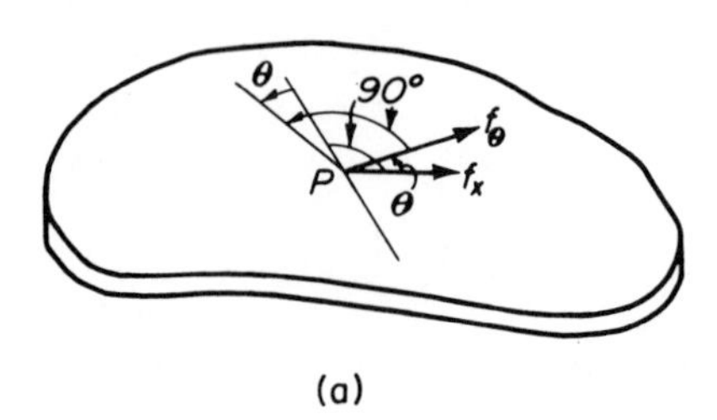

(a)

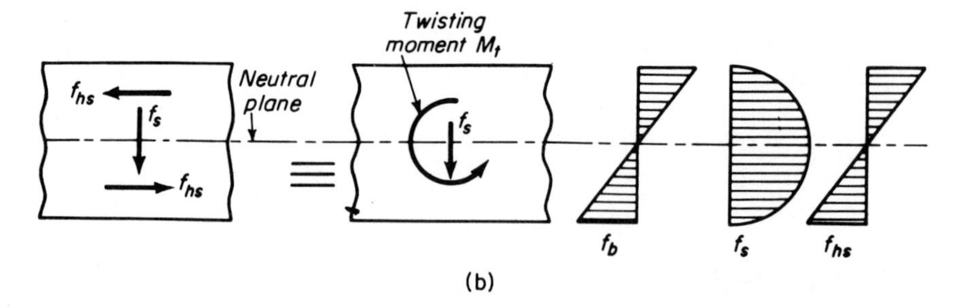

(b)

FIGURE 12.1.1

rectangular grid: the plate may be considered as a grid with an infinite number of identical beams of infinitesimal width and of depth h equal to the plate thickness. Bending and transverse shear in these perpendicular beam systems carry load in proportion to their stiffness, which is related to the beam spans only, since their cross sections are identical.

To visualize the load-carrying mechanism due to torsion, consider three identical beams 1, 2, and 3 loaded at midspan by concentrated loads W_1, W_2, W_3; their midspan deflections d_1, d_2, d_3 are proportional to the loads since the beams are identical [Fig. 12.1.2 (a)]. Let us now weld the midspan sections of the three beams so that their deflections there must become equal. The interactions between the beams are the shears S_1 and S_3 [Fig. 12.1.2 (b)]. The total loads on the identical beams are now $W_1 + S_1$, $W_2 - S_1 + S_3$, $W_3 - S_3$, and, since they produce identical deflections, they must be identical and equal to $(\frac{1}{3})(W_1 + W_2 + W_3)$. Hence, we obtain:

$$W_1 + S_1 = \tfrac{1}{3}(W_1 + W_2 + W_3), \qquad W_3 - S_3 = \tfrac{1}{3}(W_1 + W_2 + W_3),$$

from which:

$$S_1 = \tfrac{1}{3}(W_3 + W_2 - 2W_1), \qquad S_3 = \tfrac{1}{3}(2W_3 - W_2 - W_1).$$

The shears acting on beam 2 [Fig. 12.1.2 (c)] may be considered as the superposition of two symmetrical upward shears $(S_1 - S_3)/2$ and two anti-symmetrical shears $(S_1 + S_3)/2$. Hence, beam 2 also carries a total load:

$$W = W_2 - 2\frac{S_1 - S_3}{2} = \frac{1}{3}(W_1 + W_2 + W_3)$$

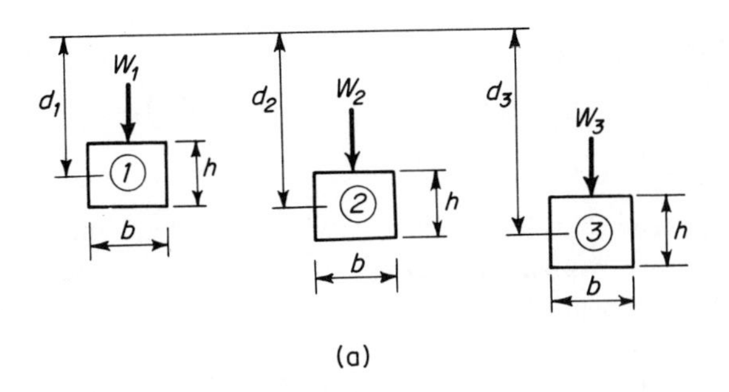

(a)

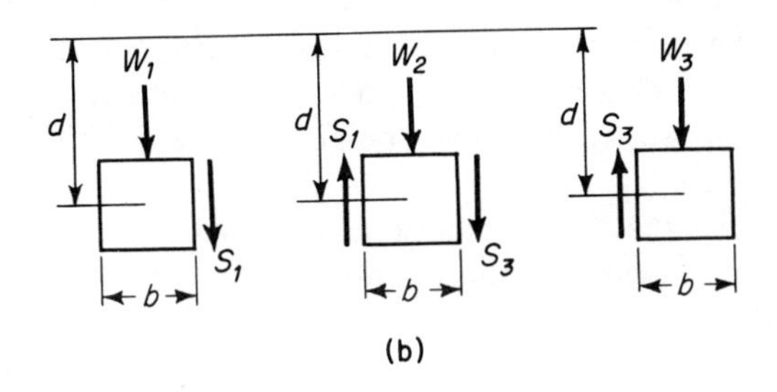

(b)

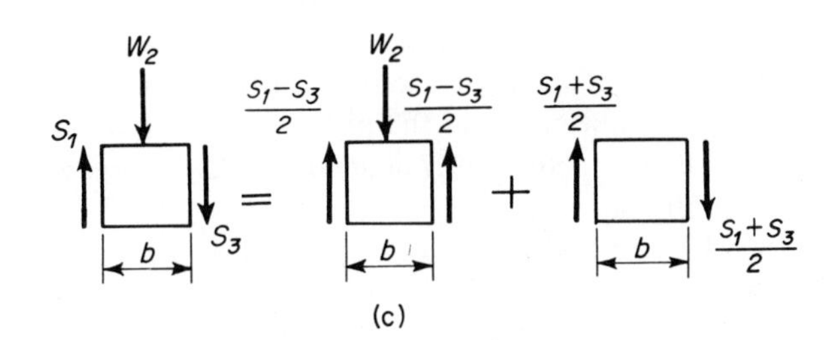

(c)

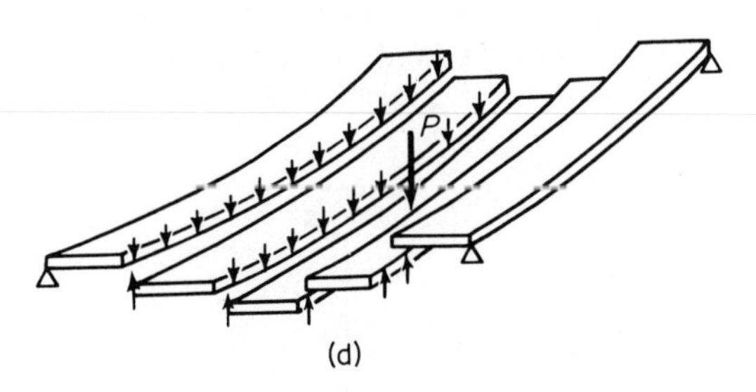

(d)

FIGURE 12.1.2

and is twisted by a torque:

$$M_t = \frac{(S_1 + S_3)}{2}b = \frac{(W_3 - W_1)}{2}b,$$

where b is its width.

It is thus seen that the additional twisting mechanism, due to continuity *across* the beams, transfers *laterally* the loads $(S_1 - S_3)/2 = \frac{1}{6}(-W_3 + 2W_2 - W_1)$ from beam 2 to beams 1 and 3, which in turn are twisted by torques $S_1b/2$ and $S_3b/2$.

Similarly, in a plate some of the load on a longitudinal strip is transferred laterally to two adjoining strips, and from these to the next strips, etc., until it reaches the supports [Fig. 12.1.2 (d)]. In the plate, the shears S vary along the length of each strip: their symmetrical components produce transverse shears f_s, their antisymmetrical components the twisting moments, i.e., the horizontal shears f_{hs}.

Moreover, this shear action takes place in two perpendicular directions, since the two sets of beams of the fictitious rectangular grid act in bending in two directions.

In conclusion, a plate transfers loads to the supports longitudinally by bending and shear in two orthogonal directions, and laterally by torsion and shear in the two directions perpendicular to the bending directions.

The share of the load carried in bending and in torsion depends on the load, the support conditions, and the plate shape. For example, in a simply supported square plate, uniformly loaded, the two mechanisms share the load almost equally.

The two-dimensional character of plates also gives them an additional secondary carrying capacity, which will be illustrated by the behavior of a simply supported circular plate of radius a (Fig. 12.1.3).

Let us assume that the simple plate support will allow radial displacements. Under the action of the load the plate deflects by δ at the center, and

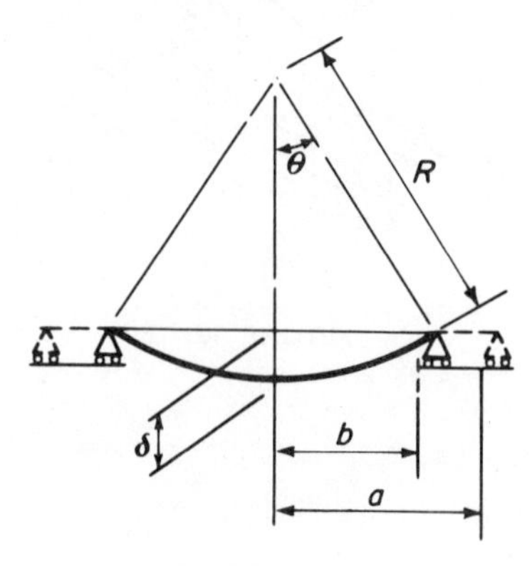

FIGURE 12.1.3

its boundary moves inward by $a - b$. Approximating its deflection by the arc of a circle of radius R, subtending an arc 2θ small enough for $\sin\theta$ to be approximated by $\theta - \theta^3/6$ and for $\cos\theta$ to be approximated by $1 - \theta^2/2$, we obtain:

$$\delta = R(1 - \cos\theta) = \frac{R\theta^2}{2}, \qquad a = R\theta,$$
$$b = R\sin\theta = R\left(\theta - \frac{\theta^3}{6}\right). \tag{a}$$

The plate boundary must shorten to move from the radius a to the radius b, and its strain is:

$$\epsilon_\theta = \frac{2\pi b - 2\pi a}{2\pi a} = \frac{b - a}{a} = \frac{(R\theta - R\theta^3/6 - R\theta)}{R\theta} = -\frac{\theta^2}{6}.$$

Substituting θ^2 in terms of δ, from (a):

$$\epsilon_\theta = -\frac{1}{6}\left(\frac{2\delta}{R}\right) = -\frac{\delta}{3R}.$$

Ignoring Poisson's effect, it is found that the plate boundary is under a compressive stress f_θ in the circumferential direction:

$$f_\theta = E\epsilon_\theta = -\frac{E\delta}{3R}.$$

Since every circular fiber in the plate is shortened by the deflection, the entire plate develops not only bending stresses in the radial and circumferential directions, but also compressive stresses in the circumferential direction which vanish only at the center of the plate. If, instead, the plate boundary is prevented from moving inward, the plate develops radial tensile stresses. These circumferential and radial stresses are capable of carrying some load and are called *membrane stresses*.

The development of membrane stresses makes the plate stiffer than it would be otherwise: the deflection evaluated by taking into account membrane stresses is smaller than the deflection evaluated neglecting them. A beam can only deflect without strain of its middle axis, i.e., without development of cable action, provided one of its ends is free to move relative to the other. A plate, instead, will generally develop membrane stresses even if its boundary is free to move because it cannot deflect without acquiring a curved shape and, in general, a two-dimensional element cannot acquire such a shape without stretching or shrinking, slightly, its middle surface. (Only a cylindrical deflection can take place in a plate without stretch of its middle surface; such a deflection is called *developable*.) Hence, most of the load on the plate is carried by bending and torsional stresses and, usually, a small amount by membrane stresses. The latter are the two-dimensional equivalent of the cable stress developed by a beam with anchored ends (see Example 5.2.2).

The load-carrying action by membrane stresses is dealt with more thoroughly in the discussion of shell action in Chapters 15 to 17. (See, in particular, Example 15.3.8 for the load-carrying capacity of plates through membrane stresses.)

12.2 RECTANGULAR HOMOGENEOUS PLATES [10.4]

The exact evaluation of stresses and deflections in a plate requires mathematical knowledge beyond that expected of the reader of this book.

The following tabulated stress and deflection coefficients for uniformly loaded plates are presented without derivation and discussed from a physical point of view so as to be of use in preliminary design.

Table 12.2.1 gives the coefficients for the maximum deflection δ_{max}, the maximum bending moments M_x, M_y in the x- and y-directions, the maximum transverse shears V_x, V_y, and the maximum reactions R_x, R_y on the sides a, b parallel to x, y of a rectangular, *simply supported* plate under a uniform load w (Fig. 12.2.1).* The maximum moments occur at the center of the plate.

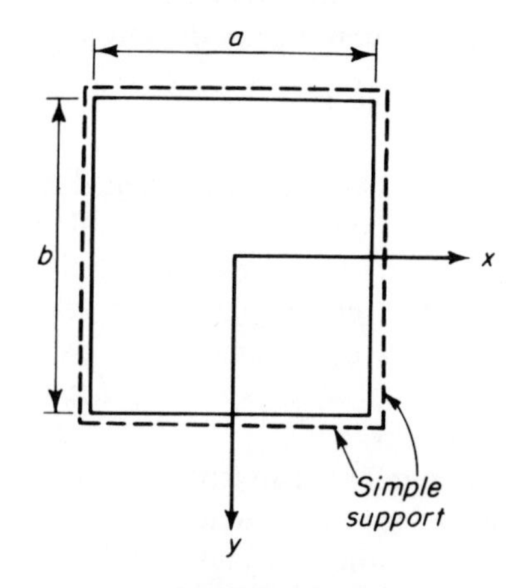

FIGURE 12.2.1

All quantities in Table 12.2.1 are *per unit of plate width*, so that the moments have units of force (for example, kN·m/m = kN, ft·lb/ft = lb), and the forces have units of force per unit length (for example, kN/m, lb/ft).

It is apparent from Table 12.2.1 that as soon as b/a is greater than 2 for all practical purposes, the two-way action of the plate vanishes and one-way beam-action takes place in the short direction a. No plate action is usually considered for $b/a > 1.5$.

*The coefficients have been evaluted for Poisson's ratio $\nu = 0.3$, but are not too sensitive to the value of ν.

Table 12.2.1 SIMPLY SUPPORTED PLATE

$\frac{b}{a}$	$\delta_{max} = dwa^4/Eh^3$	$M_{x,\,max} = m_x wa^2$	$M_{y,\,max} = m_y wa^2$	$V_{x,\,max} = v_x wa$	$V_{y,\,max} = v_y wa$	$R_{x,\,max} = r_x wa$	$R_{y,\,max} = r_y wa$
1.0	0.049	0.0479	0.0479	0.338	0.338	0.420	0.420
1.1	0.058	0.0554	0.0493	0.360	0.347	0.440	0.440
1.2	0.068	0.0627	0.0501	0.380	0.353	0.455	0.453
1.3	0.077	0.0694	0.0503	0.397	0.357	0.468	0.464
1.4	0.085	0.0755	0.0502	0.411	0.361	0.478	0.471
1.5	0.093	0.0812	0.0498	0.424	0.363	0.486	0.480
2.0	0.122	0.1017	0.0464	0.465	0.370	0.503	0.496
3.0	0.147	0.1189	0.0406	0.493	0.372	0.505	0.498
∞	0.156	0.1250	0.0375	0.500	0.372	0.500	0.500

It is readily seen from Table 12.2.1 that, because of torsion, plate action differs from grid action. Assuming, for example, that the load w on the median strips of a plate with $a = b$ is equally shared by beams of length a and width 1 at right angles to each other, the maximum moment in these beams would be:

$$M_x = \frac{1}{8}\left(\frac{w}{2}\right)a^2 = 0.0625wa^2,$$

or 30% higher than the moment in the plate. Similarly, the maximum deflections of the two orthogonal beams would be:

$$\delta_{max} = \frac{5}{384}\frac{(w/2)a^4}{E(\frac{1}{12} \times 1 \times h^3)} = 0.078\frac{wa^4}{Eh^3},$$

or 59% higher than the plate deflection. On the other hand, two diagonal fixed-end beams of unit width connecting the plate corners would have a fixed-end moment:

$$M_d = -\frac{1}{12}\left(\frac{w}{2}\right)(\sqrt{2}\ a)^2 = -0.083\,3wa^2,$$

while the moment in the diagonal direction at the plate corner, which is negative due to the rigidity of the corners, may be proved to be $-0.0325\ wa^2$, or only 39% of the beam moment.

For a plate with $b/a = 2$, equating deflections at midspan for two beams of length a and $2a$, under uniform load, it is found that the sharing is in the ratio of $(\frac{1}{2})^4$ to 1. Hence, the maximum moments in the two beams are:

$$M_x = \left(\frac{16}{17}w\right)\left(\frac{a^2}{8}\right) = 0.118\,0\,wa^2,$$

$$M_y = \left(\frac{1}{17}w\right)\left(\frac{a^2}{8}\right) = 0.007\,4\,wa^2,$$

with errors of -16% and $+84\%$ in comparison with the plate moments, while the maximum beam deflection is:

$$\delta_{\max} = \frac{5}{384} \frac{(\frac{16}{17}w)a^4}{E(\frac{1}{12} \times 1 \times h^3)} = 0.15\frac{wa^4}{Eh^3},$$

with an error of $+23\%$ in comparison with the plate deflection.

Table 12.2.2 gives the maximum deflection, the maximum positive moments M_x, M_y, and the maximum negative moments M_x and M_y in a uniformly loaded rectangular plate with *fixed* sides (Fig. 12.2.2). The maximum negative moments occur at the middle of the long sides, A, the maximum positive moments at the plate center, O.

Rectangular plates may be supported differently on each of their sides. For example, when two opposite long sides of a plate are simply supported and its two opposite short sides are fixed (Fig. 12.2.3), the maximum deflection

Table 12.2.2 FIXED PLATE

$\frac{b}{a}$	$\delta_{\max} = dwa^4/Eh^3$	$+M_{x,\max} = m_x wa^2$	$+M_{y,\max} = m_y wa^2$	$-M_{x,\max} = m_x wa^2$	$-M_{y,\max} = m_y wa^2$
1.0	0.0138	0.0231	0.0231	−0.0513	−0.0513
1.1	0.0165	0.0264	0.0231	−0.0581	−0.0538
1.2	0.0189	0.0299	0.0228	−0.0639	−0.0554
1.3	0.0210	0.0327	0.0222	−0.0687	−0.0563
1.4	0.0228	0.0349	0.0212	−0.0726	−0.0586
1.5	0.0242	0.0368	0.0203	−0.0757	−0.0570
2.0	0.0279	0.0412	0.0158	−0.0829	−0.0571
∞	0.0286	0.0417	0.0125	−0.0833	−0.0571

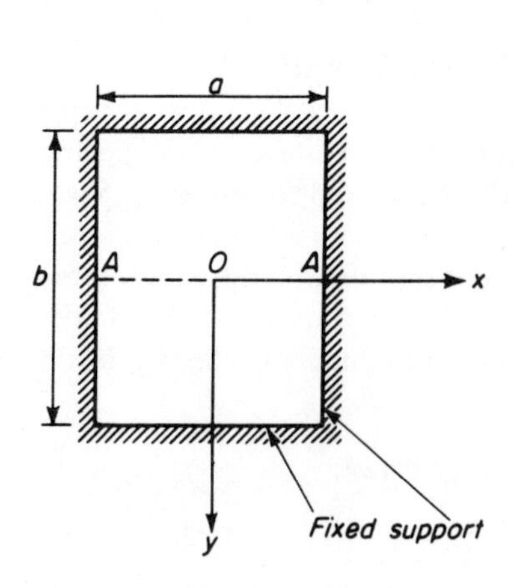

FIGURE 12.2.2

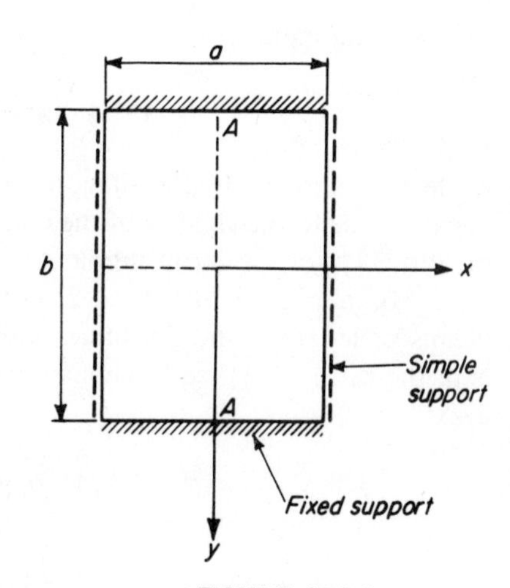

FIGURE 12.2.3

and moments are given by Table 12.2.3. The maximum negative moment occurs at the middle of the short sides, *A*.

Table 12.2.4 gives the positive moments at the center of the plate and the greater negative moment at the middle points *A*, *B* of the fixed sides in a plate with two adjacent sides fixed and the other two simply supported (Fig. 12.2.4). The maximum positive moments M_x, M_y occur at point *P*, 0.4*a* and 0.4*b* from the simply supported edges and are less than 10% above the moments at the plate center *O* given in the table.

Table 12.2.5 gives the maximum negative moment in uniformly loaded rectangular plates fixed on two *adjacent* sides and free on the other two (Fig. 12.2.5), the so-called *corner-balcony plates.* The maximum value of the moments occurs at the roots *A* and *B* of the free sides and is practically the same at the roots of the long and of the short side, whatever the ratio of sides. The positive moments are negligible for such plates.

Table 12.2.3 PLATES WITH LONG SIDES *b* SIMPLY SUPPORTED, SHORT SIDES *a* FIXED

$\frac{b}{a}$	$\delta_{max} = dwa^4/Eh$	$+M_{x,\,max} = m_x wa^2$	$+M_{y,\,max} = m_y wa^2$	$-M_{y,\,max} = m_y wa^2$
1.0	0.0210	0.0244	0.0332	−0.0697
1.1	0.0275	0.0307	0.0371	−0.0787
1.2	0.0350	0.0376	0.0400	−0.0868
1.3	0.0426	0.0446	0.0426	−0.0938
1.4	0.0505	0.0514	0.0448	−0.0998
1.5	0.0583	0.0585	0.0460	−0.1049
2.0	0.0925	0.0869	0.0474	−0.1191
∞	0.1430	0.1250	0.0375	−0.1250

Table 12.2.4 PLATES WITH TWO ADJACENT SIDES SIMPLY SUPPORTED AND THE OTHER TWO FIXED

$\frac{b}{a}$	$+M_{x,\,max} = m_x wa^2$	$-M_{x,\,max} = m_x wa^2$	$+M_{y,\,max} = m_y wa^2$	$-M_{y,\,max} = m_y wa^2$
1.0	0.0281	−0.0678	0.0281	−0.0678
1.1	0.0330	−0.0766	0.0283	−0.0709
1.2	0.0376	−0.0845	0.0279	−0.0736
1.3	0.0416	−0.0915	0.0270	−0.0754
1.4	0.0451	−0.0975	0.0260	−0.0765
1.5	0.0481	−0.1028	0.0248	−0.0772
2.0	0.0574	−0.1180	0.0191	−0.0787
∞	0.0703	−0.1250	—	—

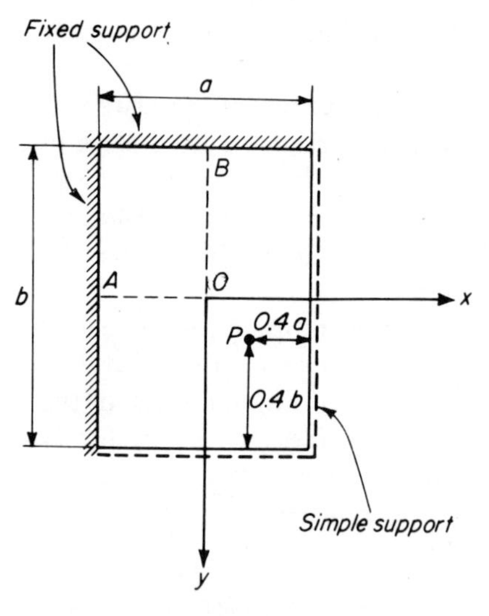

FIGURE 12.2.4

FIGURE 12.2.5

Table 12.2.5 CORNER-BALCONY PLATES

b/a	1	4/3	2	8/3	4	8	∞
$M_{x,\,\max} = M_{y,\,\max} = mwa^2$	−0.29	−0.35	−0.43	−0.45	−0.48	−0.49	−0.50

The bending stresses in a homogeneous plate are obtained by noticing that the section modulus S of a rectangular section of width 1 and depth h is $\frac{1}{12}(1)h^3/(h/2) = h^2/6$:

$$f_b = \frac{M}{S} = \frac{6M}{h^2}. \tag{12.2.1}$$

The transverse shear stresses in the same plate may be evaluated by assuming a parabolic stress distribution across the thickness with a maximum equal to 1.5 times the average stress, as is done in rectangular cross-section beams:

$$f_s = 1.5\frac{V}{A} = 1.5\frac{V}{h}. \tag{12.2.2}$$

Example 12.2.1. Compare the thicknesses h_a, h_b, h_c of two plates uniformly loaded with *aspect ratio* (ratio of sides) 1.5/1 and 1/1: (a) simply supported; (b) fixed; (c) simply supported along the long sides and fixed along the short ones, with the thickness h_d of a one-way simply supported slab of span a, indicating by F_b the allowable stress in bending of its homogeneous material.

By Table 12.2.1 and (12.2.1):

Aspect ratio 1.5/1:

$$(1)\quad F_b = 0.081\,2\,\frac{6wa^2}{h_a^2} = 0.487w\left(\frac{a}{h_a}\right)^2 \quad \therefore \quad h_a = \sqrt{0.487}\sqrt{\frac{w}{F_b}}\,a,$$

$$(2)\quad F_b = 0.075\,7\,\frac{6wa^2}{h_b^2} = 0.453w\left(\frac{a}{h_b}\right)^2 \quad \therefore \quad h_b = \sqrt{0.453}\sqrt{\frac{w}{F_b}}\,a,$$

$$(3)\quad F_b = 0.104\,9\,\frac{6wa^2}{h_c^2} = 0.629w\left(\frac{a}{h_c}\right)^2 \quad \therefore \quad h_c = \sqrt{0.629}\sqrt{\frac{w}{F_b}}\,a,$$

$$(4)\quad F_b = 0.125\,0\,\frac{6wa^2}{h_d^2} = 0.750w\left(\frac{a}{h_d}\right)^2 \quad \therefore \quad h_d = \sqrt{0.750}\sqrt{\frac{w}{F_b}}\,a,$$

$$\frac{h_a}{h_d} = \sqrt{\frac{0.487}{0.750}} = 0.81, \quad \frac{h_b}{h_d} = \sqrt{\frac{0.453}{0.750}} = 0.78, \quad \frac{h_c}{h_d} = \sqrt{\frac{0.629}{0.750}} = 0.92.$$

Aspect ratio 1/1:

$$(a)\quad \frac{h_a}{h_d} = \sqrt{\frac{0.288}{0.750}} = 0.62, \qquad (b)\quad \frac{h_b}{h_d} = \sqrt{\frac{0.307}{0.750}} = 0.64,$$

$$(c)\quad \frac{h_c}{h_d} = \sqrt{\frac{0.417}{0.750}} = 0.75.$$

These thickness ratios are also valid for balanced, reinforced concrete slab sections.

Example 12.2.2. Find the thickness ratios in Example 12.2.1 if the plate thickness is determined in all cases by the stresses due to the maximum positive bending moments.

Aspect ratio 1.5/1:

$$(a)\quad \frac{h_a}{h_d} = \sqrt{\frac{0.487}{0.750}} = 0.81, \qquad (b)\quad \frac{h_b}{h_d} = \sqrt{\frac{0.220}{0.750}} = 0.56,$$

$$(c)\quad \frac{h_c}{h_d} = \sqrt{\frac{0.351}{0.750}} = 0.68.$$

Aspect ratio 1/1:

$$(a)\quad \frac{h_a}{h_d} = \sqrt{\frac{0.288}{0.750}} = 0.62, \qquad (b)\quad \frac{h_b}{h_d} = \sqrt{\frac{0.138}{0.750}} = 0.43,$$

$$(c)\quad \frac{h_c}{h_d} = \sqrt{\frac{0.200}{0.750}} = 0.52.$$

These thickness ratios may be used for reinforced concrete slabs where the negative moment stress is absorbed by the steel in the upper part of the section.

Example 12.2.3. A rectangular hall 15 × 22.5 m (50 ft × 75 ft) is roofed by a grid of identical steel beams spaced 1.5 m (5 ft) apart. The total roof load is 7.2 kN/m² (150 psf). Design the beams of the grids, assuming them simply supported and identical.

The evaluation of the reactions of one beam system on the other would require the solution of a system of 35 equations in 35 unknowns (Fig. 12.2.6). A

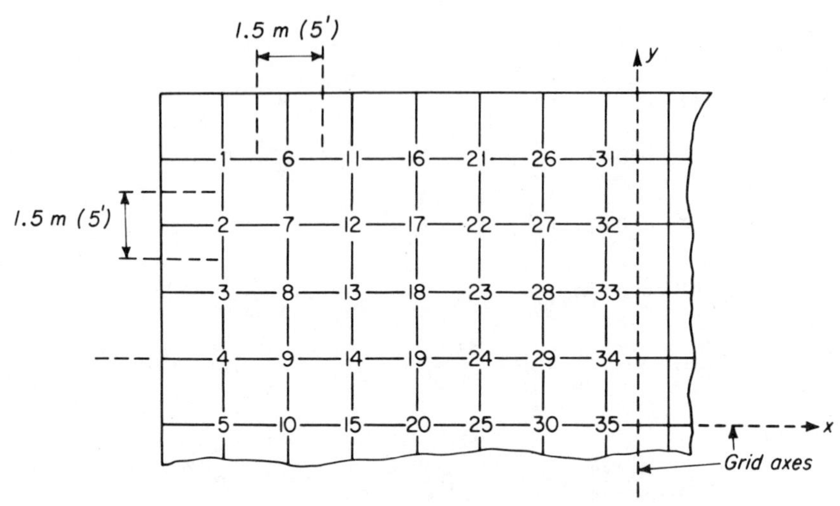

FIGURE 12.2.6

solution of sufficient practical accuracy may be easily obtained, instead, by assuming that the roof acts as a plate, by determining the maximum moments per unit width of the plate, and by designing each beam to absorb the moment of a 1.5 m width of plate.

From Table 12.2.1, for $b/a = 1.5$:

$$M_{x,\max} = 0.081 wa^2.$$

Hence, the moment for a width $c = 1.5$ m of plate is:

$$M_x = 0.081 wa^2 c = 0.081 \times 7.2 \times 15^2 \times 1.5 = 196.8 \text{ kN·m (152 k·ft)}.$$

The required section modulus in steel for an allowable stress $F_b = 165$ MPa (24 ksi) is:

$$S = \frac{M_x}{F_b} = \frac{196.8 \times 1\,000^2}{165} = 1\,193 \times 10^3 \text{ mm}^3 \text{ (72.77 in.}^3\text{)}.$$

A W460 × 74 (W18 × 50) has $S = 1\,457 \times 10^3$ mm³ and weighs 0.73 kN/m. The dead load of the beams per unit area is the weight of two beams 1.5 m in length, divided by the contributory area $1.5 \times 1.5 = 2.25$ m², i.e.,

$$\frac{3 \times 0.73}{2.25} = 0.97 \text{ kN/m}^2 \text{ (20 psf)}.$$

Example 12.2.4. Determine the thickness of a homogeneous rectangular balcony plate, fixed along two adjacent sides and free along the other two, uniformly loaded, and of aspect ratio 1.5.

From Table 12.2.5, interpolating linearly between the values for $b/a = 1.33$ and $b/a = 2$, one obtains:

$$\frac{m - (-0.35)}{1.50 - 1.33} = \frac{-0.43 - (-0.35)}{2.00 - 1.33} \quad \therefore \quad m = -0.37.$$

With an allowable stress F_b, the thickness of the plate is given by:

$$F_b = \pm\frac{M_{max}}{S} = \pm\frac{6 \times 0.37 \times wa^2}{h^2} = \pm 2.22w\frac{a^2}{h^2} \quad \therefore \quad h = 1.49\sqrt{\frac{w}{F_b}}\,a.$$

Example 12.2.5. To improve their structural efficiency, plates are often stiffened by ribs in two orthogonal directions. Such plates are called *waffle slabs.* Each rib of a waffle slab with an adjoining participating or *effective width* of plate acts as a T-beam (Fig. 12.2.7). By code, the effective width of a plate on either side of the rib must be less than eight times the plate thickness and not more than the rib spacing c.

Determine the load sharing between the rib systems of a uniformly loaded, ribbed plate of reinforced concrete 75 mm (3 in.) thick, covering a rectangular area 6 m $\times$ 9 m (20 ft $\times$ 30 ft), with equal ribs 125 mm $\times$ 250 mm (5 in. $\times$ 10 in.) spaced 750 mm (30 in.) o.c.

The effective width of the plate is the rib spacing $c = 750$ mm, since the rib spacing (Fig. 12.2.8) is less than $2 \times 8 \times 75 + 125 = 1\,325$ mm. Hence, the grid

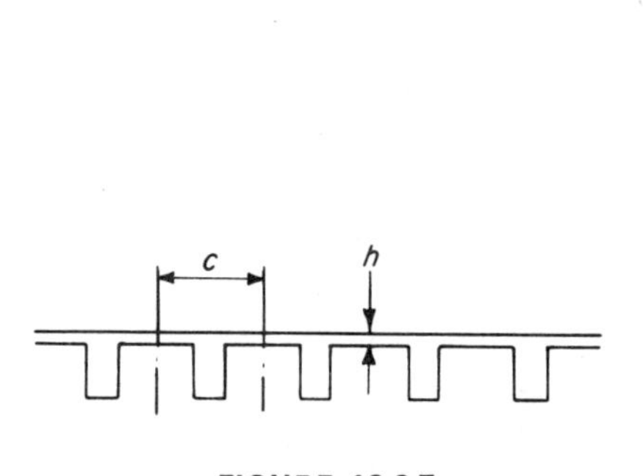

FIGURE 12.2.7

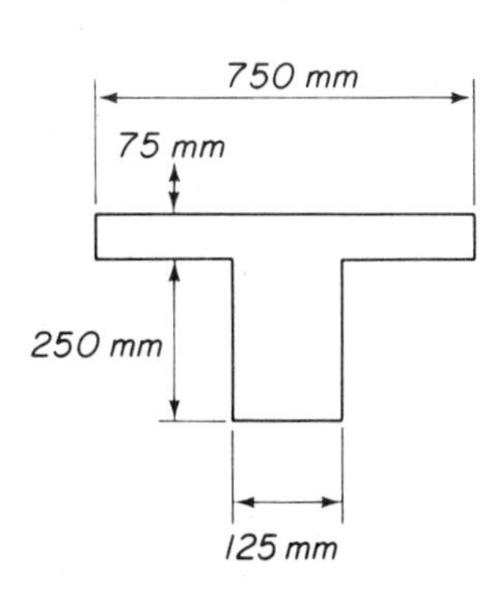

FIGURE 12.2.8

of T-beams consists of identical beams and the sharing is in proportion to the fourth power of the spans: the short ribs will carry $1.5^4/(1.5^4 + 1^4) = 0.83$ of the load, and the long ribs 0.17 of the load. Unequal spacing of the ribs may be used to reduce the difference in load sharing of the two rib systems.

12.3 TWO-WAY AND FLAT SLAB DESIGN IN REINFORCED CONCRETE

The analysis of *two-way slabs* of reinforced concrete is based on elastic plate theory, modified to account for the plastic redistribution of stress shown by experiments, which can be illustrated by considering the grid of Fig. 12.2.6 and its equivalent plate.

By Table 12.2.1, with $b/a = 1.5$, the maximum moment in the center 1.5 m (5 ft) strip of the equivalent plate is:

$$M_x = 1.5 \times 0.081\,wa^2. \tag{a}$$

The maximum moments developed by the two strips on each side of the center strip are smaller. But, if the center strip is overloaded to the point of yielding, the adjacent understressed strips will pick up the excess load without failure of the plate. Since this process of load redistribution can take place over a large part of the center section of the plate, the elastic design moments in plates can be safely reduced to take advantage of this plastic redistribution.

For instance, for the plate of Fig. 12.2.6 experiments show that the *design moment* may be assumed to be:

$$M_x = 1.5 \times 0.072\, wa^2,$$

which is 11% less than the theoretical elastic moment (a). For a square plate, this reduction is as much as 25%. Moreover, the full design moment is assumed by code* to act over the middle half of the slab, called the *middle strip* (Fig. 12.3.1) and to diminish linearly to one-third of this value at the edge over what is called the *column strip*. A comparison of the theoretical and the design moments (Fig. 12.3.1) shows the averaging procedure followed by the code.

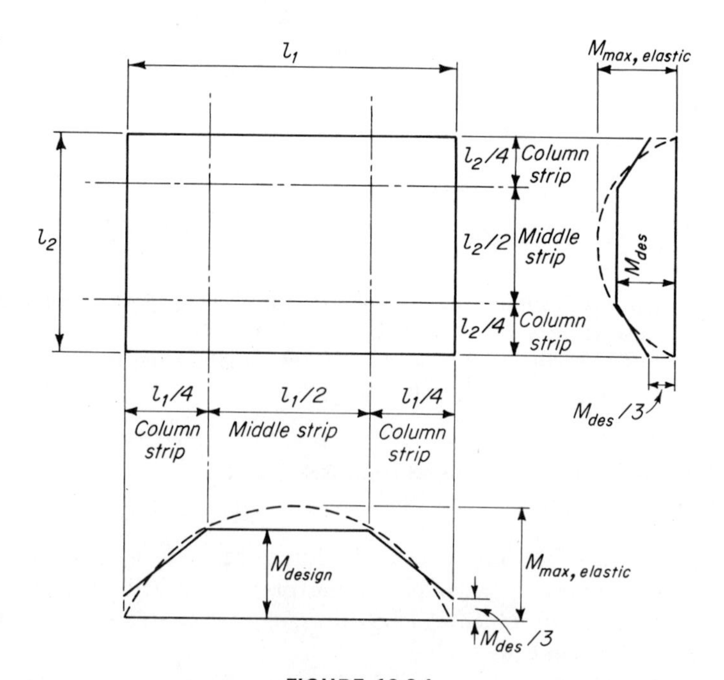

FIGURE 12.3.1

*ACI 318, American Concrete Institute, Building Code Requirements for Reinforced Concrete.

When all panels of a plate continuous over many supports are loaded, e.g., by the dead load, rotation over the supports is practically zero. If one panel is loaded by live loads and the adjacent panels are not, a rotation will occur at the support, with a reduction of fixity and higher positive moments. However, for preliminary calculations this difference may often be neglected.

Coefficients for positive and negative moments in rectangular plates are given in Tables 12.3.1, 12.3.2, and 12.3.3. The load carried by the supporting beams or walls is obtained by assuming contributory areas as shown in Fig. 12.3.2.

Table 12.3.1 COEFFICIENTS FOR *Live Load* POSITIVE MOMENTS IN SLABS*

$$\left.\begin{aligned} M_{a\,\text{pos}} &= C_a \times w \times a^2 \\ M_{b\,\text{pos}} &= C_b \times w \times b^2 \end{aligned}\right\} \text{ where } w = \text{total uniform live load}$$

Ratio $m = \frac{a}{b}$		Case 1	Case 2	Case 3	Case 4	Case 5	Case 6	Case 7	Case 8	Case 9
1.00	C_a	0.036	0.027	0.027	0.032	0.032	0.035	0.032	0.028	0.030
	C_b	0.036	0.027	0.032	0.032	0.027	0.032	0.035	0.030	0.028
0.95	C_a	0.040	0.030	0.031	0.035	0.034	0.038	0.036	0.031	0.032
	C_b	0.033	0.025	0.029	0.029	0.024	0.029	0.032	0.027	0.025
0.90	C_a	0.045	0.034	0.035	0.039	0.037	0.042	0.040	0.035	0.036
	C_b	0.029	0.022	0.027	0.026	0.021	0.025	0.029	0.024	0.022
0.85	C_a	0.050	0.037	0.040	0.043	0.041	0.046	0.045	0.040	0.039
	C_b	0.026	0.019	0.024	0.023	0.019	0.022	0.026	0.022	0.020
0.80	C_a	0.056	0.041	0.045	0.048	0.044	0.051	0.051	0.044	0.042
	C_b	0.023	0.017	0.022	0.020	0.016	0.019	0.023	0.019	0.017
0.75	C_a	0.061	0.045	0.051	0.052	0.047	0.055	0.056	0.049	0.046
	C_b	0.019	0.014	0.019	0.016	0.013	0.016	0.020	0.016	0.013
0.70	C_a	0.068	0.049	0.057	0.057	0.051	0.060	0.063	0.054	0.050
	C_b	0.016	0.012	0.016	0.014	0.011	0.013	0.017	0.014	0.011
0.65	C_a	0.074	0.053	0.064	0.062	0.055	0.064	0.070	0.059	0.054
	C_b	0.013	0.010	0.014	0.011	0.009	0.010	0.014	0.011	0.009
0.60	C_a	0.081	0.058	0.071	0.067	0.059	0.068	0.077	0.065	0.059
	C_b	0.010	0.007	0.011	0.009	0.007	0.008	0.011	0.009	0.007
0.55	C_a	0.088	0.062	0.080	0.072	0.063	0.073	0.085	0.070	0.063
	C_b	0.008	0.006	0.009	0.007	0.005	0.006	0.009	0.007	0.006
0.50	C_a	0.095	0.066	0.088	0.077	0.067	0.078	0.092	0.076	0.067
	C_b	0.006	0.004	0.007	0.005	0.004	0.005	0.007	0.005	0.004

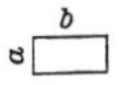

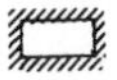

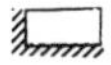

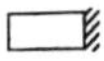

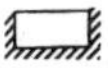

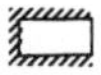

†A cross-hatched edge indicates that the slab continues across or is fixed at the support; an unmarked edge indicates a simple support at which torsional resistance is negligible.

*ACI 318–63, American Concrete Institute, Building Code Requirements for Reinforced Concrete.

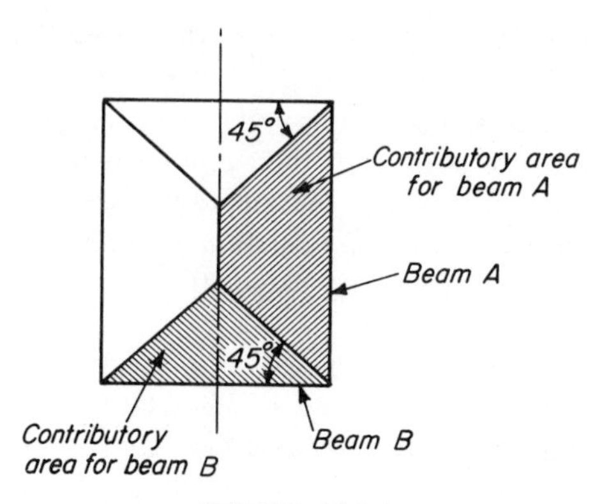

FIGURE 12.3.2

Table 12.3.2 COEFFICIENTS FOR *Dead Load* POSITIVE MOMENTS IN SLABS*

$$\left.\begin{aligned} M_{a\text{ pos}} &= C_a \times w \times a^2 \\ M_{b\text{ pos}} &= C_b \times w \times b^2 \end{aligned}\right\} \text{where } w = \text{total uniform dead load}$$

Ratio $m = \frac{a}{b}$		Case 1	Case 2	Case 3	Case 4	Case 5	Case 6	Case 7	Case 8	Case 9
1.00	C_a	0.036	0.018	0.018	0.027	0.027	0.033	0.027	0.020	0.023
	C_b	0.036	0.018	0.027	0.027	0.018	0.027	0.033	0.023	0.020
0.95	C_a	0.040	0.020	0.021	0.030	0.028	0.036	0.031	0.022	0.024
	C_b	0.033	0.016	0.025	0.024	0.015	0.024	0.031	0.021	0.017
0.90	C_a	0.045	0.022	0.025	0.033	0.029	0.039	0.035	0.025	0.026
	C_b	0.029	0.014	0.024	0.022	0.013	0.021	0.028	0.019	0.015
0.85	C_a	0.050	0.024	0.029	0.036	0.031	0.042	0.040	0.029	0.028
	C_b	0.026	0.012	0.022	0.019	0.011	0.017	0.025	0.017	0.013
0.80	C_a	0.056	0.026	0.034	0.039	0.032	0.045	0.045	0.032	0.029
	C_b	0.023	0.011	0.020	0.016	0.009	0.015	0.022	0.015	0.010
0.75	C_a	0.061	0.028	0.040	0.043	0.033	0.048	0.051	0.036	0.031
	C_b	0.019	0.009	0.018	0.013	0.007	0.012	0.020	0.013	0.007
0.70	C_a	0.068	0.030	0.046	0.046	0.035	0.051	0.058	0.040	0.033
	C_b	0.016	0.007	0.016	0.011	0.005	0.009	0.017	0.011	0.006
0.65	C_a	0.074	0.032	0.054	0.050	0.036	0.054	0.065	0.044	0.034
	C_b	0.013	0.006	0.014	0.009	0.004	0.007	0.014	0.009	0.005
0.60	C_a	0.081	0.034	0.062	0.053	0.037	0.056	0.073	0.048	0.036
	C_b	0.010	0.004	0.011	0.007	0.003	0.006	0.012	0.007	0.004
0.55	C_a	0.088	0.035	0.071	0.056	0.038	0.058	0.081	0.052	0.037
	C_b	0.008	0.003	0.009	0.005	0.002	0.004	0.009	0.005	0.003
0.50	C_a	0.095	0.037	0.080	0.059	0.039	0.061	0.089	0.056	0.038
	C_b	0.006	0.002	0.007	0.004	0.001	0.003	0.007	0.004	0.002

*A cross-hatched edge indicates that the slab continues across or is fixed at the support; an unmarked edge indicates a simple support at which torsional resistance is negligible.

Table 12.3.3 COEFFICIENTS FOR NEGATIVE MOMENTS IN SLABS*

$$\left.\begin{aligned} M_{a\,\text{neg}} &= C_a \times w \times a^2 \\ M_{b\,\text{neg}} &= C_b \times w \times b^2 \end{aligned}\right\}\ \text{where } w = \text{total uniform dead plus live load}$$

Ratio $m = \frac{a}{b}$		Case 1	Case 2	Case 3	Case 4	Case 5	Case 6	Case 7	Case 8	Case 9
1.00	C_a		0.045		0.050	0.075	0.071		0.033	0.061
	C_b		0.045	0.076	0.050			0.071	0.061	0.033
0.95	C_a		0.050		0.055	0.079	0.075		0.038	0.065
	C_b		0.041	0.072	0.045			0.067	0.056	0.029
0.90	C_a		0.055		0.060	0.080	0.079		0.043	0.068
	C_b		0.037	0.070	0.040			0.062	0.052	0.025
0.85	C_a		0.060		0.066	0.082	0.083		0.049	0.072
	C_b		0.031	0.065	0.034			0.057	0.046	0.021
0.80	C_a		0.065		0.071	0.083	0.086		0.055	0.075
	C_b		0.027	0.061	0.029			0.051	0.041	0.017
0.75	C_a		0.069		0.076	0.085	0.088		0.061	0.078
	C_b		0.022	0.056	0.024			0.044	0.036	0.014
0.70	C_a		0.074		0.081	0.086	0.091		0.068	0.081
	C_b		0.017	0.050	0.019			0.038	0.029	0.011
0.65	C_a		0.077		0.085	0.087	0.093		0.074	0.083
	C_b		0.014	0.043	0.015			0.031	0.024	0.008
0.60	C_a		0.081		0.089	0.088	0.095		0.080	0.085
	C_b		0.010	0.035	0.011			0.024	0.018	0.006
0.55	C_a		0.084		0.092	0.089	0.096		0.085	0.086
	C_b		0.007	0.028	0.008			0.019	0.014	0.005
0.50	C_a		0.086		0.094	0.090	0.097		0.089	0.088
	C_b		0.006	0.022	0.006			0.014	0.010	0.003

*A cross-hatched edge indicates that the slab continues across or is fixed at the support; an unmarked edge indicates a simple support at which torsional resistance is negligible.

Example 12.3.1. A concrete roof slab 3.5 × 4.75 m (12 ft × 16 ft), supported on four sides by beams, is loaded by a snow load of 2 kN/m² (40 psf) and a roofing and insulation load of 0.5 kN/m² (10 psf). Determine the slab thickness for $f'_c = 30$ MPa (4,350 psi).

The ratio of sides m is $3.5/4.75 \simeq 0.75$. From Table 12.3.1, Case 1, and Table 12.3.2, Case 1, $C_a = 0.061$ and $C_b = 0.019$. Assuming a 100 mm (4 in.) thick slab weighing 2.28 kN/m² (50 psf), the total load is $0.5 + 2 + 2.28 \simeq 4.8$ kN/m². The moments are:

$$M_a = C_a w a^2 = 0.061 \times 4.8 \times (3.5)^2 = 3.59\ \text{kN}\cdot\text{m}\ (880\ \text{ft}\cdot\text{lb}),$$

$$M_b = C_b w b^2 = 0.019 \times 4.8 \times (4.75)^2 = 2.06\ \text{kN}\cdot\text{m}\ (480\ \text{ft}\cdot\text{lb}).$$

The resisting moment of a concrete slab is [see (b′) of Section 4.2]:

$$M_R = \tfrac{1}{6} f_c b d^2.$$

From Table 2.4.2, for $f'_c = 30$ MPa, $f' = 13.5$ MPa. With a width $b =$ 1 000 mm, the maximum moment M_a requires a slab depth:

$$d = \sqrt{\frac{6M_a}{f_c b}} = \sqrt{\frac{6 \times 3.59 \times 1\,000^2}{13.5 \times 1\,000}} = 40 \text{ mm (1.6 in.)},$$

and, hence, a thickness:

$$t = 40 + 25 = 65 \text{ mm (2.6 in.)}.$$

The minimum slab thickness required by code is 1/180 of the slab perimeter. Applying this criterion, $t = 16.5 \times 1\,000/180 = 91.7$ mm (4.0 in.), which governs over the requirement for bending.

A slab supported directly on columns without connecting beams is called a *flat slab*. Normally, such slabs have uniform thickness although at times, to reduce the shear stresses around the column, the slab is thickened there by means of a so-called *drop panel*. In addition, to reduce bending stresses as well as shear stresses within the slab, the column is often flared at the top to form a *capital* (Fig. 12.3.3). Finally, the slab itself may be formed as a waffle to obtain greater depth without correspondingly increasing its weight.*

Although the rigorous analysis of a flat slab is complex, an approximate evaluation of its stresses may be obtained by analogy with a two-way slab, by dividing the slab into *column* and *middle* strips (Fig. 12.3.4). The loads

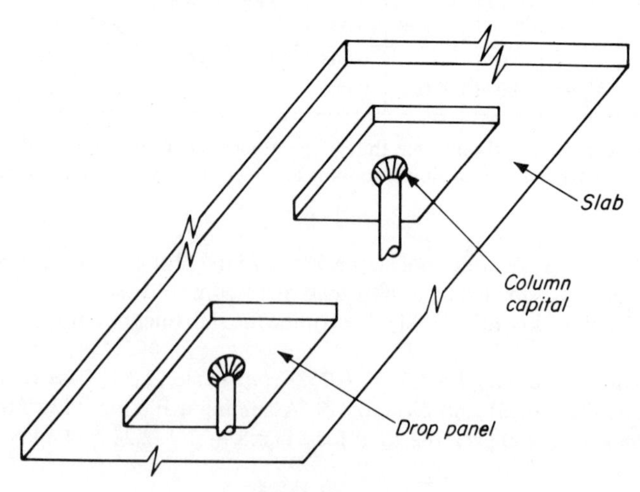

FIGURE 12.3.3

*Tables for the design of standard reinforced concrete waffle slabs appear in the *Concrete Reinforcing Steel Institute Handbook*, under the heading "Dome Slabs."

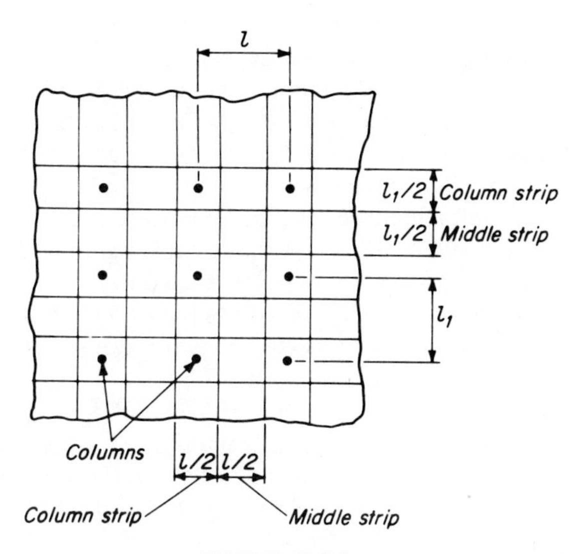

FIGURE 12.3.4

on the middle strip are assumed to be transferred first to the column strip in a direction perpendicular to it, and then to the columns along the column strip. The action of the column strip is thus equivalent to that of a supporting boundary beam in a two-way slab. Neglecting the reduction in span resulting from the use of a capital, the total (positive plus negative) moments in the directions l_1, l of an $l_1 \times l$ rectangular panel are theoretically equal to:

$$M_{l_1} = \tfrac{1}{8}Wl_1, \qquad M_l = \tfrac{1}{8}Wl, \tag{12.3.1}$$

where the total load on the panel:

$$W = wl_1 l, \tag{12.3.2}$$

is assumed to be carried first in one direction and then in the perpendicular direction.

It has been determined analytically and experimentally that the *total design moments* for a flat slab are about 25% less than those given by (12.3.1) and that (neglecting the effect of the capital) they may be taken equal to:

$$M_{l_1} = 0.09\ Wl_1, \qquad M_l = 0.09\ Wl \qquad (W = wl_1 l). \tag{12.3.3}$$

The apportioning of these total moments between column and middle strips as positive and negative moments is given in the ACI 318 Code for various conditions of continuity. For an interior panel with $(l_1/l) < 1.33$, these percentages are:

Column strip positive moment	22%
Column strip negative moment	46%
Middle strip positive moment	16%
Middle strip negative moment	16%
	100%

This distribution varies when drop panels are used because of the additional stiffness of the slab around the column.

Detailed requirements for minimum slab thickness, critical shear areas, size of drop panels and capitals, etc., are also given in the code. The reader is referred to these requirements for the detailed design of a flat slab.

For preliminary design purposes, approximate values of the slab thickness and of the reinforcing steel required for various spans and uniform loadings are shown in Fig. 12.3.5. Although this figure is derived from a square grid of columns, it can be shown that the error in steel quantities for a rectangular grid with an aspect ratio of less than 1.5 is of the order of 2%, if the equivalent square-slab span is chosen as the average of the two spans of the rectangular slab.

Example 12.3.2. Determine the minimum thickness of a flat slab, used in an apartment building, with bay sizes of 3.5 × 4.5 m (12 ft × 15 ft), using $f_c' =$ 20 MPa (2,900 psi). The minimum thickness specified in the ACI 318 Code for a slab without drop panels is 1/36 of the longer span, or 125 mm (5 in.).

The loads to be carried are (see Chapter 1):

	kN/m²	psf
125 mm slab (assumed)	2.9	60
25 mm plaster ceiling	0.5	10
Wood flooring	0.2	4
Partitions	1.0	20
Live load	1.9	40
Total	$w = 6.5$	134

The total panel load is $W = 6.5 \times 3.5 \times 4.5 = 102.4$ kN (24.2 k). By (12.3.3), the total design moments in the two directions l and l_1 are:

$$\begin{aligned} M_l &= 0.09 \times 102.4 \times 3.5 = 32.3 \text{ kN·m (26.0 ft·k)}, \\ M_{l_1} &= 0.09 \times 102.4 \times 4.5 = 41.5 \text{ kN·m (32.5 ft·k)}. \end{aligned} \quad \text{(a)}$$

The maximum moments occur over the columns (negative moments) and are 46% of the values given by (a). Since the width of the column strip is one-half the panel width, the negative moments per unit width of column strips are:

$$M_l = \frac{0.46 \times 32.3}{4.5/2} = -6.6 \text{ kN·m/m. } (-1.6 \text{ ft·k/ft}),$$

$$M_{l_1} = -\frac{0.46 \times 41.5}{3.5/2} = -10.9 \text{ kN·m/m. } (-2.5 \text{ ft·k/ft}).$$

The required depth and thickness, using the maximum moment of 10.9 kN·m/m, are:

$$d = \sqrt{\frac{6 \times 10.9 \times 1\,000^2}{9.0 \times 1\,000}} \simeq 85 \text{ mm (3.4 in.)},$$

$$t = 85 + 25 = 110 \text{ mm},$$

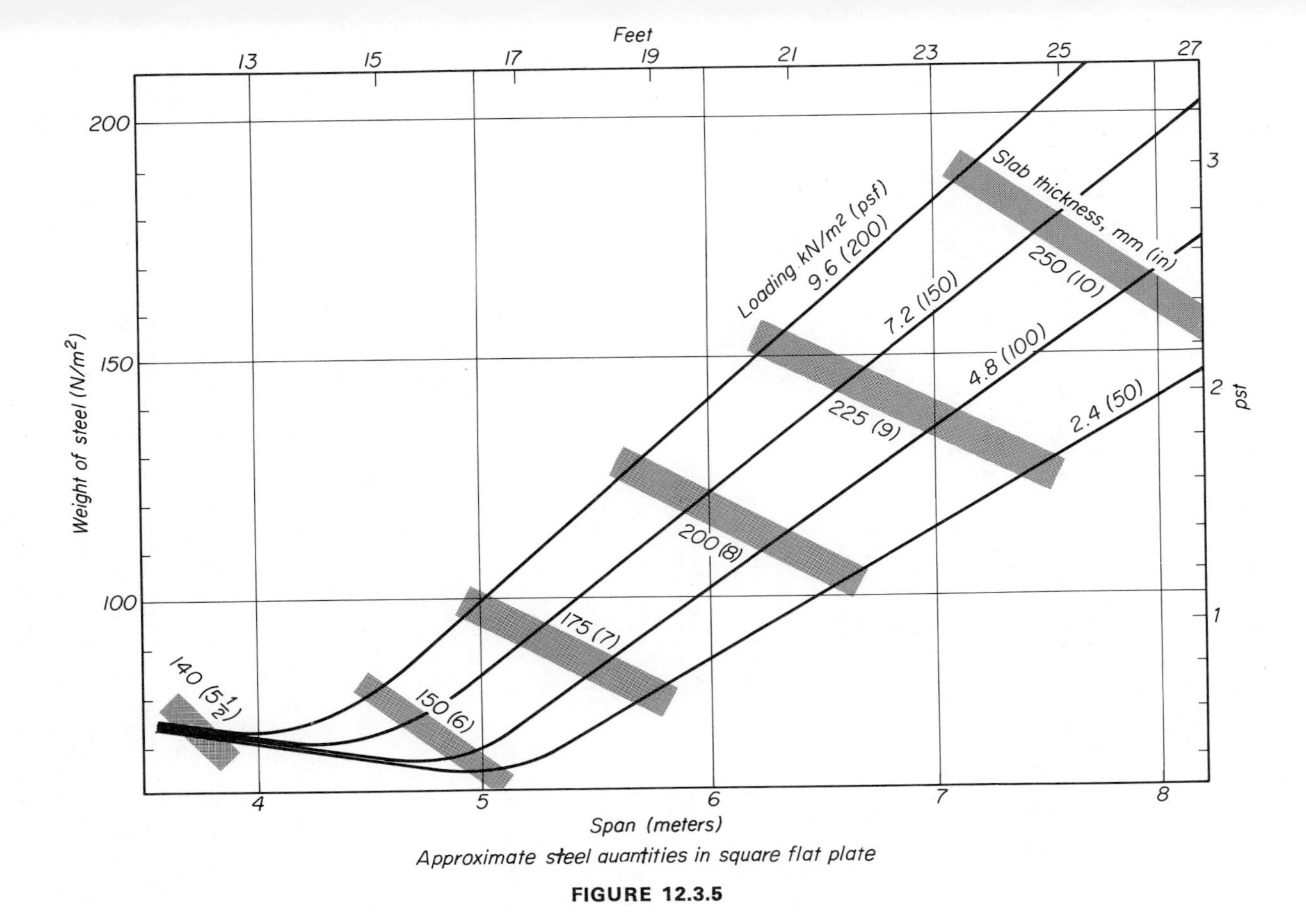

Approximate steel quantities in square flat plate

FIGURE 12.3.5

allowing 20 mm ($\frac{3}{4}$ in.) cover for reinforcing, or 25 mm (1 in) to the center of the bar.

The slab thickness is therefore governed by the minimum of 125 mm (5 in.) specified by the ACI 318 Code.

Figure 12.3.5 gives for $l = (3.5 + 4.5)/2 = 4$ m a slab thickness of approximately 140 mm ($5\frac{1}{2}$ in.) and a steel quantity of about 30 N/m^2 (0.5 psf).

Since in many cases maximum span-depth ratios are governed by specifications or limited by acceptable deflections, these are summarized in Table 12.3.4.

Table 12.3.4

	l/t
Beams	20–26*
One-way slabs	25–35*
One-way joists	24
Flat slabs	36–40†
Waffle slabs	28
Two-way slabs	45–54‡

*Depends on end conditions.
†Without or with drop panel.
‡Depends on ratio of sides.

12.4 CIRCULAR AND POLYGONAL PLATES [10.4]

Circular plates are often used to cover circular areas. The maximum bending moment M_r in the radial direction in a simply supported plate of diameter a under a uniform load w occurs at the center of the plate and is given by:

$$M_{r,\max} = 0.051\ wa^2, \tag{12.4.1}$$

while a fixed plate has a maximum negative radial moment at the boundary:

$$M_{r,\min} = -0.031\ wa^2, \tag{12.4.2}$$

and a maximum positive moment at the center:

$$M_{r,\max} = 0.016\ wa^2. \tag{12.4.3}$$

Comparison of (12.4.1) with the maximum moment in a simply supported square plate of side a ($m = 0.048$), shows that the moment in the circular plate may be approximated by that in the square plate circumscribed to it [Fig. 12.4.1 (a)]. This is physically understandable, since the stiff corners of the square plate, for all practical purposes, do not deflect. Similarly, by (12.4.3) and Table 12.2.2, the maximum positive and negative moments in a fixed square plate of side $\frac{4}{5}a$ [Fig. 12.4.1 (b)]:

$$M_{x,\min} = -0.051\ w(\tfrac{4}{5}a)^2 = -0.033\ wa^2,$$
$$M_{x,\max} = 0.023\ w(\tfrac{4}{5}a)^2 = 0.015\ wa^2,$$

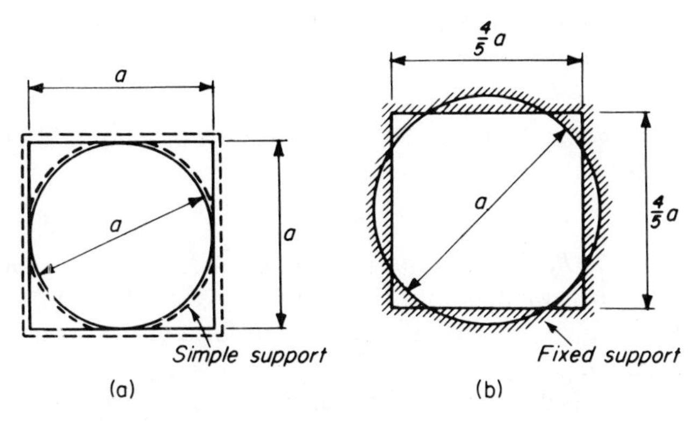

FIGURE 12.4.1

are approximately equal to those in the fixed circular plate of diameter a.

Maximum moments in simply supported or fixed polygonal plates are well approximated by those in the circumscribed simply supported circular plate and the inscribed fixed circular plate, respectively.

The maximum moment in a *ring plate* bounded by two circular boundaries of radii a and b (Fig. 12.4.2), when a is not small in comparison with b, is practically the same as that in a one-way slab of span $l = b - a$, since the ratio of the average plate circumference, $2\pi(b + a)/2 = \pi(b + a)$, to the span $(b - a)$ is:

$$\frac{\pi(b + a)}{b - a} = \pi\frac{1 + (a/b)}{1 - (a/b)} > \pi = 3.14 > 2,$$

and two-way action is lost in plates of aspect ratio larger than 2.

In many modern buildings the floors are built as plates with re-entrant corners, *fixed* on an inner core and *simply supported* on an outer wall or a set of columns (Fig. 12.4.3), with equal spans a in the directions parallel to

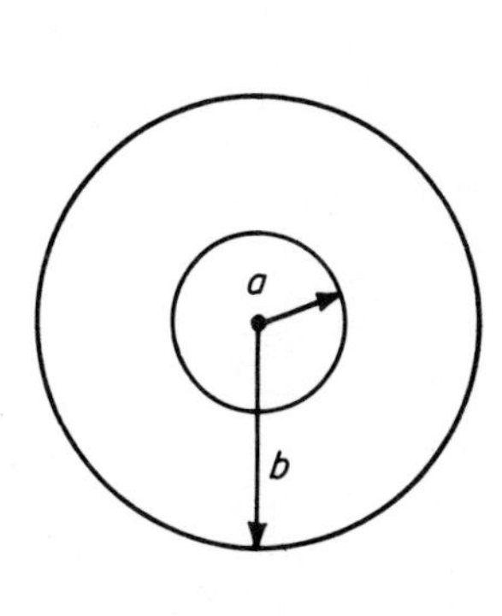

FIGURE 12.4.2

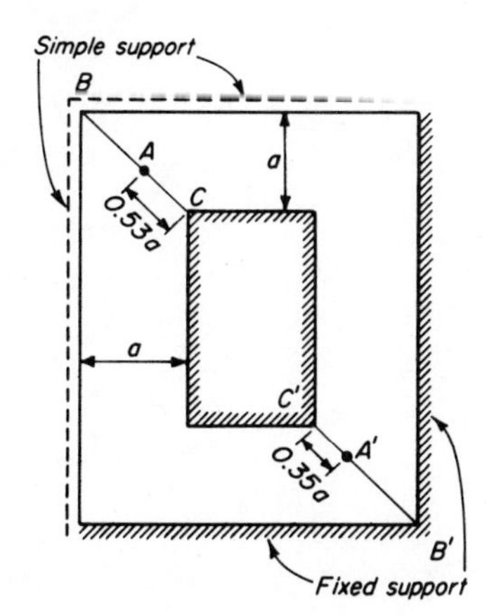

FIGURE 12.4.3

the walls. The maximum moment for support conditions of this type in a so-called *corner plate:*

$$M_{max} = 0.070\ wa^2 \tag{12.4.4}$$

occurs in the diagonal direction BC at a point A a distance $0.53a$ from the re-entrant corner C. The maximum negative moment:

$$M_{min} = -0.083\ wa^2 \tag{12.4.5}$$

occurs at the corner B. The moments (12.4.4), (12.4.5) may be compared with the moments $+0.070\ wa^2$ and $-0.125\ wa^2$ given by Table 4.1.2 (13) for a strip of unit width, simply supported at one end and fixed at the other, in a one-way slab of span a.

When both the core-supported boundary and the outer boundary of the plate are fixed, the maximum positive moment:

$$M_{max} = 0.04\ wa^2 \tag{12.4.6}$$

occurs at a point A' a distance $0.35a$ from the corner C', while the maximum negative moment:

$$M_{min} = -0.10\ wa^2 \tag{12.4.7}$$

occurs at the corner B'. The moments (12.4.6), (12.4.7) may be compared with the moments $+0.042$ and -0.083 given by Table 4.1.2 (9) for a strip of unit width in a one-way slab of span a fixed at both ends.

12.5 FOLDED PLATES [10.8]

It is shown in Example 12.2.5 how the flexural rigidity of a thin plate may be increased by means of ribs. The same results may also be economically obtained, whenever feasible, by folding the plate, i.e., by giving the plate, for example, one of the cross sections shown in Fig. 12.5.1.

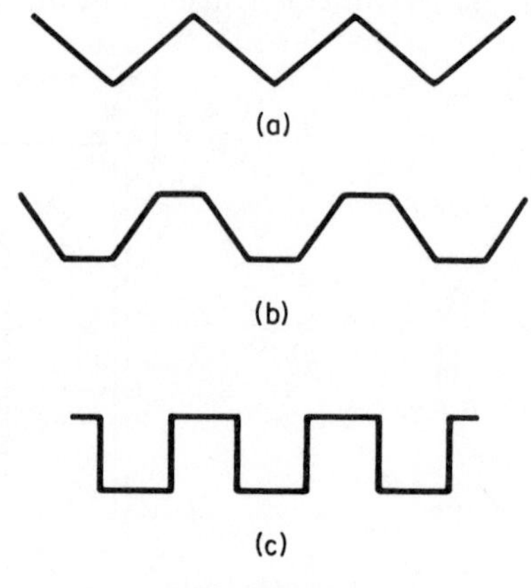

FIGURE 12.5.1

Each slab of a *folded plate* has a length l much larger than its width a [Fig. 12.5.2 (a)]. Hence, bending occurs only across its width (see Section 12.2); i.e., the slabs develop one-way bending action in the transverse direction. In a folded plate with numerous folds under uniform vertical load, all slabs deflect identically, except those near the plate boundaries, which are weakened by the lack of transverse support from adjoining slabs. Since there is no relative displacement between the interior folds, each transverse unit strip of an interior slab [Fig. 12.5.2 (a)] behaves in bending as *a continuous beam on rigid supports* acted upon by the *normal* component of the load [Fig. 12.5.2 (b)].

When all the slabs are identical, by symmetry, the folds do not rotate, and since the angle β between the slabs is maintained by continuity, the moment over the supports, i.e., at the folds, is equal to the fixed-end moment for a beam of unit width and length a. For example, the folded plate of Fig. 12.5.3 (a) consists of identical slabs of thickness t and width a, inclined at an angle α to the horizontal, and loaded by a dead load w kN/m²(psf). Hence [Fig. 12.5.3 (b)], the transverse bending moments at the folds and mid-span between the folds, for a unit strip of plate, equal [see Table 4.1.2 (9)]:

$$M_1 = \frac{-(w \cos \alpha)a^2}{12}, \qquad M_2 = \frac{+(w \cos \alpha)a^2}{24}. \tag{12.5.1}$$

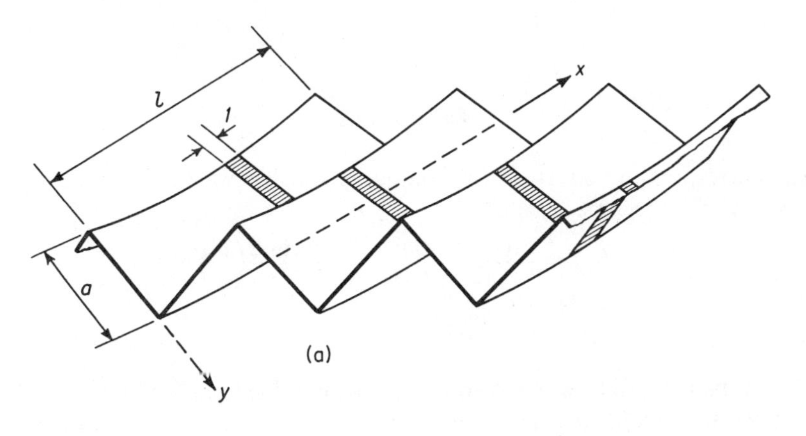

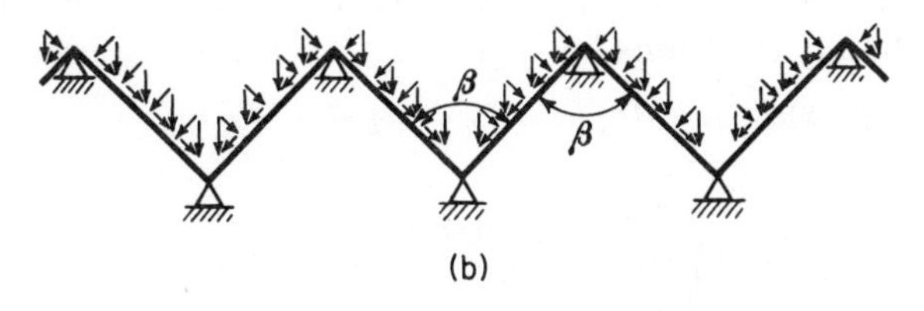

FIGURE 12.5.2

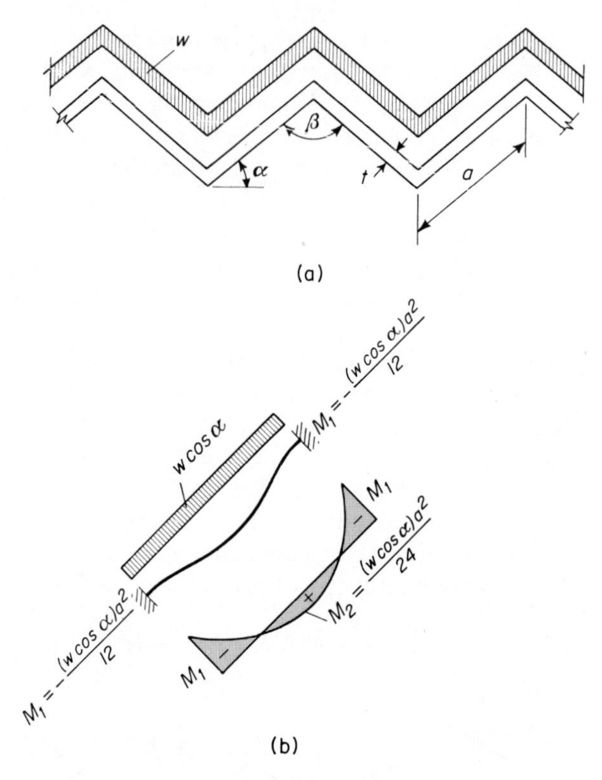

FIGURE 12.5.3

The corresponding stresses in a homogeneous slab with a section modulus $S = 1 \times t^2/6$ become:

$$f_{1,y} = -\frac{(w\cos\alpha)a^2}{12(t^2/6)} = -\frac{w\cos\alpha}{2}\left(\frac{a}{t}\right)^2,$$

$$f_{2,y} = +\frac{w\cos\alpha}{4}\left(\frac{a}{t}\right)^2.*$$

When the slabs are not equal, as in the folded plate of Fig. 12.5.4 (a), all the interior folds deflect equally, but they also rotate in order to maintain the angle β between the slabs of different widths. Hence, the slabs behave transversely, as the continuous beam of Fig. 12.5.4 (b). Dropping the common factor $1/EI$, the continuity equation (4.6.1) gives in this case (see Section 4.6):

$$\tfrac{1}{24}(w\cos\alpha)a_1^3 + \tfrac{1}{2}Ma_1 = -(\tfrac{1}{24}wa_2^3 + \tfrac{1}{2}Ma_2),$$

from which:

$$M = -\frac{wa_2^2}{12}\frac{1 + (a_1/a_2)^3\cos\alpha}{1 + (a_1/a_2)}. \qquad (12.5.3)$$

*These are also the stresses in a reinforced concrete slab with a so-called *balanced section*, in which yield in the steel and crushing of the concrete occur simultaneously.

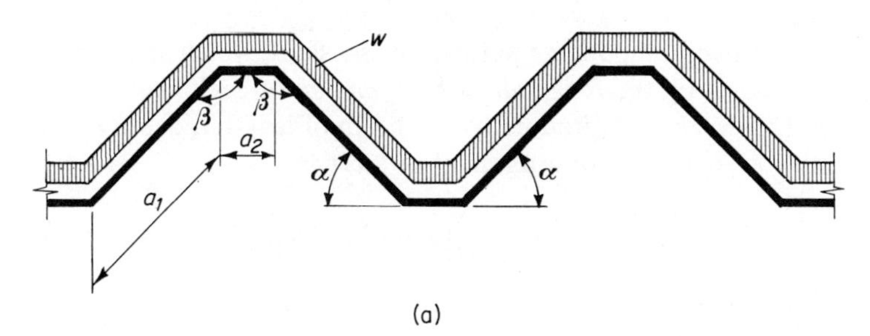

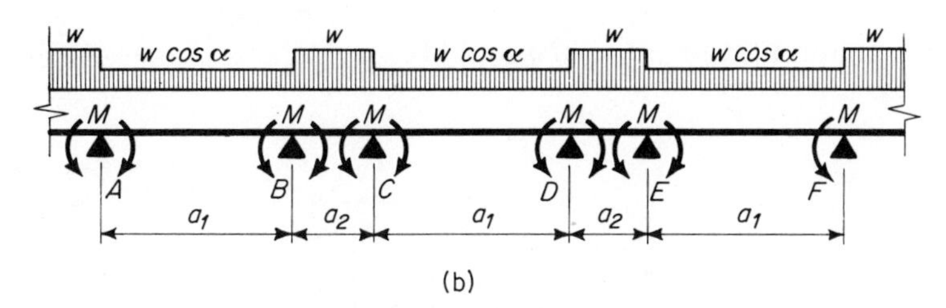

FIGURE 12.5.4

The transverse bending action of the slabs transfers to the folds the reactions R due to the normal component $p_n = wa \cos \alpha$ of the load on the slab, while the tangential component p_t of the load $p_t = wa \sin \alpha$ is transferred to the folds by direct stress along the slab [Fig. 12.5.5 (a)]. Hence, the folds carry the total load $p = wa$ per unit length. This total load p is split at

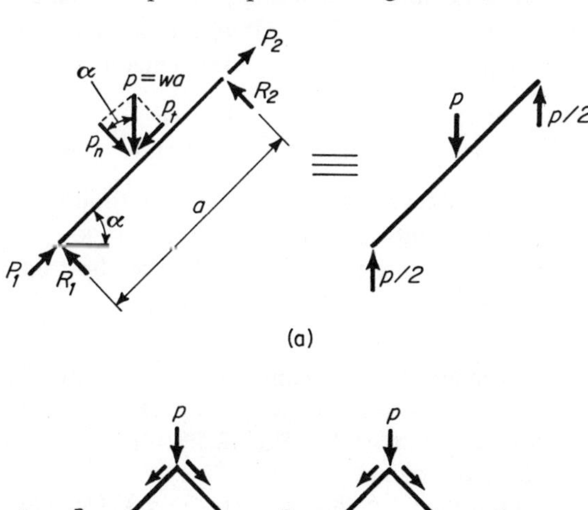

FIGURE 12.5.5

the folds into two components parallel to the slabs, which are transferred to the end supports by the *longitudinal beam action* of each slab. Hence, each slab (Fig. 12.5.6) acts longitudinally as a beam of length l, depth $h = a \sin \alpha$, and width $b = t/\sin \alpha$, with a moment of inertia:

$$I = \frac{1}{12} bh^3 = \frac{1}{12}\left(\frac{t}{\sin \alpha}\right)(a \sin \alpha)^3 = \frac{1}{12} ta^3 \sin^2 \alpha, \qquad (12.5.4)$$

and a section modulus:

$$S = \frac{I}{h/2} = \frac{\frac{1}{12} ta^3 \sin^2 \alpha}{(a \sin \alpha)/2} = \frac{1}{6} ta^2 \sin \alpha, \qquad (12.5.5)$$

under a uniform load $p = wa$ kN/m (lb/ft).

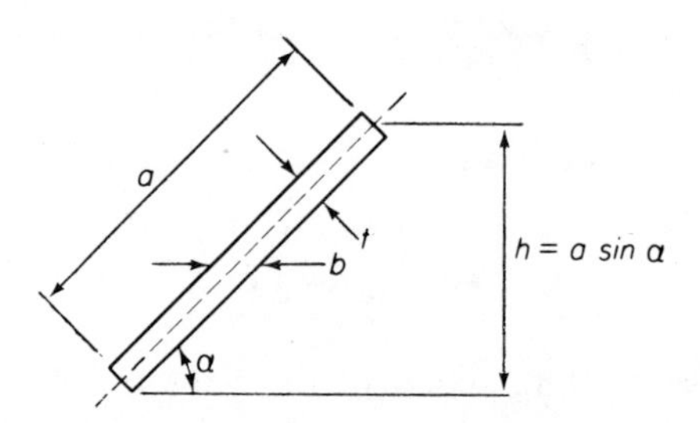

FIGURE 12.5.6

Folded plates are usually supported at their ends on stiffeners, which are rigid vertically and flexible horizontally; hence, the slabs act longitudinally as simply supported beams. Their maximum moment is $pl^2/8 = wal^2/8$, and the maximum longitudinal stress in a homogeneous slab becomes:

$$f_{x,\max} = \frac{pl^2/8}{S} = \frac{wa\, l^2/8}{ta^2 \sin \alpha/6} = \frac{3}{4} \frac{wl^2}{ta \sin \alpha} = \frac{3}{4} \frac{l^2}{th} w. \qquad (12.5.6)$$

When the folded plates are continuous in the longitudinal direction, their longitudinal bending moments are those of a continuous beam on rigid supports with a moment of inertia equal to that of the folded plate cross section.

Example 12.5.1. Determine the maximum longitudinal stress $f_{x,\max}$ due to the dead load w in a homogeneous folded plate with the cross section of Fig. 12.5.7. The folded plate has two spans of length l [Fig. 12.5.7 (b)] and is supported on three stiffeners which are flexible horizontally.

The load per unit of plate length is $p = w(a_1 + a_2)$. The maximum moment in the continuous beams occurs over the middle support and equals $-\frac{1}{8}pl^2$ [see Table 4.1.2 (13)]. The moment of inertia of the beam equals:

$$I = 2\left(\frac{1}{2} a_2\right)(t)\left(\frac{h}{2}\right)^2 + \frac{1}{2} \frac{t}{\sin \alpha} h^3 = \frac{th^2}{4}\left(a_2 + \frac{a_1}{3}\right),$$

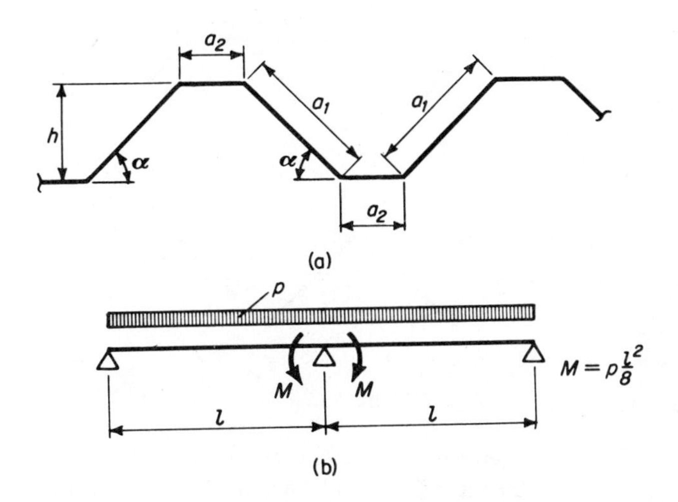

FIGURE 12.5.7

and its section modulus is:

$$S = \frac{I}{h/2} = \frac{th}{2}\left(a_2 + \frac{a_1}{3}\right).$$

The maximum longitudinal stress in a homogeneous plate is, thus:

$$f_{x,\max} = \pm\frac{w(a_1 + a_2)l^2/8}{th(a_2 + a_1/3)/2} = \pm\frac{1}{4}\left(\frac{a_2 + a_1}{a_2 + a_1/3}\right)\left(\frac{l^2}{th}\right)w. \tag{12.5.7}$$

Similarly, with $\cos\alpha = \sqrt{a_1^2 - h^2}/a_1 = \sqrt{1 - (h/a_1)^2}$, the maximum transverse moment (12.5.3) gives the maximum bending stress:

$$f_{y,\max} = \pm\frac{wa_2^2}{12}\,\frac{1 + (a_1/a_2)^3\cos\alpha}{1 + a_1/a_2}\Bigg/\frac{t^2}{6}$$

$$= \pm\frac{1 + (a_1/a_2)^3\sqrt{1 - (h/a_1)^2}}{2(1 + a_1/a_2)}\left(\frac{a_2}{t}\right)^2 w. \tag{12.5.8}$$

For example, for $a_1 = \sqrt{2}a$, $a_2 = a$, $\alpha = 45°$, $h = a$, (12.5.7) and (12.5.8) give:

$$f_{x,\max} = \pm\frac{1}{4}\,\frac{1 + \sqrt{2}}{1 + \sqrt{2}/3}\,\frac{l^2}{ta}w = \pm 0.410\frac{l^2}{ta}w = \pm 0.410\left(\frac{l}{a}\right)^2\left(\frac{a}{t}\right)w,$$

$$f_{y,\max} = \pm\frac{1 + (\sqrt{2})^3\sqrt{2}/2}{2(1 + \sqrt{2})}\left(\frac{a}{t}\right)^2 w = \pm 0.622\left(\frac{a}{t}\right)^2 w.$$

The exterior slabs near the longitudinal boundaries deflect more than the interior slabs, and an exact evaluation of the load transferred to their folds may be obtained by considering the transverse unit strips as supported on elastic rather than rigid supports. In order to equalize the deflections, and hence the load carried by the slabs, the external slabs are often stiffened by vertical beams or slabs (Fig. 12.5.8).

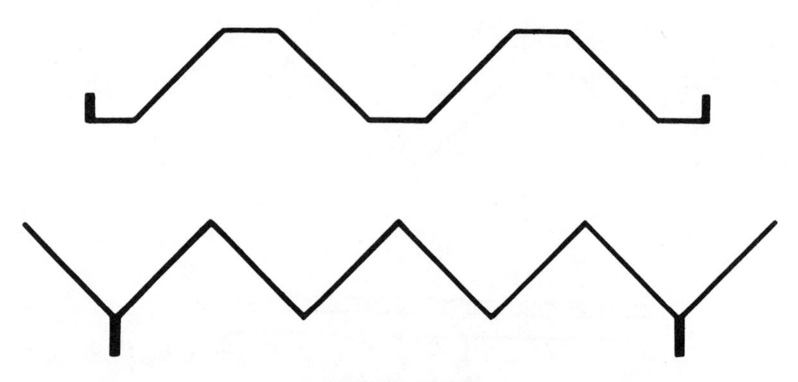

FIGURE 12.5.8

The one-way transverse bending action in the slabs is based on the fact that $l/a \gg 1$. On the other hand, the slabs are usually built-in into the end stiffeners, so that in a small area near the stiffeners the slabs develop local bending moments at right angles to their short sides as well. The maximum value of this moment is the value of the negative moment at the middle of the short side a of a rectangular plate of sides l and a, with $l \gg a$, fixed all around the boundary and loaded by a uniform load $w \cos \alpha$. From Table 12.2.2 this value is:

$$M_y = -0.057(w \cos \alpha)a^2, \tag{12.5.9}$$

and the corresponding stress in a homogeneous plate is:

$$f_{y,\text{bending}} = \pm 0.343(w \cos \alpha)\left(\frac{a}{t}\right)^2. \tag{12.5.10}$$

The formulas of this section allow the preliminary determination of the folded plate moments, which can be used to check the longitudinal and transverse stresses in homogeneous as well as in reinforced concrete plates of balanced design. A more rigorous analysis is required for final design.

Folded plates may be built of wood, aluminum, or steel, but are ideally suited to reinforced concrete construction because of simplicity of formwork, ease of placement of the reinforcement, and variety of shapes. Reinforced concrete folded plates are used mostly as roofs, but may also be economically employed as walls to resist lateral pressures due to wind, earth, or water.

Example 12.5.2. Determine the reinforcing required in a single-span, concrete folded plate with a span of 20 m (67 ft) and the cross section shown in Fig. 12.5.5, where $a = 2.8$ m (9.4 ft), $t = 150$ mm (6 in.), and $\alpha = 45°$.

The dead load of the folded plate, based on a concrete weight of 22 kN/m^3 (140 pcf) is:

$$w = 0.15(22) = 3.3 \text{ kN/m}^2 \text{ (70 psf)}.$$

The longitudinal stress is, by (12.5.6):

$$f_{x,\max} = \frac{3}{4}\,\frac{3.3(20)^2 1\,000}{150(2.8 \times 1\,000) \sin 45^\circ} = 3.33 \text{ MPa (488 psi)}.$$

This stress is tensile at the bottom of the folded plate and varies from a maximum of 3.33 MPa to zero at mid-height. The total tensile force T in one plate is:

$$T = \frac{3.33}{1\,000}(150)\left(\frac{2\,800}{2}\right)\frac{1}{2} = 350 \text{ kN (78.75 k)},$$

and with an allowable steel stress of 138 MPa (20 ksi), the total steel area required is

$$A_s = \frac{350 \times 1\,000}{138} = 2\,533 \text{ mm}^2 \text{ (3.9 in.}^2\text{)},$$

which is equivalent to six 25 mm bars having a total area of 6(490) = 2 940 mm². Most of this steel is concentrated at the bottom, where the required steel area is:

$$A_s = \frac{3.33}{138}(150)(1\,000) = 3\,620 \text{ mm}^2/\text{m},$$

and the corresponding minimum spacing of 25 mm bars is:

$$\frac{490}{3\,620}(1\,000) = 135 \text{ mm (5.3 in.)}.$$

The maximum bending stress, which is obtained from (12.5.10):

$$f_{y,\text{bending}} = \pm 0.343(3.3 \cos 45^\circ)\left(\frac{2\,800}{150}\right)^2 = \pm 278.8 \text{ kN/m}^2$$

$$= \pm 0.279 \text{ MPa (40 psi)},$$

is seen to be small in comparison with the longitudinal stress.

PROBLEMS

12.1 A simply supported A36 steel plate is used to cover a 1.5 × 2.1 m (5 ft × 7 ft) pit. Determine the plate thickness if the live load is 10 kN/m² (200 psf).

12.2 Determine the maximum uniformly distributed load q which can be applied to a 1.2 m × 2.4 m (4 ft × 8 ft) sheet of 12 mm ($\frac{1}{2}$ in.) plywood: (a) simply supported along its edges, (b) fixed on all sides, for F_b = 10 MPa (1,500 psi). Obtain results on the basis of bending strength only.

12.3 A removable floor in a computer room consists of 600 mm × 600 mm (2 ft × 2 ft) aluminum panels of 6061-T6 alloy simply supported along the edges. For a 5 kN/m² (100 psf) live load, determine the thickness of the panels.

12.4 Determine the thickness of a 1.8 m × 2.4 m (6 ft × 8 ft) apartment-house balcony slab fixed along two adjacent sides, to be constructed in concrete for a superimposed load of 3 kN/m² (60 psf). Assume f_c' = 30 MPa (4,350 psi) and $k = \frac{3}{8}$.

12.5 A simply supported, two-way concrete slab 4.8 m $\times$ 6.0 m (16 ft $\times$ 20 ft) supports a live load of 12.5 kN/m^2 (250 psf). Determine the slab thickness if $f_c' = 20$ MPa (2,900 psi).

12.6 What is the maximum size of a simply supported, two-way square concrete slab 150 mm (6 in.) thick, carrying a live load of 5 kN/m^2 (100 psf) if $f_c' =$ 30 MPa (4,350 psi)?

12.7 A simply supported, concrete flat slab is specified for a warehouse floor with a 12.5 kN/m^2 (250 psf) live load. Determine the slab thickness if the bays are 5.5 m $\times$ 6.75 m (18 ft $\times$ 22 ft) and $f_c' = 20$ MPa (2,900 psi).

12.8 A simply supported, two-way concrete waffle slab spans a 7.8 m $\times$ 9 m (26 ft $\times$ 30 ft) bay. The ribs are 150 mm (6 in.) wide and are spaced 750 mm (2 ft 6 in.) o.c. The slab thickness is 75 mm (3 in.) and the total rib depth is 430 mm (17 in.). Determine the maximum live load that can be carried by the slab if $f_c' = 20$ MPa (2,900 psi).

12.9 A flat slab floor is supported by a row of columns uniformly spaced 6 m (20 ft) o.c. in one direction and alternately spaced at 5.5 m (18 ft) and 6.75 m (22 ft) in the other. If the specified live load is 7.5 kN/m^2 (150 psf), determine the design moments, the slab thickness, and the reinforcing steel in both middle and column strips for $f_c' = 20$ MPa (2,900 psi).

12.10 An office building with a central core and closely spaced perimeter columns fixed to the slab, as shown in Fig. 12.4.3, has a span $a = 7.5$ m (25 ft). Determine the slab thickness with $f_c' = 30$ MPa (4,350 psi) and a superimposed load of 5 kN/m^2 (100 psf).

12.11 Determine the maximum longitudinal stress and transverse bending moment in a folded plate with two equal spans of total length 30 m (100 ft). The cross section of the plate is shown in Fig. 12.5.2, with $a = 1.5$ m (5 ft) and $\beta = 90°$. The slab thickness is assumed to be 100 mm (4 in.).

12.12 The steel reinforcement parallel to the slab sides in a simply supported concrete rectangular slab is proportional to the maximum bending moments in the directions of its sides. Compare the amount of total bending steel reinforcement in plates with ratios $a/b = 1$, 1.5, and ∞, under uniform load.

12.13 A very long concrete rectangular plate of width a is supported on transverse beams equally spaced by b and is uniformly loaded. Compare the total bending steel in a plate bay for $b/a = 1$, 1.5, and 2. Assume the negative steel to extend over one-half the span b.

12.14 A concrete flat slab under uniform load is designed by code. Compare the total bending steel in the slab for $l/l_1 = 1$, 1.5, and 2.

12.15 A concrete circular slab of radius 4.5 m (15 ft) is uniformly loaded with 5 kN/m^2 (100 psf). Determine its thickness and reinforcement: (a) if it is simply supported, (b) if it is fixed. (The concrete strength is 30 MPa (4,350 psi) and the steel is grade 60.)

12.16 The folded slab of Fig. 12.5.2 has $a = 3$ m (10 ft) and $\alpha = 45°$, spans 45 m (150 ft), and is simply supported. It carries a live load of 2 kN/m² (40 psf) in horizontal projection. Determine the slab thickness on the basis of its longitudinal action, and its transverse reinforcement on the basis of its transverse action, assuming all folds to deflect equally. (Assume a 50 MPa (7,250 psi) concrete and an allowable steel stress of 138 MPa (20 ksi).)

12.17 A concrete foundation mat with $f'_c = 20$ MPa (2,900 psi) is designed as an inverted flat slab carrying the soil reaction. Determine the thickness of a mat carrying columns on 7.2 m (24 ft) square bays and supported on a 200 kN/m² (4,000 psf) soil.

12.18 A square concrete foundation slab 600 mm (2 ft) thick must resist the hydrostatic upward pressure of 3 m (10 ft) of water. The slab is continuous over beams forming 6 m × 7.2 m (20 ft × 24 ft) bays. Determine its positive and negative reinforcement for Grade 40 Steel.

12.19 Determine the maximum live load that could be carried by the slab of Problem 12.8 if the same amount of concrete were used to build it of uniform thickness.

12.20 A square concrete slab 9 m (30 ft) on a side and 250 mm (10 in.) thick is simply supported and carries a live load of 2 kN/m² (40 psf). How much would its maximum moment be reduced if the slab were supported by a column at its center? Calling a the plate side, the deflection due to a concentrated load P at the center of the plate equals $0.14\, Pa^2/Eh^3$, and the maximum moment due to P at the plate center equals approximately $0.3\, P$.

13. Stress Redistribution and Plastic Design

13.1 PRINCIPAL STRESSES [9.2]

In the preceding chapters stresses in one-dimensional structural elements were evaluated on cross sections perpendicular to the axis of the element. We now wish to investigate stresses on cross sections which make an angle different from 90° with the axis of the element. To start with, we consider the elementary case of a straight, rectangular cross section bar of area A under a tensile load T (Fig. 13.1.1).

The tensile stress f_0 on a section AB at right angles to the bar axis is:

$$f_0 = \frac{T}{A}. \tag{13.1.1}$$

A cross section $A'B'$ at an angle φ to AB has an area:

$$A' = \frac{A}{\cos \varphi}. \tag{a}$$

To obtain the stresses on $A'B'$ the tensile force T is split into a *normal* component:

$$T' = T \cos \varphi, \tag{b}$$

and a *shear* component:

$$S' = T \sin \varphi. \tag{c}$$

The corresponding normal and shear stresses on $A'B'$ are, thus:

$$f' = \frac{T'}{A'} = \frac{T \cos \varphi}{A/\cos \varphi} = \frac{T}{A} \cos^2 \varphi = f_0 \cos^2 \varphi, \tag{13.1.2}$$

$$f'_s = \frac{S'}{A'} = \frac{T \sin \varphi}{A/\cos \varphi} = \frac{T}{A} \sin \varphi \cos \varphi = \frac{f_0}{2} \sin 2\varphi. \tag{13.1.3}$$

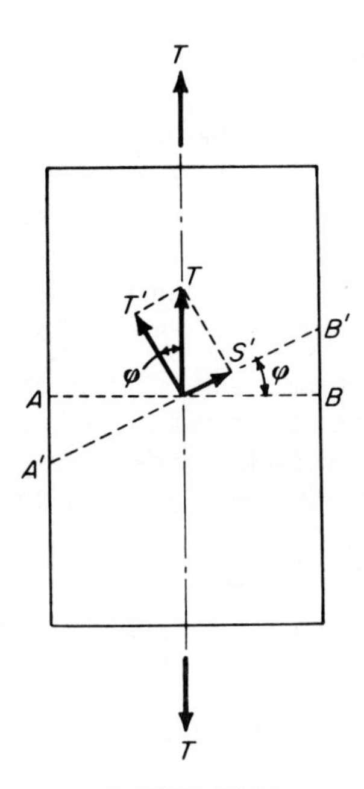

FIGURE 13.1.1

Equations (13.1.2) and (13.1.3) show that as the section $A'B'$ rotates, i.e., as φ goes from 0° to 90°, the normal stress f' acquires its maximum value f_0 at $\varphi = 0°$ and its minimum value, zero, at $\varphi = 90°$, while the shear stress is zero at $\varphi = 0°$ and $\varphi = 90°$ and acquires its maximum and minimum value, $\pm f_0/2$, at $\varphi = 45°$ and $\varphi = 135°$. Figure 13.1.2, showing the stresses on the faces of a rectangular element cut out of the bar at various inclinations to the bar axis, points out the following essential features:

1. The sections on which the normal stress f' becomes respectively maximum and minimum ($\varphi = 0°$ and $\varphi = 90°$) are at right angles.

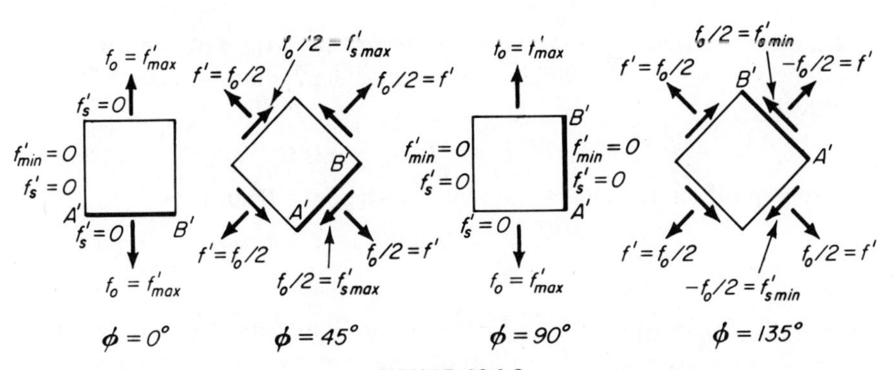

FIGURE 13.1.2

2. On the sections for which the normal stress is respectively maximum and minimum, the shear stresses are zero.

3. The sections on which the shear stress becomes maximum and minimum ($\varphi = 45°, 135°, f'_s = \pm f_0/2$) are also at right angles to each other.

4. The sections of maximum and minimum shear stresses are at 45° to the sections of maximum and minimum normal stresses.

5. The maximum shear stress equals half the difference between the maximum (f_0) and the minimum (0) normal stresses.

The importance of these results consists of the fact that the above-mentioned five features may be proved to hold, in general, for *any* state of plane stress. For example, consider the elementary state of stress in which a plate is acted upon by a tensile stress f_1 in the y-direction and a tensile stress f_2 in the x-direction (Fig. 13.1.3). By (13.1.2), and (13.1.3), the stress f_1 develops on the section $A'B'$ stresses:

$$f'_1 = f_1 \cos^2 \varphi, \qquad f'_{1s} = \tfrac{1}{2} f_1 \sin 2\varphi. \tag{d}$$

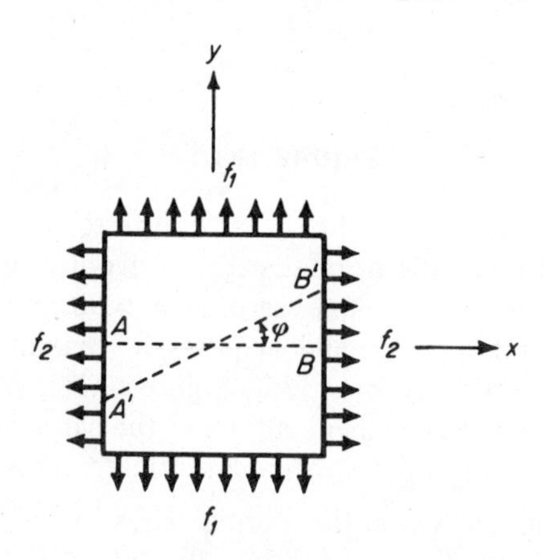

FIGURE 13.1.3

The stresses due to f_2 are obtained from (d) by changing φ into $\varphi + 90°$, i.e., $\cos \varphi$ into $\cos(\varphi + 90°) = -\sin \varphi$ and $\sin 2\varphi$ into $\sin 2(\varphi + 90°) = -\sin 2\varphi$:

$$f'_2 = f_2 \sin^2 \varphi, \qquad f'_{2s} = -\tfrac{1}{2} f_2 \sin 2\varphi. \tag{e}$$

By superposition, the total stresses on the section $A'B'$ are:

$$f' = f'_1 + f'_2 = f_1 \cos^2 \varphi + f_2 \sin^2 \varphi, \tag{13.1.4}$$

$$f'_s = f'_{1s} + f'_{2s} = \tfrac{1}{2}(f_1 - f_2) \sin 2\varphi. \tag{13.1.5}$$

If $f_1 < f_2$, $f'_{max} = f_1$ at $\varphi = 0°$, $f'_{min} = f_2$ at $\varphi = 90°$, and, at $\varphi = 0°$, $\varphi = 90°$, $f'_s = 0$; also $f'_{s\,\substack{max\\min}} = \pm\tfrac{1}{2}(f_1 - f_2)$ at $\varphi = 45°$ and $\varphi = 135°$. In the particular case in which $f_2 = -f_1$, i.e., f_2 is a compression equal to the tension f_1 and at right angles to it:

$$f'_{max} = f_1, \qquad f'_{min} = f_2 = -f_1, \tag{f}$$

$$f'_{s,max} = \tfrac{1}{2}(f_1 - f_2) = f_1, \qquad f'_{s,min} = -\tfrac{1}{2}(f_1 - f_2) = -f_1. \tag{g}$$

Equations (f) and (g) prove that a state of tension and compression at right angles ($\varphi = 0°$, $\varphi = 90°$) is equivalent to a state of pure shear at 45° to the normal stresses (Fig. 13.1.4).

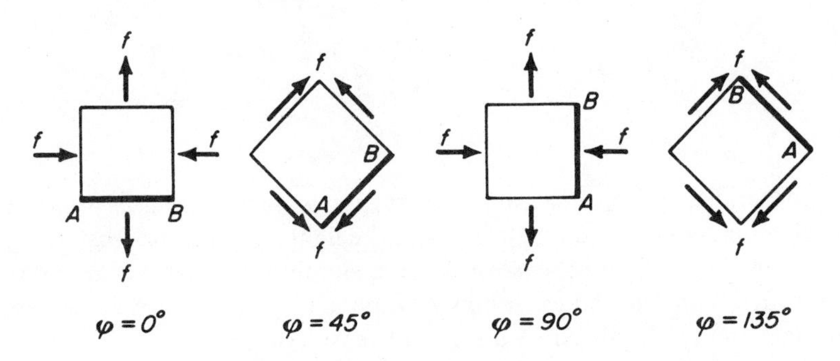

FIGURE 13.1.4

The maximum and minimum values of the normal stresses are called their *principal values;* the *principal stress directions* are often easily spotted by locating the directions in which the shear stresses vanish.

13.2 LINES OF PRINCIPAL STRESS [9.2]

In states of plane stress, like those developed in the web of an I beam or in a rectangular cross-section beam, the values and the directions of the principal stresses usually vary from point to point. If the principal stress directions are determined (mathematically or experimentally) at a number of points, one may indicate these perpendicular directions by a small cross at each point (Fig. 13.2.1). The lines connecting these crosses are the lines along which the principal stresses "flow," and are called *principal stress lines,* or *isostatics* (Fig. 13.2.1).

The isostatics allow the visualization of the stress pattern in complicated situations. For example, consider the isostatics of a rectangular cantilever beam acted upon by a concentrated load at its tip. Since the upper and lower boundaries of the beam are free of stress, and, in particular, free of shear stress, they are isostatics. We know that the upper boundary of the beam is in tension and the lower in compression, and that the neutral axis is free of normal stress. Hence, the isostatics will start at right angles to one boundary

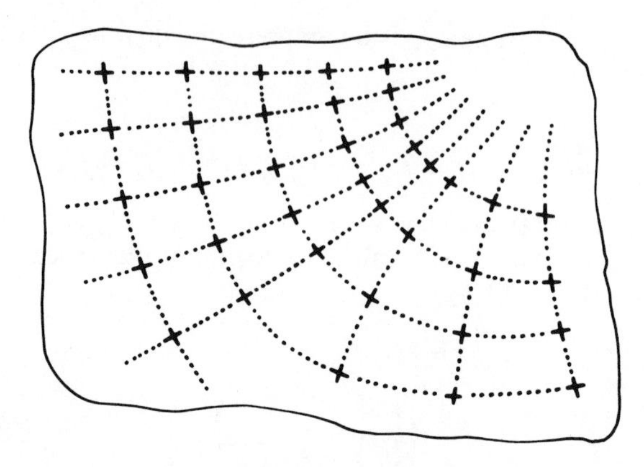

FIGURE 13.2.1

and end parallel to the other and will appear as in Fig. 13.2.2, where the heavy lines are compression isostatics and the light lines tension isostatics.

Figure 13.2.3 shows the isostatics in a simply supported beam under a concentrated load and indicates how a beam acts as a series of compressed (funicular) arches supported by a series of tensile cables.

Figure 13.2.4 shows the isostatics of a bar in tension with a centered hole: the light tensile lines, which start evenly spaced away from the hole, are deviated and crowded in order to go around the hole. The compression lines, at right angles to the tension lines, meet the hole at right angles, since the hole is free of shear stress and hence is an isostatic line.

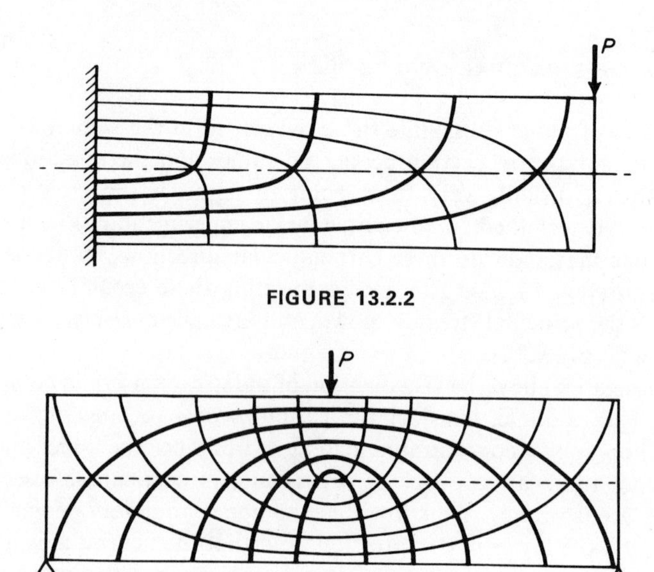

FIGURE 13.2.2

FIGURE 13.2.3

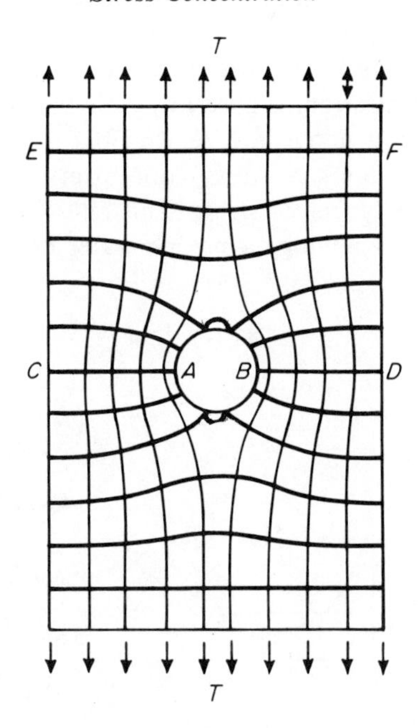

FIGURE 13.2.4

13.3 STRESS CONCENTRATION [9.2]

The crowding of the isostatic lines around the hole in the tension bar of Fig. 13.2.4 indicates that the uniform distribution of tensile stress is disturbed by the hole and that the tensile stress is higher around the points A, B. This phenomenon is known as *stress concentration* and may be understood by a simple analogy. The flow of tensile stress in the bar may be compared to the flow of water in a river. When the current finds no obstructions, the water has the same speed at all points of the river cross section. If a round pile obstructs the river cross section, the water acquires a greater speed, in order to allow the same amount of water to flow through. It is noticeable, moreover, that the water speed is greater next to the pile than near the river banks. Similarly, in Fig. 13.2.4, the tensile stresses are greater in the reduced sections CA, BD than in section EF, and the stress at A and B is greater than at C and D. The stress distribution across the sections CA, BD due to a tensile force T is shown in Fig. 13.3.1, where $f_{av} = T/A'$ is the average stress across the reduced section of area A'. Not only is f_{av} greater than $f_0 = T/A$, since the full cross-sectional area A is greater than A', but the peak stress f_{max} is three times f_{av}. The ratio of f_{max} to f_{av} is called the *stress-concentration factor* and is indicated by k. A round hole in a narrow tension field produces a stress-concentration factor $k = 4$, in a wide tension field a stress concentration factor $k = 3$. (In practice, tension fields with holes are wide.)

In general, any abrupt change in section, any notch or even scratch produces stress concentration. Figure 13.3.2 shows the crowding of the tension isostatics around a notch in a tension bar and indicates how the re-entrant corner A produces a stress concentration while the protruding corners B are practically free of stress. This stress behavior may be understood by the water-flow analogy, since no water flows in the dead corners,

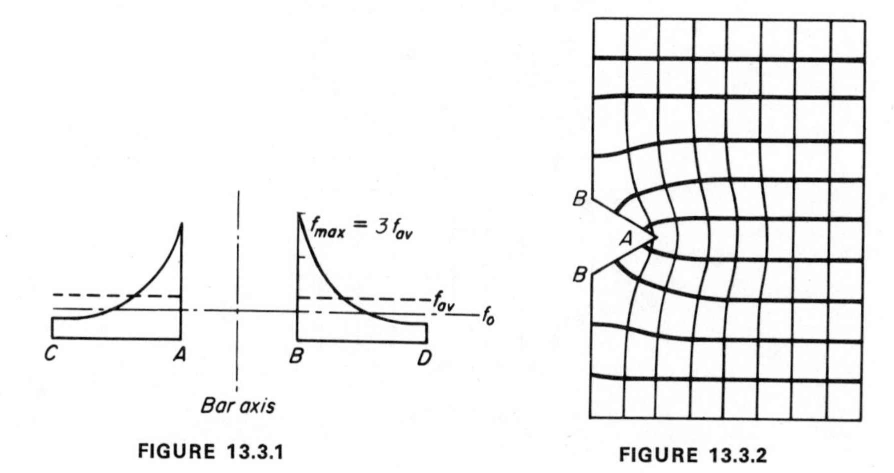

FIGURE 13.3.1 **FIGURE 13.3.2**

while the water circulates at high velocity around the re-entrant corner. Figure 13.3.3 gives the factors for the most commonly encountered cases of stress concentration.

Stress concentration is particularly dangerous in brittle materials, which behave elastically up to failure. Thus, a small scratch on a piece of glass produces a stress concentration high enough for the material to fail in tension when bent. This is why sheet glass is scratched with a diamond to cut it along a given line. On the other hand, materials capable of plastic flow (see Sections 2.1 and 13.4) adapt to stress concentration by allowing the material to flow wherever the stresses are above the yield stress f_y, i.e., by effecting a *redistribution of stress.* For example, if f_{max} in Fig. 13.3.1 were to be above the yield stress, the actual stress in the material could not reach this value; since the highest actual stress in the material equals f_y, in order to develop the required total tension the understressed part of the section will develop higher stresses than in Fig. 13.3.1, exhibiting the stress distribution indicated by the solid line in Fig. 13.3.4. The areas under the solid and the dashed lines are equal (and equal to the tensile load T in the bar when multiplied by the bar thickness), but the plastic stress line represents a more even distribution of stress. Thus, plastic stress redistribution contributes to the safety of structures by smoothing out peak stresses wherever they occur. This is one essential reason why all structural materials must have a plastic range of stress—without it most structures would fail due to stress concentration.

Stress concentration factors

			k
Square corner			2
Corner with fillet		$\frac{r}{d}$	
		$1/16$	1.75
		$1/8$	1.50
		$1/4$	1.20
		$1/2$	1.10
Hole in tension field		Narrow	4
		Wide	3
Hole under pure shear			4
Surface scratch			1.2

FIGURE 13.3.3

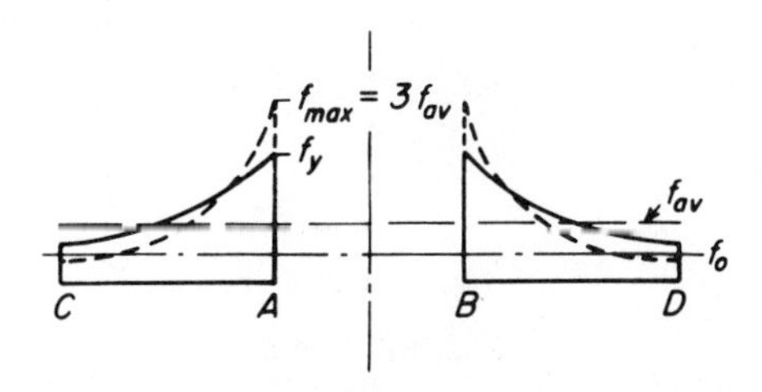

FIGURE 13.3.4

13.4 PLASTIC HINGES IN BENDING [9.3]

Plastic redistribution of stress is particularly significant in beam bending. Consider a material like steel behaving identically in tension and compression, with a well defined yield stress f_y and with such a small amount of strain hardening (see Section 2.2) that its elastic limit f_e virtually coincides with both f_y and its ultimate stress f_u (Fig. 13.4.1). The distribution of bending

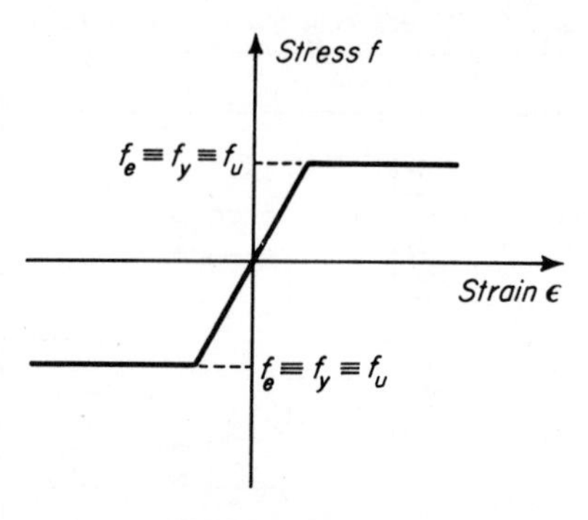

FIGURE 13.4.1

stress in a cross section of a beam made out of this *elasto-plastic material* is linear up to a value of the extreme fiber stress equal to f_e [Fig. 13.4.2 (a)]. If the loads on the beam were to produce fiber stresses greater than $f_e = f_y$, the overstressed fibers would yield, and the stresses in fibers nearer the neutral axis would grow up to f_y, so that the stress distribution would become that shown in Fig. 13.4.2 (b). The maximum contribution to the load-carrying capacity of the bent beam occurs when *all* the fibers of the section have developed the maximum stress f_y [Fig. 13.4.2 (c)]. At this point all the fibers yield, and the section develops a *plastic hinge*, which allows the relative rotation of the adjoining beam segments without an increase in load.

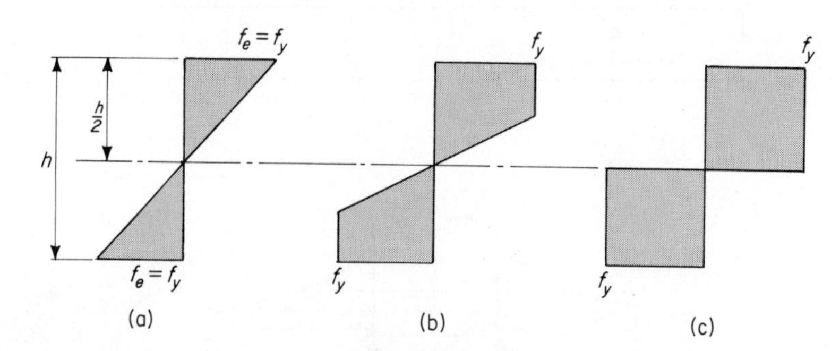

FIGURE 13.4.2

Let us consider a simply supported elasto-plastic beam of rectangular cross section $b \times h$ loaded by a concentrated load at midspan (Fig. 13.4.3). The maximum bending moment, $M = \frac{1}{4}Pl$, and hence the maximum bending stresses, occur at midspan. The bending moment M of the elastic internal stresses, when the extreme fiber stress is f (Fig. 13.4.4) equals:

$$M = F\left(\frac{2}{3}h\right) = \left(\frac{b}{2}\right)\left(\frac{h}{2}\right)(f)\left(\frac{2}{3}h\right) = \frac{1}{6}(bh)hf = \frac{Ah}{6}f.$$

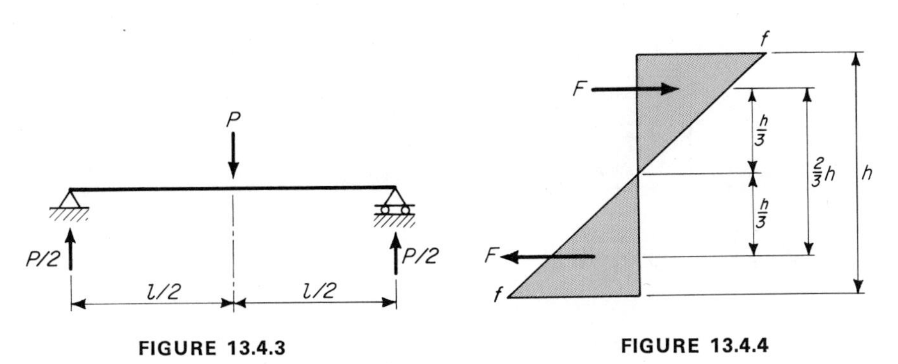

FIGURE 13.4.3

FIGURE 13.4.4

When $f = f_y$, the corresponding moment M_y equals:

$$M_y = \frac{Ah}{6} f_y. \tag{13.4.1}$$

The moment of the internal stresses equilibrates the external moment $Pl/4$; hence:

$$\frac{Pl}{4} = \frac{Ah}{6} f \quad \therefore \quad P = \frac{2}{3}\left(\frac{h}{l}\right) Af.$$

If $f = f_e = f_y$:

$$P_y = \frac{2}{3}\left(\frac{h}{l}\right) Af_y \tag{a}$$

is the maximum load the beam can carry by elastic stresses, i.e., without yielding.

The *capacity moment* M_p, i.e., the bending moment of the internal stresses when *every* fiber in the midspan section develops plastic stresses f_y (Fig. 13.4.5), equals:

$$M_p = F_y\left(\frac{h}{2}\right) = \left(\frac{bh}{2} f_y\right)\left(\frac{h}{2}\right) = \frac{bh}{4} hf_y = \frac{Ah}{4} f_y. \tag{13.4.2}$$

Equating M_p to the external moment M, one obtains the maximum or *ultimate load* P_u, carried by the beam when the midspan section is plasticized;

$$\frac{P_u l}{4} = \frac{Ah}{4} f_y \quad \therefore \quad P_u = \left(\frac{h}{l}\right) Af_y. \tag{b}$$

Comparison of (b) with (a) shows that $P_u = \frac{3}{2} P_y$; i.e., that before collapsing the beam can carry a load 50% greater than the elastic-stress limit load P_y. As soon as $P \geq P_u$, a plastic hinge develops at midspan and the beam collapses because a three-hinge beam becomes a mechanism (Fig. 13.4.6).

The same reserve of strength is developed by an elasto-plastic, simply supported beam under a uniform load w. Equating the maximum external

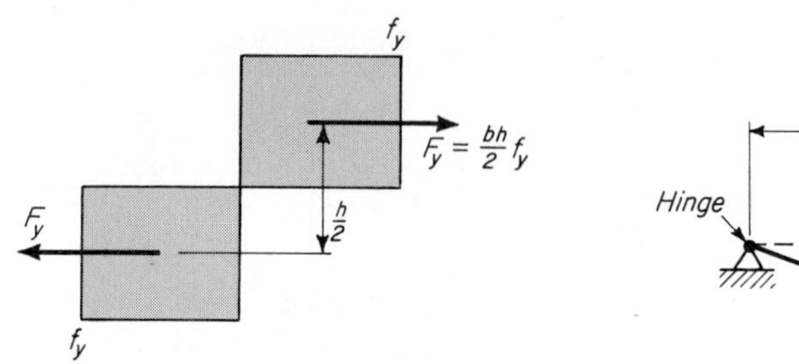

FIGURE 13.4.5

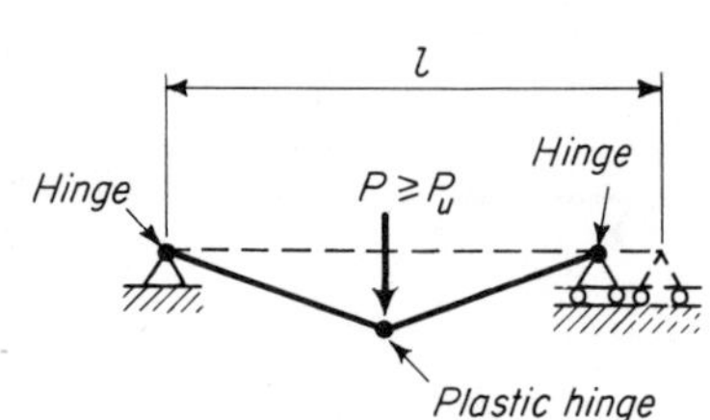

FIGURE 13.4.6

moment $M = wl^2/8$ to M_y and to M_u, we find:

$$\frac{wl^2}{8} = \frac{Ah}{6} f_y \quad \therefore \quad w_y l = \frac{4}{3} \frac{Ah}{l} f_y,$$

$$\frac{wl^2}{8} = \frac{Ah}{4} f_y \quad \therefore \quad w_u l = 2 \frac{Ah}{l} f_y,$$

$$w_u l = \tfrac{3}{2} w_y l.$$

The reserve of strength in a cantilever of rectangular cross section is also 50%, since, whatever the loading condition, the ratio of M_u to M_y is 3/2. In general, we may conclude that the reserve of strength in any statically determinate, elasto-plastic beam of rectangular cross section is 50%.

A statically determinate I-beam, instead, has almost no reserve of strength. In fact, calling A_1 the area of one flange, the bending moment developed by the flanges when the stress in the flanges has reached the value f_y equals:

$$M_F = 2(A_1 f_y)\left(\frac{h}{2}\right) = A_1 h f_y,$$

while, indicating by A_2 the area of the web, the moment developed by the web equals, by (13.4.1):

$$M_W = \frac{A_2 h}{6} f_y,$$

so that the total moment M_y equals:

$$M_y = M_F + M_W = \left(A_1 + \frac{A_2}{6}\right) h f_y. \tag{13.4.3}$$

Since $A_2/6$ is small in comparison with A_2, M_y is well approximated by the M_y of an *ideal I-section* with a web of zero area:

$$M_y \approx A_1 h f_y. \tag{13.4.4}$$

When the external moment is greater than M_y, the only fibers capable of increasing their stress are those in the web, since the flange stresses have already reached their maximum value f_y. By (13.4.2) the plastic moment developed by the web, when it is entirely plasticized, equals:

$$M_{W,u} = \frac{A_2 h}{4} f_y.$$

so that the *ultimate moment* M_u becomes:

$$M_u = M_F + M_{W,u} = \left(A_1 + \frac{A_2}{4}\right) h f_y \equiv z f_y, \tag{13.4.5}$$

where z is the *plastic section modulus.*

The difference $(\frac{1}{4} - \frac{1}{6}) A_2 h f_y$ between M_y and M_u is usually small. For example, standard wide flange sections have ratios M_u/M_y varying between 1.10 and 1.23.

The reserve of strength of statically indeterminate, elasto-plastic beams is, in general, substantially greater than that of statically determinate beams. Consider, for example, a fixed-end beam of rectangular cross section carrying a uniform load w [Fig. 13.4.7(a)]. The bending moments at midspan and at the ends, by Table 4.1.2 (9), are:

$$M_C = \frac{w\, l^2}{24}, \qquad M_A = M_B = -\frac{w\, l^2}{12}. \tag{c}$$

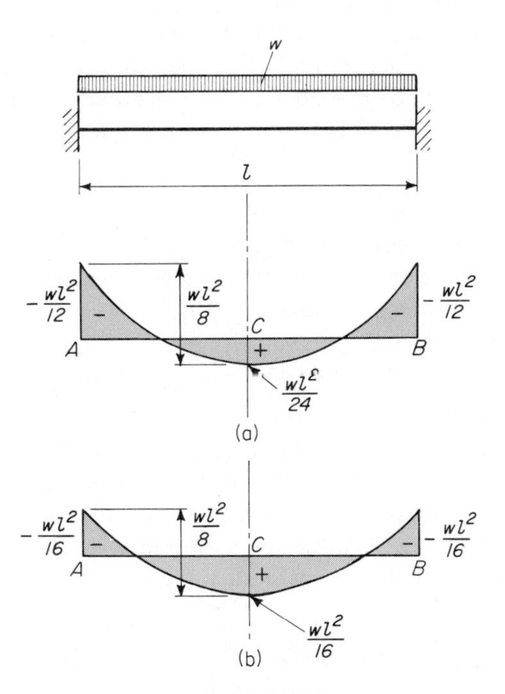

FIGURE 13.4.7

Equating the maximum moment $wl^2/12$ to the elastic internal moment (13.4.1), we obtain:

$$\frac{w_y l^2}{12} = \frac{Ah}{6} f_y \quad \therefore \quad w_y l = 2\left(\frac{h}{l}\right) A f_y. \tag{d}$$

As the load w grows, the end sections plasticize first, since the moment is greater at the ends. But after the end sections have entirely plasticized, there is still a reserve of strength in the midspan section, and it plasticizes next. The beam carries its ultimate load when *all three sections* are plasticized. At this point the end moments and the midspan moment are equal, and, since the sum of the end moment and the midspan moment must always add up to the simply supported beam moment, $wl^2/8$, each equals $wl^2/16$ [Fig. 13.4.7(b)]. Hence, equating $wl^2/16$ to the capacity moment (13.4.2):

$$\frac{w_u l^2}{16} = \frac{Ah}{4} f_y \quad \therefore \quad w_u l = 4\left(\frac{h}{l}\right) A f_y. \tag{e}$$

A comparison of (d) and (e) shows that the reserve of strength is 100%. This increase of strength reserve is due to the fact that, after redistributing stresses at its end sections, the statically indeterminate, fixed-end beam redistributes stresses at its midspan section, thus changing the bending moment diagram of the beam. This change of bending moment diagram cannot occur in a statically determinate beam.

It is now clear that a statically indeterminate wide flange beam may also have a substantial reserve of strength. Consider, for example, a fixed-end wide flange beam under a uniform load. The load w_y which develops stresses f_y at the end sections of the beam is well approximated by (13.4.4):

$$\frac{w_y l^2}{12} = A_1 h f_y \quad \therefore \quad w_y l = 12\left(\frac{h}{l}\right) A_1 f_y. \tag{f}$$

As soon as this load is reached, the end sections plasticize and rotate. The midspan moment increases, and eventually it also reaches the capacity value $A_1 h f_y$. The midspan moment and the end moments are now equal; hence, their value is $wl^2/16$ and:

$$\frac{w_u l^2}{16} = A_1 h f_y \quad \therefore \quad w_u l = 16\left(\frac{h}{l}\right) A_1 f_y. \tag{g}$$

A comparison of (f) and (g) shows that the fixed-end wide flange beam has a reserve of strength of $\frac{4}{12}$, or 33%. Thus, the reserve of strength of a statically indeterminate beam depends on its cross section, the type of load, and the boundary conditions. For example, a fixed-end wide flange beam under a concentrated load at midspan has a small reserve of strength, since from Table 4.1.2 (8):

$$M_A = M_B = M_C = \frac{Pl}{8},$$

so that the end sections and the midspan section plasticize at the same time without change in the bending moment diagram.

Example 13.4.1. A rectangular cross section beam with fixed ends carries a load P at the third point (Fig. 13.4.8). Determine its reserve of strength.

By Table 4.1.2 (8), the maximum moment occurs at the end A and equals $\frac{12}{81}Pl$. Hence:

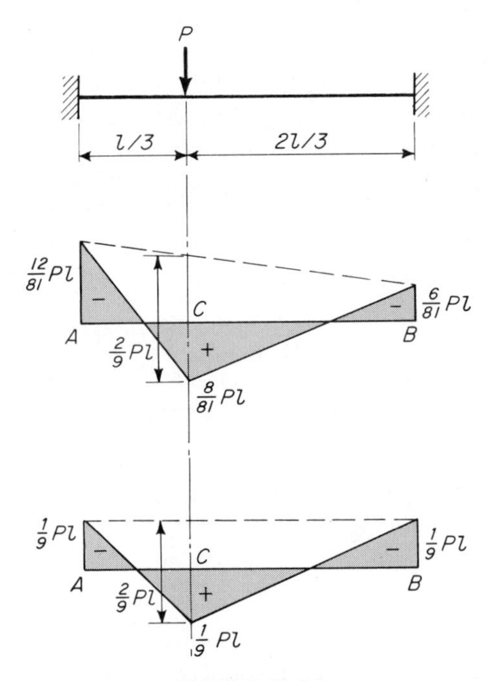

FIGURE 13.4.8

After complete plasticization of sections A, B, and C, the moments at these sections are identical and equal to one-half the maximum, simply supported beam moment $\frac{2}{9}Pl$ [Table 4.1.2 (4)]. Hence, by (13.4.2):

$$\frac{1}{9}P_u l = \frac{Ah}{4}f_y \quad \therefore \quad P_u = \frac{9}{4}\left(\frac{h}{l}\right)Af_y = 2P_y.$$

The reserve of strength equals 100%.

13.5 ULTIMATE DESIGN OF STEEL STRUCTURES

The design of a structure can be based on the load it can carry by elastic stresses or on its ultimate load. Both types of design are conventional, since they are based on simplifying theoretical assumptions. *Plastic* or *ultimate design* is often adopted for steel structures because:

1. Steel behaves as an almost ideal elasto-plastic material.

2. The design is often simplified by eliminating the calculation of statically indeterminate unknowns.

3. Some economy of material may be achieved following the ultimate-design code specifications.

The ultimate load P_u of a structure is written as the allowable (working) load P_w multiplied by a *load factor* F:

$$P_u = FP_w. \tag{13.5.1}$$

Safety in plastic design is assured by the use of the load factor, which plays in ultimate design the role played by the safety factor in elastic design.

The codes require that two design conditions be examined: (a) dead load + live load; (b) dead load + live load + wind *or* earthquake. The load factor in (a) is the product of the ratio of the yield stress to the allowable stress and the *shape factor* of the member, which equals the ratio M_p/M_y for a statically determinate beam. In building design the first ratio is taken to be:

$$\frac{f_y}{f_w} = 1.65.$$

The average value of the shape factor for steel beams is 1.12. Therefore:

$$F = 1.65 \times 1.12 = 1.85. \tag{13.5.2a}$$

In the case (b) including wind or earthquake loads, for which it is usual to allow an increase in working stress of 33%:

$$F = \frac{1.85}{1.33} = 1.40. \tag{13.5.2b}$$

As seen in Section 13.4, the plastic design of statically indeterminate systems does not require the derivation of the elastic moment diagrams, but only the satisfaction of three conditions at ultimate loading:

1. Equilibrium must exist between the given ultimate loads and the external reactions.

2. The structure must be on the verge of becoming a mechanism.

3. The moments produced by the ultimate loads must not exceed the plastic capacity moment at any point in the structure.

For relatively simple structures and loading conditions, these three requirements can be satisfied using equilibrium equations and a semigraphical solution. The following examples illustrate this approach.

Example 13.5.1. Derive the ultimate value of the loads for a continuous beam with two equal spans, loaded by concentrated loads at the center of each span [Fig. 13.5.1(a)].

The structure is first made statically determinate by introducing a hinge at the center support. The maximum determinate bending moment M_s in each span is then equal to $Pl/4$. A diagram of the restraining bending moment M_r due to the condition of continuity is superimposed on the determinate moment diagram [Fig. 13.5.1(b)]. The moment at any point is the difference in the ordinates of the two diagrams. The magnitude of these moment differences at the points where they are maximum is then increased until they become equal to M_p. Inspection of the structure with the plastic hinges thus formed indicates that it is on the verge of becoming a mechanism [Fig. 13.5.1(c)]. From the geometry of the diagrams:

$$M_p = M_s - \frac{M_p}{2}, \qquad M_p = \frac{2}{3}M_s = \frac{P_u l}{6}, \quad \therefore \quad P_u = 6\frac{M_p}{l}.$$

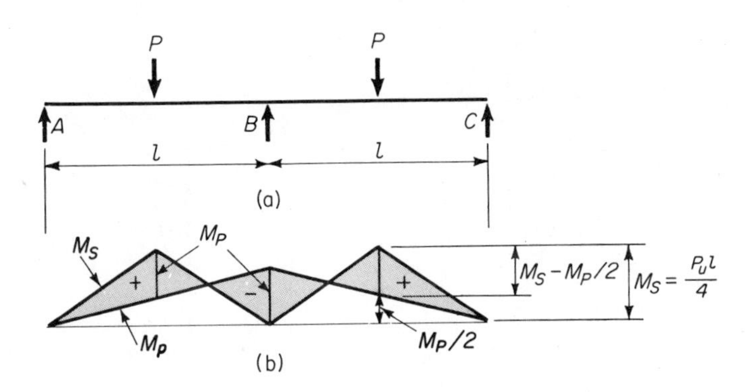

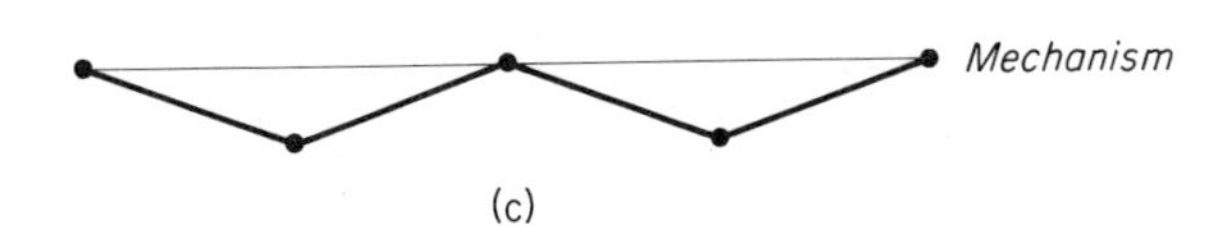

FIGURE 13.5.1

Example 13.5.2. Derive the ultimate load for a hinged frame uniformly loaded on its beam of span l. The beam and the columns are wide flange sections with plastic section moduli z_1 and z_2, respectively.

The capacity moment of the beam and columns are, respectively, $M_{p_1} = f_y z_1$ and $M_{p_2} = f_y z_2$. The capacity moment of the column is usually smaller than that of the beam and, hence, is the capacity moment at the beam end. Therefore:

$$\frac{w_u l^2}{8} = M_{p_1} + M_{p_2} \quad \therefore \quad w_u l = \frac{8}{l}(z_1 + z_2) f_y.$$

Example 13.5.3. Derive the ultimate bending moment diagram for the hinged frame of Fig. 13.5.2, and design its members.

The frame is made statically determinate by assuming the hinge D to be on a horizontal roller, so that the hinge A absorbs the entire thrust P and the moment at B becomes $M_s = Ph$, while it is zero at C. The frame deformation is antisym-

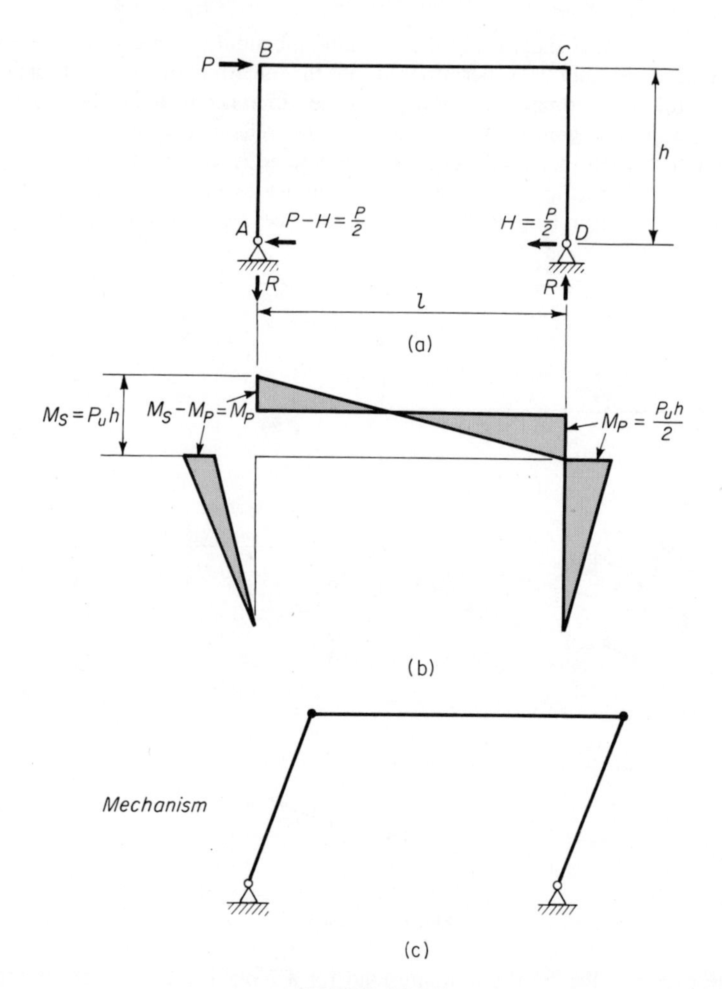

FIGURE 13.5.2

metrical, so that the end moments of the beam, M_p at C and $M_s - M_p$ at B, are equal and opposite. By Fig. 13.5.2 (b):

$$M_p = M_s - M_p \quad \therefore \quad M_p = \frac{M_s}{2} = \frac{P_u h}{2}.$$

With $P = 50$ kN (12 k), $F = 1.85$, $h = 5$ m (16 ft), $f_y = 248$ MPa (36 ksi):

$$P_u = 50 \times 1.85 = 92.5 \text{ kN}, \quad z = \frac{M_p}{f_y} = \frac{92.5 \times (5 \times 1\,000) \times 1\,000}{2 \times 248}$$

$$= 933 \times 10^3 \text{ mm}^3 \text{ (56.9 in.}^3\text{)}.$$

A W410 × 54 (W16 × 36) with $z = 1\,049 \times 10^3$ mm³* is the most economical size.

*See Appendix A.

Example 13.5.4. A wide flange beam, built-in at one end and simply supported at the other, carries a uniform load w (Fig. 13.5.3). Determine its ultimate bending moment diagram, and design the beam.

The beam is made statically determinate by introducing a hinge at A, so that [Table 4.1.2 (5)]:

$$M_s(x) = \frac{wl^2}{2}\left(\frac{x}{l} - \frac{x^2}{l^2}\right). \tag{a}$$

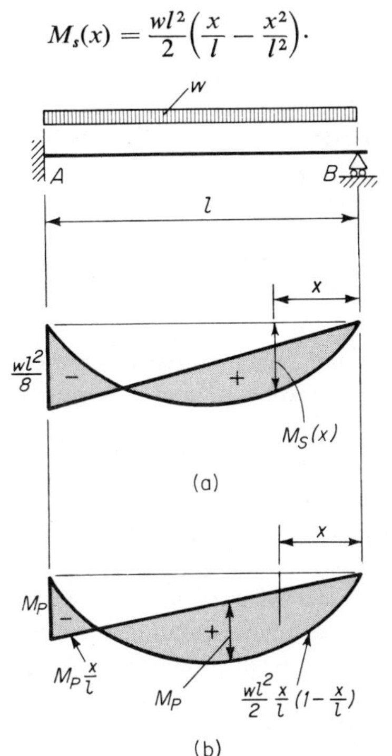

FIGURE 13.5.3

The restraining moment $-M_p$ at A produces a bending moment diagram $-M_p(x/l)$. Letting $\xi = x/l$, the total moment becomes:

$$M(\xi) = \frac{wl^2}{2}(\xi - \xi^2) - M_p\xi. \tag{b}$$

$M(\xi)$ has a positive maximum at a section $\bar{\xi}$, where:

$$\frac{dM}{d\xi} = \frac{wl^2}{2}(1 - 2\xi) - M_p = 0 \quad \therefore \quad \bar{\xi} = \frac{1-\alpha}{2} \qquad \left(\alpha = \frac{M_p}{wl^2/2}\right) \tag{c}$$

By (b) and (c), the maximum positive moment equals:

$$M(\bar{\xi}) = \frac{wl^2}{2}(\bar{\xi} - \bar{\xi}^2 - \alpha\bar{\xi}) = \frac{wl^2}{2}[(1-\alpha)\bar{\xi} - \bar{\xi}^2] = \frac{wl^2}{2}\frac{(1-\alpha)^2}{4}.$$

Equating $M(\bar{\xi})$ to M_p and solving for α, we obtain:

$$M_p = \alpha\frac{wl^2}{2} = \frac{wl^2}{2}\frac{(1-\alpha)^2}{4}, \qquad \alpha = \frac{1}{4}(1-\alpha)^2, \tag{d}$$

from which:

$$\alpha^2 - 6\alpha + 1 = 0 \quad \therefore \quad \alpha = 3 - \sqrt{9 - 1} = 0.172.$$

Hence, by (c), $\bar{\xi} = (1 - 0.172)/2 = 0.414$ and, by (d):

$$M_p = \frac{(1 - 0.172)^2}{4} \frac{wl^2}{2} = 0.086wl^2. \tag{e}$$

For $w = 15$ kN/m (1 k/ft), $F = 1.85$, $l = 7.5$ m (25 ft), and $f_y = 248$ MPa (36 ksi):

$$M_p = 0.086(15 \times 1.85)(7.5)^2 = 134 \text{ kN·m } (99 \text{ k·ft}).$$

$$z = \frac{134 \times 1\,000}{248} = 540 \times 10^3 \text{ mm}^3 \ (32.97 \text{ in.}^3),$$

and the most economical beam size is a W360 × 33 (W14 × 22) with $z = 542.5 \times 10^3$ mm³ (33.1 in.³).

The ultimate design of concrete structures, based on somewhat analogous principles, has been developed and has also been introduced in the codes.*

13.6 THE PLASTIC BEHAVIOR OF PLATES [10.6]

The plate action described in Chapter 12 assumes that the plate develops only elastic stresses and, hence, small deflections. Under loads exceeding its elastic stress capacity, large deflections and plastic redistribution of stress take place in the plate. It is important to realize that, because of its two-dimensional action, a plate has a greater reserve of strength than a one-dimensional structure like a beam.

Consider a simply supported square plate of uniform material and of side a under a uniform load w. The maximum stress in this plate occurs at the center and, by Table 12.2.1, equals:

$$f = \frac{0.048wa^2}{h^2/6} = 0.288w\left(\frac{a}{h}\right)^2.$$

When f reaches the yield value f_y, the total load W_y on the plate is:

$$W_y = wa^2 = \frac{h^2}{0.288} f_y = 3.47h^2 f_y. \tag{13.6.1}$$

As w grows, the sections at the plate center plasticize, but the plate does *not* collapse because a *point hinge* does not transform it into a mechanism.

As the load continues to grow, additional sections plasticize and the plate load reaches its ultimate value when *hinge* or *yield lines* are formed capable of allowing an indefinite deformation of the plate. Experiments show

*See ACI 318-77 code.

that, in a rectangular plate, hinge lines develop along its diagonals, so that the plate is transformed into a mechanism consisting of four hinged triangles (Fig. 13.6.1). The ultimate load for a square plate $W_u = w_u a^2$ is computed as follows. When the hinge lines are just formed, each triangle carries one-quarter of the ultimate load W_u, and the bending moment *per unit width* of plate along the sides OA, OB of each triangle (Fig. 13.6.2) is the capacity moment for a rectangular cross section of unit width and depth h, $M_p = (h^2/4) f_y$ [see (13.4.2)]. For equilibrium, the moment of the load about

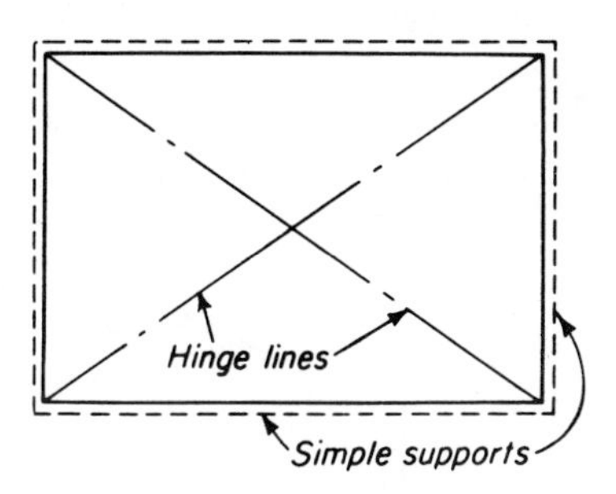

FIGURE 13.6.1

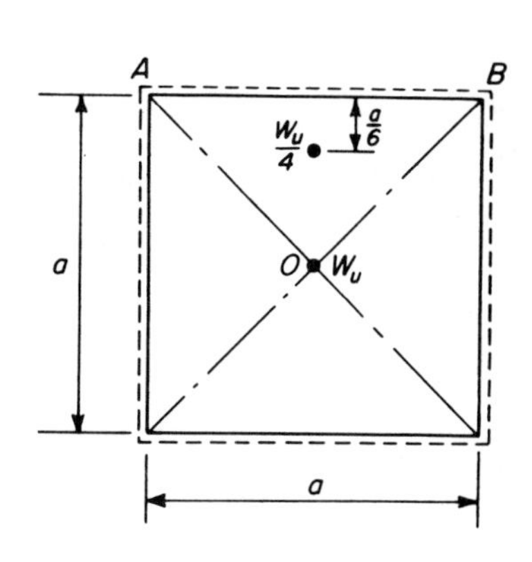

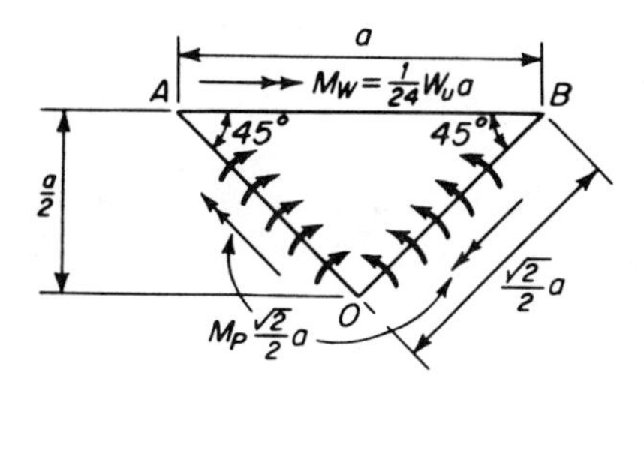

FIGURE 13.6.2

the side AB of the plate must be equal and opposite to the component about the same side of the moments along the diagonal sides OA, OB. The moment of the load about AB equals $\frac{1}{4}W_u(a/6) = \frac{1}{24}W_u a$. The total moment along each diagonal side equals $(\sqrt{2}/2)aM_p$, and its component about the plate side equals:

$$\frac{\sqrt{2}}{2} M_p a \cos 45° = \frac{\sqrt{2}}{2} M_p a \frac{\sqrt{2}}{2} = \frac{1}{2} M_p a.$$

Hence:

$$\tfrac{1}{24} W_u a = 2(\tfrac{1}{2} M_p a) \quad \therefore \quad W_u = 24 M_p = 6h^2 f_y. \tag{13.6.2}$$

It is thus seen by (13.6.1) and (13.6.2) that:

$$\frac{W_u}{W_y} = \frac{6}{3.47} \quad \therefore \quad W_u = 1.73W_y. \tag{13.6.3}$$

The simply supported square plate has a reserve of strength of 73%.

For a square plate fixed along the boundary, the ultimate load is reached when hinge lines develop along the sides as well as along the diagonals. Hence, for equilibrium of one of the triangles (Fig. 13.6.3):

$$2(\tfrac{1}{2}M_pa) + M_pa = \tfrac{1}{24}W_ua \quad \therefore \quad W_u = 48M_p = 12h^2f_y. \tag{13.6.4}$$

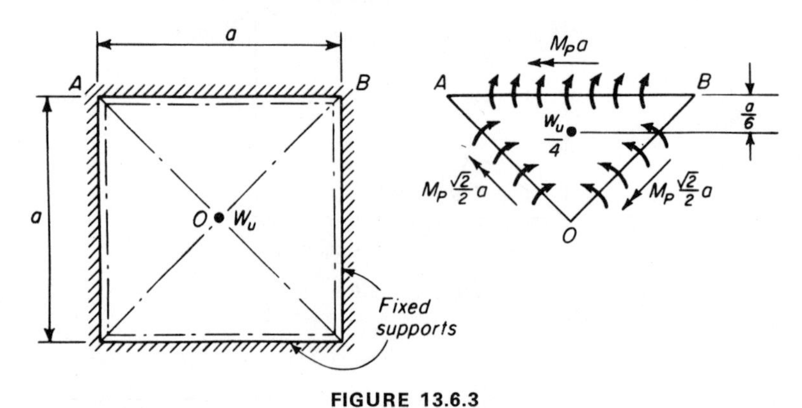

FIGURE 13.6.3

From Table 12.2.2, the maximum stress in a square plate with fixed sides, which occurs at the middle of the sides, is:

$$f = -\frac{0.051wa^2}{h^2/6} = -0.307w\left(\frac{a}{h}\right)^2,$$

from which, when $f = f_y$:

$$W_y = wa^2 = \frac{h^2}{0.307}f_y = 3.26h^2f_y. \tag{13.6.5}$$

Hence, by (13.6.4) and (13.6.5):

$$\frac{W_u}{W_y} = \frac{12}{3.26} \quad \therefore \quad W_u = 3.69W_y. \tag{13.6.6}$$

The fixed plate has a reserve of strength of 269%.

In the case of a simply supported circular plate, every radial line must become a hinge line for the plate to flow plastically. From the equilibrium of a very small circular sector, the component of the moments along the radial lines M_pR in the direction of the tangent to the support (Fig. 13.6.4) is:

$$2M_pR \sin\theta \approx 2M_pR\theta; \tag{a}$$

the ultimate load on the sector, W_u, is to the total load on the plate, $w_u\pi R^2$,

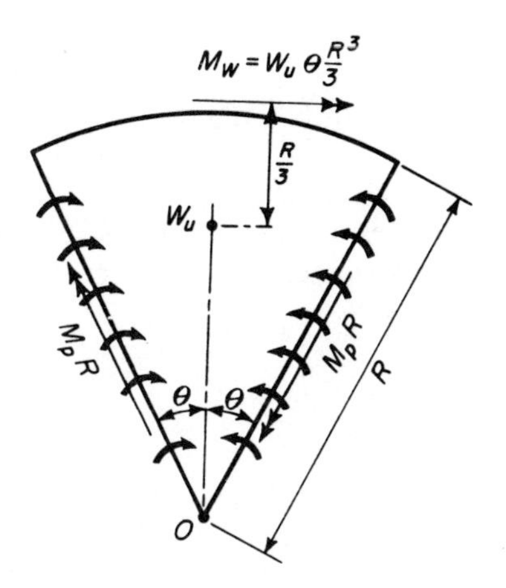

FIGURE 13.6.4

as θ is to π: $W_u = w_u\theta R^2$, and the moment of this load about the tangent to the support is:

$$w_u\theta R^2\left(\frac{R}{3}\right) = \frac{w_u\theta R^3}{3}. \tag{b}$$

Equating (a) to (b), we obtain:

$$w_uR^2 = 6M_p$$

$$\therefore \quad W_u \equiv \pi w_uR^2 = 6\pi M_p = 6\pi\frac{h^2f_y}{4} = 4.72h^2f_y. \tag{13.6.7}$$

By (12.4.1), where $a = 2R$, the maximum stress in the plate equals $f_r = 0.051\ wa^2/(h^2/6) = 0.31\ w(a/h)^2$, from which, setting $f_r = f_y$ and solving for w_ya^2, the maximum load producing elastic stresses is:

$$W_y \equiv \pi w_y\left(\frac{a}{2}\right)^2 = \frac{\pi}{4}\frac{h^2}{0.31}f_y = 2.53h^2f_y. \tag{13.6.8}$$

Hence:

$$W_u = \frac{4.27}{2.53}W_y = 1.87W_y, \tag{13.6.9}$$

with a reserve of strength of 87%.

For a fixed circular plate, one obtains, similarly:

$$2M_pR\theta + M_pR(2\theta) = \frac{w_u\theta R^3}{3},$$

from which:

$$w_uR^2 = 12M_p \quad \therefore \quad W_u = \pi w_uR^2 = 9.44h^2f_y. \tag{13.6.10}$$

By (12.4.2), $f_{r,\min} = -0.031\ wa^2/(h^2/6) = -0.19\ w(a/h)^2$, from which:

$$W_y = \frac{\pi}{4}\frac{h^2}{0.19}f_y = 4.12h^2f_y, \tag{13.6.11}$$

and hence:

$$W_u = \frac{9.44}{4.12} W_y = 2.29 W_y, \tag{13.6.12}$$

with a reserve of strength of 129%.

The above given values of the reserve of strength for plates of uniform materials can be used approximately for reinforced concrete slabs of balanced design.

PROBLEMS

13.1 Determine the maximum shear stress in a 300 mm × 300 mm (12 in. × 12 in.) concrete column with an axial force of 1 280 kN (288 k).

13.2 A concrete wall under horizontal loading in a high-rise building is subject to shearing forces on its vertical and horizontal edges. Determine the maximum possible shear stress in the wall if the maximum allowable tension is 0.34 MPa (50 psi). Neglect the compression in the wall.

13.3 Sketch the lines of principal stress in a triangular vertical plate uniformly supported along its lower edge and loaded by a vertical force at its top.

13.4 Sketch the lines of principal stress in a square plate loaded at each corner by equal, externally directed forces along the square's diagonals.

13.5 What is the maximum stress concentration in a bolted splice of A36 steel if the average tensile stress is 165 MPa (24 ksi)? What is the actual maximum tensile stress?

13.6 Derive the ultimate bending-moment diagram for a continuous beam with two equal spans l loaded by a force P at the center of the left span.

13.7 A W410 × 60 (W16 × 40) beam of A36 steel spans continuously over two spans of 9 m (30 ft). What is the ultimate value of two equal loads, placed at the center of each span?

13.8 Determine the ultimate uniform load that can be carried by the beam of Problem 13.7.

13.9 What is the ultimate concentrated load that can be carried at the center of the left span of the beam of Problem 13.7?

13.10 The frame shown in Fig. 13.5.2 with $h = 2.4$ m (8 ft) and $l = 3.5$ m (12 ft) consists of W360 × 33 (W14 × 22) members of A36 steel. What is the ultimate horizontal load P that can be carried by the frame?

13.11 Determine by the yield-line method the ultimate uniform load that can be carried by a 1.2 m × 1.2 m (4 ft × 4 ft) square plate, 10 mm ($\frac{3}{8}$ in.) thick of A36 steel if: (a) the edges are simply supported, (b) the edges are fixed.

13.12 Determine the ultimate loads for the inscribed and circumscribed circular plates used to approximate the fixed-edge square plates in Problem 13.11.

13.13 Determine the radii R_1 and R_2 of a simply supported and a fixed circular plate which have the same ultimate loads as those of the simply supported square plates of Problem 13.11 (a) and (b).

13.14 A marble column of square cross section supports a load P. If the load P grows in intensity, will the column fail in compression or in shear?

13.15 A cantilever beam of span l and width b supports a distributed load w. The beam depth is h, and its material is homogeneous. Determine the maximum shear stress at the cantilever root: (a) at the neutral axis, (b) at the top fiber, and (c) at the bottom fiber of the beam. (Notice that the load w produces a vertical compression w/b at the top of the beam.)

13.16 A flat A36 tension bar is connected to a plate by circular fillets of radius r. Determine the width d of the bar capable of carrying a load of 90 kN (20 k) if its thickness is 6 mm ($\frac{1}{4}$ in.) for $r/d = \frac{1}{2}$ and $r/d = \frac{1}{8}$.

13.17 Determine by the yield-line method the ultimate load of an equilateral triangular plate of side a and thickness h, carrying a concentrated load P at its center: (a) when the plate is simply supported, (b) when the plate is fixed.

13.18 Determine by the yield-line method the ultimate load on a uniformly loaded hexagonal plate of side a: (a) when simply supported, (b) when fixed.

13.19 A beam of rectangular cross section $b \times h$ has a yield moment $M_y = Fh/2$, where $F = f_y bh/2$. Determine its yield moment M_y' if the beam is initially under a tensile force T. (Indicate by c_1 the distance of the neutral axis from the upper extreme fiber of the beam; locate c_1 by an equilibrium equation in the direction of the beam axis.)

14. The Geometry of Shell Surfaces

14.1 LOCAL PROPERTIES OF SURFACES [12.2]

In order to obtain a thorough understanding of the structural behavior of two-dimensional curved structures, such as membranes and shells, it is essential to become familiar first with the geometrical properties of their surfaces.

These properties can be divided into two categories: (a) *local properties*, which define the geometry of the surface in the immediate neighborhood of a point, and (b) *general properties*, which refer to the shape of the surface as a whole.

Given a point O on a surface and the plane tangent to the surface at O (Fig. 14.1.1), the perpendicular to the tangent plane at O is called the *normal* to the surface at O.

The curves, intersected on the surface by planes containing the normal at O, are called the *normal sections* of the surface. The *local* geometrical properties of a surface are determined by the variation of the properties of its normal sections as the intersecting plane rotates around the normal.

Let us choose the normal pointing *into* the surface at O as the positive z-axis, and the intersection of a given normal plane with the tangent plane as the x-axis (Fig. 14.1.1). The normal section cut by the given plane is then a curve $z = f(x)$. The slope $m_x = \tan \alpha$ of the curve $z = f(x)$ at a point P_1 of coordinates $(x_1, 0, z_1)$ is called the *slope of the surface in the x-direction* at P_1. It is measured by:

$$m_x = \left.\frac{df}{dx}\right]_{x=x_1} = \left.\frac{\partial z}{\partial x}\right]_{\substack{x=x_1\\y=0}}, \qquad (14.1.1)$$

where $z = z(x, y)$ is the equation of the surface.

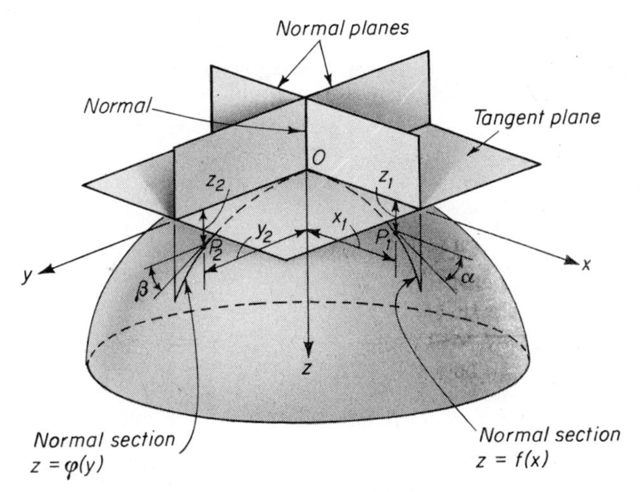

FIGURE 14.1.1

The slope of the surface at O is zero, since the x-axis is tangent to the curve, and the slope changes as one moves away from O along the x-axis. The *rate of change at O of the slope m_x in the direction of the slope*, i.e., in the x-direction, is called the *curvature of the surface in the x-direction at O*, and is measured by:

$$c_x = \left.\frac{d^2f}{dx^2}\right]_{x=0} = \left.\frac{\partial^2 z}{\partial x^2}\right]_{\substack{x=0\\y=0}}. \tag{14.1.2}$$

When c_x is positive, the curve $z = f(x)$ is concave inward, i.e., in the $+z$-direction. When c_x is negative, the curve is concave outward, i.e., in the $-z$-direction; when $c_x = 0$, the curve either is a straight line or changes from $+$ to $-$ curvature and is said to have an *inflection point* at $x = 0$ (Fig. 14.1.2). A small arc of the curve $z = f(x)$ in the immediate neighborhood of O may be approximated by an arc of the circle of radius R_x (Fig. 14.1.3), called the *osculating circle*, which is the nearest to the curve among the tangent circles. The change in slope between O and P_1 is $\Delta m_x = \alpha - 0 = \alpha$; for small α,

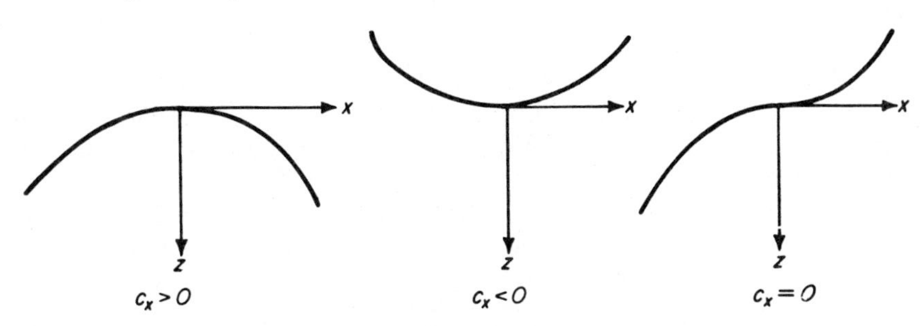

FIGURE 14.1.2

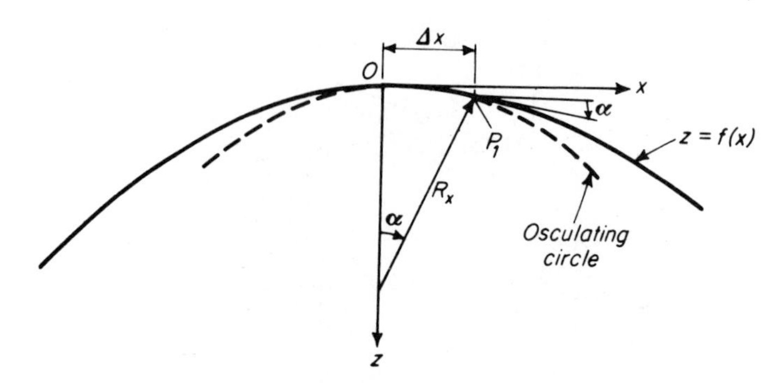

FIGURE 14.1.3

the distance $OP_1 = \Delta x = R_x\alpha$; hence, the rate of change of the slope, i.e., the curvature, is given by:

$$c_x = \frac{\Delta m_x}{\Delta x} = \frac{\alpha}{R_x\alpha} = \frac{1}{R_x}. \tag{14.1.3}$$

R_x is called the *radius of curvature* of the surface in the x-direction at O.

A plane through O perpendicular to the (x, y)-plane cuts the tangent plane along the y-axis and cuts the surface along a normal section $z = \varphi(y)$ (Fig. 14.1.1). The slope m_y of $z = \varphi(y)$ at a point $P_2(O, y_2, z_2)$ (Fig. 14.1.1), given by:

$$m_y = \frac{d\varphi}{dy}\Big]_{y=y_2} = \frac{\partial z}{\partial y}\Big]_{\substack{x=0\\y=y_2}}, \tag{14.1.4}$$

is the slope of the surface in the y-direction at P_2. The rate of change of m_y in the y-direction at O, given by:

$$c_y = \frac{d^2\varphi}{dy^2}\Big]_{y=0} = \frac{\partial^2 z}{\partial y^2}\Big]_{\substack{x=0\\y=0}}, \tag{14.1.5}$$

is the *curvature of the surface in the y-direction at O*. The curvature c_y is the reciprocal of the radius of curvature R_y of the osculating circle in the y-direction:

$$c_y = \frac{1}{R_y}. \tag{14.1.6}$$

A last quantity of geometrical (and structural) significance is the *rate of change of the slope at right angles to the slope* or the *geometrical twist* of the surface. Consider a point P_2 on the curve $z = \varphi(y)$ and the slope $m_x = \alpha_2$ of the surface in the x-direction at P_2 (Fig. 14.1.4). The change in the slope m_x as one moves from O to P_2 is $\Delta m_x = \alpha_2 - 0$; the distance $OP_2 = \Delta y$ for a small angle β equals $R_y\beta$; hence, the twist t_{xy} at O is given by:

$$t_{xy} = \frac{\Delta m_x}{\Delta y} = \frac{\alpha_2}{R_y\beta} = \frac{\Delta(\partial z/\partial x)}{\Delta y} = \frac{\partial^2 z}{\partial y\, \partial x}\Big]_{\substack{x=0\\y=0}}. \tag{14.1.7}$$

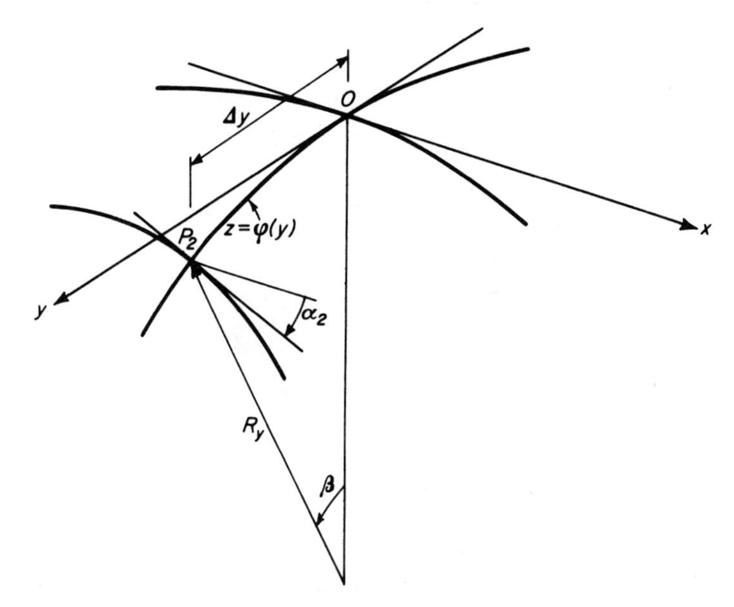

FIGURE 14.1.4

Similarly:

$$t_{yx} = \frac{\Delta m_y}{\Delta x} = \frac{\alpha_1}{R_x \alpha} = \frac{\Delta(\partial z/\partial y)}{\Delta x} = \left.\frac{\partial^2 z}{\partial x\, \partial y}\right]_{\substack{x=0\\y=0}},$$

and, since $(\partial^2 z/\partial x\, \partial y) = (\partial^2 z/\partial y\, \partial x)$ for any continuous function $z(x, y)$ with continuous first derivatives:

$$t_{xy} = t_{yx}. \tag{14.1.8}$$

As the cutting plane rotates about the normal to the surface at a point, the curvature of the normal sections usually changes value, and it may be proved that, in general:

1. There are two directions *at right angles to each other* for which the curvature acquires, respectively, maximum and minimum values. These are called the *principal values* c_1, c_2 of the curvature at that point; the directions in which they occur are called the *principal directions of curvature*. For example, at the top of an elliptical dome (Fig. 14.1.5), the principal curvatures occur in the direction of the smallest and largest semi-axis, a and b, respectively. In Fig. 14.1.5, the maximum curvature $c_1 = c_x$, the minimum $c_2 = c_y$.

2. The twist t_{12} in the principal directions is zero. At the top of the elliptical dome, the slope $m_x = m_1 = 0$, and it remains zero if one moves in the y-direction, i.e., along the curve 2. Similarly, the slope $m_y = m_2 = 0$ at the top of the dome, and it remains zero if one moves along the curve 1. The directions of principal curvature are often easily spotted as those for which there is no twist.

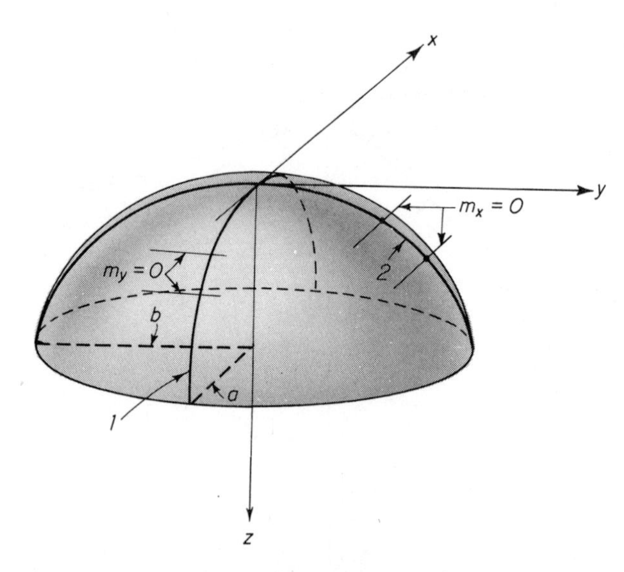

FIGURE 14.1.5

3. There are two more directions at right angles to each other for which the twist becomes maximum and minimum. These directions bisect the principal directions of curvature.

As was done with lines of principal stress (see Section 13.2), indicating by crosses the directions of principal curvature at various points of a surface, one may draw its *lines of principal curvature* which are parallel to the crosses at each point. Figure 14.1.6 shows the lines of principal curvature for an ellipsoid.

Surfaces are divided into three distinct categories according to the variation of their curvature around a point:

1. When the curvature at a point is of the same sign in all directions, the surface is called *synclastic* at that point. An elliptic and a spherical dome are synclastic surfaces at all points; they are everywhere and in all directions

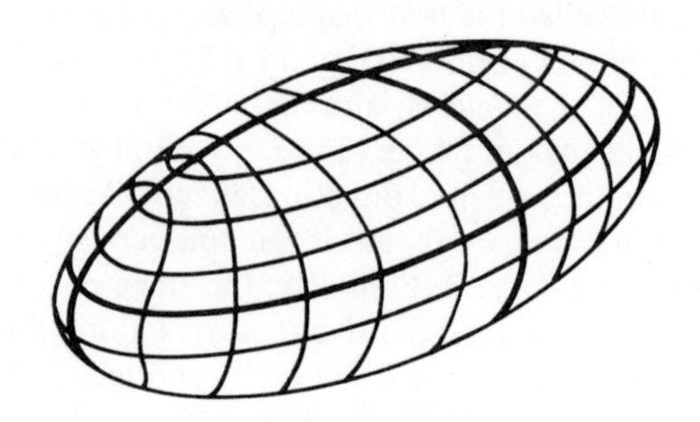

FIGURE 14.1.6

concave into the dome (downward) (Fig. 14.1.5). A dish-shaped tensile roof is also synclastic; its curvature at any point is upward in all directions.

In a synclastic surface, the principal curvatures c_1 and c_2 have the same sign, their product is positive and, since $t_{12} = 0$:

$$c_1c_2 - t_{12}^2 > 0. \tag{14.1.9}$$

Synclastic surfaces are also called surfaces of *positive Gaussian curvature* because the quantity $c_1c_2 - t_{12}^2$ is called the *Gaussian curvature* of the surface at a point.

2. When the curvature at a point is of the same sign in all but one direction, in which it is *zero*, the surface is called *developable* at that point. If c is never negative, its minimum value is $c_2 = 0$; if c is never positive, its maximum value is $c_1 = 0$. Hence, for a developable surface:

$$c_1c_2 - t_{12}^2 = 0 \tag{14.1.10}$$

and a developable surface has zero Gaussian curvature.

A cylinder is a developable surface at all points, with zero curvature in the direction of the generatrices (Fig. 14.1.7). A developable surface may be flattened into a plane *without stretching or shrinking it.*

3. When the curvature at a point is positive in certain directions and negative in others, the surface is called an *anticlastic* or a *saddle surface* at that point. In this case, $c_1 > 0$, $c_2 < 0$, and:

$$c_1c_2 - t_{12}^2 < 0, \tag{14.1.11}$$

so that the Gaussian curvature is negative.

Figure 14.1.8 shows, projected on the x, y-plane, the surface sectors of amplitude γ_1 in which the curvature is positive, and those of amplitude

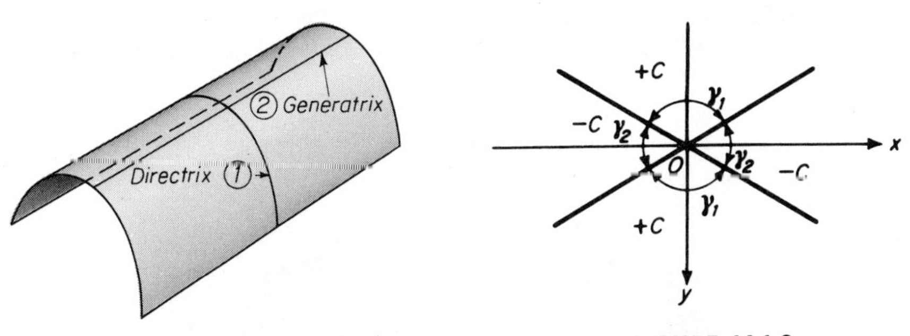

FIGURE 14.1.7

FIGURE 14.1.8

γ_2 in which the curvature is negative around the point O. Along the lines dividing these sectors, the curvature must be zero, since it changes from positive to negative values or vice versa. Hence, two lines of no curvature, i.e., two straight lines, lie on the saddle surface at O. The hyperbolic paraboloid (Fig. 14.1.9) is a saddle surface at all points.

Surfaces in which the curvature changes sign more than four times around a particular point may be readily conceived: they are "scalloped" and have three or more straight lines lying on them at that particular point. A surface with three straight lines on it at a particular point is called a "monkey saddle" (Fig. 14.1.10).

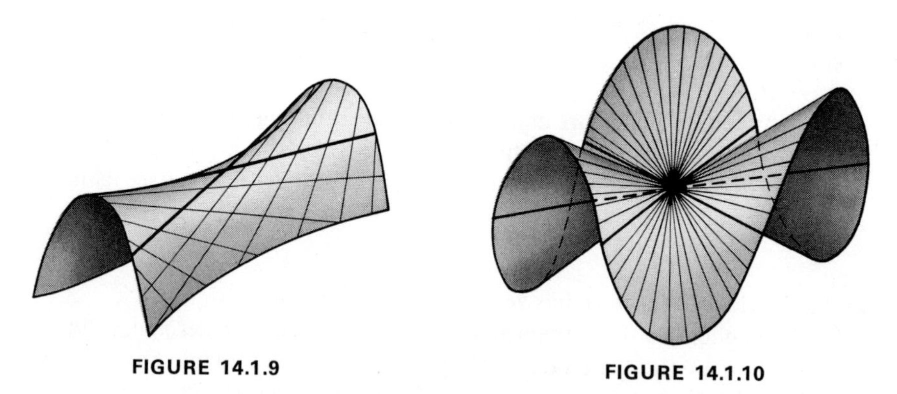

FIGURE 14.1.9 FIGURE 14.1.10

Neither synclastic nor anticlastic surfaces may be opened up into planes without stretching or shrinking them. This resistance to flattening makes them stiffer than developable surfaces.

14.2 GENERATION OF SURFACES [12.3, 12.4, 12.5]

Most of the geometrically defined surfaces used in structures are generated by one of two basic processes, the rotation or the translation of a plane curve. In the first process, the curve rotates around a line, called its *rotational axis*, and generates a *rotational surface* (Fig. 14.2.1). In the second

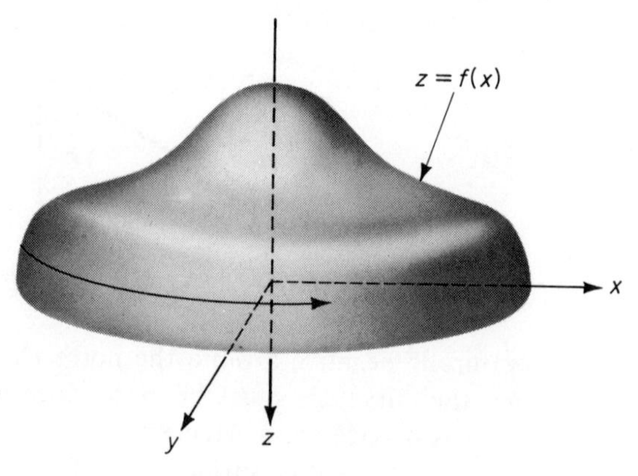

FIGURE 14.2.1

process, the curve moves both at an angle to, and in its vertical plane, while remaining parallel to itself, and generates a *translational surface* (Fig. 14.2.2).

When the axis of a rotational surface is vertical and the curve intersects the axis, the surface is a *dome* or a *dish*. The rotating curve is the *meridian* of the surface; the horizontal sections are its *parallels* (Fig. 14.2.3). The meridians are lines of principal curvature, since there is no change in the horizontal slope of the parallels as one moves at right angles to them along the meridians; their radius of curvature R_1 at a point P is, therefore, one

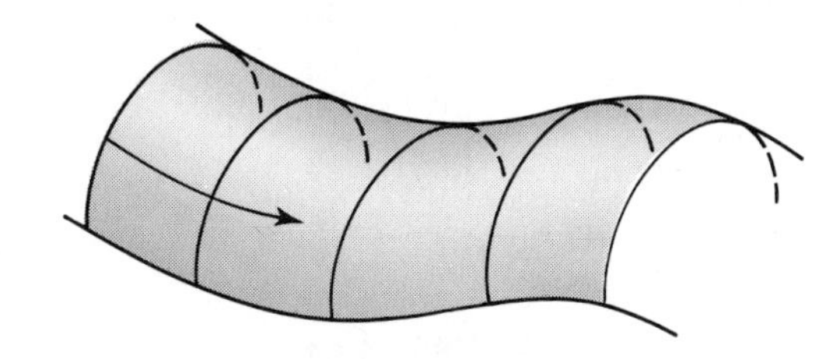

FIGURE 14.2.2

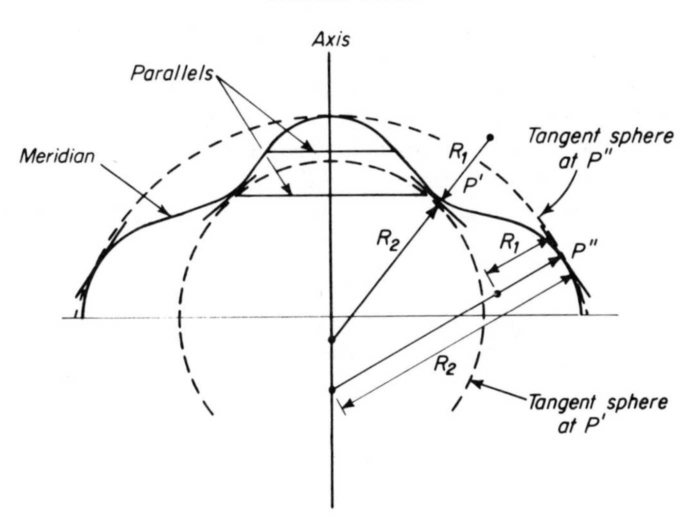

FIGURE 14.2.3

of the principal radii of curvature. The second principal direction of curvature is at right angles to the meridian; the corresponding principal radius R_2 (in a plane at right angles to the meridional plane) is the radius of the tangent sphere nearest to the surface. By symmetry, R_2 is the distance, along the normal to the surface, from the point on the surface to the axis of rotation (Fig. 14.2.3).

Any curve may be used as a meridian (Fig. 14.2.4): a circle gives rise to a sphere (a), an ellipse to an ellipsoid of revolution (b), a parabola to a rota-

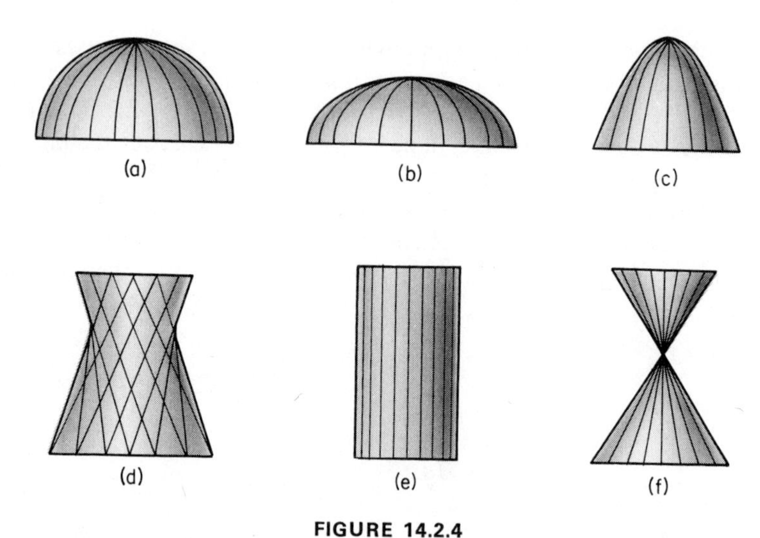

FIGURE 14.2.4

tional paraboloid (c), a hyperbola to a rotational hyperboloid (d). A straight line gives rise to a circular cylinder (e), a rotational hyperboloid (d), or a circular cone (f), depending on whether the line is parallel to the axis, is inclined to it without intersecting it, or intersects it. In Fig. 14.2.4, surfaces (a), (b), and (c) are synclastic; (e) and (f) developable; and (d) anticlastic.

The surface generated by the rotation of any *closed* curve around an axis which it does not intersect is called a *torus* [Fig. 14.2.5 (a)]. The most

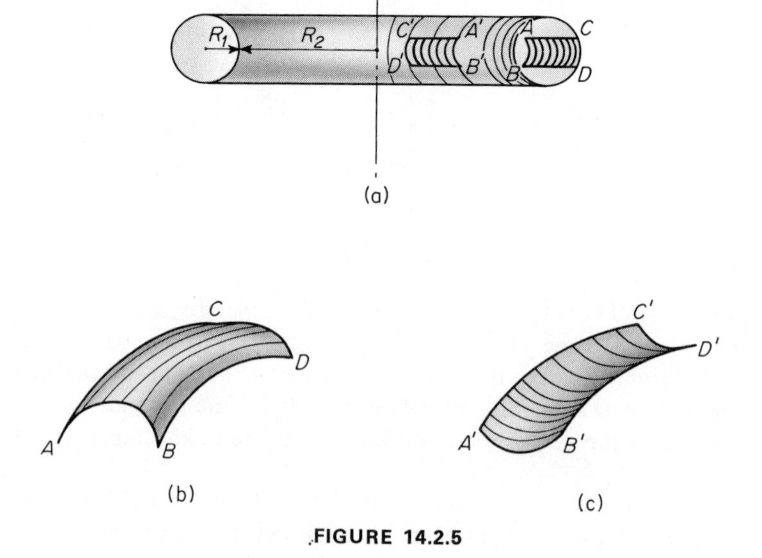

FIGURE 14.2.5

commonly used torus is circular. Sectors of the same torus may have totally different appearances: a sector cut out near the outer portion of the torus [Fig. 14.2.5 (b)] is a synclastic surface similar to an elliptic paraboloid (see Fig. 14.2.8), a sector cut out near its inner portion is a saddle surface similar to the hyperbolic paraboloid (see Fig. 14.2.9).

If the meridional curve is represented in the x, z-plane by the equation (Fig. 14.2.1):

$$z = f(x), \tag{14.2.1}$$

where z is the axis of rotation, the equation of the rotational surface in cartesian coordinates is (Fig. 14.2.6):

$$z = f(\sqrt{x^2 + y^2}). \tag{14.2.2}$$

A surface is generated by *translation* when a plane curve 1 slides, remaining parallel to itself, on another plane curve 2, usually at right angles to the first (Fig. 14.2.7). Since any combination of curves may be used, a large variety of surfaces is obtainable by translation.

Translating a plane curve 1 on a straight line 2, one obtains cylinders (circular, parabolic, elliptic, catenary, etc.). Translating a parabola 1 with downward curvature on another parabola 2 also with downward curvature, one obtains an *elliptic paraboloid* whose horizontal sections are ellipses (Fig. 14.2.8). (Synclastic surfaces similar to elliptic paraboloids may be obtained by translating an arc of circle on another arc of circle.) A parabola 1

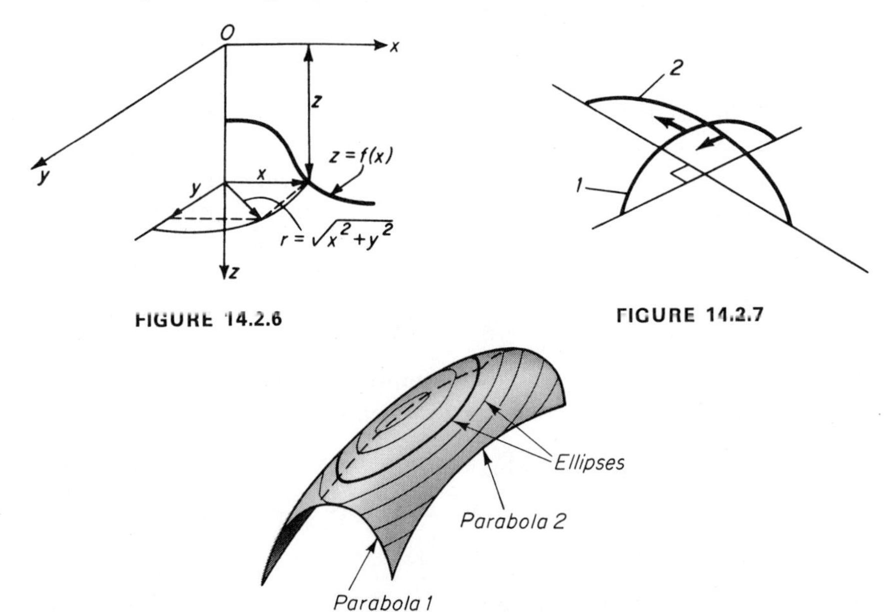

FIGURE 14.2.6

FIGURE 14.2.7

FIGURE 14.2.8

with downward curvature translated on another parabola 2 with upward curvature generates a saddle surface, the *hyperbolic paraboloid*, whose horizontal sections are hyperbolas (Fig. 14.2.9).

If the equations of the two curves at right angles are:

$$z = f_1(x), \qquad z = f_2(y), \tag{14.2.3}$$

the equation of the translational surface is:

$$z = f_1(x) + f_2(y). \tag{14.2.4}$$

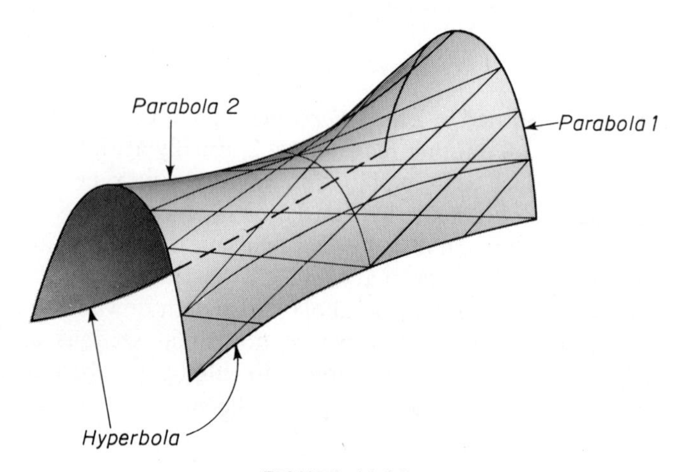

FIGURE 14.2.9

In the translational surface (14.2.4), $t_{xy} = 0$, since the slope at a point P of curve 1 does not change as curve 1 is moved at right angles along curve 2. Hence, x and y are the principal directions of curvature at any point. By (14.2.4), the hyperbolic paraboloid generated by two parabolas:

$$z_a = k_a x^2, \qquad z_b = -k_b y^2$$

has the equation:

$$z = k_a x^2 - k_b y^2. \tag{14.2.5}$$

A *rectangular* hyperbolic paraboloid, a translational surface commonly used in shell construction, is generated by two identical parabolas:

$$z_a = kx^2, \qquad z_b = -ky^2. \tag{14.2.6}$$

By (14.2.4), the surface of the rectangular hyperbolic paraboloid is:

$$z = k(x^2 - y^2) = k(x - y)(x + y). \tag{14.2.7}$$

Along the 45° lines through the origin OA, OB (Fig. 14.2.10), of equations:

$$y = \pm x \quad \text{or} \quad y - x = 0, \qquad y + x = 0,$$

$z = 0$, so that the paraboloid is horizontal along OA and OB. Along any 45° line MN, parallel to OB (Fig. 14.2.10), of equations:

$$y = y_M + \frac{1}{\sqrt{2}}y', \qquad x = x_M - \frac{1}{\sqrt{2}}y',$$

the ordinate z of the paraboloid:

$$z = k\left[\left(x_M - \frac{1}{\sqrt{2}}y'\right)^2 - \left(y_M + \frac{1}{\sqrt{2}}y'\right)^2\right]$$
$$= k[x_M^2 - y_M^2 - 2\sqrt{2}(x_M + y_M)y']$$

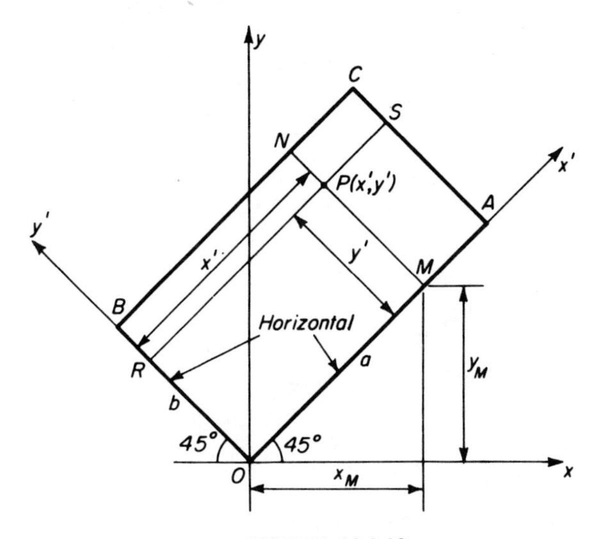

FIGURE 14.2.10

varies *linearly* with y'. Similarly, z varies linearly with x' along any 45° line RS parallel to OA. Hence, the rectangular hyperbolic paraboloid is also described by a straight line MN sliding on the two skew lines OA, BC, or by a straight line RS sliding on the two skew lines OB, AC. Since OA and OB are perpendicular, the equation of the rectangular hyperbolic paraboloid referred to (x', y')-axes is of the form:

$$z = cx'y'. \tag{a}$$

Indicating by h the *rise* of the hyperbolic paraboloid at C, where $x' = a$ and $y' = b$, one obtains:

$$h = cab \quad \therefore \quad c = \frac{h}{ab},$$

and the equation of the paraboloid referred to the new axes becomes:

$$z' = h\left(\frac{x'}{a}\right)\left(\frac{y'}{b}\right). \tag{14.2.8}$$

Conoidal surfaces are generated by translating a line AB, called the *generatrix*, on two vertical plane curves called the *directrices* (Fig. 14.2.11). When one of the two directrices is a horizontal straight line, the surface is anticlastic and called a *conoid* (Fig. 14.2.12). When the two directrices are identical, the surface is developable and is a cylinder. When the two directrices

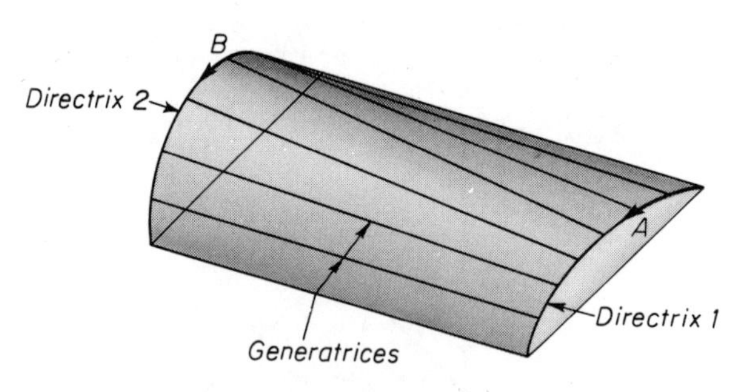

FIGURE 14.2.11

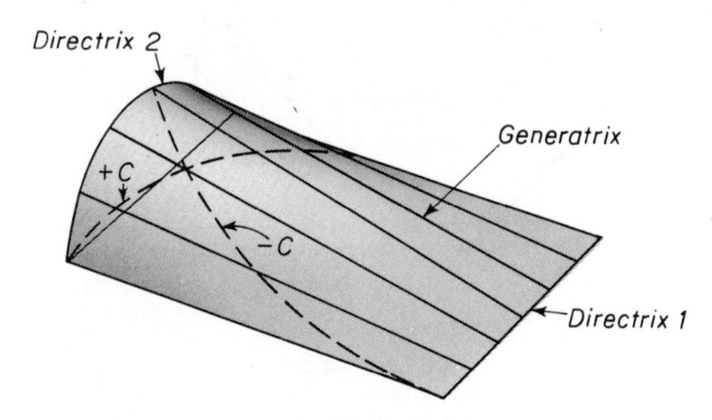

FIGURE 14.2.12

are two skew straight lines, the anticlastic surface is a hyperbolic paraboloid (Fig. 14.2.13).

If the directrices of a conoid are located at $x = 0$, $x = l$ and have equations (Fig. 14.2.14):

$$z_1 = f_1(y_1), \qquad z_2 = f_2(y_2), \tag{14.2.9}$$

the conoid has the equation:

$$z = f_1(y_1) - \frac{x}{l}[f_1(y_1) - f_2(y_2)] \tag{14.2.10 a}$$

with:

$$y_2 = \frac{b_2}{b_1}y_1, \qquad y = \left[1 - \left(1 - \frac{b_2}{b_1}\right)\left(\frac{x}{l}\right)\right]y_1. \tag{14.2.10 b}$$

It is seen from this brief description of surface generation that the same surface may be obtained in a variety of ways. The cylinder can be generated

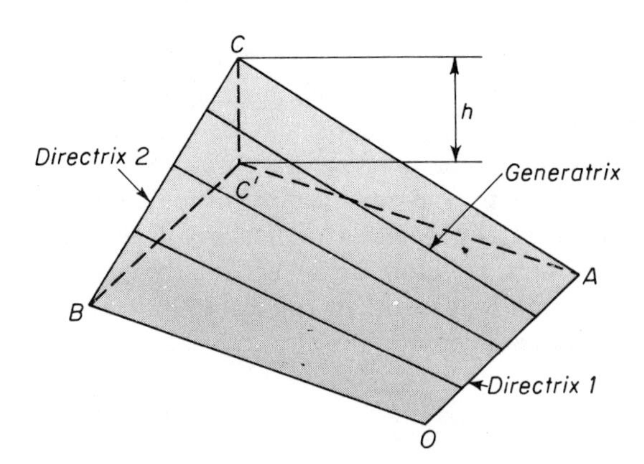

FIGURE 14.2.13

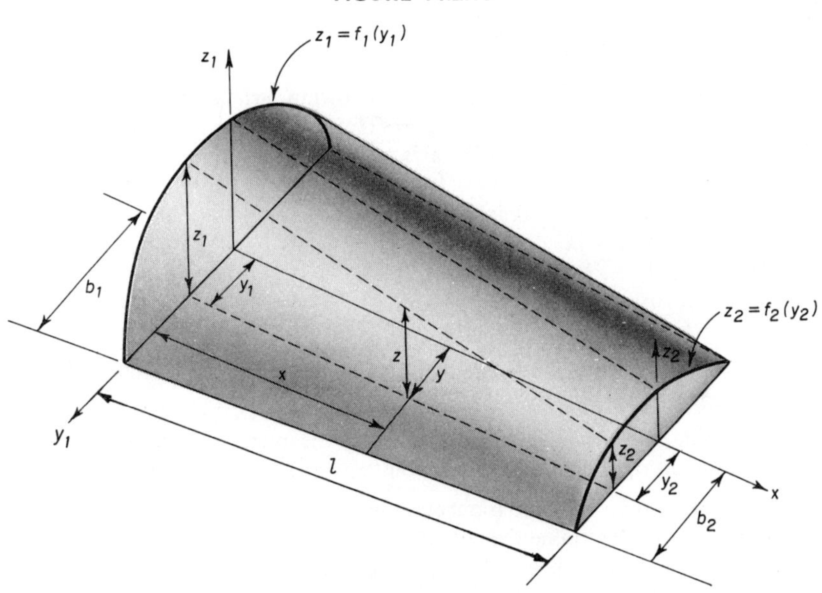

FIGURE 14.2.14

as a rotational surface, a translational surface, or a conoidal surface. The rotational hyperboloid may be obtained by rotating either a hyperbola or an inclined line around an axis. The hyperbolic paraboloid is generated either by translating its principal sections (the parabolas) on one another, or by translating a straight line on two skew straight lines (the straight lines are the lines of zero curvature which lie on any saddle surface).

Any surface described by the motion of a straight line is called a *ruled surface*. The cylinder, the cone, the rotational hyperboloid, and the hyperbolic paraboloid are ruled surfaces.

Surfaces which cannot be generated by a simple geometrical procedure or are not represented by a simple equation may still be well adapted to structural use; their structural analysis may be obtained by approximate methods quite satisfactory for practical purposes. On the other hand, the shape of a curved surface to be used structurally should be chosen essentially on the basis of its structural properties, particularly if destined to span large distances. It is important, therefore, to understand the mechanisms by which curved surfaces transmit loads to their support and, hence, to the ground. Such mechanisms are considered in the next two chapters.

PROBLEMS

14.1 The meridian in the (x, z)-plane of the following surfaces has the given equation $z = f(x)$. Write the equation $z = f(x, y)$ of the corresponding rotational surfaces having the z-axis as the axis of rotation:
(a) $z = kx/a$ (cone),
(b) $z = kx^2/a^2$ (paraboloid),
(c) $z = k\sqrt{1 - x^2/k^2}$ (sphere),
(d) $z = k\sqrt{1 - x^2/a^2}$ (ellipsoid),
(e) $z = k \cosh (x/a)$ (catenoid),
(f) $z = ka/x$ (hyperboloid).

14.2 Determine the slope along the meridian at the point x, z of the surfaces of Problem 14.1.

14.3 Determine the principal curvature along the meridian and the principal curvature at right angles to it at a point x, z of the surfaces of Problem 14.1.

14.4 The equation of a spheroid (nonrotational ellipsoid) is given by $x^2/a^2 + y^2/b^2 + z^2/c^2 = 1$. Determine the curvatures of the spheroid in the x and y directions at $x = 0$, $y = 0$. Are these principal curvatures?

14.5 The equations of cylinders with axes parallel to the y-axis are given by the equations (b) to (e) of Problem 14.1. Determine the ratio of sag h to span l for these cylinders.

14.6 Classify the following surfaces according to their Gaussian curvature at the point $x = 0$, $y = 0$:
(a) $z = k(x^2 - y^2)$,
(b) $z = kxy$,
(c) $z = k(x^2 - 1)y^2$,
(d) $z = k(x^2 - 2)(y^2 - 1)$,
(e) $z = k \cosh \sqrt{x^2 + y^2}$,
(f) $z = k\sqrt{2 - x^2 - y^2}$,
(g) $z = kxy^2$.

14.7 Write the equations of the nine translational surfaces obtained by translating the curves (a), (b), (c) on the curves (d), (e), (f). Classify these surfaces according to their Gaussian curvature at $x = 0$, $y = 0$.

(a) $z = kx$,
(b) $z = k\sqrt{1 - x^2/a^2}$,
(c) $z = kx^2$,
(d) $z = c\sqrt{1 - y^2/b^2}$,
(e) $z = -ky^2$,
(f) $z = c$.

14.8 Write the equation of the conoid with directrices $f_1(y_1) = k\sqrt{1 - y_1^2/a^2}$, $f_2(y_2) = 0$, and $b_2 = b_1 = a$. Classify this surface according to its Gaussian curvature.

14.9 Determine the directions and values of the principal curvatures of the surface $z = x^2y^2/2$ at $x = 1, y = 0$. Classify this surface at points x, y. Are all points of the same type?

14.10 Determine the values of k and a in the curves of Problem 14.1 (a) to (e) which result in a given ratio r of span l to rise h, and evaluate the slope of these curves at $x = l/2$.

15. Membranes

15.1 MEMBRANE ACTION [11.1, 11.2]

An *ideal membrane* is a sheet of material so thin in comparison with its lateral dimensions that it can only develop tension.

Calling h the membrane thickness, its flexural rigidity per unit width is measured by $EI = E[1 \times (h^3/12)]$; hence, for a small enough h, the bending rigidity is bound to become negligible, and bending and transverse shear, which are proportional to EI, must vanish. In other words, although a membrane is a two-dimensional structural element, its "plate action" is negligible.

The resistance to compression of a membrane is also negligible, since, due to its thinness, it is bound to buckle under very small compressive stresses. Hence, an ideal membrane can only carry loads by tension in all directions and can only be built out of materials with good tensile resistance. Such materials include sheet metals, prestressed concrete, reinforced plastics, and fabrics—in particular, plastic fabrics, such as those made of nylon or reinforced by glass fibers.

A membrane is the two-dimensional equivalent of a cable and, in order to carry loads *by tension only*, at times it must adapt its shape to the load. But because of their two-dimensional behavior, membranes are not as unstable as cables: their shape, which is always the *funicular surface* of the load, may be funicular *for a variety of loadings*, rather than for only one loading, as is the case for cables.

The greater inherent stability of membranes is due to their geometry and to the types of stresses they develop under load. Consider a rectangular element of sides a, b parallel to x, y, cut out of a curved membrane with radii of curvature $R_x = 1/c_x$, $R_y = 1/c_y$ and twist t_{xy} in the x, y-directions [Fig. 15.1.1 (a)]. Let the load per unit area on the element be q, and indicate

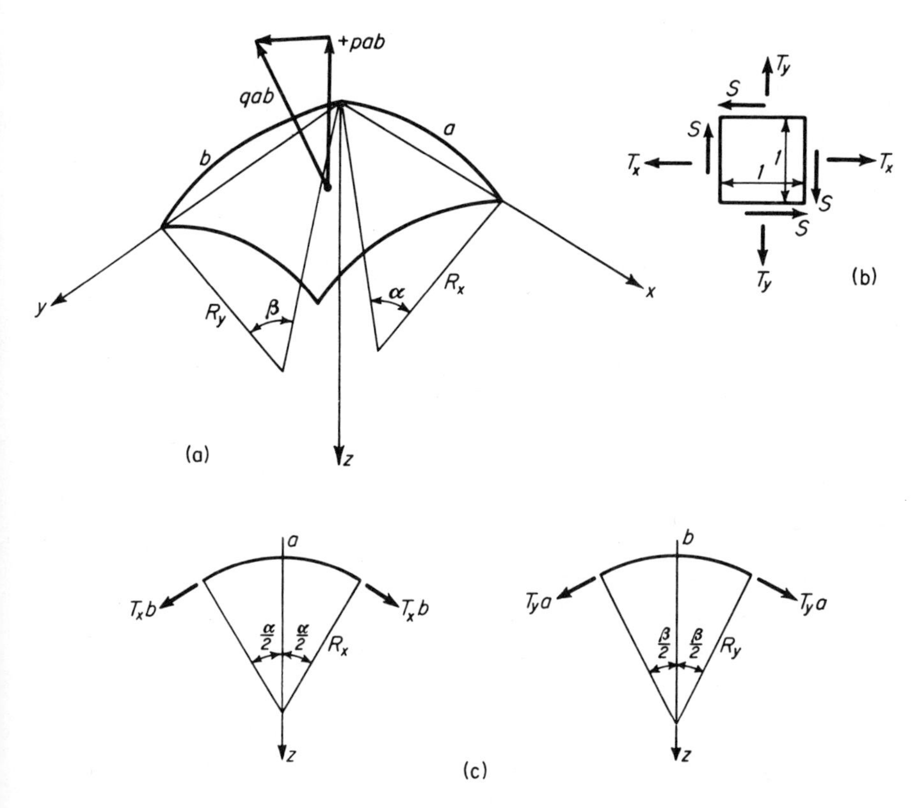

FIGURE 15.1.1

by p its component in the *negative* direction of the normal z to the element. The forces on the sides of the element are a tension T_x *per unit length* on the sides b, a tension T_y per unit length on the sides a, and shears S per unit length on all sides [Fig. 15.1.1 (b)]. T_x, T_y, and S have dimensions of force per unit length (kN/m, lb/in.). The *tangential shears* S act in the plane tangent to the membrane and, in the absence of bending, are evenly distributed across the membrane thickness, as are T_x and T_y. The three *membrane stresses* T_x, T_y, S can always be determined by considering the equilibrium of a loaded membrane element in the x-, y-, and z-directions. Hence, their values do not depend on the membrane deformation: membrane stresses are *internally statically determinate.* (The support conditions may, at times, require consideration of the membrane deformations and thus make the membrane statically indeterminate externally.)

The tensions T_x and T_y carry load by what might be called *two-way cable action.* If a, b are small in comparison with R_x, R_y so that the angles α, β [Fig. 15.1.1 (a)] are so small that $\sin \alpha \approx \alpha$ and $\sin \beta \approx \beta$, the sides of the element may be approximated by arcs of circle:

$$a = R_x\alpha, \qquad b = R_y\beta.$$

The resultants in the z-direction of the tensile forces on the sides b and a equal, respectively [Fig. 15.1.1 (c)]:

$$2T_x b \sin\frac{\alpha}{2} \approx 2T_x b\frac{\alpha}{2} = T_x\frac{ba}{R_x},$$

$$2T_y a \sin\frac{\beta}{2} \approx 2T_y a\frac{\beta}{2} = T_y\frac{ab}{R_y}.$$

The sum of these two resultants can equilibrate a normal load $p'ab$ in the *negative* z-direction given by:

$$p'ab = T_x\frac{ab}{R_x} + T_y\frac{ab}{R_y} \quad \therefore \quad p' = \frac{T_x}{R_x} + \frac{T_y}{R_y}. \tag{a}$$

p' is the normal load or pressure per unit area carried by cable action in the x- and y-directions through the curvatures $c_x = 1/R_x$ and $c_y = 1/R_y$.

The shears S carry load by a mechanism which has no counterpart in the cable mechanism because it is essentially a two-dimensional action. Indicate by s_x the slope in the x-direction of the membrane along the side AB of the element [Fig. 15.1.2 (a)] and by s'_x the slope along the opposite

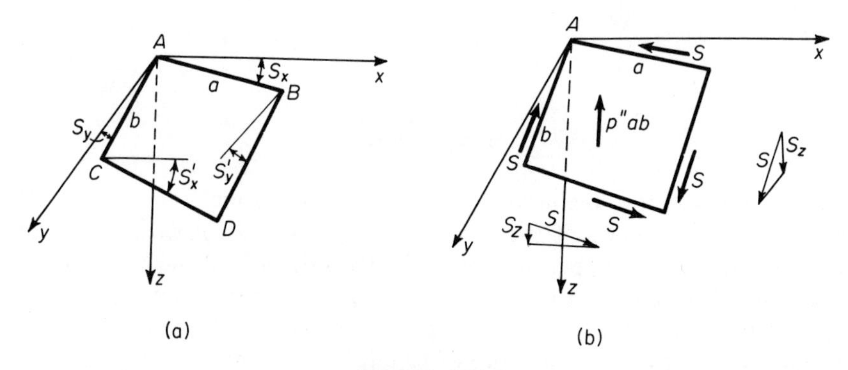

FIGURE 15.1.2

side CD. Since t_{xy} is the rate of change of the slope s_x in the y-direction, and since the side AC has a length b:

$$s'_x = s_x + t_{xy}b.$$

Similarly:

$$s'_y = s_y + t_{yx}a = s_y + t_{xy}a.$$

The resultant in the z-direction of the shears S [Fig. 15.1.2 (b)] is thus:

$$\begin{aligned} &-Sas_x + Sas'_x - Sbs_y + Sbs'_y \\ &\quad = -Sas_x + Sa(s_x + t_{xy}b) - Sbs_y + Sb(s_y + t_{xy}a) = 2St_{xy}ab. \end{aligned}$$

Hence, the normal load $p''ab$ in the negative z-direction that can be equilibrated by the shears S is:

$$p''ab = 2St_{xy}ab \quad \therefore \quad p'' = 2St_{xy}. \tag{b}$$

p'' is the load per unit area carried by the tangential shears through the geometrical twist [Fig. 15.1.2 (b)]. Introducing the reciprocal of the twist:

$$R_{xy} \equiv \frac{1}{t_{xy}}, \tag{15.1.1}$$

the total *normal* load $p = p' + p''$ per unit area carried by the membrane stresses is given by:

$$p = \frac{T_x}{R_x} + \frac{T_y}{R_y} + 2\frac{S}{R_{xy}}. \tag{15.1.2}$$

The carrying capacity of the membrane is seen to depend essentially on its geometric characteristics, the curvatures and the twist. A flat membrane, with no curvatures and no twist, i.e., with $R_x = R_y = R_{xy} = \infty$, cannot carry normal loads.

When the directions x, y are the principal directions of curvature 1, 2 of the membrane surface (see Section 14.1), the twist $t_{12} = 0$, shear action vanishes, and the normal load becomes:

$$p = \frac{T_1}{R_1} + \frac{T_2}{R_2}. \tag{15.1.3}$$

For a cylindrical or a conical surface $R_1 = \infty$, and the normal load becomes:

$$p = \frac{T_2}{R_2}; \tag{15.1.4}$$

in this case, the tension T_2 is often referred to as a *hoop force*.

The load q per unit area on a membrane has, in general, besides the component p *normal* to the shell, a component acting in the plane tangent to the shell. The *tangential* component of q is also carried by direct and shear stresses lying in the tangent plane. Hence, the entire load q is carried by membrane stresses, provided compression is not developed in any direction.

The greater stability of membranes can now be intuitively grasped. Whenever the membrane shape does not allow the entire load to be carried by cable action in two directions, shear action carries the excess load without requiring a change of membrane shape, provided no compression is developed.

15.2 MEMBRANE STRESSES IN SURFACES OF REVOLUTION [11.2]

It was seen in Section 14.2 that the principal directions of curvature in a surface of revolution coincide with the meridional direction and the direction at right angles to the meridian. Moreover, if a rotational membrane is loaded and supported axisymmetrically, the shears S on the meridional section vanish by symmetry, since there is no tendency for one meridional section to slide relatively to another. Thus, the directions of principal stress coincide with those of principal curvature. Therefore, letting $T_1 = T_\varphi$ denote the meridional force and $T_2 = T_\theta$ the circumferential or parallel force *per unit length*, R_1 the meridional radius, and R_2 the principal radius at right angles to the meridian (Fig. 15.2.1), by (15.1.3) the normal component p of the total load q becomes:

$$p = \frac{T_\varphi}{R_1} + \frac{T_\theta}{R_2}. \tag{15.2.1}$$

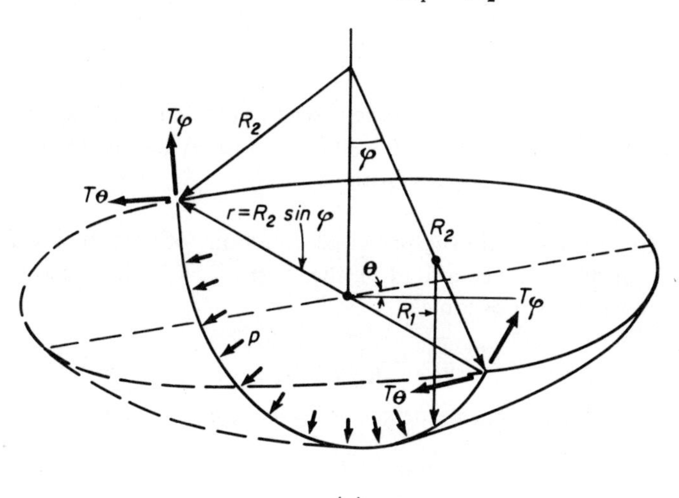

(a)

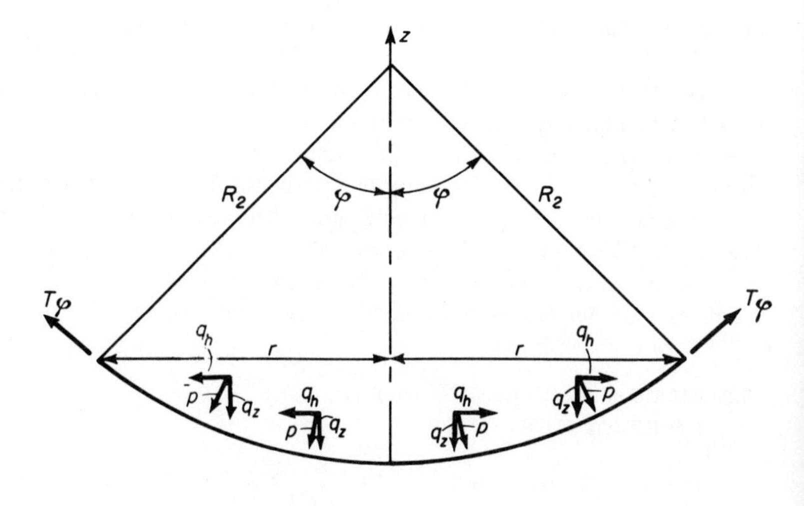

(b)

FIGURE 15.2.1

In particular, for a sphere $R_1 = R_2 = R$ and:

$$p = \frac{1}{R}(T_\varphi + T_\theta). \tag{15.2.2}$$

In order to determine T_φ and T_θ separately, we use another equation, which states the equilibrium in the *vertical* direction of the entire sector of membrane below the parallel defined by the colatitude angle φ. Due to the axisymmetric stress condition in the membrane, the horizontal components of the load q balance each other [Fig. 15.2.1 (b)]. Hence, the resultant of the load q is a vertical force Q_z along the membrane axis. The vertical resultant of the force T_φ applied to the circular boundary of the sector of radius r [Fig. 15.2.1 (a)] equals:

$$T_\varphi(2\pi r) \sin \varphi = 2\pi T_\varphi R_2 \sin^2 \varphi. \tag{a}$$

Equating (a) to Q_z:

$$T_\varphi = \frac{Q_z}{2\pi R_2 \sin^2 \varphi}, \tag{15.2.3}$$

and, by (15.2.3) and (15.2.1):

$$T_\theta = R_2\left(p - \frac{T_\varphi}{R_1}\right). \tag{15.2.4}$$

In particular, for a spherical membrane with $R_1 = R_2 = R$:

$$T_\varphi = \frac{Q_z}{2\pi R \sin^2 \varphi}, \qquad T_\theta = pR - T_\varphi. \tag{15.2.5}$$

Example 15.2.1. Determine the tensile stress f_t in a spherical membrane of radius 30 m (100 ft) inflated by a pressure of 0.001 N/mm² (0.15 psi).

By (15.2.2), with $T_\varphi = T_\theta$ due to symmetry:

$$T_\varphi = T_\theta = \tfrac{1}{2} \times 0.001(30 \times 1\,000) = 15 \text{ N/mm}.$$

From Table 2.7.2, the ultimate tensile strength of a vinyl-coated nylon balloon is 70 N/mm. The factor of safety is, therefore:

$$\tfrac{70}{15} = 4.7 > 4,$$

which is adequate for a fabric structure.

Example 15.2.2. A spherical membrane carries a vertical snow load $q_z = q$ kN/m² (psf), uniformly distributed in *horizontal projection* (Fig. 15.2.2). Determine its meridional and parallel tensions T_φ, T_θ.

The vertical equilibrium of the sector requires that:

$$Q_z = q\pi r^2 = q\pi R^2 \sin^2 \varphi = 2\pi(R \sin \varphi)(T_\varphi \sin \varphi)$$

$$\therefore \quad T_\varphi = \tfrac{1}{2} qR. \tag{15.2.6}$$

The normal component p of the load, per unit area of membrane, is the component of q in the normal direction, $p' = q \cos \varphi$, divided by an area A' of membrane

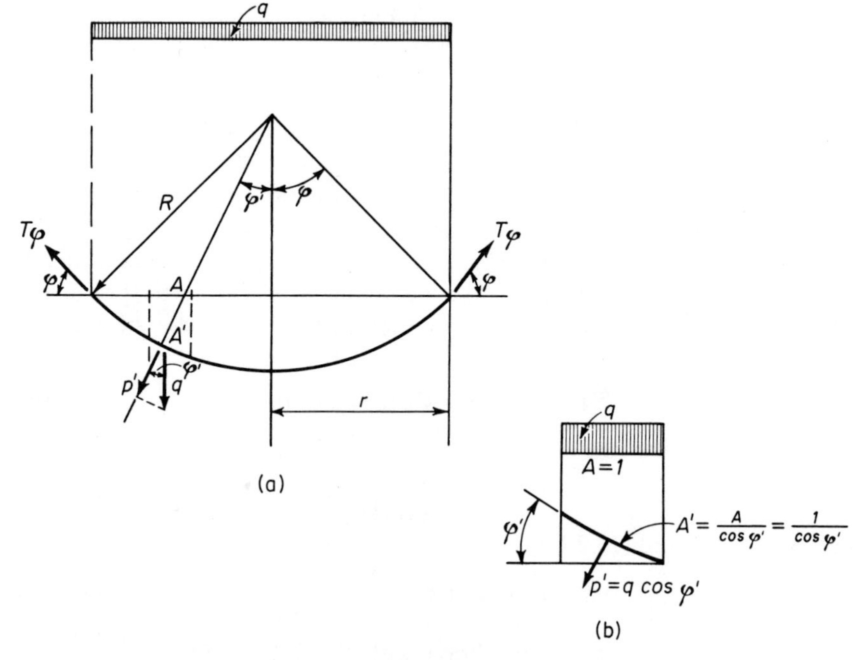

FIGURE 15.2.2

which has a projection $A = 1$ on a horizontal plane: $A' = A/\cos\varphi = 1/\cos\varphi$ [Fig. 15.2.2 (b)]. Hence:

$$p = \frac{p'}{A'} = q\cos^2\varphi, \tag{b}$$

and, by (15.2.5), and (15.2.6):

$$T_\theta = qR(\cos^2\varphi - \tfrac{1}{2}) = \tfrac{1}{2}qR\cos 2\varphi. \tag{15.2.7}$$

It is seen from (15.2.7) that if:

$$\cos^2\varphi < \frac{1}{2}, \quad \text{i.e.,} \quad \cos\varphi < \frac{1}{\sqrt{2}} \quad \therefore \quad \varphi > 45°,$$

T_θ becomes negative, i.e., is a compressive force. Hence, when $\varphi_0 > 45°$, in order to support the snow load without developing compressive parallel or *hoop stresses*, the membrane must change shape. The membrane is unstable for an opening angle $\varphi_0 > 45°$.

Example 15.2.3. A spherical membrane carries its own dead load w kN/m² (psf). Determine its meridional and parallel stresses.

The weight W of a membrane sector of opening angle φ can be computed by adding the weights of an infinite number of infinitesimal parallel rings of width $R\,d\varphi'$ (Fig. 15.2.3):

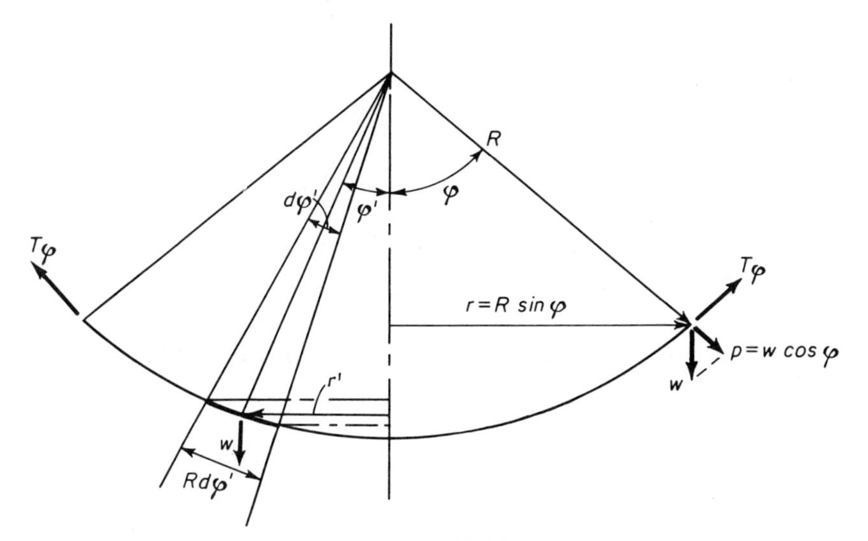

FIGURE 15.2.3

$$dW = (2\pi r')(R\,d\varphi')w = 2\pi w R^2 \sin\varphi'\,d\varphi',$$

$$W = \int_0^{\varphi} dW = 2\pi R^2 w \int_0^{\varphi} \sin\varphi'\,d\varphi' = 2\pi w R^2(1 - \cos\varphi). \qquad \text{(c)}$$

Vertical equilibrium requires that:

$$2\pi(R\sin\varphi)T_\varphi \sin\varphi = 2\pi w R^2(1 - \cos\varphi),$$

$$\therefore \quad T_\varphi = wR\frac{1-\cos\varphi}{\sin^2\varphi} = wR\frac{1-\cos\varphi}{1-\cos^2\varphi}$$

$$= wR\frac{1-\cos\varphi}{(1-\cos\varphi)(1+\cos\varphi)} = \frac{wR}{1+\cos\varphi}. \qquad (15.2.8)$$

T_φ is tensile everywhere, is maximum at $\varphi = 90°$, where it equals wR, and minimum at $\varphi = 0°$, where it equals $\frac{1}{2}wR$.

By (15.2.5), with $p = w\cos\varphi$, and (15.2.8):

$$T_\theta = wR\left(\cos\varphi - \frac{1}{1+\cos\varphi}\right). \qquad (15.2.9)$$

T_θ is tensile when:

$$\cos\varphi \geq \frac{1}{1+\cos\varphi}, \quad \text{i.e.,} \quad \cos^2\varphi + \cos\varphi - 1 \geq 0,$$

$$\cos\varphi \geq -\frac{1}{2} + \sqrt{\left(\frac{1}{2}\right)^2 + 1} = -\frac{1}{2} + \frac{\sqrt{5}}{2} = 0.63 \quad \therefore \quad \varphi \leq 52°.$$

Examples 15.2.2 and 15.2.3 show that, provided the opening angle $\varphi_0 < 45°$, a spherical membrane can carry both a uniform live load and a uniform dead load without changing shape.

Example 15.2.4. A conical membrane of opening angle α carries a load W at its vertex (Fig. 15.2.4). Determine $T_\varphi \equiv T_x$ and T_θ.

By vertical equilibrium:

$$(2\pi r)T_x \cos\alpha = (2\pi x \sin\alpha)T_x \cos\alpha = W,$$

$$\therefore \quad T_x = \frac{W}{2\pi x \sin\alpha \cos\alpha} = \frac{1}{\pi x}\frac{W}{\sin 2\alpha}. \tag{15.2.10}$$

By (15.1.4), with $p = 0$ (since there is no distributed load on the membrane) and $R_2 = x\tan\alpha$:

$$T_2 \equiv T_\theta = px\tan\alpha = 0.$$

T_x is tensile and becomes theoretically infinite at $x = 0$, since W is a finite load supported at a point, i.e., on "zero area."

Example 15.2.5. A conical membrane of opening α carries its own dead load w kN/m² (psf). Determine T_x and T_θ (Fig. 15.2.5).

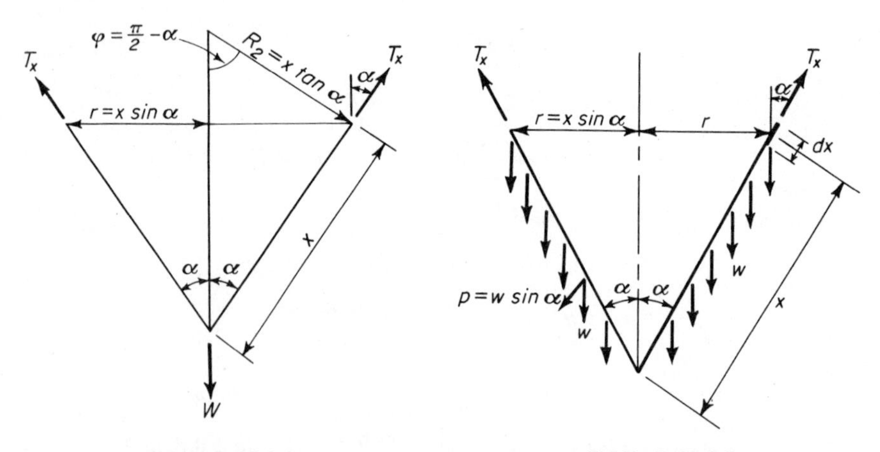

FIGURE 15.2.4 FIGURE 15.2.5

The weight of the cone up to a parallel x equals:

$$W = w\int_0^x 2\pi r\,dx = 2\pi w\int_0^x x\sin\alpha\,dx = 2\pi w\frac{x^2}{2}\sin\alpha = \pi w x^2 \sin\alpha.$$

Vertical equilibrium requires that:

$$(2\pi r)T_x\cos\alpha = 2\pi x\sin\alpha\cos\alpha\,T_x = \pi w x^2\sin\alpha,$$

$$\therefore \quad T_x = \frac{1}{2\cos\alpha}wx. \tag{15.2.11}$$

By (15.1.4), with $p = w\sin\alpha$, $R_2 = x\tan\alpha$:

$$T_2 \equiv T_\theta = (x\tan\alpha)(w\sin\alpha) = \frac{\sin^2\alpha}{\cos\alpha}wx. \tag{15.2.12}$$

Both T_x and T_θ are tensile everywhere.

Example 15.2.6. A shallow, spherical membrane of radius R, loaded by an internal pressure p, has a base radius r_0 and a rise h (Fig. 15.2.6). Determine the total weight of the membrane material needed to resist the pressure p if the material unit weight is γ and its allowable stress is F_t.

Equation (15.2.2), with $T_\varphi = T_\theta$ gives:

$$T_\varphi = T_\theta = T = \tfrac{1}{2}pR. \tag{d}$$

As shown by (d), the tension T in the membrane is the same in all directions. The area of material per unit length (Fig. 15.2.7) required to resist this force is T/F_t, and the volume of material per unit area is $1 \times T/F_t$. Hence, its weight w_m per unit area is:

$$w_m = \frac{1}{2}pR\left(\frac{\gamma}{F_t}\right).$$

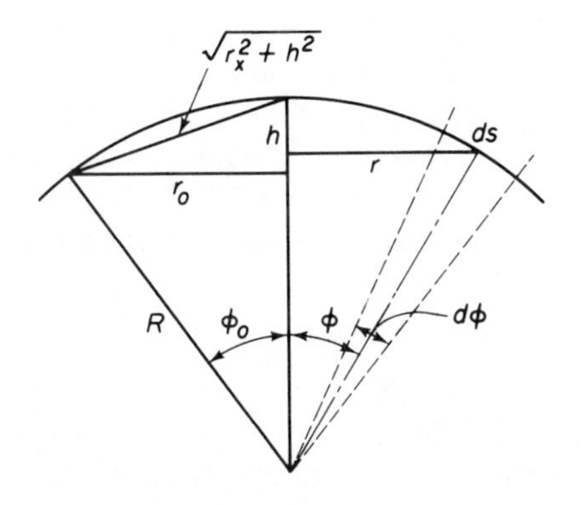

FIGURE 15.2.6

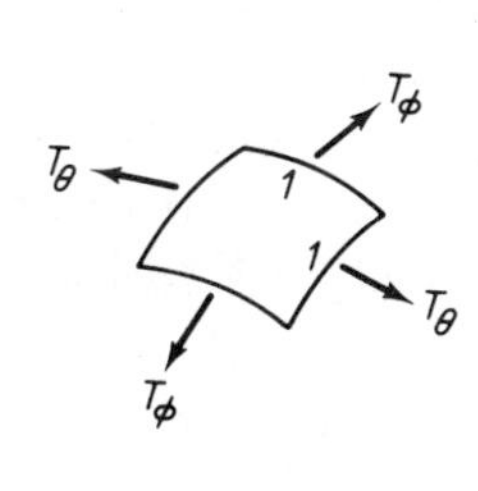

FIGURE 15.2.7

The total weight of the membrane W_m is:

$$W_m = \frac{1}{2}pR\frac{\gamma}{F_t}A_s,$$

where A_s is the area of the spherical sector; by integration, we obtain (Fig. 15.2.6):

$$A_s = \int_0^{\varphi_0} 2\pi rR\,d\varphi = \int_0^{\varphi_0} 2\pi(R\sin\varphi)R\,d\varphi = 2\pi R^2\int_0^{\varphi_0}\sin\varphi\,d\varphi$$
$$= 2\pi R^2(1 - \cos\varphi_0) = 2\pi R\cdot R(1 - \cos\varphi_0) = 2\pi Rh.$$

But, by Pythagoras' theorem,

$$R^2 = r_0^2 + (R - h)^2 = r_0^2 + R^2 - 2Rh + h^2 \quad \therefore \quad R = \frac{r_0^2 + h^2}{2h},$$

so that:

$$A_s = 2\pi\frac{r_0^2 + h^2}{2h}h = \pi(r_0^2 + h^2). \tag{e}$$

The total weight of the membrane is, therefore:

$$W_m = \frac{1}{2}p\frac{\gamma}{F_t}\frac{r_0^2 + h^2}{2h}\pi(r_0^2 + h^2) = \frac{\pi}{4}\frac{p}{F_t}\gamma r_0^3\left(\frac{r_0}{h} + 2\frac{h}{r_0} + \frac{h^3}{r_0^3}\right)$$
$$\approx \frac{\pi}{4}\frac{p}{F_t}\gamma r_0^3\left(\frac{r_0}{h} + 2\frac{h}{r_0}\right). \tag{f}$$

If instead of a uniform membrane, we consider the case of a cable-reinforced fabric, the cable material is needed separately in the two directions φ and θ to resist the forces given by (d) in the two directions. The total weight of cable W_c required is therefore twice that given by (f), where F_t is now the allowable tensile stress in the cables and γ_c their unit weight:

$$W_c = \frac{\pi}{2}\frac{p}{F_t}\gamma_c r_0^3\left(\frac{r_0}{h} + 2\frac{h}{r_0}\right). \tag{g}$$

15.3 STABILIZED MEMBRANES [11.1]

Membranes can be stabilized by *pretensioning*, i.e., by stretching them to such initial tensile stresses that the maximum compressive stresses due to the load will always be smaller than the initial tensile stresses. Pretensioning may be obtained by external forces applied to the boundary in open membranes or by internal pressure in closed membranes. It is readily seen that pretensioning by external forces on the membrane boundary can only be applied to *saddle* surfaces.

In a pretensioned unloaded membrane, the external normal load p is zero; hence, by (15.1.3):

$$\frac{T_1}{R_1} + \frac{T_2}{R_1} = 0 \quad \therefore \quad \frac{T_1}{T_2} = -\frac{R_1}{R_2}. \tag{15.3.1}$$

Since T_1 and T_2 are both tensile, i.e., positive, the left-hand member of (15.3.1) is positive and, for the right-hand member to be also positive, the signs of R_1 and R_2 must be opposite; one of the principal curvatures must be upward and the other downward.

Example 15.3.1. A translational *saddle* surface is obtained by translating on one another two identical, vertical arcs of circles of radius R at right angles to each other. Its surface is flat enough for its vertical dead and live loads to be practically normal to the surface, i.e., for $w + q$ to be considered a normal load p. Determine the pretensioning force T necessary to stabilize the membrane.

With $R_1 = -R_2 = R$, the total load carried by the unit forces T_1', T_2' in the unprestressed membrane is given by (15.1.3):

$$p = \frac{1}{R}(T_1' - T_2'). \tag{a}$$

By symmetry $T_2' = -T_1'$, i.e., the principal unit forces are equal and of opposite sign, so that:

$$T_1' = -T_2' = \tfrac{1}{2}pR. \tag{15.3.2}$$

Hence, the unprestressed saddle surface cannot carry any load without developing a compression $T_2' = -\frac{1}{2}pR$.

By (15.3.1), with $R_1 = -R_2$, the prestressing unit forces T_1'', T_2'' are equal:

$$T_1'' = T_2'' = T''. \tag{b}$$

The minimum prestressing tension capable of wiping out the compression due to the load is:

$$T'' = \tfrac{1}{2}pR, \tag{15.3.3}$$

and the membrane forces after prestressing become:

$$T_1 = T_1' + T'' = pR, \qquad T_2 = T_2' + T'' = 0. \tag{15.3.4}$$

For a given, allowable force T_1, it is seen that half the tensile capacity of the membrane is used to prestress it and half to carry the load. In order to have a safety margin against instability, T'' must be made greater than $\frac{1}{2}pR$.

Pretensioning by internal pressure may be used whenever the membrane encloses a space. Pool and tennis court enclosures, and domes covering radar antennas (radomes) or large stadiums are often built as membranes stabilized by internal pressure.

Example 15.3.2. A hemispherical dome of dead load w kN/m² (psf) with a live load q uniform in horizontal projection must be stabilized by an internal pressure p. Determine the value of p with a safety factory of $\frac{4}{3}$.

The tensile forces due to p are, by (15.2.2), with $T_\varphi = T_\theta$:

$$T_\varphi' = \tfrac{1}{2}pR, \qquad T_\theta' = \tfrac{1}{2}pR. \tag{c}$$

Due to its *downward* curvatures, the dome develops maximum membrane forces due to w, equal and *opposite* to those given in (15.2.8) and (15.2.9):

$$\begin{aligned} T_\varphi'']_{\varphi=90^\circ} &= -wR, \qquad & T_\varphi'']_{\varphi=0^\circ} &= -\tfrac{1}{2}wR, \\ T_\theta'']_{\varphi=90^\circ} &= wR, \qquad & T_\theta'']_{\varphi=0^\circ} &= -\tfrac{1}{2}wR. \end{aligned} \tag{d}$$

Similarly, the maximum compressive forces due to q are equal and opposite to those given by (15.2.6), (15.2.7):

$$\begin{gathered} T_\varphi''']_{\varphi=0^\circ} = T_\varphi''']_{\varphi=90^\circ} = -\tfrac{1}{2}qR, \qquad T_\theta''']_{\varphi=0^\circ} = -\tfrac{1}{2}qR, \\ T_\theta''']_{\varphi=90^\circ} = +\tfrac{1}{2}qR. \end{gathered} \tag{e}$$

Hence, the maximum required prestressing pressure with a safety factor of $\frac{4}{3}$ is given by the condition of no compressive T_φ at $\varphi = 90^\circ$:

$$\tfrac{1}{2}pR - \tfrac{4}{3}(wR + \tfrac{1}{2}qR) = 0 \quad \therefore \quad p = \tfrac{4}{3}(2w + q). \tag{15.3.5}$$

The maximum tensile forces in the membrane are:

$$\begin{aligned} T_\varphi]_{\varphi=0^\circ} &= \tfrac{1}{2} \times \tfrac{4}{3}(2w + q)R - \tfrac{1}{2}wR - \tfrac{1}{2}qR = \tfrac{1}{6}(5w + q)R, \\ T_\theta]_{\varphi=90^\circ} &= \tfrac{1}{2} \times \tfrac{4}{3}(2w + q)R + wR + \tfrac{1}{2}qR = \tfrac{7}{6}(2w + q)R. \end{aligned}$$

For example, a membrane covering a circular area of 15 m (50 ft) radius, and supporting a snow load $q = 2$ kN/m² (40 psf), and a negligible dead load, requires a prestressing pressure of:

$$p = \tfrac{4}{3}(2) = 2.7 \text{ kN/m}^2 = 0.002\,7 \text{ MPa (0.39 psi)}$$

and develops a maximum tensile force:

$$T_\theta]_{\varphi=90^\circ} = \tfrac{7}{6}(2 \times 15) = 35 \text{ kN/m} = 35 \text{ N/mm (200 lb/in.)}.$$

From Table 2.7.2, the ultimate strength of Teflon-coated fiberglass is 122.5 N/mm (700 lb/in.). This results in a factor of safety for the membrane of:

$$\frac{122.5}{35} = 3.5.$$

Example 15.3.3. A balloon built by means of two flat, identical, spherical sectors with a radius of 50 m (160 ft) and an opening angle of 30°, supports a live load $q = 1.5$ kN/m² (30 psf) and is to be prestressed with a safety factor of $\frac{4}{3}$ (Fig. 15.3.1). The two halves of the balloon are tied to a compression ring by radial cables. Determine the prestressing pressure, the maximum tension in the balloon, and the compression in the ring, neglecting the dead load of the balloon.

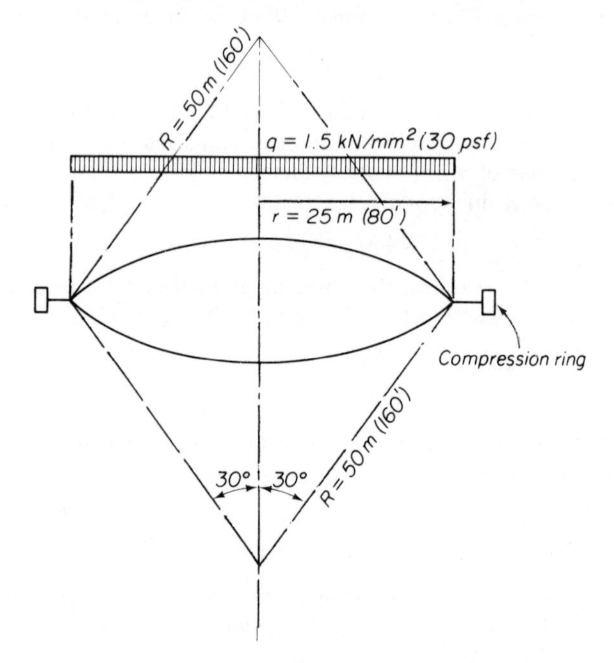

FIGURE 15.3.1

By (15.2.6), (15.2.7), and (15.2.2) the maximum forces in the loaded upper half of the balloon are:

$$T_\varphi = -\tfrac{1}{2}qR, \qquad T_\theta]_{\max} = T_\theta]_{\varphi=0^\circ} = -\tfrac{1}{2}qR,$$
$$T'_\varphi = +\tfrac{1}{2}pR, \qquad T'_\theta = +\tfrac{1}{2}pR.$$

Hence, the condition of no tensile force at $\varphi = 0^\circ$ requires that:

$$\tfrac{1}{2}pR = (\tfrac{4}{3})(\tfrac{1}{2})qR \quad \therefore \quad p = \tfrac{4}{3}q.$$

The maximum tension in the unloaded sector is, by (15.2.2):

$$T'_\varphi = +\tfrac{1}{2}(\tfrac{4}{3}q)R = +\tfrac{2}{3}qR.$$

The resultant of the horizontal components of the two T'_φ's at the rims of the upper and lower sectors is maximum in the absence of live load and equals:

$$H = 2T'_\varphi \cos 30° = 2(\tfrac{2}{3}qR)\cos 30° = 1.15qR.$$

The compression in the ring of radius r_0 (Fig. 15.3.2) is given by:

$$C = Hr_0 = 1.15qR^2 \sin \varphi_0.$$

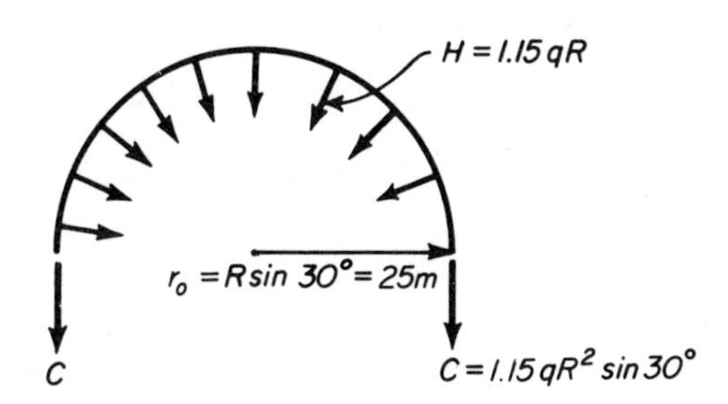

FIGURE 15.3.2

For the given balloon:

$$p = \tfrac{4}{3}(1.5) = 2 \text{ kN/mm}^2 = 0.002 \text{ MPa } (0.28 \text{ psi}),$$

$$T'_\varphi]_{\max} = \tfrac{2}{3} \times 1.5 \times 50 = 50 \text{ kN/m } (3{,}200 \text{ lb/ft}),$$

$$C = 1.15 \times 1.5 \times 50^2 \times 0.5 = 2\,156 \text{ kN } (442 \text{ k}).$$

As shown by the following examples, stabilized membranes may be used as structural elements, like beams and columns. They offer the advantage of a minimum dead load when built out of plastic-coated fabrics only a few millimeters thick, with an ultimate tensile strength of as much as 122.5 N/mm (700 lb/in.).*

Example 15.3.4. A membrane balloon in the shape of a circular cylinder of radius R, with a hemispherical head at one end, is used as a cantilever beam of length l to support a lighting fixture weighing W kg (lb). The membrane thickness is h. Determine the prestressing pressure in the balloon with a factor of safety of 2.

Neglecting the weight of the balloon, the maximum beam moment is $M = Wl$, and the maximum compressive stress at the bottom of the fixed-end with $S = \pi h R^2$ [see Table 4.1.1 (10)] is:

$$f_c = \frac{1}{\pi}\frac{Wl}{hR^2}. \tag{f}$$

The longitudinal stress f_t due to a pressure p in the spherical head of the balloon is obtained from (15.2.2):

$$T = f_t h = \frac{1}{2}pR \quad \therefore \quad f_t = \frac{pR}{2h}. \tag{g}$$

*See Table 2.7.2.

The prestressing pressure with a safety factor of 2 is given by the condition:

$$\frac{pR}{2h} = 2\frac{1}{\pi}\frac{Wl}{hR^2} \quad \therefore \quad p = \frac{4}{\pi}\frac{Wl}{R^3}, \tag{15.3.6}$$

and is independent of h.

For example, for $W = 50$ kg (100 lb), $l = 1.5$ m (5 ft), $R = 0.15$ m ($\frac{1}{2}$ ft):

$$p = \frac{4}{\pi} \times \frac{50 \times 9.807 \times 1.5}{(0.15)^3} = 277\,500 \text{ N/m}^2 = 0.278 \text{ MPa (35 psi)},$$

and, by (15.1.4), the hoop stress in a ballon with $h = 1$ mm (0.02 in.) equals:

$$f_t]_{\text{hoop}} = p\frac{R}{h} = \frac{277\,500 \times 150}{1.0} = 41\,700\,000 \text{ N/m}^2 = 41.70 \text{ MPa (10.6 ksi)}.$$

The maximum stress in the spherical head equals:

$$f_{t,s} = \frac{1}{2}\frac{277\,500 \times 150}{1.0} = 20\,850\,000 \text{ N/m}^2 = 20.85 \text{ MPa (5.3 ksi)}.$$

Example 15.3.5. A membrane balloon of cylindrical shape of radius R and thickness h, closed at each end by a plate, is to be used as a column to carry a centered load W. Determine the prestressing pressure p with a safety factor of 2.

The internal pressure p produces an axial tensile force $p\pi R^2$ and a tensile stress in the cylinder:

$$f_t = \frac{p\pi R^2}{2\pi Rh} = \frac{pR}{2h}. \tag{h}$$

The load W produces a compressive stress:

$$f_c = -\frac{W}{2\pi Rh}. \tag{i}$$

Hence, the prestressing pressure with a safety factor of 2 is given by the condition:

$$\frac{pR}{2h} = 2\frac{W}{2\pi Rh} \quad \therefore \quad p = \frac{2W}{\pi R^2}. \tag{15.3.7}$$

For example, for $W = 500$ kg (1,000 lb), $R = 0.15$ m ($\frac{1}{2}$ ft), $h = 0.3$ mm (0.01 in.):

$$p = \frac{2 \times 500 \times 9.807}{3.14 \times 0.15^2} = 139\,000 \text{ N/m}^2 = 0.139 \text{ MPa (17.6 psi)},$$

$$f_t]_{\text{long}} = \frac{2W - W}{2\pi Rh} = \frac{500 \times 9.807}{2\pi \times 0.15 \times 1\,000 \times 0.3} = 17.35 \text{ MPa (2.67 ksi)},$$

$$f_t]_{\text{hoop}} = \frac{0.139 \times 0.15 \times 1\,000}{0.3} = 69.5 \text{ MPa (10.58 ksi)}.$$

A membrane was shown to carry load if curved, but an initially flat prestressed (or unprestressed) membrane deflects under load, and the deflection introduces curvatures sufficient to allow the membrane to carry the load, as shown by the following examples.

Example 15.3.6. Determine the uniform pressure p carried by a circular, flat membrane of radius r, initially stretched by a uniform tension T, assuming the deflection δ at the center of the plate due to p to be small in comparison with r.

Under the pressure p, the membrane deflects into a surface which can be approximated by a sector of a sphere of radius R (Fig. 15.3.3). For very small angles θ:

$$r = R \sin \theta \approx R\theta, \qquad \delta = R(1 - \cos \theta) \approx R\frac{\theta^2}{2}, \tag{j}$$

so that:

$$\theta = \frac{r}{R}, \qquad \delta = \frac{R}{2}\left(\frac{r}{R}\right)^2 = \frac{r^2}{2R} \quad \therefore \quad R = \frac{r^2}{2\delta}. \tag{k}$$

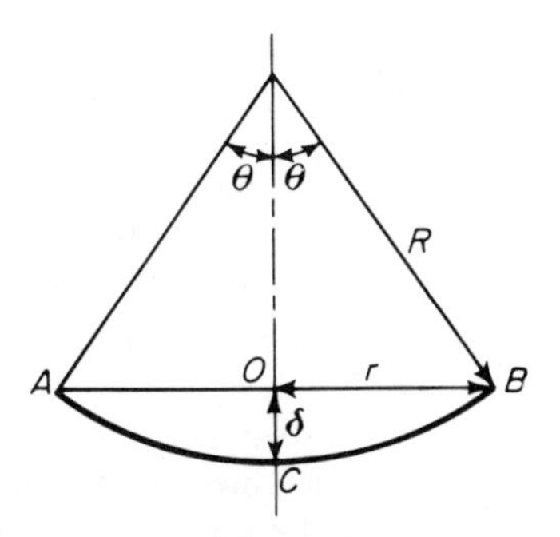

FIGURE 15.3.3

Neglecting the *increase* in T due to the stretching of the membrane into a shallow nondevelopable surface, (15.2.2), with $T_\varphi = T_\theta$, gives:

$$T_\varphi = T_\theta = T = \frac{1}{2}pR = \frac{1}{2}p\frac{r^2}{2\delta} = p\frac{r^2}{4\delta}.$$

The pressure p is thus given by:

$$p = \frac{4T}{r^2}\delta. \tag{15.3.8}$$

and is proportional to δ.

Equation (15.3.8) may also be used to evaluate approximately the uniform load q carried by the membrane, provided $\delta \ll r$.

Example 15.3.7. The membrane of Example 15.3.6 is originally unstretched and fixed at its boundary. Evaluate the pressure p carried when the membrane center displaces by δ.

In this case, the membrane stresses are due exclusively to the stretching of the membrane into a shallow nondevelopable surface (Fig. 15.3.3). The average strain in the radial direction ϵ_φ is due to the stretching of the radius $OB = r =$

$R \sin \theta$ into the arc $CB = R\theta$. Approximating $\sin \theta$ by $\theta - (\theta^3/6)$:

$$\epsilon_\varphi = \frac{R\theta - R\sin\theta}{R\theta} = \frac{\theta - (\theta - \theta^3/6)}{\theta} = \frac{\theta^2}{6},$$

or, with $\theta^2 = 2\delta/R$ as given by (j):

$$\epsilon_\varphi = \frac{\delta}{3R}. \tag{l}$$

The radial tension T_φ is thus:

$$T_\varphi = f_\varphi h = E\epsilon_\varphi h = \frac{Eh\delta}{3R}.$$

The stretching in the circumferential direction is negligible, since the points of the membrane move essentially downward and not horizontally. Hence, by (k) and (15.2.2), with $T_\theta = 0$:

$$p = \frac{T_\varphi}{R} = \frac{Eh\delta}{3R^2} = \frac{Eh\delta}{3(r^2/2\delta)^2} = \frac{4}{3}\frac{Eh}{r^4}\delta^3 \tag{15.3.9}$$

In this case, p is proportional to δ^3. Equation (15.3.9) may also be used to approximate the uniform load q carried by the membrane, provided $\delta \ll r$.

Example 15.3.8. The middle plane of a plate deflecting under load into a non-developable surface is bound to stretch (see Sections 12.1 and 14.1) and, hence, to develop membrane stresses. For the case of a circular plate under a uniform load q with a boundary restrained from moving inward, the radial strain ϵ_φ in the middle surface of the plate is given by (l), and a fraction q_m of the load q, approximately given by (15.3.9), is carried by membrane stresses.

To evaluate the ratio of q_m to the load q_b carried by bending stresses, we notice that the center deflection δ of the plate due to bending action under the load q_b (for Poisson's ratio equal to zero) may be shown to equal:

$$\delta = \frac{15}{16}\frac{q_b r^4}{Eh^3}, \tag{m}$$

from which:

$$q_b = \frac{16}{15}\frac{Eh^3}{r^4}\delta. \tag{n}$$

By (15.3.9), the load q_m carried by membrane stresses is approximately:

$$q_m = \frac{4}{3}\frac{Eh}{r^4}\delta^3. \tag{o}$$

The ratio of q_m to q_b thus becomes:

$$\frac{q_m}{q_b} = \frac{5}{4}\left(\frac{\delta}{h}\right)^2. \tag{15.3.10}$$

Since (n) is only valid for $\delta < h$, (15.3.10) also is only valid for $\delta < h$. For example, for $\delta/h = \frac{1}{2}$, $q_m/q_b = \frac{5}{16}$; for $\delta/h = \frac{1}{5}$, $q_m/q_b = \frac{1}{20}$.

When a plate is stressed beyond the elastic limit, its deflections are large, and the fraction of load carried by membrane stresses is larger than that given by (15.3.10).

PROBLEMS

15.1 A spherical membrane spans a distance of 30 m (100 ft) with a sag of 3 m (10 ft). It carries a snow load of 1.5 kN/m² (30 psf) and a dead load of 0.8 kN/m² (15 psf). Determine the maximum T_φ and T_θ unit forces in the membrane. If a radial and circumferential cable roof is substituted for the membrane, what are the required sections for the meridional and parallel cables at the boundary, if both sets of cables are spaced by 3 m (10 ft) there? What is the cross-sectional area A_c of the concrete compression ring needed to support this roof for $f'_c = 30$ MPa (4,350 psi) assuming the ring is continuously supported vertically?

15.2 The roof of Problem 15.1 has a parabolic shape. Determine $T_{\varphi_{max}}$, $T_{\theta_{max}}$, and A_c.

15.3 A cylindrical wood silo 3 m (10 ft) in diameter and 9 m (30 ft) high is filled with wheat. Determine the areas of the four A36 steel hoops, spaced by 3 m (10 ft), needed to equilibrate the outward pressure of the wheat, assuming wheat to weigh 6.6 kN/m² (42 lb/cu ft) and the outward pressure to be equal to one-third of the vertical wheat pressure. Assume the wood staves to be simply supported between hoops.

15.4 A vertical concrete tank, 3 m (10 ft) in radius and 6 m (20 ft) high is filled with water. Determine the required hoop reinforcing at 1.5 m (5-ft) intervals from the bottom. Assume grade 40 steel.

15.5 An inverted conical roof is supported on an A36 steel ring 6 m (20 ft) in radius and carries a total load of 2 kN/m² (40 psf). It has a roof sag of 3 m (10 ft) and a central hole 1.5 m (5 ft) in radius. Determine the maximum T_x and T_θ in the roof and the cross-sectional area of the upper support ring.

15.6 A plastic membrane roof covers a circular area and has an elliptic cross section. The radius of the covered area is 6 m (20 ft); the height of the roof is 4.5 m (15 ft). The roof is supported by an internal pressure of 0.01 kN/m² (0.2 psf) and carries no external load. Determine the maximum T_φ and T_θ in the membrane, and establish its thickness for $F_t = 40$ MPa (6,000 psi).

15.7 A nylon balloon in the shape of a half sphere of radius 4.5 m (15 ft) is used as a formwork to build a concrete hemispherical dome 75 mm (3 in.) thick. Determine the internal pressure in the balloon required to support the concrete before it sets, with a safety factor of 2.5.

15.8 A nylon balloon is used as a form to build a 75 mm (3 in.) thick cylindrical concrete roof, spanning 9 m (30 ft) with a rise of 3 m (10 ft). Determine the internal pressure of the balloon with a safety factor of 2.5.

15.9 A tent in the shape of a hyperbolic paraboloid is hung from two poles 6 m (20 ft) high, set at opposite corners of a square 12 m (40 ft) on a side. The tent is tied down at the other two opposite corners of the square. Determine the tensile forces at the tied-down corners if the tent carries a load of 1 kN/m² (20 psf) of horizontal projection. (Assume the tensile forces to be the resultants of uniform shears along the sides of the tent.)

15.10 A circular membrane roof 30 m (100 ft) in radius and supported along its edge is stretched by a uniform radial tension of 1.5 kN/m (1 k/ft). Determine its sag under a uniform load of 0.8 kN/m² (15 psf) if the tension is assumed to remain constant.

15.11 The roof of Problem 15.10 is originally unstretched. Its material has a modulus $E = 13\,800$ MPa (2,000 ksi). Determine its sag if the material is 3 mm ($\frac{1}{8}$ in.) thick.

15.12 A closed cylindrical balloon 3 m (10 ft) long, 300 mm (1 ft) in diameter, and 0.1 mm (0.004 in.) thick is inflated to a pressure of 0.02 MPa (3 psi). Determine its hoop and longitudinal stresses. What uniform wind pressure would collapse the balloon if it were cantilevered in a vertical position?

15.13 A rubber raft is supported by two longitudinal cylindrical balloons 300 mm (1 ft) in diameter and 2.4 m (8 ft) long. Determine the load carried by the raft when the balloons are barely submerged under the water surface, the internal pressure needed to stabilize the balloons, and their thickness if $F_t = 13.8$ MPa (2,000 psi).

15.14 A sail of a square rigged sailing ship is tied to top and bottom wood yards 6 m (20 ft) apart. Under a wind of 64 km/hr (40 mph) the sail bellies out 1.2 m (4 ft) and takes a parabolic shape. Determine the force per meter (foot) exerted by the sail on the yards and the yards' cross section if they are 9 m (30 ft) long and tied to the mast at midspan.

15.15 Three membranal rotational roofs, one spherical, one parabolic, and one conical, have the same radius and sag and carry the same load per square foot of horizontal projection. List the roofs in order of increasing stresses T_φ and T_θ.

15.16 A spherical membrane spans a distance l and carries a load q per square meter (square foot) in horizontal projection, with a coefficient of safety of 3 when its sag is $l/3$. What is its coefficient of safety if its sag is reduced to $l/4$? For what sag will the membrane fail?

15.17 A spherical membrane roof of radius r and sag h carries a load q kN/m² (psf) in horizontal projection and is supported by a compression ring of area A with a coefficient of safety of 2.5. What reduction in sag will produce the collapse of the ring in compression?

15.18 A circular horizontal membrane of radius r is originally unstretched and carries a load p kN/m² (psf) with a center deflection δ. What additional uniform tension T must be added to the membrane to cut the center deflection in half?

15.19 A simply supported circular plate spanning 6 m (20 ft) is uniformly loaded by a load of 7.5 kN/m² (150 psf). By code, its center deflection must be at most

1/360 of the span; its thickness is usually 1/40 of the span. What share of the load is carried by membrane stresses and what share by bending stresses?

15.20 A load of 90 kN (20 k) hangs from the vertex of a conical membrane which is supported by a compression ring. The ring is 6 m (20 ft) in diameter and the cone vertex is 1.5 m (5 ft) below the ring. Determine T_φ and T_θ in the membrane. Substitute, for the membrane, steel wires equally spaced by 0.75 m (2.5 ft) around the ring and determine their diameter for $F_t = 550$ MPa (80,000 psi). How much can the point of support of the load be lifted before the cables snap if $F_u = 1\ 380$ MPa (200,000 psi)?

16. Membrane Stresses in Shells

16.1 THIN SHELLS [12.1]

A thin shell is a curved membrane thin enough to develop negligible bending stresses over most of its surface but thick enough not to buckle under small compressive stresses, as an ideal membrane would. Under load, a thin shell develops *membrane stresses*, i.e., tensile, compressive, and *tangential* shear stresses. A thin shell is *stable* under any smooth load which does not overstress it, since it does not have to change shape to avoid the development of compressive stresses.

Thin shells are made of materials such as metals, wood, and plastics, capable of resisting compressive and, at times, tensile stresses, but reinforced concrete is an ideal material for thin shell construction because of the ease with which concrete is poured or sprayed into curved shapes.

In order to develop membrane stresses over most of its surface, a thin shell must be *properly supported.* A proper support is one which: (a) develops *membrane* reactions, i.e., reactions acting in the plane tangent to the shell at the boundary, and (b) allows at the boundary of the shell membrane displacements, i.e., the displacements developed by the strains due to the membrane stresses. If the support reactions are not tangent to the shell or if membrane displacements are prevented by the supports, the shell develops bending stresses (usually in the neighborhood of the boundary), which are referred to as bending *boundary disturbances.* If the shell shape and the support conditions are both incorrectly chosen, the shell may develop bending stresses all over its surface. Such an improperly designed "shell" cannot act as a thin shell, i.e., does not support most of the load by membrane stresses.

An illustration of the influence of the support conditions on thin shell stresses is given by the behavior of a cylindrical barrel under its own dead load. When a long cylindrical barrel, i.e., a barrel with a length at least twice its width, is supported at its ends on stiff frames [Fig. 16.1.1 (a)], it acts essentially as a beam of semicircular cross section and develops longitudinal stresses f_x distributed linearly across its depth, tangential shears $f_{x\varphi}$, and hoop stresses f_φ which vanish along its longitudinal boundary [Fig. 16.1.1 (b)]. The membrane stresses f_x, f_φ, and $f_{x\varphi}$ are capable of carrying the load without the development of bending stresses and lateral thrust. Moreover, the tangential shear mechanism (see Section 15.1) makes the barrel funicular for any smooth load.

The same barrel supported along its longitudinal boundaries (Fig. 16.1.2) acts, instead, as a series of identical semicircular arches and, since the circle is not the funicular of the dead load (see Section 9.1), the shell develops a thrust and, besides the hoop stress f_φ, bending stresses f_b and *transverse* shear stresses f_{sx} all over its surface. Even if the longitudinally supported barrel has a catenary cross section which is funicular for the dead load, it cannot be funicular for any other load and is bound to develop bending stresses under, for example, a snow load.

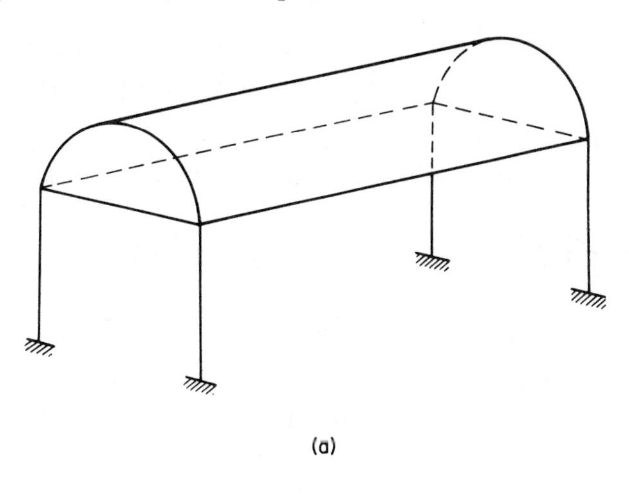

(a)

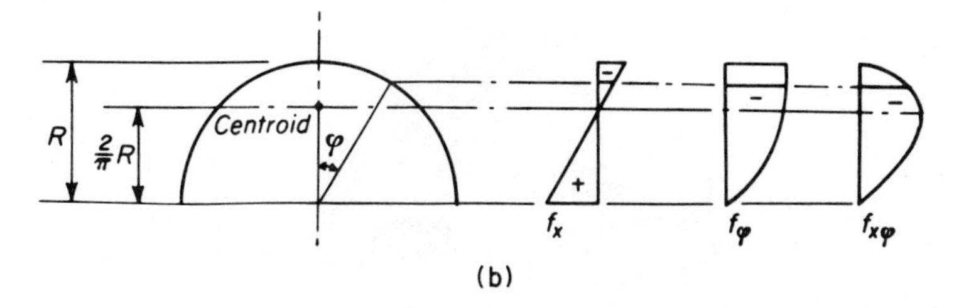

(b)

FIGURE 16.1.1

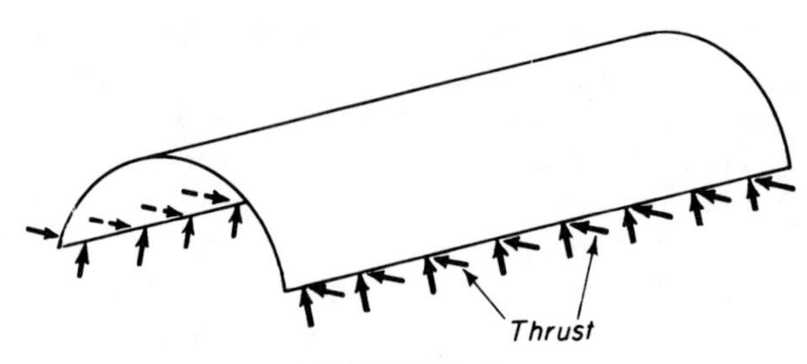

FIGURE 16.1.2

The bending stresses developed by the prevention of boundary displacements are illustrated by the behavior of a hemispherical dome, which under its dead load has a tendency to expand at its equator [Fig. 16.1.3 (a)]. The prevention of this radial displacement requires a radial horizontal reaction H (N/mm, lb/in.) which is not tangent to the shell and, hence, develops bending stresses around its boundary [Fig. 16.1.3 (b)]. Finally, a thin shell cannot carry concentrated loads by membrane stresses, since the deformation of the shell under the load involves local curvatures and, hence, bending stresses (Fig. 16.1.4).

It is thus seen that a thin shell will act properly, i.e., carry most of its load by membrane stresses only, if it is *thin*, *properly shaped*, and *correctly supported.*

In general, membrane stresses are so small that in most cases the thickness of a shell is determined by the bending boundary disturbances. Even

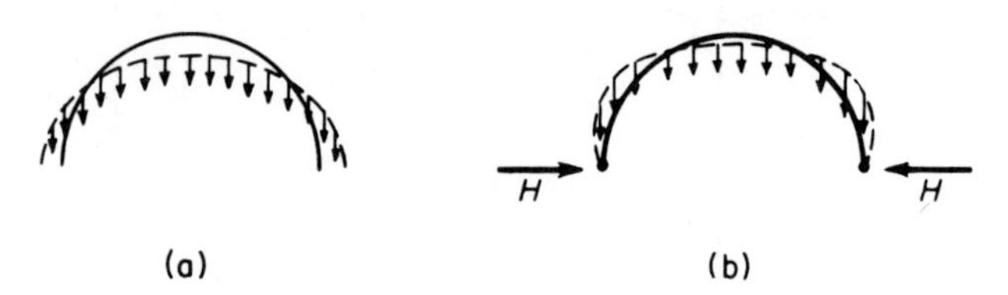

FIGURE 16.1.3

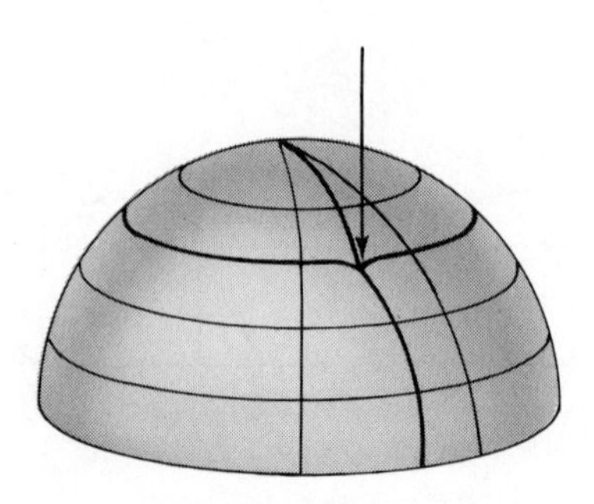

FIGURE 16.1.4

so, one must evaluate membrane stresses in order to: (a) determine where tensile stresses may develop and provide adequate tensile reinforcement if the shell is made out of an essentially compressive material, (b) determine the highest compressive stress and check buckling, (c) determine membrane boundary displacements and the bending stresses developed by their partial or total prevention.

16.2 MEMBRANE STRESSES IN AXISYMMETRICAL SHELLS [12.7]

The membrane forces per unit length in a shell of revolution, symmetrically loaded and symmetrically supported with respect to its axis, are analogous to those in a rotational membrane.

The tangential shears S in the meridional and parallel directions vanish since, by symmetry, there is no tendency for adjoining sections to slide in those directions. The meridional and parallel unit forces are indicated by N_φ and N_θ, respectively (Fig. 16.2.1). (These unit forces are sometimes referred to as membrane stresses, although they have dimensions of force per unit length.) The membrane unit forces N_φ and N_θ are obtained by the membrane equation (15.2.1):

$$\frac{N_\varphi}{R_1} + \frac{N_\theta}{R_2} = p, \tag{16.2.1}$$

and by (15.2.3), stating the vertical equilibrium of the shell sector above the parallel φ (Fig. 16.2.1):

$$N_\varphi = \frac{Q_z}{2\pi R_2 \sin^2\varphi}, \tag{16.2.2}$$

where p is the normal component of the load q (MPa, psi), Q_z (kN, lb) is the resultant of all loads above the parallel φ (which is vertical by symmetry), and N_φ, N_θ (kN/mm, lb/in.) are assumed *positive when compressive* (Fig.16.2.1).

From the results in Section 15.2 we obtain the following unit forces, which are identical to those in ideal membranes but which may now have either positive (compressive) or negative (tensile) values.

A. *Spherical Shell* (Fig. 16.2.2):

1. Dead load w:

$$N_\varphi = \frac{wR}{1+\cos\varphi}, \qquad N_\theta = wR\left(\cos\varphi - \frac{1}{1+\cos\varphi}\right), \tag{16.2.3}$$

$$N_\varphi]_{\max} = wR, \qquad N_\theta]_{\max} = \tfrac{1}{2}wR, \qquad N_\theta]_{\min} = -wR. \tag{16.2.4}$$

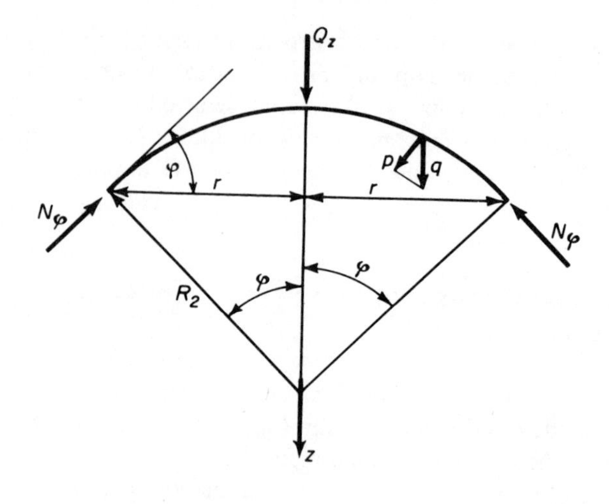

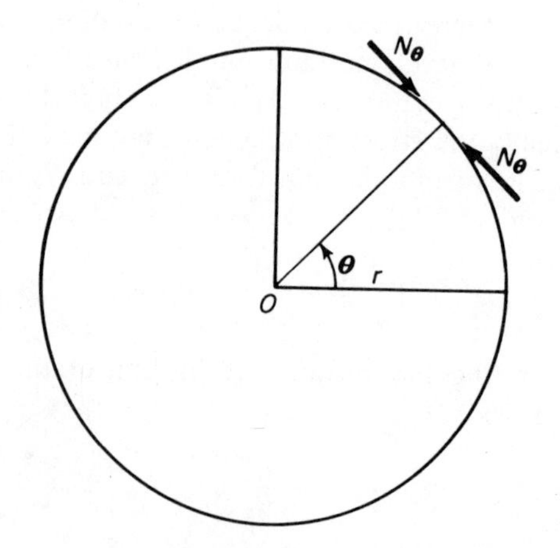

FIGURE 16.2.1

2. Live load q:

$$N_\varphi = \tfrac{1}{2}qR, \qquad N_\theta = qR(\cos^2\varphi - \tfrac{1}{2}) = \tfrac{1}{2}qR\cos 2\varphi, \tag{16.2.5}$$

$$N_\theta]_{\max} = \tfrac{1}{2}qR, \qquad N_\theta]_{\min} = -\tfrac{1}{2}qR. \tag{16.2.6}$$

3. External pressure p:

$$N_\varphi = N_\theta = \tfrac{1}{2}pR. \tag{16.2.7}$$

B. *Conical Shell* (Fig. 16.2.3):

1. Dead load w:

$$N_\varphi \equiv N_x = \frac{wx}{2\cos\alpha}, \tag{16.2.8a}$$

$$N_\theta = \frac{\sin^2\alpha}{\cos\alpha}wx. \tag{16.2.8b}$$

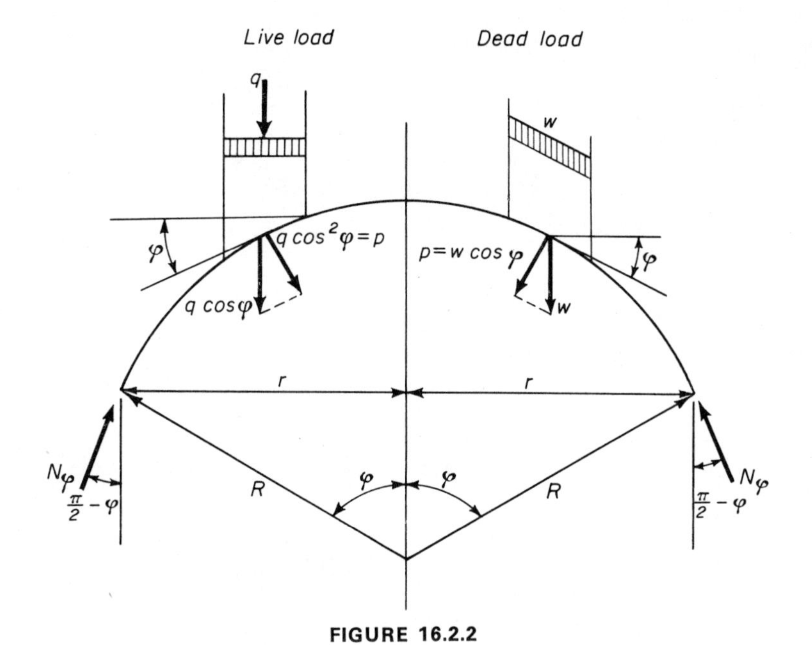

FIGURE 16.2.2

FIGURE 16.2.3

2. Live load q:
By (16.2.2):

$$\pi(x \sin \alpha)^2 q = N_x(2\pi x \sin \alpha) \cos \alpha \quad \therefore \quad N_x = \tfrac{1}{2} qx \tan \alpha. \tag{16.2.9a}$$

By (16.2.1), with $R_1 = \infty$, $R_2 = x \tan \alpha$; $p = q \sin^2 \alpha$:

$$N_\theta = pR_2 = \frac{\sin^3 \alpha}{\cos \alpha} qx. \tag{16.2.9b}$$

C. *Paraboloid of Revolution* (Fig. 16.2.4):

The equation of the meridian is:

$$z = f\frac{x^2}{r^2} = \frac{1}{c}x^2 \qquad \left(c = \frac{r^2}{f}\right). \tag{16.2.10}$$

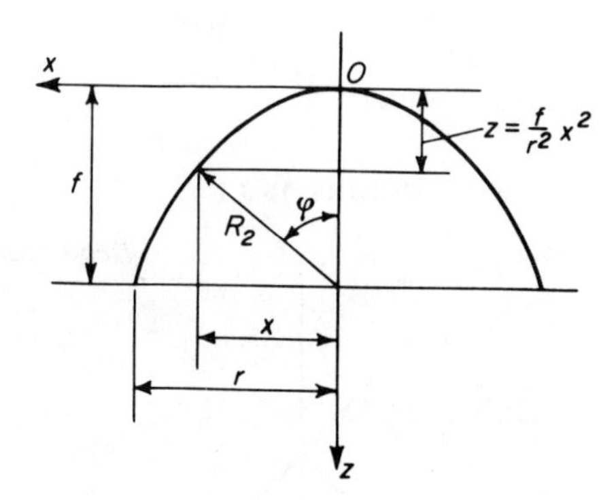

FIGURE 16.2.4

The stresses may be shown to be given by the following formulas:

1. Dead load w:

$$N_\varphi = \frac{wc}{6k^2}\left[(1 + k^2)^2 - \sqrt{1 + k^2}\right], \qquad k = \frac{2x}{c}, \tag{16.2.11}$$

$$N_\theta = \frac{wc}{2} - \frac{N_\varphi}{\sqrt{1 + k^2}}. \tag{16.2.12}$$

2. Live load q:

$$N_\varphi = \frac{qc}{4}\sqrt{1 + k^2}, \tag{16.2.13}$$

$$N_\theta = \frac{qc}{4}\Big/\sqrt{1 + k^2}. \tag{16.2.14}$$

Example 16.2.1. A hemispherical shell 75 mm (3 in.) thick, of radius $R = 12$ m (40 ft) is made of concrete. Determine the maximum compressive and maximum tensile stress due to the dead load and a snow load of 1.5 kN/m² (30 psf). For the examples in this chapter the weight of concrete is assumed to be 22 kN/m³ (140 lb/cu ft) (see Table 1.2.1).

By (16.2.4) and (16.2.6):

$$N_\varphi]_{\max} = 22 \times (0.075) \times 12 + \tfrac{1}{2} \times 1.5 \times 12 = 28.8 \text{ kN/m } (2{,}000 \text{ lb/ft}),$$

$$f_\varphi]_{\max} = \frac{28.8 \times 1\,000}{75 \times 1\,000} = 0.384 \text{ MPa } (56 \text{ psi}),$$

$$N_\theta]_{\min} = -22(0.075) \times 12 - \tfrac{1}{2} \times 1.5 \times 12 = -28.8 \text{ kN/m } (-2{,}000 \text{ lb/ft}),$$

$$f_\theta]_{\min} = -0.384 \text{ MPa } (-56 \text{ psi}).$$

For an allowable steel stress of 165 MPa (24 ksi), the maximum tension at the equator of 28.8 kN/m (2,000 lb/ft) can be taken by 175 mm² of steel per meter (0.08 sq in. of steel per foot).

Example 16.2.2. A conical shell has a radius of 6 m (20 ft) at the base and a rise of 3 m (10 ft); it is made out of wood planks 25 mm (1 in.) thick and carries a snow load of 2 kN/m² (40 psf). Determine the maximum meridional stress in the shell. Wood weighs ≈ 7.9 kN/m³ (50 lb/cu ft).

The angle α has a tangent equal to $\frac{6}{3} = 2$; hence, $\alpha = 63°$, $\sin\alpha = 0.89$, $\cos\alpha = 0.45$. The maximum value of x equals $\sqrt{3^2 + 6^2} = 6.71$ m. Hence, by (16.2.8) and (16.2.9):

$$N_x]_{\max} = \frac{7.9 \times 0.025 \times 6.71}{2 \times 0.45} + \frac{1}{2} \times 2 \times 6.71 \times 2 = 14.74 \text{ kN/m } (1{,}000 \text{ lb/ft}),$$

$$f_x]_{\max} = \frac{14.74}{25} = 0.60 \text{ MPa } (83 \text{ psi}).$$

Example 16.2.3. A concrete shell in the shape of a rotational paraboloid has a radius $r = 6$ m (20 ft) at the base and a rise $f = 15$ m (50 ft). The shell is 50 mm (2 in.) thick and carries a snow load of 2 kN/m² (40 psf). Determine the maximum and minimum stresses in the shell.

With $f = 15$ m, $r = 6$ m, $c = 6^2/15 = 2.4$ m, the maximum compressive N_φ occurs at the base, i.e., at $x = r = 6$ m, and, by (16.2.11) and (16.2.13) with $k = 2 \times \frac{6}{2.4} = 5$.

$$N_\varphi]_{\max} = \frac{(0.05 \times 22) \times 2.4}{6 \times 5^2}\left[(1 + 5^2)^2 - \sqrt{1 + 5^2}\right] + \frac{2 \times 2.4}{4}\sqrt{1 + 5^2}$$

$$= 11.81 + 6.12 = 17.93 \text{ kN/m } (1.24 \text{ k/ft}),$$

$$f_\varphi]_{\max} = \frac{17.93}{50} = 0.36 \text{ MPa } (52 \text{ psi}).$$

The maximum tensile N_θ also occurs at the base and, by (16.2.12) and (16.2.14), equals:

$$N_\theta]_{\max} = \frac{(0.05 \times 22) \times 2.4}{2} - \frac{17.93}{5.1} = 1.32 - 3.52$$

$$= -2.20 \text{ kN/m } (-150.4 \text{ lb/ft}).$$

This tensile force can be absorbed by less than 13.5 mm² of steel per meter (0.006 in.² per foot).

Example 16.2.4. What would be the maximum span of a spherical shell made out of 75 mm (3 in.) thick concrete capable of carrying its own dead load with an allowable stress of 6.9 MPa (1,000 psi), if one could ignore the danger of buckling?

The weight of a cubic meter of concrete is 22 kN; the weight per square meter of a shell of thickness h is $w = 22h$. Hence, by (16.2.4):

$$N_\varphi]_{\max} = hf_\theta]_{\max} = 75 \times 6.9 \times 1\,000 = 22 \times \frac{75}{1\,000}R$$

$$\therefore \quad R = \frac{6.9 \times 1\,000^2}{22\,000} = 313.64 \text{ m (1,029 ft)}.$$

The span of the shell, i.e., its diameter, equals 627 m (2,057 ft).

16.3 DISPLACEMENTS IN AXISYMMETRICAL SHELLS

The horizontal displacements induced by membrane stresses in axisymmetrical shells can be readily evaluated. The evaluation of the vertical displacements is mathematically more complicated.

For example, the outward displacement δ of the equator of a spherical shell under its own dead load is obtained from (16.2.4) as follows:

$$N_\theta]_{\min} = -wR, \qquad f_\theta = \frac{N_\theta}{h} = -\frac{wR}{h}, \qquad \epsilon_\theta = -\frac{f_\theta}{E} = \frac{wR}{Eh},$$

$$\delta = \epsilon_\theta R = \frac{wR^2}{Eh}, \tag{16.3.1}$$

where δ is *positive when outward.*

The vertical displacement δ_0, positive downward, at the top of the shell may be shown to equal:

$$\delta_0 = 1.76\frac{wR^2}{Eh}, \tag{16.3.2}$$

assuming Poisson's ratio ν equal to zero.

The rotation α of the shell boundary, *positive when outward,* may be shown to equal:

$$\alpha = 2\frac{wR}{Eh}, \tag{16.3.3}$$

for $\nu = 0$.

Example 16.3.1. A spherical dome is made out of concrete, is 100 mm (4 in.) thick, and has a radius of 9 m (30 ft). Evaluate the displacement of the top of the shell, the outward displacement of its boundary, and its boundary rotation due to its dead load.

$w = 22 \times (0.1) = 2.2 \text{ kN/m}^2$ (47 psf),

$$\delta_0 = 1.76\frac{(2.2 \times 1\,000)(9)^2}{23\,000 \times 100} = 0.137 \text{ mm } (0.006 \text{ in.}),$$

$$\delta = \frac{(2.2 \times 1\,000)(9)^2}{23\,000 \times 100} = 0.077 \text{ mm } (0.003\,5 \text{ in.}),$$

$$\alpha = 2\frac{2.2 \times 9}{23\,000 \times 100} = 1.72 \times 10^{-5} \text{ rad} = 3.55 \text{ sec of a degree } (4.03 \text{ sec}).$$

These results clearly indicate the minuteness of membrane displacements, i.e., the exceptional stiffness of thin shells.

Example 16.3.2. Derive the equatorial displacement δ due to a snow load q in a spherical shell.

$$N_\theta]_{\varphi=90^\circ} = -\frac{1}{2}qR, \quad f_\theta = -\frac{1}{2}\frac{qR}{h}, \quad \epsilon_\theta = \frac{1}{2}\frac{qR}{Eh}, \quad \delta = \frac{1}{2}\frac{qR^2}{Eh}. \tag{16.3.4}$$

It may be shown that the shell boundary does not rotate under a uniform live load and that the vertical displacement δ_0 of the top of the shell is equal to:

$$\delta_0 = \frac{3}{2}\frac{qR^2}{Eh}. \tag{16.3.5}$$

Example 16.3.3. Derive the equatorial displacement for a conical shell under dead load.

Indicate by a the radius of the base, by f the rise, and by $l = \sqrt{a^2 + f^2}$ the side of the cone (Fig. 16.2.3). Noticing that $\tan\alpha = a/f$, $\sin\alpha = a/l$, $\cos\alpha = f/l$, we obtain, by (16.2.8b):

$$N_\theta]_{\max} = \frac{wl\sin^2\alpha}{\cos\alpha}, \qquad f_\theta = \frac{wl}{h}\frac{\sin^2\alpha}{\cos\alpha},$$

$$\epsilon_\theta = -\frac{wl}{Eh}\frac{\sin^2\alpha}{\cos\alpha}, \tag{16.3.6}$$

$$\delta = -\frac{wla}{Eh}\frac{\sin^2\alpha}{\cos\alpha} = -\left(\frac{a}{f}\right)\frac{wa^2}{Eh}.$$

δ is an inward (negative) displacement.

The vertical displacement of the top of the cone may be shown to equal:

$$\delta_0 = \frac{1}{2}\left(\frac{l}{a}\right)^2\frac{wa^2}{Eh}. \tag{16.3.7}$$

For example, for $a/f = 2$, $l/f = \sqrt{1 + (a/f)^2} = 2.24$, $l/a = 1.12$:

$$\delta = -2\frac{wa^2}{Eh}, \quad \delta_0 = 0.63\frac{wa^2}{Eh}.$$

16.4 WIND STRESSES IN SHELLS OF REVOLUTION [12.7]

An approximate evaluation of the maximum membrane stresses induced by the wind in a spherical shell may be obtained by statics, assuming the normal stress N_φ and the shear stress $N_{\theta\varphi}$ at the equator to vary respectively as $\cos\theta$ and $\sin\theta$, where $\theta = 0$ is the direction of the wind (Fig. 16.4.1):

$$N_\varphi = N\cos\theta, \qquad N_{\theta\varphi} = S\sin\theta. \tag{16.4.1}$$

Indicating by A the area of the vertical cross section of the shell, the total horizontal force due to the wind pressure p_w equals:

$$P = p_w A. \tag{16.4.2}$$

The resultant in the direction $\theta = 0$ [Fig. 16.4.1 (c)] of the shears $N_{\theta\varphi}$ is:

$$\int_0^{2\pi} (S\sin\theta)\sin\theta\, r\, d\theta = S\pi r. \tag{16.4.3}$$

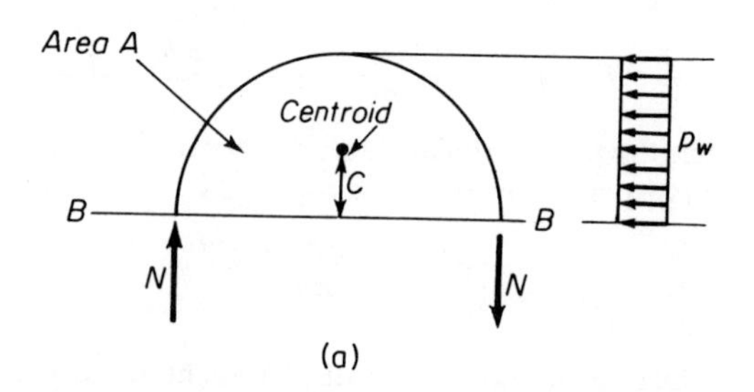

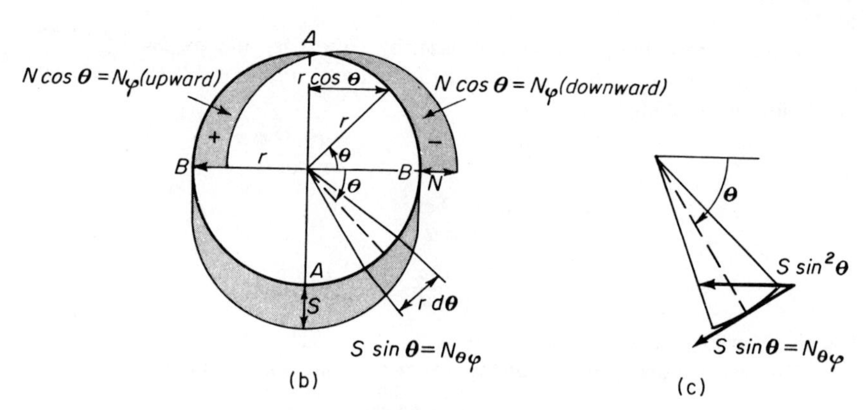

FIGURE 16.4.1

Equating (16.4.3) to (16.4.2), we obtain:

$$S = \frac{1}{\pi}\frac{p_w A}{r}. \tag{16.4.4}$$

Indicating by c the height over the equator of the centroid of the shell cross section, the moment of the wind force about the A-A axis is $p_w A c$. The moment of the N_φ about the same axis is:

$$4\int_0^{\pi/2} (N\cos\theta)(r\cos\theta)\, r\, d\theta = 4Nr^2\left(\frac{\pi}{4}\right) = \pi N r^2.$$

Equating moments, we obtain:

$$N = \frac{1}{\pi}\frac{p_w A c}{r^2}. \tag{16.4.5}$$

For example, for a spherical dome with $r = R$, the cross-sectional area and the centroidal heights are:

$$A = \frac{1}{2}\pi R^2, \qquad c = \frac{4}{3\pi}R, \tag{a}$$

by means of which (16.4.4) and (16.4.5) give:

$$S = \frac{1}{2}p_w R, \qquad N = \frac{2}{3\pi}p_w R. \tag{16.4.6}$$

For a vertical cylinder of radius R and height H, $A = 2RH$, $c = H/2$, and:

$$S = \frac{2}{\pi}p_w H, \qquad N = \frac{1}{\pi}\left(\frac{H}{R}\right)p_w H. \tag{16.4.7}$$

The wind stresses are usually of minor importance. A dome 30 m (100 ft) in radius under a wind of 1.5 kN/m² (30 psf) develops a maximum normal unit force:

$$N = \frac{2}{3 \times 3.14} 1.5 \times 30 = 9.55 \text{ kN/m} \ (0.64 \text{ k/ft})$$

and a maximum shear unit force:

$$S = \tfrac{1}{2} \times 1.5 \times 30 = 22.50 \text{ kN/m} \ (1.54 \text{ k/ft}),$$

which, with a thickness of 100 mm (4 in.), gives:

$$f_{\substack{\max\\ \min}} = \pm\frac{9.55}{100 \times 1\,000} = \pm 1 \times 10^{-4} \text{ kN/mm}^2 = \pm 0.10 \text{ MPa} \ (\pm 13.3 \text{ psi}),$$

$$f_{s,\max} = \frac{22.50}{100 \times 1\,000} = 2.25 \times 10^{-4} \text{ kN/mm}^2 = 0.225 \text{ MPa} \ (31.2 \text{ psi}).$$

16.5 MEMBRANE STRESSES IN BARRELS [12.9]

Long circular barrels, supported at their ends on stiffeners, arches, or diaphragms rigid in the vertical direction and flexible horizontally, behave like simply supported beams of circular cross section. The barrels are supported by the tangential shear forces $N_{x\varphi}$ developing between the end stiffeners and the barrel and do not exert any longitudinal action on the end stiffeners (Fig. 16.5.1). Their longitudinal stresses N_x and shear stresses $N_{x\varphi}$ due to the *live load* are easily evaluated by simple beam theory.

Indicating the weight per unit area of the semicircular shell by w (kN/m², psf), its radius by R, its thickness by h, and its length by l, its weight per unit

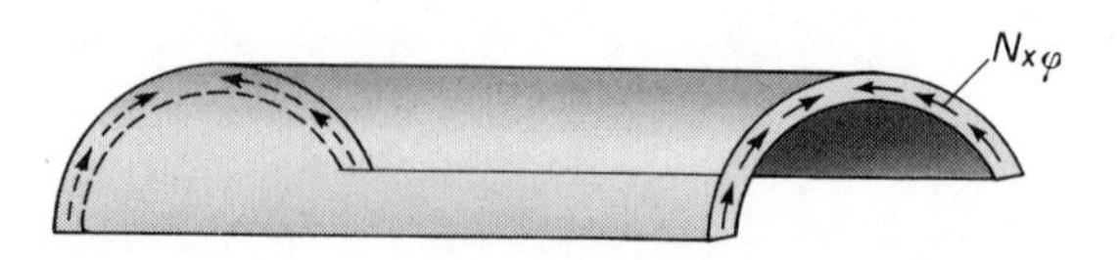

FIGURE 16.5.1

length is πwR, and the maximum bending moment in the barrel is:

$$M_{\max} = \tfrac{1}{8}(\pi wR)l^2 = 0.393wRl^2.$$

By Table 4.1.1 (12), the maximum compressive and tensile unit forces due to $M_{\max}$ are:

$$\begin{aligned} N_x]_{\max} &= f_{x,\max}h = \frac{0.393wRl^2}{0.83hR^2}h = 0.47\left(\frac{l}{R}\right)wl, \\ N_x]_{\min} &= f_{x,\min}h = -\frac{0.393wRl^2}{0.47hR^2}h = -0.83\left(\frac{l}{R}\right)wl. \end{aligned} \tag{16.5.1}$$

The unit shear force $N_{x\varphi}$ is evaluated by first determining the beam shear stresses f_s:

$$f_s = \frac{VS_y}{bI}, \tag{a}$$

where V is the beam shear, b the *horizontal* beam width, I its moment of inertia, and S_y the static moment of the section above the fiber where f_s is evaluated. The maximum f_s occurs at the neutral axis of the end sections, where (Fig. 16.5.2):

$$V = \tfrac{1}{2}(\pi wR)l = 1.57wRl, \qquad \sin\alpha = \frac{2}{\pi} = 0.64,$$

$$\cos\alpha = 0.77, \qquad \alpha = 40^\circ, \qquad \beta = 50^\circ = 0.88 \text{ rad},$$

$$b = \frac{2h}{\cos\alpha} = 2.60\,h,$$

$$S_y = 2\int_0^{\beta}(hR\,d\varphi)R(\cos\varphi - \cos\beta) = 2R^2h(\sin\beta - \beta\cos\beta) = 0.42hR^2,$$

so that, by (a) and Table 4.1.1 (12):

$$f_s]_{\max} = \frac{(1.57wRl)(0.21hR^2)}{(2.60h)(0.30hR^3)} = 0.84\left(\frac{l}{h}\right)w.$$

The beam shear stress f_s is vertical. The tangential shear $f_{x\varphi}$ [Fig. 16.5.2 (c)] is thus given by:

$$f_{x\varphi}]_{\max} = \frac{f_s}{\cos\alpha} = 1.10\left(\frac{l}{h}\right)w,$$

and the unit shear force by:

$$N_{x\varphi}]_{\max} = f_{x\varphi}h = 1.10wl. \tag{16.5.2}$$

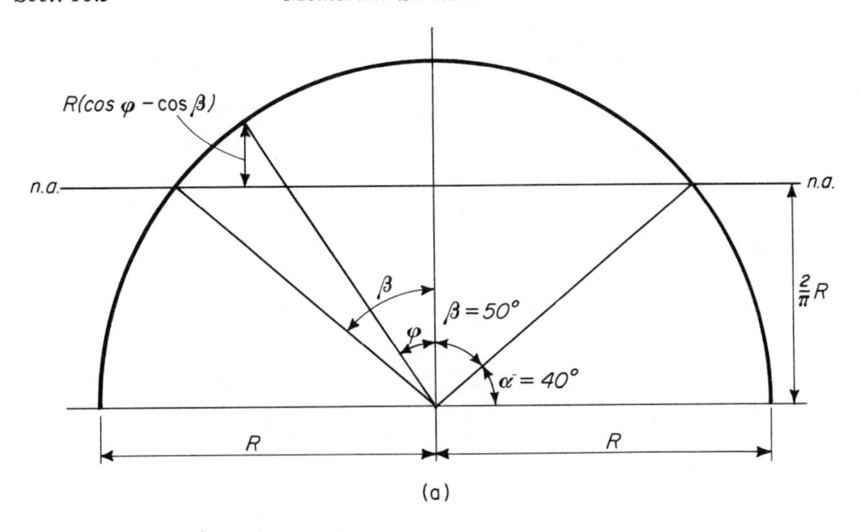

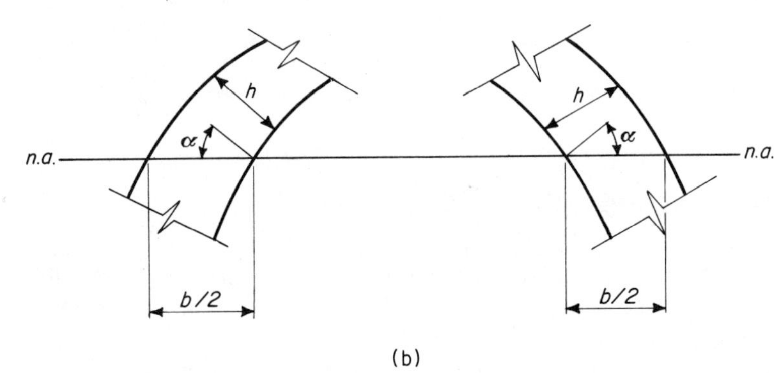

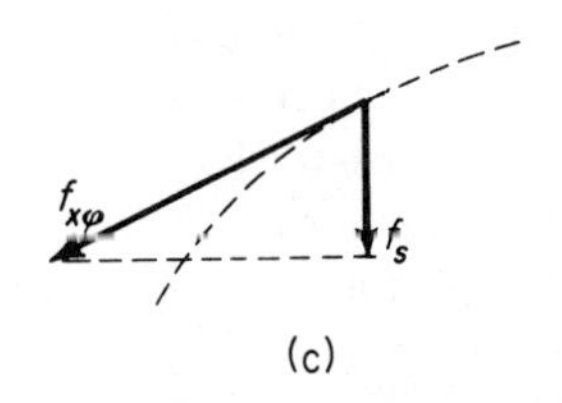

FIGURE 16.5.2

The meridional unit force N_φ is given by (16.2.1), with $R_1 = R$, $R_2 = \infty$ and $p = w \cos \varphi$:

$$N_\varphi = wR \cos \varphi, \tag{16.5.3}$$

and vanishes at $\varphi = 90°$ on the longitudinal boundaries.

The N_x and $N_{x\varphi}$ due to a *live load* q (kN/m², psf), equivalent to a load $2qR$ per unit of beam length, are in the ratio of $2/\pi$ to those due to the dead load, so that:

$$N_{x,\max} = 0.30\left(\frac{l}{R}\right)ql, \qquad N_{x,\min} = -0.53\left(\frac{l}{R}\right)ql,$$
$$N_{x\varphi}]_{\max} = 0.70ql. \tag{16.5.4}$$

The normal component of a live load q is given by $p = q\cos^2\varphi$ (Fig. 16.5.3) and:

$$N_\varphi = qR\cos^2\varphi \quad \therefore \quad N_\varphi]_{\max} = qR. \tag{16.5.5}$$

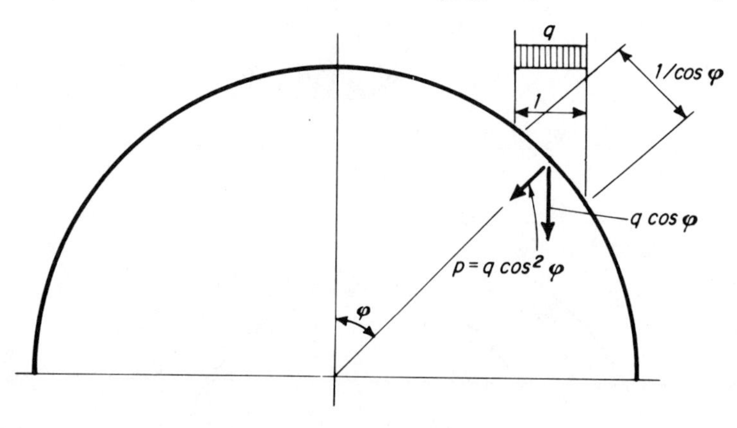

FIGURE 16.5.3

The N_x and $N_{x\varphi}$ for barrels continuous over three or more stiffeners are similarly evaluated by means of the maximum bending moment and shears in the corresponding continuous beam.

All the above formulas assume that the cross section of the barrel remains undistorted under load, as is the case for full-section beams. The cross sections of a thin circular barrel, instead, change shape under load. The change in stress due to this effect is negligible for long barrels, with $l > 4R$, but is very marked in short barrels, with $l < 2R$. In short barrels the distribution of stress is not linear across the depth of the barrel and may present tension both at the top and at the bottom of the shell, with compressive stresses in the interior fibers (Fig. 16.5.4).

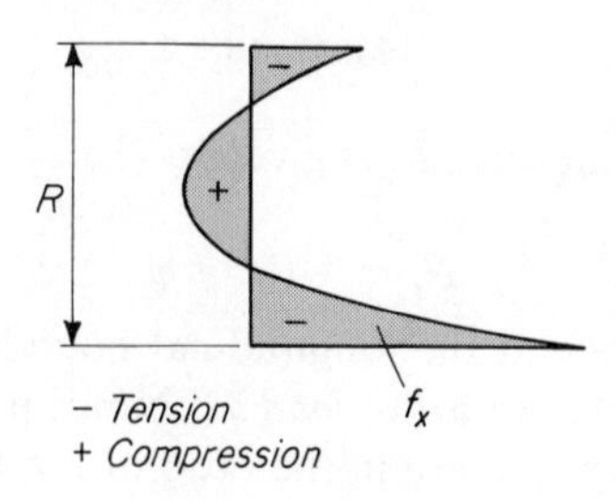

FIGURE 16.5.4

Example 16.5.1. A concrete barrel with two spans $l = 12$ m (40 ft), $R = 3$ m (10 ft), $h = 100$ mm (4 in.) carries its own dead load and a snow load of 1.5 kN/m² (30 psf). The barrel is continuous over the middle stiffener. Determine its maximum stresses.

From Table 4.1.2 (13), the maximum (negative) moment over the middle support is $-(\frac{1}{8})(\pi wR + 2qR)l^2$; the maximum positive moment occurs at $\frac{3}{8}l$ from the outer supports and equals $(\frac{9}{128})(\pi wR + 2qR)l^2$; the maximum shear occurs over the support and equals $(\frac{5}{8})(\pi wR + 2qR)l$. Hence, at the middle support, by (16.5.1) and (16.5.4):

$$N_x]_{\min} = -0.47\left(\frac{l}{R}\right)wl - 0.30\left(\frac{l}{R}\right)ql$$

$$= -0.47(\tfrac{12}{3})(22 \times 0.1) \times 12 - 0.30(\tfrac{12}{3})1.5 \times 12 = -71.23 \text{ kN/m } (3.64 \text{ k/ft}),$$

$$N_x]_{\max} = 0.83(\tfrac{12}{3})(22 \times 0.1) \times 12 + 0.53(\tfrac{12}{3})1.5 \times 12 = 125.81 \text{ kN/m } (6.46 \text{ k/ft}).$$

At the section $\frac{3}{8}l$ from the end supports, the moments have opposite signs and are in the ratio of $(\frac{9}{128})/(\frac{1}{8}) = 0.56$ to the moment over the support, hence:

$$N_x]_{\max} = 0.56 \times 71.23 = 39.89 \text{ kN/m } (2.04 \text{ k/ft}),$$

$$N_x]_{\min} = -0.56 \times 125.81 = -70.45 \text{ kN/m } (3.62 \text{ k/ft}).$$

The maximum shear is $(\frac{5}{8})/(\frac{1}{2}) = \frac{5}{4}$ of the maximum shear in a simply supported beam: hence, by (16.5.2) and (16.5.4):

$$N_{x\varphi}]_{\max} = \tfrac{5}{4}(1.10wl + 0.70ql)$$

$$= 1.25[0.55(22 \times 0.1) \times 12 + 0.35 \times 1.5 \times 12]$$

$$= 26.03 \text{ kN/m } (1.81 \text{ k/ft}).$$

The maximum N_φ at the shell top equals:

$$N_\varphi = wR + qR = (22 \times 0.1 + 1.5) \times 3 = 11.1 \text{ kN/m } (0.77 \text{ k/ft}).$$

A particular type of barrel, used mostly in industrial buildings and called a *north-light shell*, consists of an incomplete barrel with a vertical or an inclined glazed surface, which usually faces north (Fig. 16.5.5). The incomplete barrel is stiffened at the top and at the bottom by longitudinal edge beams that are connected by inclined struts.

The rigorous evaluation of membrane stresses in north-light shells is

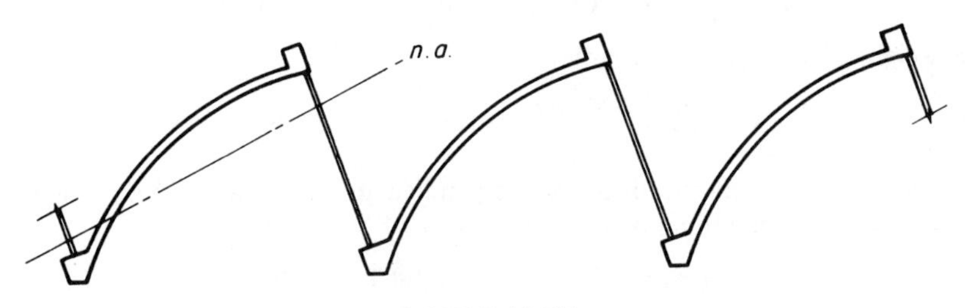

FIGURE 16.5.5

a nonelementary task, but approximate values of these stresses in long shells may be obtained by beam theory, noticing that the neutral axis of the beam cross section is inclined to the horizontal and that bending occurs about both the neutral axis and at right angles to it.

Example 16.5.2. A simply supported, reinforced concrete north-light shell consists of an arc of circle with an opening angle of 90° stiffened by two identical edge beams. Its dimensions are given in Fig. 16.5.6 (a). The shell span l is 12 m (40 ft). Determine the membrane stresses due to its dead load.

By symmetry, the neutral axis y-y is at 45° to the horizontal. The centroid G lies on the perpendicular to y-y through O. Its vertical ordinate c is determined by the ratio of the static moment S_x of the shell's cross section about the x-axis to its area A [Fig. 16.5.6 (b), (c)]:

$$A = \frac{\pi}{2}(1.5 \times 1\,000) \times 100 + 2 \times (300 \times 250) = 385\,500 \text{ mm}^2 \text{ (616 in.}^2\text{)},$$

$$\begin{aligned} S_x &= \int_0^{\pi/2} (hR\,d\varphi)R\cos\varphi + A_1R = R^2h\int_0^{\pi/2}\cos\varphi\,d\varphi + A_1R \\ &= R^2h + A_1R = (1.5 \times 1\,000)^2 \times 100 + (250 \times 300)(1.5 \times 1\,000) \\ &= 337\,500\,000 \text{ mm}^3 \text{ (21,600 cu in.)}, \end{aligned}$$

$$c = \frac{S}{A} = \frac{337\,500\,000}{385\,500} = 875.5 \text{ mm (35 in.)}.$$

Hence:

$$OG = \frac{875.5}{\cos 45^\circ} = 1\,238 \text{ mm}, \qquad OM = 1\,550 \text{ mm}$$

$$GM = 1\,550 - 1\,238 = 312 \text{ mm}$$

$$ON' = 1\,500\sqrt{2}/2 = 1\,061 \text{ mm}, \qquad GN' = OG - ON' = \bar{z} = 177 \text{ mm}$$

$$GN = GN' + N'N = 177 + 250\cos 45^\circ = 354 \text{ mm}.$$

The moment of inertia of the cross section [Fig. 16.5.6 (c)] is given approximately by:

$$\begin{aligned} I_y &= I_{y1} - A\bar{z}^2 = 2\int_0^{\pi/4} hR\,d\varphi[R(\cos 45^\circ - \cos\varphi)]^2 - A\bar{z}^2 \\ &= 2hR^3\left(\frac{\pi}{8} - 1 + \frac{\pi}{8} + \frac{1}{4}\right) - A\bar{z}^2 = 0.071hR^3 - A\bar{z}^2 \\ &= 0.071 \times 100 \times (1500)^3 - 385\,500 \times 177^2 = 11.82 \times 10^9 \text{ mm}^4. \end{aligned}$$

The weight of shell per unit length is:

$$w = 22 \times \frac{385\,500}{1\,000^2} = 8.48 \text{ kN/m (600 lb/ft)}.$$

Its components w_y in the direction perpendicular to y-y is $(\sqrt{2}/2) \times 8.48 =$ 6 kN/m. The maximum moment M_y equals:

$$M_y = \tfrac{1}{8}w_yl^2 = \tfrac{1}{8}(6 \times 12^2) = 108 \text{ kN·m (85 k ft)},$$

In a balanced design the maximum shell compression f_c equals:

$$f_c = \frac{(108 \times 1\,000) \times 312}{11.82 \times 10^9} = 0.002\,85 \text{ kN/mm}^2 = 2.85 \text{ MPa (420 psi).}$$

The tension in the steel reinforcement at midspan, of area A_s, equilibrates the moment M_y with a lever arm larger than $N'G$, since the resultant of the compressive stresses in the concrete is above the neutral axis. Hence, conservatively, with

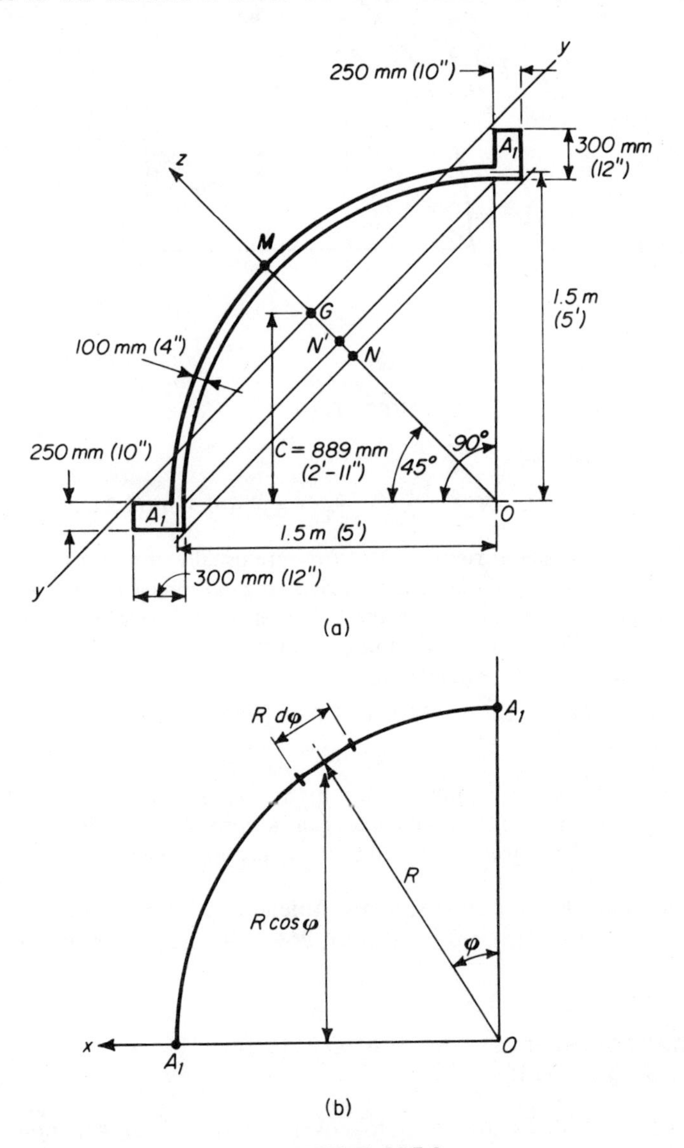

FIGURE 16.5.6

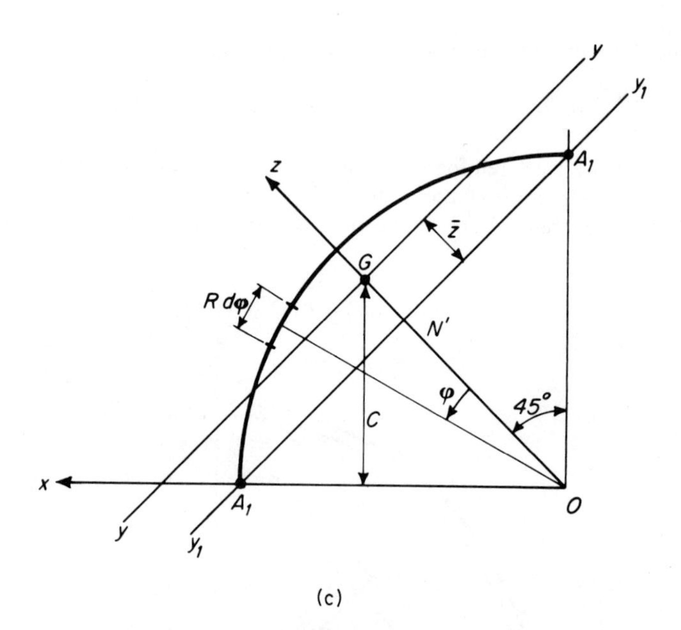

(c)

FIGURE 16.5.6

$F_s = 138$ MPa (20,000 psi):

$$M_y = F_s A_s \bar{z} \quad \therefore \quad A_s = \frac{108 \times 1\,000^2}{138 \times 177} = 4\,410 \text{ mm}^2 \text{ (7.2 in.}^2\text{)}.$$

The bending stresses at right angles to *y-y* are usually of minor importance, since the moment of inertia about an axis at right angles to *y-y* is much larger than the moment about *y-y*. Since, by symmetry, the moment M_z about the *z*-axis equals M_y, with $F_s = 138$ MPa (20,000 psi), it may be absorbed by a steel area A'_s in the lower edge beam, given approximately by:

$$108 \times 1\,000 = 2A'_s \times 138 \times (1\,500 \times \sqrt{2}/2) = 292\,700\,A'_s$$

$$\therefore \quad A'_s = 0.369 \text{ mm}^2 \text{ (0.60 in.}^2\text{)}.$$

The compressive force in the upper edge beam equals $138 \times 0.369 = 50.9$ N. With $F_c = 6.9$ MPa (1 ksi), the concrete can absorb $250 \times 300(6.9/1\,000) = 517.5$ kN (120 k), and no steel is required in the upper edge beam.

Circular barrels are also frequently stiffened by longitudinal boundary beams. Their approximate stress analysis follows the calculations of Example 16.5.2.

16.6 MEMBRANE STRESSES IN TRANSLATIONAL SHELLS [12.11, 12.12]

Translational shells are used to cover rectangular areas; their most common shapes are the elliptic paraboloid and the hyperbolic paraboloid (see Section 14.2).

The *elliptic paraboloid* is a synclastic surface described by a parabola $z_a = c_a(x/a)^2$ of span $2a$ and rise c_a sliding, at right angles to it, on a parabola $z_b = c_b(y/b)^2$ of span $2b$ and rise c_b [Fig. 16.6.1 (a)]. The shell is supported on four boundary stiffeners by tangential shears S only [Fig. 16.6.1 (b)], since the thin vertical stiffeners (diaphragms or arches) are unable to react out of their own plane. We wish to compute the unit forces N_x, N_y and S due to a live load q on the elliptic paraboloid.

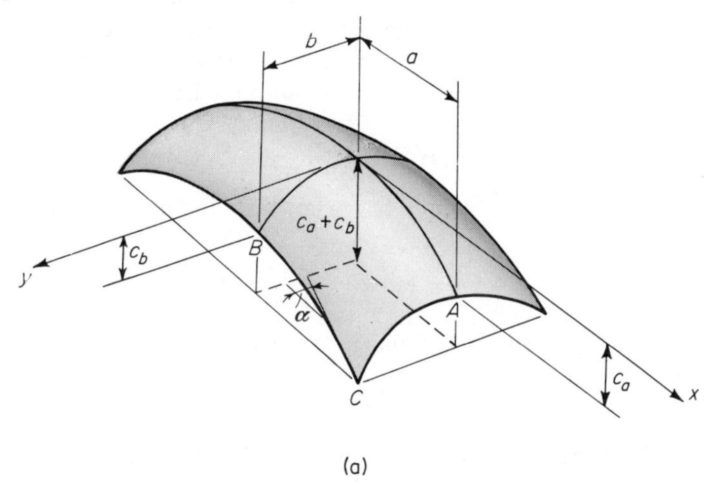

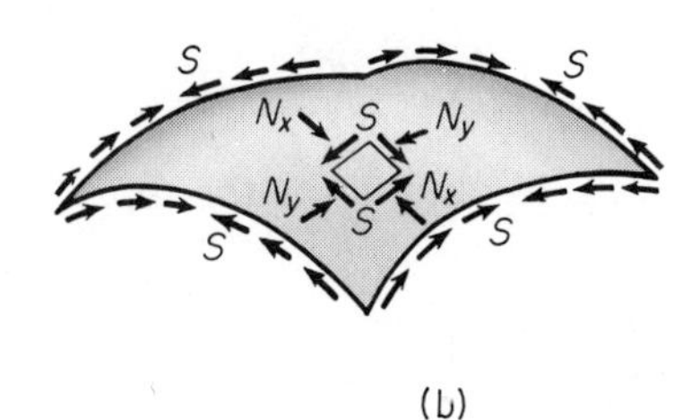

FIGURE 16.6.1

The radii of curvature of the parabolas at the top of the shell are:

$$R_x = \frac{1}{d^2z_a/dx^2} = \frac{a^2}{2c_a}, \qquad R_y = \frac{1}{d^2z_b/dy^2} = \frac{b^2}{2c_b}. \tag{16.6.1}$$

When $b = a$ and $c_a = c_b = c$, the radii are equal, $R_x = R_y = R = a^2/2c$. Since, moreover, at the top of the shell by symmetry, $S = 0$ and $N_x = N_y$, the values of N_x and N_y at the top of the shell, where $p = q$ because q is normal to the shell, are given by (15.1.2):

$$q = \frac{N_x}{R_x} + \frac{N_y}{R_y} = \frac{2N_x}{R} \quad \therefore \quad N_x = N_y = N_{\min} = \frac{1}{2}qR = \frac{qa^2}{4c}. \tag{16.6.2a}$$

At the boundary point B [Fig. 16.6.1(a)] of coordinates $x = 0$, $y = b$, $N_y = 0$, the twist is zero, i.e., $R_{xy} = \infty$, and (15.1.2) gives:

$$N_{x,\max} = pR = \frac{pa^2}{2c} \approx \frac{qa^2}{2c}, \tag{16.6.2b}$$

since the normal component p of the load q may be conservatively considered equal to the total load q.

The boundary shears S are zero at the top of the boundary arches by symmetry and are maximum at the corners of the shell (where theoretically they become infinite as is shown at the end of this section). Calling S_0 their maximum value at the corner, their variation may be approximated by:

$$S = S_0\left(\frac{x}{a}\right)^2. \tag{16.6.3a}$$

The slope m of the parabola $z_a = c(x/a)^2$ on the boundary $y = b$ [Fig. 16.6.1(a)] equals:

$$m = \tan\alpha = \frac{2c}{a}\left(\frac{x}{a}\right), \tag{a}$$

and, for $c \ll a$, the vertical component of S may be approximated by:

$$S_v = S\sin\alpha \approx S\tan\alpha = \left(\frac{2c}{a}\right)(S_0)\left(\frac{x}{a}\right)^3.$$

The resultant of the S_v around the shell must equal the total load $4qa^2$:

$$4qa^2 = 8\int_0^a S_v\,dx = 16cS_0\int_0^a\left(\frac{x}{a}\right)^3 d\left(\frac{x}{a}\right) = 4cS_0,$$

so that:

$$S_{\max} = S_0 = \frac{qa^2}{c}. \tag{16.6.3b}$$

When $a \neq b$ and $c_a \neq c_b$, N_x and N_y at the top of the shell may be shown to be in the ratio:

$$\frac{N_y}{N_x} = \left(\frac{b}{a}\right)^2\sqrt{\frac{c_a}{c_b}} = r. \tag{16.6.4}$$

With $N_y = rN_x$, (15.1.2) gives:

$$q = \frac{N_x}{R_x} + \frac{N_y}{R_y} = N_x\left(\frac{1}{R_x} + \frac{r}{R_y}\right)$$

$$\therefore \quad N_{x,\min} = \frac{q}{(1/R_x) + (r/R_y)}, \qquad N_{y,\min} = \frac{rq}{(1/R_x) + (r/R_y)}. \tag{b}$$

By (16.6.1) and (16.6.4), equations (b) become:

$$\begin{aligned} N_{x,\min} &= \frac{q}{(2c_a/a^2) + (b^2/a^2)\sqrt{c_a/c_b}\,(2c_b/b^2)} = \frac{qa^2/\sqrt{c_a c_b}}{2(1 + \sqrt{c_a/c_b})}, \\ N_{y,\min} &= q\frac{(b^2/a^2)\sqrt{c_a/c_b}}{(2c_a/a^2) + (b^2/a^2)\sqrt{c_a/c_b}\,(2c_b/b^2)} = \frac{qb^2/\sqrt{c_a c_b}}{2(1 + \sqrt{c_b/c_a})}, \end{aligned} \tag{16.6.5}$$

while S_{max} becomes:

$$S_{max} = \frac{qab}{\sqrt{c_a c_b}}. \tag{16.6.6}$$

At the top of the boundary arches parallel to x or y the load is carried by N_x or N_y only, so that:

$$N_{x,max} = qR_x = \frac{qa^2}{2c_a}, \qquad N_{y,max} = qR_y = \frac{qb^2}{2c_b}. \tag{16.6.7}$$

The stresses in shells of translation generated by arcs of circles or of other curves having the same spans and rises as parabolic arches do not differ substantially from those for the elliptic paraboloid. It is useful to remember that for often-used circular arches:

$$R_x = \frac{a^2 + c_a^2}{2c_a}, \qquad R_y = \frac{b^2 + c_b^2}{2c_b}. \tag{c}$$

The formulas derived above for a uniform live load q may be used in connection with a dead load w, provided the rises c_a and c_b be less than one-quarter of the corresponding sides $2a$ and $2b$.

Example 16.6.1. Determine the maximum stresses in a concrete elliptic paraboloid, 100 mm (4 in.) thick, with $a = 9$ m (30 ft), $b = 12$ m (40 ft), $c_a = 1.75$ m (6 ft), $c_b = 2.5$ m (8 ft), and carrying a snow load of 1.5 kN/m² (30 psf).

The total load per unit area on the shell is:

$$w = 22(0.1) + 1.5 = 3.7 \text{ kN/m}^2 \text{ (77 psf)},$$

so that, by (16.6.7):

$$N_{x,max} = 3.7 \times \frac{9^2}{2 \times 1.75} = 85.63 \text{ kN/m},$$

$$f_{x,max} = \frac{85.63}{1\,000 \times 100} = 0.000\,856 \text{ kN/mm}^2 = 0.856 \text{ MPa (120 psi)},$$

$$N_{y,max} = 3.7 \times \frac{12^2}{2 \times 2.5} = 106.56 \text{ kN/m},$$

$$f_{y,max} = \frac{106.56}{1\,000 \times 100} = 0.001\,066 \text{ kN/mm}^2 = 1.066 \text{ MPa (160 psi)},$$

by (16.6.6):

$$S_{max} = 3.7 \times \frac{9 \times 12}{\sqrt{1.75 \times 2.5}} = 191 \text{ kN/m},$$

$$f_{s,max} = \frac{191}{1\,000 \times 100} = 0.001\,91 \text{ kN/mm}^2 = 1.91 \text{ MPa (277 psi)},$$

and, by (16.6.5):

$$N_{x,min} = 3.7 \times \frac{(9^2/\sqrt{1.75 \times 2.5})}{2(1 + \sqrt{1.75/2.5})} = 39 \text{ kN/m},$$

$$f_{x,min} = \frac{39}{1\,000 \times 100} = 0.000\,39 \text{ kN/mm}^2 = 0.39 \text{ MPa (56 psi)},$$

$$N_{y,\min} = 3.7 \times \frac{(12^2/\sqrt{1.75 \times 2.5})}{2(1 + \sqrt{2.5/1.75}} = 58 \text{ kN/mm},$$

$$f_{y,\min} = \frac{58}{1\,000 \times 100} = 0.000\,58 \text{ kN/mm}^2 = 0.58 \text{ MPa (86 psi)}.$$

The high value of $f_{s,\max}$ requires diagonal steel reinforcement (see Example 16.6.3).

Example 16.6.2. Evaluate $N_{x,\max}$, $N_{y,\max}$ for a circular translational shell with the same rises and sides as the shell in Example 16.6.1.

$$R_x = \frac{9^2 + 1.75^2}{2 \times 1.75} = 24.02 \text{ m}; \qquad R_y = \frac{12^2 + 2.5^2}{2 \times 2.5} = 30.05 \text{ m},$$

By (16.6.7),

$$N_{x,\max} = 3.7 \times 24.02 = 88.87 \text{ kN/m}; \qquad N_{y,\max} = 3.7 \times 30.05 = 111.19 \text{ kN/m}.$$

These unit forces are about 4% higher than those of Example 16.6.1.

Example 16.6.3. The maximum tensile stress in an elliptic paraboloid occurs at the corners as the resultant of the shears S (Fig. 16.6.2). Since $N_x = N_y = 0$ at the corner, an element at the corner is under pure shear, and the diagonal tensile force T equals the shear S_0; hence:

$$T_{\max} = S_0 = \frac{qab}{\sqrt{c_a c_b}}. \tag{16.6.8}$$

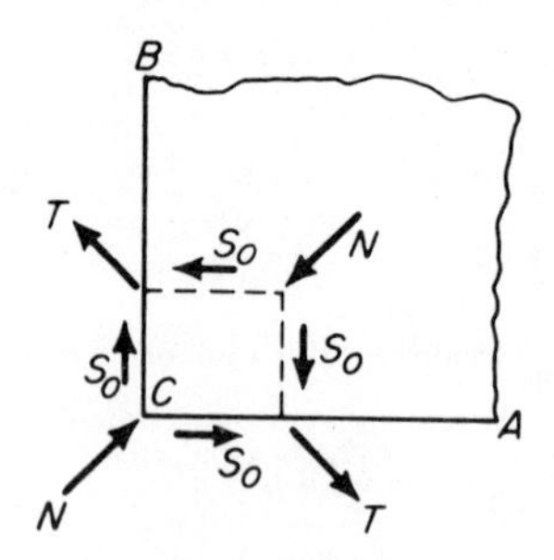

FIGURE 16.6.2

In the shell of Example 16.6.1, $T_{\max} = 191$ kN/m and requires 1 384 mm² (2.1 sq in.) of steel per meter for an allowable stress in the steel of 138 MPa (20 ksi).

The state of stress at the corner, actually, cannot be a membrane state of stress. In fact, (Fig. 16.6.2), N_x and N_y are zero, respectively, along the boundaries $x = a$ and $y = b$ because the boundary stiffeners cannot react in a direction perpendicular to their planes, and, hence, are both zero at the corner C ($x = a, y = b$). Moreover, the geometrical twist in the directions x, y is zero (see Section 14.2), i.e., $R_{xy} = \infty$, so that, by (15.1.2), the load carried by the membrane stresses at C is:

$$p = \frac{0}{R_x} + \frac{0}{R_y} + \frac{S_0}{\infty}$$

and is zero unless S_0 is infinite. As membrane stresses cannot carry the load, bending stresses are unavoidable at the corner. Bending stresses along the boundary of the elliptic paraboloid are considered in Section 17.4.

The membrane stresses in a shallow *rectangular hyperbolic paraboloid*, of rises c_a and c_b, small in comparison with a and b (Fig. 16.6.3), are determined by (15.1.2), in which p is approximately equal to the vertical load w, the radii of curvature in the principal directions x, y are practically constant and equal, $R_x = -R_y = R = a^2/2c_a = b^2/2c_b$ [see (16.6.1)], the twist is zero (see Section 14.2), $N_x = C$ is compressive, $N_y = T$ is tensile, and $C = -T$, since the two parabolas in the x- and y-directions are identical. Thus:

$$w = \frac{C}{R} + \frac{-C}{-R} = \frac{2C}{R} \quad \therefore \quad C = -T = \frac{1}{2}wR = \frac{1}{4}\frac{wa^2}{c_a} = \frac{1}{4}\frac{wb^2}{c_b}.$$

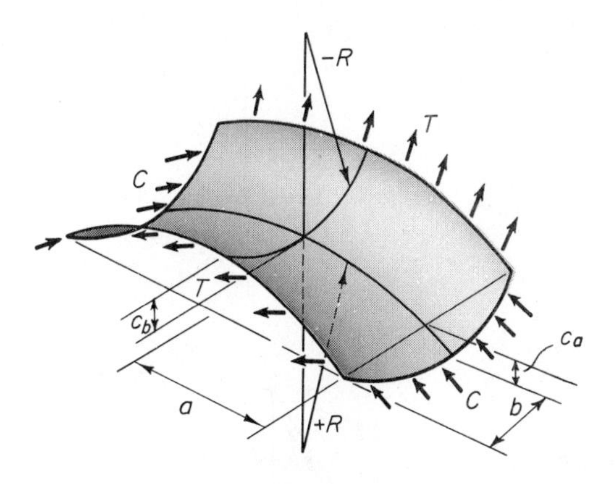

FIGURE 16.6.3

Referring the paraboloid to axes x', y' at 45° to x, y, and calling c the rise at $x' = a$, $y' = b$ (Fig. 16.6.4), the radius R equals ab/c, by means of which the membrane forces in the x- and y-directions become:

$$C = w\frac{ab}{2c}, \qquad T = -w\frac{ab}{2c}. \qquad (16.6.9)$$

These equations show that the load on a *shallow* hyperbolic paraboloid is supported by funicular arch action in the x-direction and by funicular cable action in the y-direction through membrane stresses which are constant over the entire shell surface. Hence, the membrane stresses developed by the shell, which are also constant through its thickness, have a constant value at *all* points of the shell and the material is ideally utilized.

The equal compression and tension in the x- and y-directions are equivalent to identical shears at 45° to the x-, y-axes (see Section 13.1); hence, the

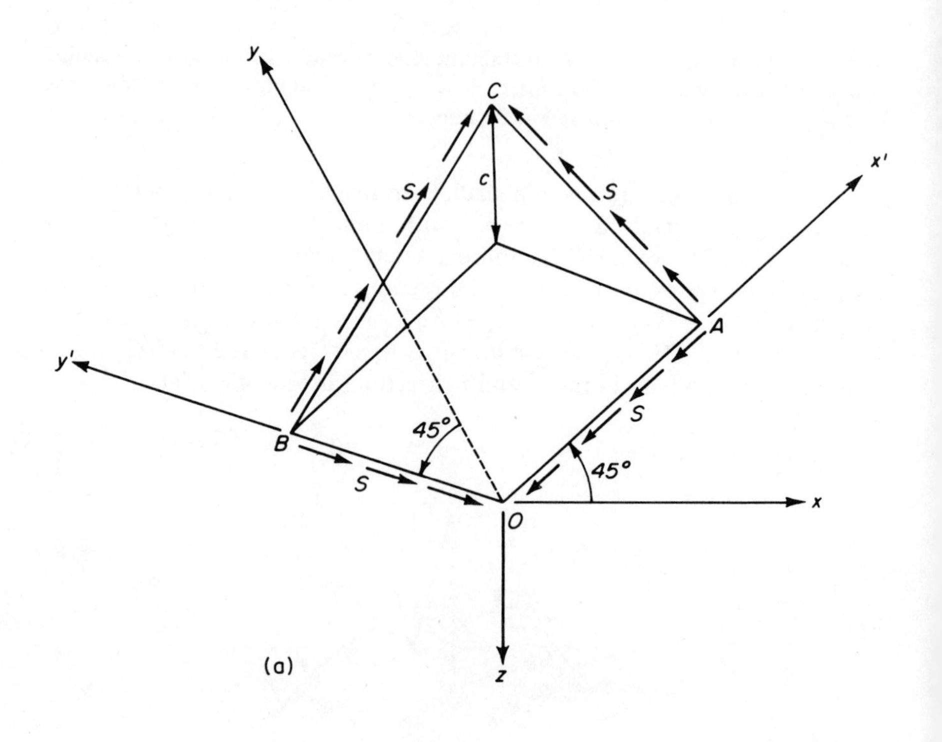

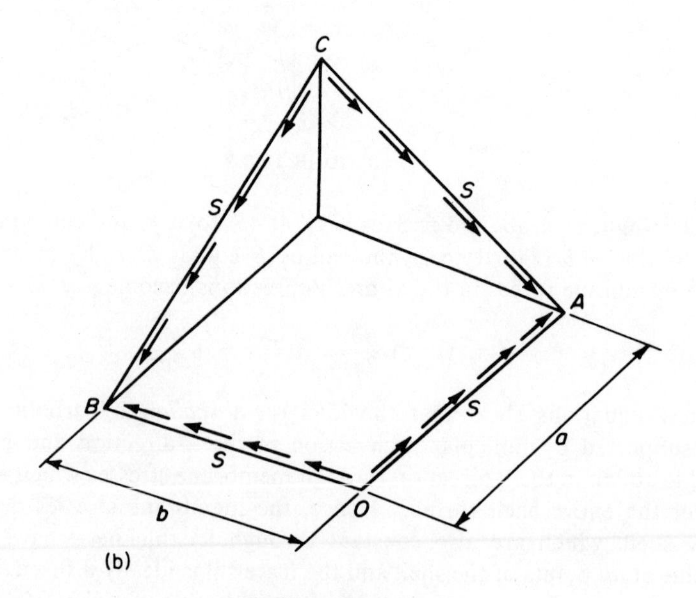

FIGURE 16.6.4

membrane shears S are also given by:

$$S = \pm w\frac{ab}{2c}. \tag{16.6.10}$$

If the hyperbolic paraboloid is to be supported along arches parallel to x and to y at $x = \pm a$ and $y = \pm b$, the supports should be able to supply to the shell boundary reactions equal and opposite to C and T, respectively, and thus be capable of reacting at right angles to their own plane (Fig. 16.6.3). Since support stiffeners are usually incapable of providing such reactions, it is customary to support the paraboloid along straight lines at 45° to x and y, along which the unit forces are the pure shears S of (16.6.10). The shears on the boundary of an element of paraboloid with sides a, b at 45° to the x-, y-axes are shown in Fig. 16.6.4 (a). Such shears are provided by boundary struts [Fig. 16.6.4 (b)], which are thus acted upon by shears equal and opposite to those at the shell boundary: the struts are compressed by the accumulation of shears from O to A and O to B, and from C to A and C to B. The maximum compressive forces in the struts OA, CB, and OB, CA are given, respectively, by:

$$F'_c = C\sqrt{a^2 + c^2} = w\frac{ab}{2c}\sqrt{a^2 + c^2}, \qquad F''_c = w\frac{ab}{2c}\sqrt{b^2 + c^2}. \tag{16.6.11}$$

It is now apparent that the element of paraboloid shown in Fig. 16.6.4 (a) can only be in equilibrium if external forces F'_c, F''_c equal to those in (16.6.11) are provided to equilibrate the struts. This is usually done by adjoining four identical paraboloidal elements and by providing tie rods so that the forces in each element balance those in the adjoining elements, as shown in Fig. 16.6.5 (a).

Figure 16.6.5 (b) shows a balanced paraboloidal shell, a hyperbolic paraboloid umbrella, in which the horizontal struts are tensed and the inclined struts compressed by the accumulation of boundary shears. (In both cases the weight of the horizontal struts hangs from the shell, introducing some bending.)

It must finally be noted that one is not free to choose the boundary conditions at the edges of a hyperbolic paraboloid, if the load is to be equilibrated by membrane stresses only. In fact, starting at a boundary point A of a square element (Fig. 16.6.6), and splitting the shear S into its components C and T, it is seen that C is transmitted by arch action and T by catenary action to the adjoining boundaries so as to meet at the same point. Their resultant is a shear S parallel to the boundary on all four sides of the element. For example, in order to free the edge HL of stresses, an equal and opposite S should be applied at A, and its components $-C$, $-T$ would be transmitted by arch and catenary action to the points B, C, and D, wiping out the boundary shear *on all four edges*. Hence, in trying to have one edge free of stress we would free all edges of stress, and the paraboloid could not carry

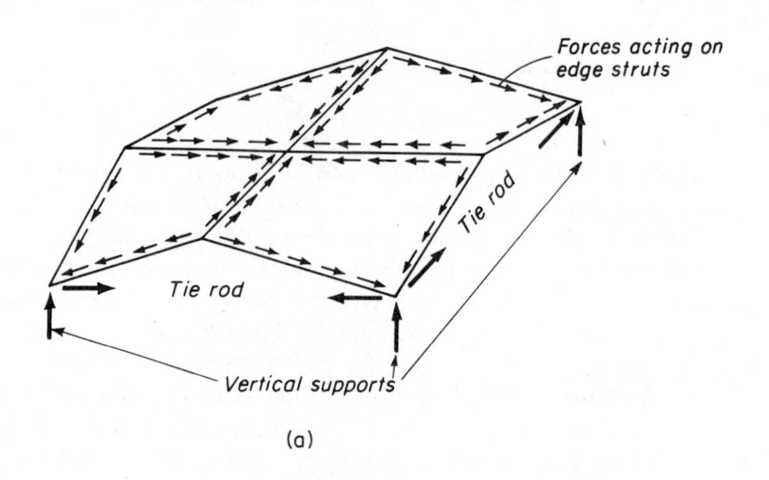

(a)

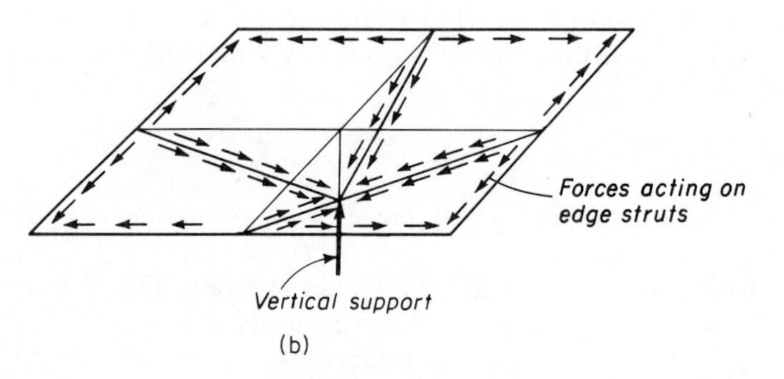

(b)

FIGURE 16.6.5

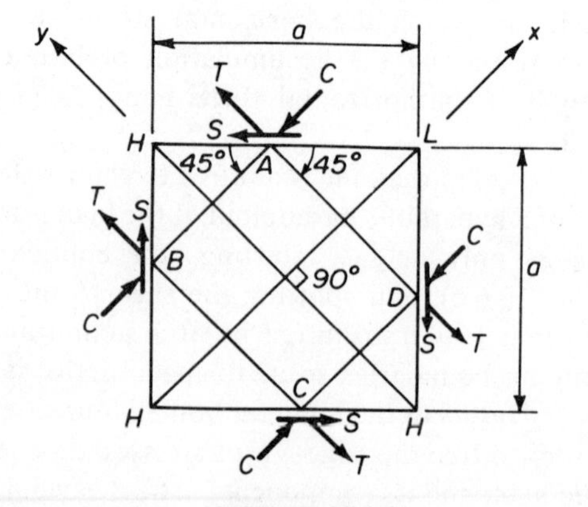

FIGURE 16.6.6

load for lack of equilibrating boundary forces. Since, in a hyperbolic paraboloid with one free edge, membrane stresses alone cannot carry load, bending stresses are bound to arise if such edge condition actually exists.

A hyperbolic paraboloid is seldom supported along its parabolic boundaries, since two of them have upward curvature. If this is done and the boundary stiffeners react with shears only, the paraboloid acts like two intersecting cylinders of opposite curvature supported at their ends.

Example 16.6.4. A concrete roof made out of four hyperbolic paraboloid sectors [Fig. 16.6.5 (a)], with a thickness $h = 100$ mm (4 in.), spans 30 m (100 ft) by 24 m (80 ft) and has a rise c of 6 m (20 ft). It is supported on its four corners and carries a snow load of 1.5 kN/m² (30 psf). Determine the stresses in the paraboloid and the dimensions of its boundary beams for $F_c = 6.9$ MPa (1 ksi).

$$w = 22 \times 0.1 + 1.5 = 3.7\ \text{kN/m}^2, \qquad a = 15\ \text{m}, \qquad b = 12\ \text{m},$$

$$S = C = -T = 3.7 \times \frac{15 \times 12}{2 \times 6} = 55.50\ \text{kN/m (3.85 k/ft)},$$

$$f_c = \frac{55.5}{100} = = 0.555\ \text{MPa}, \qquad A_s = \frac{55.5 \times 1\,000}{138} \approx 402\ \text{mm}^2/\text{m (0.19 in.}^2/\text{ft)},$$

$$F'_c = C\sqrt{a^2 + c^2} = 55.5 \times 16.16 = 896\ \text{kN (208 k)},$$

$$A' = \frac{896 \times 1\,000}{6.9} = 129\,900\ \text{mm}^2 \approx 300\ \text{mm} \times 450\ \text{mm (12 in.} \times 18\ \text{in.)}.$$

$$F''_c = C\sqrt{b^2 + c^2} = 55.5 \times 13.42 = 745\ \text{kN (173 k)},$$

$$A'' = \frac{745 \times 1\,000}{6.9} = 108\,000\ \text{mm}^2 = 250\ \text{mm} \times 450\ \text{mm (10 in.} \times 18\ \text{in.)}.$$

16.7 SUMMARY

The unit membrane forces in the most commonly used types of thin shells are summarized in Table 16.7.1.

PROBLEMS

16.1 A concrete spherical shell spans 45 m (150 ft) with a rise of 7.5 m (25 ft). It is 100 mm (4 in.) thick and carries a snow load of 1.5 kN/m² (30 psf). Evaluate the maximum compressive stress in the shell, the membrane thrust exerted by the shell on its support, and the reinforcement in mm²/m (sq in./ft) at the shell boundary. Use grade 60 steel.

Table 16.7.1

Type	Loading	Unit forces		Figure
Spherical	Dead	$N_\varphi = \dfrac{wR}{1+\cos\varphi}$	$N_\theta = wR\left(\cos\varphi - \dfrac{1}{1+\cos\varphi}\right)$	16.2.2
	Live	$N_\varphi = \frac{1}{2}qR$	$N_\theta = \frac{1}{2}qR\cos 2\varphi$	
	External pressure	$N_\varphi = \frac{1}{2}pR$	$N_\theta = \frac{1}{2}pR$	
Conical	Dead	$N_x = \dfrac{wx}{2\cos\alpha}$	$N_\theta = wx\dfrac{\sin^2\alpha}{\cos\alpha}$	16.2.3
	Live	$N_x = \frac{1}{2}qx\tan\alpha$	$N_\theta = qx\dfrac{\sin^3\alpha}{\cos\alpha}$	
	External pressure	$N_x = \frac{1}{2}px\tan\alpha$	$N_\theta = px\tan\alpha$	
Paraboloid of revolution	Dead	$N_\varphi = \dfrac{wc}{6k^2}[(1+k^2)^2 - \sqrt{1+k^2}]$	$N_\theta = \dfrac{wc}{2} - \dfrac{N_\varphi}{\sqrt{1+k^2}}$	16.2.4
	Live	$N_\varphi = \dfrac{qc}{4}\sqrt{1+k^2}$	$N_\theta = \dfrac{qc}{4}\dfrac{1}{\sqrt{1+k^2}}$	
		$\left(k = \dfrac{2x}{c}\right)$		

Cylindrical	Dead	$N_{x,\,\max} = 0.47\left(\frac{l}{R}\right)wl$		16.5.2
		$N_{x,\,\min} = -0.83\left(\frac{l}{R}\right)wl$	$N_{x\varphi,\,\max} = 1.10\,wl$	
		$N_{\varphi} = wR\cos\varphi$		
	Live	$N_{x,\,\max} = 0.30\left(\frac{l}{R}\right)ql$		16.5.3
		$N_{x,\,\min} = -0.53\left(\frac{l}{R}\right)ql$	$N_{x\varphi,\,\max} = 0.70ql$	
		$N_{\varphi,\,\max} = qR$		
Hyperbolic paraboloid (shallow)	Dead or live	$N_x = C = w\frac{ab}{2c}$		16.6.4
		$N_y = T = -w\frac{ab}{2c}$		
		$S = \pm\, w\frac{ab}{2c}$		

16.2 A concrete spherical shell spans 30 m (100 ft) with a rise of 6 m (20 ft). It is 75 mm (3 in.) thick and carries its dead load only, with a safety factor of 3. By how much should we decrease its rise to produce the collapse of the dome in compression? Use $f'_c = 20$ MPa (2,900 psi).

16.3 Compare the values of the maximum compressive unit forces in three rotational concrete domes with circular, parabolic, and straight-line meridians, spanning 30 m (100 ft), with a rise of 4.5 m (15 ft), 75 mm (3 in.) thick, and carrying no live load.

16.4 A hemispherical dome of radius R carries a dead load w and a live load $q = w$. Determine by trial and error the value of φ for which the hoop stress vanishes.

16.5 A 3 m (10 ft) high concrete conical shell 75 mm (3 in.) thick, of radius 9 m (30 ft), carries a live load of 1 kN/m^2 (20 psf). Determine its maximum compressive and tensile stress and the tensile reinforcement in mm^2/m (sq in./ft) at its boundary. Use grade 60 steel.

16.6 What is the ultimate span of a concrete spherical shell 100 mm (4 in.) thick under dead load, if the span-to-rise ratio is 5, and the ultimate strength of the concrete in compression is 35 MPa (5,000 psi) and the shell is prevented from buckling?

16.7 Determine the value of x for which a paraboloid of revolution with $f/r = 2$ develops zero hoop stresses under dead load. (See Fig. 16.2.4.)

16.8 Does a paraboloid of revolution with $f/r = 1$ ever develop tensile hoop stresses under a live load equal to its dead load?

16.9 What is the smallest value of f/r for which a paraboloid of revolution will not develop tensile hoop stresses under dead load?

16.10 Determine the maximum membrane unit forces under a dead load w in an inverted cone of opening angle α supported by a ring one-third of the rise from the cone vertex.

16.11 Determine the membrane unit forces under dead load in an inverted hemispherical shell sector supported on the parallel $\phi = 45°$.

16.12 A hemispherical dome of radius 6 m (20 ft) rests on a circular vertical wall. Determine the membrane reactions of the half dome on the wall if the dome is made of concrete 50 mm (2 in.) thick and carries a live load of 1.5 kN/m^2 (30 psf).

16.13 Determine the horizontal displacement at the boundary of the shell of Problem 16.1.

16.14 Determine the horizontal displacement at the boundary and the vertical displacement at the top of a hemispherical concrete shell 30 m (100 ft) in radius and 150 mm (6 in.) thick under its dead load and a 2 kN/m^2 (40 psf) snow load.

16.15 The unit forces N_ϕ in a parabolic arch of span l, rise f, thickness h, and unit width, under a live load q uniform in horizontal projection, and those of a rotational paraboloid with the same geometrical characteristics and the same load are both of the membrane type. Compare them and explain why they are different. Which structure has a greater displacement at the top?

16.16 Determine the maximum wind unit forces in a hemispherical concrete shell of radius 15 m (50 ft) and thickness 75 mm (3 in.). Are the N and S stresses within 33% of the deadload stresses? Assume $p_w = 1.5\ \text{kN/m}^2$ (30 psf).

16.17 Derive by the methods of Sections 16.4 the membrane unit forces due to wind in a conical shell.

16.18 A concrete barrel shell, supported on thin end stiffeners, spans 45 m (150 ft), has a radius of 3 m (10 ft), and a thickness of 100 mm (4 in.). Determine its maximum compressive and tensile unit forces under its dead load and a snow load of 2 kN/m² (40 psf), and the reinforcement in mm²/m (sq in./ft) to absorb the maximum tensile force. Use grade 40 steel.

16.19 Solve Problem 16.18 if the barrel is supported by three stiffeners spaced 22.5 m (75 ft) apart.

16.20 A concrete elliptic paraboloid covers a square area 18 m (60 ft) on a side, with a total rise of 3.6 m (12 ft). It is 100 mm (4 in.) thick and carries a snow load of 2 kN/m² (40 psf). Determine its maximum compressive unit force and the reinforcement to absorb the corner shears. Dimension the tie rods on the sides of the shell in A36 steel.

16.21 A concrete, square, hyperbolic paraboloid umbrella is supported on a central column. It is 9 m (30 ft) on the side, 75 mm (3 in.) thick, and carries 1.5 kN/m² (30 psf) of snow load. Determine the reinforcing steel in the four boundary beams if the rise is 1.8 m (6 ft). Use grade 40 steel.

16.22 A concrete roof made out of four rectangular hyperbolic paraboloid sectors is 18 m (60 ft) on the side, 100 mm (4 in.) thick, and carries a 2 kN/m² (40 psf) snow load. It is supported by four columns and has a rise of 4.5 m (15 ft). Determine the maximum compressive and tensile unit forces, the compressive forces in the inclined boundary beams, and the reinforcement in the horizontal boundary beams.

16.23 The dead load of a concrete roof, made out of four square hyperbolic paraboloid sectors with sides $2a$ and a rise f, is supported on four columns with a coefficient of safety of 4 in the concrete compression, of 3 in the tensile steel, and of 3 in the tie rods. What decrease in rise will produce the collapse of the shell, and for what cause?

16.24 Substitute for the shell of Problem 16.1 a space frame made of A36 steel with radial and circumferential bars. The radial bars are spaced by one-thirtieth of the circumference at the boundary and the circumferential bars by one-tenth of the radius at the boundary. Determine the cross-sectional areas of the most stressed bars by having those bars absorb the shell membrane forces of the contributory areas.

16.25 Substitute for the hyperbolic paraboloid shell of Problem 16.21 a space frame of A36 steel bars spaced by 1.5 m (5 ft) on horizontal projection and parallel to the sides of the shell. Determine the area of the bars.

17. Bending and Buckling of Shells

17.1 PRIMARY AND SECONDARY BENDING [12.7]

In Chapter 16 it was shown that thin shells can carry any smooth load by membrane stresses, and that the membrane stresses developed in a shell depend essentially on the support conditions of its boundary. We now wish to investigate what type of boundary conditions develop a state of pure membrane stresses in a shell. Two requirements must be met for this to happen:

1. The reactive forces on the shell boundary must be equal and opposite to the membrane forces at the boundary induced by the load.
2. The support must allow the shell boundary to undergo the displacements induced by the membrane strains.

Whenever either or both of these requirements are not met, bending stresses develop in the shell.

For example, a spherical dome under dead and live loads develops membrane stresses and undergoes membrane displacements. In order to have in the dome a state of pure membrane stress, the reactions must be equal and opposite to the N_φ on the boundary, and the boundary must be free to move outward by δ and to rotate by α (see Sections 16.2 and 16.3). Hence, not only must the shell equator move freely outward by δ, but the reactions must rotate by α together with the boundary in order to remain tangent to the deformed shell (Fig. 17.1.1). Since it is impractical to allow such movements and rotations, bending stresses are bound to develop. But, since the membrane stresses are capable of carrying the load, the bending stresses thus developed are not required for equilibrium. Bending stresses needed for *compatibility of displacements* but not for equilibrium are called *secondary bending stresses.*

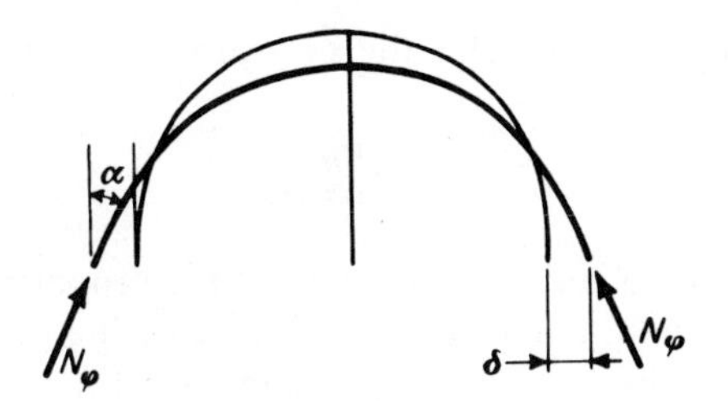

FIGURE 17.1.1

On the other hand, it is shown in Section 16.6 that, if a sector of a rectangular hyperbolic paraboloid has a free edge, the paraboloidal sector cannot be in equilibrium under membrane stresses. The bending stresses developed in the shell are required for equilibrium and are called *primary bending stresses.*

A shell is properly shaped and properly supported if it carries most of the load by membrane stresses and, hence, develops only secondary bending stresses.

17.2 AXISYMMETRICAL BENDING IN CYLINDERS [12.10]

For a large variety of shells, the secondary bending stresses due to the incompatibility of boundary displacements may be approximately evaluated in terms of the *symmetrical* bending stresses developed in a cylinder under radial pressure. This basic case will now be considered.

An infinitely long cylinder of radius R and thickness h, under a constant, outer radial pressure p (Fig. 17.2.1) develops a *membrane* (hoop) unit force identical to the hoop force in a ring [see (10.1.4)]:

$$N_\varphi = pR,$$

and, hence, has a radial *membrane* displacement, *positive inward*, given by:

$$\delta - \epsilon_\varphi R = \frac{\sigma_\varphi}{E} R = \frac{N_\varphi}{Eh} R = \frac{pR^2}{Eh}. \tag{17.2.1}$$

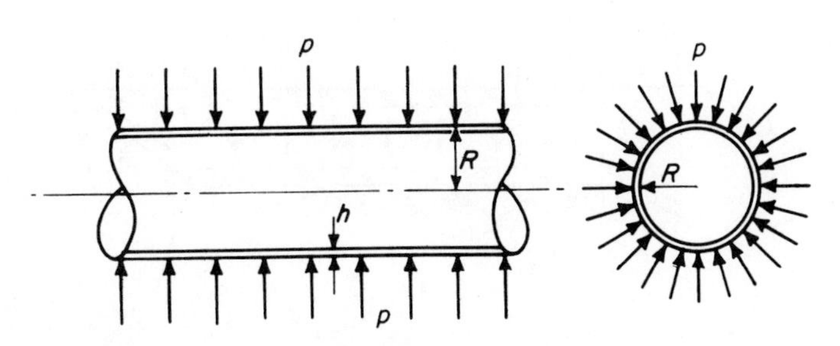

FIGURE 17.2.1

Solving (17.2.1) for p, we see that the pressure carried by a *membrane* displacement δ is equal to:

$$p = \frac{Eh}{R^2}\delta. \tag{17.2.2}$$

The load p carried by a unit area of the shell is proportional to the radial displacement δ, just as the load P carried by a spring of spring constant k is proportional to its displacement δ (see Section 4.8):

$$P = k\delta. \tag{a}$$

Comparison of (17.2.2) and (a) shows that the shell behaves like a spring of spring constant:

$$k = \frac{Eh}{R^2}, \tag{17.2.3}$$

i.e., that each longitudinal strip of shell of width 1 is supported by the rest of the shell as if it were a beam of cross section $1 \times h$ supported on an elastic foundation with a spring constant Eh/R^2 per unit length N/mm³ (lb/in.³) (Fig. 17.2.2). It is seen from (17.2.3) that the "elastic foundation" effect is due to the shell curvature: as R becomes larger, k decreases and tends to zero as R approaches infinity. Hence, a flat plate does not exhibit this effect.

Let us now consider a finite cylinder of length l, in which the end sections are prevented from displacing inward by the δ of (17.2.1). (Fig. 17.2.3). This incompatibility with membrane displacements will develop secondary bending stresses and bending displacements w to be added to the membrane displacement δ. The equation satisfied by w is obtained by noticing that the load q per unit length carried by a beam, whose deflection is w, equals $EI(d^4w/dx^4)$ [see (4.1.4 c)], and that, therefore, the bending displacement w of a cylinder strip of unit width (with a cross section $1 \times h$ and a moment of inertia $1 \times h^3/12$) will carry a load per unit area:

$$q = \frac{Eh^3}{12}\frac{d^4w}{dx^4}. \tag{b}$$

Since (Fig. 17.2.2) the external load p is carried through the membrane

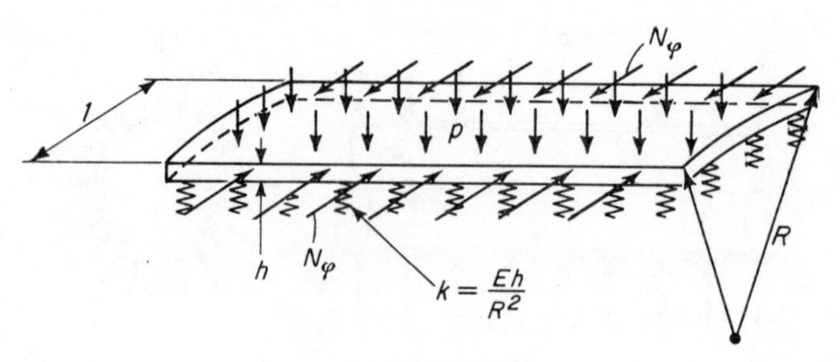

FIGURE 17.2.2

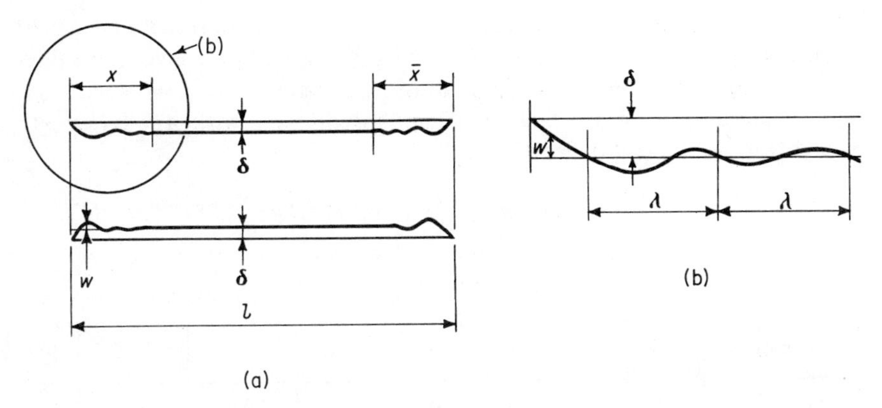

FIGURE 17.2.3

displacement δ by the membrane stress f_m:

$$f_m = \frac{N_\varphi}{h} = \frac{pR}{h}, \tag{17.2.4}$$

the bending displacements carry no external load, but support only the "elastic foundation" reaction $q = -kw$, which acts outward, i.e., is negative for an inward (positive) displacement w*, and vice versa. By (17.2.3), equation (b) becomes:

$$\frac{Eh^3}{12}\frac{d^4w}{dx^4} = -kw = -\frac{Eh}{R^2}w \quad \therefore \quad \frac{d^4w}{dx^4} + 4\frac{3}{R^2h^2}w = 0,$$

or, introducing the constant:

$$c^4 = \frac{R^2h^2}{3}, \quad c^2 = \frac{Rh}{\sqrt{3}}, \quad c = \frac{1}{\sqrt[4]{3}}\sqrt{Rh} = 0.76\sqrt{Rh}, \tag{17.2.5}$$

the equation:

$$\frac{d^4w}{dx^4} + \frac{4}{c^4}w = 0. \tag{17.2.6}$$

The bending deflections w would be expected to diminish away from the cylinder ends, where δ is prevented, and to have an oscillating behavior similar to that of a continuous beam (see Example 4.6.2), since the longitudinal strips of the cylinder are continuous over the "elastic foundation" provided by the hoops of the rest of the cylinder. These expectations are actually correct. As may be checked by substitution into the equation, the following displacements satisfy (17.2.6):

$$w_l = Ae^{-x/c}\sin\left(\frac{x}{c} + \psi\right), \qquad w_r = Be^{-\bar{x}/c}\sin\left(\frac{\bar{x}}{c} + \psi\right), \tag{17.2.7}$$

*The radial displacements δ and w in this chapter are positive inward, while those in Chapter 16 are positive outward.

where A, B, and ψ are constants and $\bar{x} = l - x$ is the coordinate measured from the right end of the cylinder (Fig. 17.2.3). Table 17.2.1 gives the values of w for $\psi = 0$, $A = 1$, $B = 1$, and of the ratios $w/w_{max} = w/w]_{x/c=\pi/4}$. The displacements w are "damped" and oscillatory, with a wave length $x = \lambda$ [Fig. 17.2.3 (b)] defined by:

$$\frac{\lambda}{c} = 2\pi \quad \therefore \quad \lambda = 2\pi c = 2\pi \times 0.76\sqrt{Rh} = 4.78\sqrt{Rh}. \quad (17.2.8)$$

Table 17.2.1

x/c or $\bar{x}/c$	0	$\pi/4$	$\pi/2$	$3\pi/4$	π	$5\pi/4$	$6\pi/4$	$7\pi/4$	2π
w	0	0.32	0.21	0.067	0	−0.014	−0.009	−0.003	0
w/w_{max}	0	1.0	0.64	0.21	0	−0.043	−0.03	−0.01	0

Each successive maximum and minimum value of w is reduced in the ratio of 1 to $e^{-\pi} = 0.043$, i.e., by 96%.

Hence, for a small thickness h the secondary bending displacements are damped out within a short distance and disturb the membrane state of stress in a small region near the shell ends. The displacements w_l and w_r do not usually interfere with each other and are called *boundary disturbances.*

The derivative of w_l with respect to x (or of w_r with respect to $\bar{x}$):

$$\frac{dw}{dx} = Ae^{-x/c}\left[-\left(\frac{1}{c}\right)\sin\left(\frac{x}{c} + \psi\right) + \left(\frac{1}{c}\right)\cos\left(\frac{x}{c} + \psi\right)\right],$$

when multiplied and divided by $-1/\sqrt{2} = -\sin(\pi/4) = -\cos(\pi/4)$, may be written as:

$$\begin{aligned}\frac{dw}{dx} &= -\sqrt{2}\left(\frac{1}{c}\right)Ae^{-x/c}\left[\sin\left(\frac{x}{c} + \psi\right)\left(\frac{1}{\sqrt{2}}\right) - \left(\frac{1}{\sqrt{2}}\right)\cos\left(\frac{x}{c} + \psi\right)\right] \\ &= -\sqrt{2}\left(\frac{1}{c}\right)Ae^{-x/c}\left[\sin\left(\frac{x}{c} + \psi\right)\cos\frac{\pi}{4} - \sin\frac{\pi}{4}\cos\left(\frac{x}{c} + \psi\right)\right] \\ &= -\left(\frac{\sqrt{2}}{c}\right)Ae^{-x/c}\sin\left(\frac{x}{c} + \psi - \frac{\pi}{4}\right) \qquad (17.2.9)\end{aligned}$$

and shows that each differentiation is equivalent to multiplying w by $-\sqrt{2}/c$ and subtracting $\pi/4$ from ψ. Hence:

$$\begin{aligned}\frac{d^2w}{dx^2} &= (2/c^2)\,Ae^{-x/c}\sin\left(\frac{x}{c} + \psi - \frac{\pi}{2}\right), \\ \frac{d^3w}{dx^3} &= -(2\sqrt{2}/c^3)\,Ae^{-x/c}\sin\left(\frac{x}{c} + \psi - 3\frac{\pi}{4}\right).\end{aligned} \qquad (17.2.10)$$

The bending moment M and shear V *per unit of shell width* are obtained by simple beam theory [see (4.1.4 a) and (4.1.4 b)], using (17.2.5):

$$M(x) = -EI\frac{d^2w}{dx^2} = -\frac{Eh^3}{12}\frac{2}{c^2}Ae^{-x/c}\sin\left(\frac{x}{c}+\psi-\frac{\pi}{2}\right)$$

$$= -\frac{1}{2\sqrt{3}}\frac{Eh^2}{R}Ae^{-x/c}\sin\left(\frac{x}{c}+\psi-\frac{\pi}{2}\right), \tag{17.2.11}$$

$$V(x) = -EI\frac{d^3w}{dx^3} = \frac{Eh^3}{12}\frac{2\sqrt{2}}{c^3}Ae^{-x/c}\sin\left(\frac{x}{c}+\psi-3\frac{\pi}{4}\right)$$

$$= \frac{1}{\sqrt{2}\sqrt[4]{3}}\frac{Eh}{R}\sqrt{\frac{h}{R}}Ae^{-x/c}\sin\left(\frac{x}{c}+\psi-3\frac{\pi}{4}\right). \tag{17.2.12}$$

A positive M induces compression in the outer fibers and tension in the inner fibers and a positive V, tension at 135° to the axis of the cylinder.

In order to obtain a complete definition of the bending stresses one must determine the angle ψ by means of the boundary conditions at the cylinder's end. When the cylinder is *fixed* so that it *cannot rotate at* $x = 0$, the slope dw/dx is zero at $x = 0$ and, by (17.2.9) and (17.2.7):

$$\left.\frac{dw}{dx}\right]_{x=0} = -\left(\frac{\sqrt{2}}{c}\right)A\sin\left(\psi-\frac{\pi}{4}\right) = 0 \quad \therefore \quad \psi = \frac{\pi}{4},$$

$$w(0) = \delta = A\sin\frac{\pi}{4} = \frac{A}{\sqrt{2}} \quad \therefore \quad A = \sqrt{2}\,\delta,$$

and by (17.2.7), (17.2.11), and (17.2.12):

$$w(x) = \sqrt{2}\,\delta\, e^{-x/c}\sin\left(\frac{x}{c}+\frac{\pi}{4}\right), \quad w(0) = \delta, \tag{17.2.13a}$$

$$M(x) = -\frac{Eh^3}{12}\frac{d^2w}{dx^2} = -\frac{Eh^2}{\sqrt{6}}\frac{\delta}{R}e^{-x/c}\sin\left(\frac{x}{c}-\frac{\pi}{4}\right),$$

$$M_0 = \left(\frac{Eh^2}{2\sqrt{3}}\right)\left(\frac{\delta}{R}\right) = 0.29Eh^2\left(\frac{\delta}{R}\right), \tag{17.2.13b}$$

$$V(x) = -\frac{Eh^3}{12}\frac{d^3w}{dx^3} = \frac{1}{\sqrt[4]{3}}Eh\sqrt{\frac{h}{R}}\left(\frac{\delta}{R}\right)e^{-x/c}\sin\left(\frac{x}{c}-\frac{\pi}{2}\right),$$

$$V_0 = -\left(\frac{Eh}{\sqrt[4]{3}}\right)\sqrt{\frac{h}{R}}\left(\frac{\delta}{R}\right). \tag{17.2.13c}$$

The bending moment M has a maximum value M_0 at $x = 0$, and vanishes first at $x/c = \pi/4$, i.e., at a distance x_p from the boundary, called the *penetration*, given by:

$$x_{p,f} = \frac{\pi}{4}c = \frac{\pi}{4\sqrt[4]{3}}\sqrt{Rh} = 0.60\sqrt{Rh}. \tag{17.2.14}$$

M becomes minimum (negative maximum) at $x = 1.2\sqrt{Rh}$, where it is -21% of M_0.

When the shell is *free to rotate* at the end $x = 0$, by (17.2.11) and (17.2.7):

$$M(0) = -\frac{1}{2\sqrt{3}}\frac{Eh^2}{R}A\sin\left(\psi - \frac{\pi}{2}\right) = 0 \quad \therefore \quad \psi = \frac{\pi}{2},$$

$$w(0) = \delta = A\sin\frac{\pi}{2} = A,$$

and by (17.2.7), (17.2.9), (17.2.11), and (17.2.12):

$$w(x) = \delta\, e^{-x/c}\sin\left(\frac{x}{c} + \frac{\pi}{2}\right), \quad w(0) = \delta, \tag{17.2.15a}$$

$$\frac{dw}{dx} = -\sqrt{2}\,\sqrt[4]{3}\left(\frac{\delta}{\sqrt{Rh}}\right)e^{-x/c}\sin\left(\frac{x}{c} + \frac{\pi}{4}\right),$$

$$\left.\frac{dw}{dx}\right]_{x=0} = -\sqrt[4]{3}\left(\frac{\delta}{\sqrt{Rh}}\right), \tag{17.2.15b}$$

$$M(x) = -\frac{Eh^2}{2\sqrt{3}}\left(\frac{\delta}{R}\right)e^{-x/c}\sin\frac{x}{c}, \quad M_0 = 0, \tag{17.2.15c}$$

$$V(x) = \left(\frac{Eh}{\sqrt{2}\,\sqrt[4]{3}}\right)\sqrt{\frac{h}{R}}\left(\frac{\delta}{R}\right)e^{-x/c}\sin\left(\frac{x}{c} - \frac{\pi}{4}\right),$$

$$V_0 = -\left(\frac{Eh}{2\sqrt[4]{3}}\right)\sqrt{\frac{h}{R}}\left(\frac{\delta}{R}\right). \tag{17.2.15d}$$

The moment is maximum at:

$$x = \left(\frac{\pi}{4}\right)c = 0.60\sqrt{Rh}, \tag{17.2.16}$$

where, by Table 17.2.1, it has the value:

$$M_{\max} = -0.32\left(\frac{Eh^2}{2\sqrt{3}}\right)\left(\frac{\delta}{R}\right) = -0.092Eh^2\left(\frac{\delta}{R}\right), \tag{17.2.17}$$

and vanishes at the penetration:

$$x_{p,s} = \pi c = 2.40\sqrt{Rh}. \tag{17.2.18}$$

Example 17.2.1. A vertical cylindrical steel tank with a radius $R = 15$ m (50 ft) and a thickness $h = 25$ mm (1 in.) is built-in into a rigid foundation. The temperature of the air is 38°C (100°F), the temperature of the foundation is 10°C (50°F). Evaluate its bending stress in the neighborhood of the foundation, and the penetration.

The outward (negative) thermal displacement of the free tank wall equals:

$$-(\gamma\,\Delta T)R, \tag{a}$$

where $\gamma = 1 \times 10^{-5}$ mm/mm per °C (0.000 006 in./in. per °F) is the coefficient of thermal expansion for steel. This membrane displacement must be wiped out at

$x = 0$ by an equal and opposite bending displacement $w(0) = +(\gamma\,\Delta T)R$; hence, by (17.2.13b):

$$M_0 = \frac{Eh^2}{2\sqrt{3}}\frac{\gamma\,\Delta T\,R}{R} = 0.29\gamma Eh^2\,\Delta T. \tag{b}$$

The corresponding bending stress:

$$f_b = \frac{M(0)}{S} = \frac{6}{1 \times h^2}\frac{1}{2\sqrt{3}}\gamma Eh^2\,\Delta T = 1.73\,\gamma E\,\Delta T,$$

is independent of the thickness. In this example the value of f_b is:

$$f_b = 1.73 \times 1 \times 10^{-5} \times 200\,000 \times 28 = 96.9 \text{ MPa (15.6 ksi)}.$$

The penetration, by (17.2.14), equals:

$$x_{p,f} = 0.60\sqrt{(15 \times 1\,000) \times 25} = 367.4 \text{ mm (14.7 in.)}.$$

Example 17.2.2. The tank of Example 17.2.1 is attached to the foundation so that it cannot expand but is free to rotate. Compute the maximum stress in the tank and evaluate the penetration.

By (17.2.17):

$$M_{\max} = -0.092Eh^2\gamma\,\Delta T = -0.32\frac{Eh^2}{2\sqrt{3}}\gamma\,\Delta T.$$

The maximum bending stress is -0.32 of the stress in Example 17.2.1, or -31 MPa (-4.9 ksi). The penetration for $R = 15$ m (50 ft), $h = 25$ mm (1 in.) is, by (17.2.18):

$$x_{p,s} = 2.40\sqrt{(15 \times 1\,000) \times 25} = 1\,470 \text{ mm (59 in.)}.$$

Example 17.2.3. A cylindrical reinforced concrete oil tank is 15 m (50 ft) high, has a radius of 15 m (50 ft) and a thickness of 150 mm (6 in.). It is built-in into a heavy foundation. Determine the maximum and minimum moments in the neighborhood of the shell boundary, and the penetration when the tank is full. Oil weighs approximately 800 kg/m^3 $\approx$ 7.9 kN/m^3 (50 lb/cu ft).

The pressure of the oil at the bottom of the tank wall of height H is:

$$p = qH = 7.9 \times 15 = 118.5 \text{ kN/m}^2 \text{ (2.5 ksf)},$$

The membrane tensile force in the tank is:

$$N_\theta = -pR = -118.5 \times 15 = -1\,778 \text{ kN/m}.$$

The outward radial membrane displacement is, by (17.2.1):

$$\delta = -\frac{pR^2}{Eh} = -\frac{(118.5/1\,000)(15 \times 1\,000)^2}{23\,000 \times 150} = -7.7 \text{ mm (0.35 in.)}.$$

The maximum moment, by (17.2.13 b), is*:

$$M_0 = \frac{Eh^2}{2\sqrt{3}}\frac{pR}{Eh} = \frac{pRh}{2\sqrt{3}} = \frac{(118.5/1\,000)(15 \times 1\,000) \times 150}{2\sqrt{3}}$$

$$= 77 \text{ N·m/m (18 k·in./in.)}.$$

The maximum negative moment is only $-0.043M_0 = -3.31$ Nm/m.

*This result ignores the small rotation due to the variation of p with depth (see Example 17.5.4).

If the shell were hinged at the base, by means of a Mesnager hinge with crossed reinforcement (Fig. 17.2.4), the maximum negative moment would be:

$$M_{\min} = -0.32 \times 77 = -24.64 \text{ N}\cdot\text{m/m } (5.76 \text{ k}\cdot\text{in./in.}).$$

The penetrations for fixed and simply supported boundaries equal:

$$x_{p,f} = 0.60\sqrt{Rh} = 0.60\sqrt{(15 \times 1\,000) \times 150} = 900 \text{ mm (36 in.)},$$

$$x_{p,s} = 2.40\sqrt{Rh} = 3\,600 \text{ mm (144 in.)}.$$

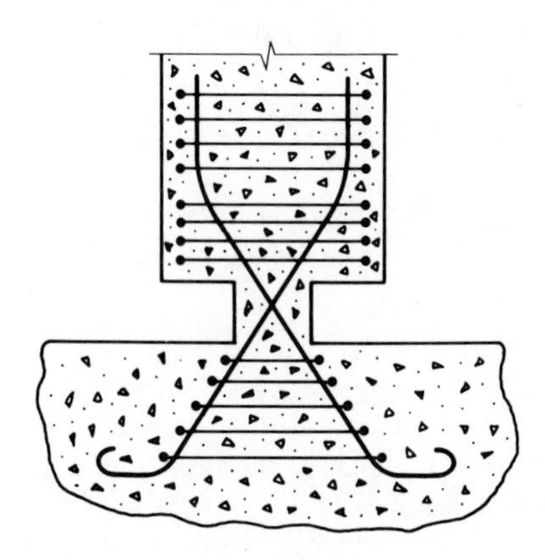

Mesnager hinge

FIGURE 17.2.4

17.3 BENDING DISTURBANCES IN BARREL SHELLS [12.10]

Bending disturbances in barrel shells occur at the intersection of the barrel with the stiffeners and along the longitudinal edges. The first kind of disturbance is due to the incompatibility of displacements between the shell and the stiffener; the second either to the incompatibility of displacement between the shell and the boundary beams, or to the lack of suitable membrane reactions along the longitudinal boundaries when boundary beams are not provided.

The bending disturbances in the neighborhood of the stiffeners may be approximated by those for axisymmetrical stresses in cylinders.

Indicating by q the dead load of the barrel, the hoop force in the barrel was found to be a *variable* force $N_\varphi = qR \cos \varphi$ [see (16.5.3)]. An approximation to the boundary disturbances is obtained by ignoring the influence of the variation of N_φ with φ, i.e., by assuming that the barrel behaves at each point φ of the meridian as if the N_φ had everywhere the value it has at that point φ. Considering the maximum value qR of N_φ, the corresponding maximum compressive strain becomes qR/Eh, and the maximum inward membrane displacement equals:

$$\delta = \frac{qR^2}{Eh}. \tag{17.3.1}$$

When the barrel is *fixed* into the stiffener (as is the case for the boundary of two barrels continuous over the stiffener), δ is wiped out by an equal and opposite displacement due to bending, and the maximum moment and the maximum shear are given by (17.2.13 b, c):

$$\begin{aligned} M_0 &= \left(\frac{Eh^2}{2\sqrt{3}}\right)\left(-\frac{qR}{Eh}\right) = -\frac{1}{2\sqrt{3}}qRh = -0.29qRh, \\ V_0 &= -\frac{Eh}{\sqrt[4]{3}}\sqrt{\frac{h}{R}}\left(-\frac{qR}{Eh}\right) = 0.76\sqrt{\frac{h}{R}}qR. \end{aligned} \tag{17.3.2}$$

When the barrel is *hinged* at the stiffener, the maximum moment occurs at $x/c = \pi/4$ and the maximum shear at $x/c = 0$; their values are given by (17.2.17) and (17.2.15 d):

$$\begin{aligned} M_{\max} &= -0.32\frac{Eh^2}{2\sqrt{3}}\left(-\frac{qR}{Eh}\right) = 0.092qRh, \\ V_0 &= -\frac{Eh}{2\sqrt[4]{3}}\sqrt{\frac{h}{R}}\left(-\frac{qR}{Eh}\right) = 0.38\sqrt{\frac{h}{R}}qR. \end{aligned} \tag{17.3.3}$$

The variation of M and V with φ is obtained by multiplying (17.3.2) and (17.3.3) by $\cos\varphi$, since N_φ varies as $\cos\varphi$.

By (17.2.14) and (17.2.18), the ratio of the maximum penetrations to the radius R are, in the two cases:

$$\frac{x_{p,f}}{R} = 0.60\sqrt{\frac{h}{R}}, \qquad \frac{x_{p,s}}{R} = 2.40\sqrt{\frac{h}{R}}. \tag{17.3.4}$$

The evaluation of the bending disturbances along the longitudinal boundaries of the barrel is a complex mathematical procedure and shows that the penetration at right angles to the barrel axis is greater than along the barrel axis. Tables for the evaluation of such disturbances simplify their calculation*, but it is seldom that such bending stresses dictate a preliminary design.

Example 17.3.1. Evaluate the maximum moments and shears in the barrel of Example 16.5.1.

Indicating by q the sum of the dead and the snow loads, $q = 22(0.1) + 1.5 = 3.7$ kN/m² (77 psf), and with $R = 3$ m (10 ft), $h = 100$ mm (4 in.) $= 0.1$ m ($\frac{1}{3}$ ft), we obtain for a fixed barrel, i.e., at the intermediate support, by (17.3.2):

$$M_0 = -0.29 \times 3.7 \times 3 \times 0.1 = -0.32 \text{ kN·m/m } (-74 \text{ lb in./in.}),$$

$$V_0 = 0.76\sqrt{\frac{0.1}{3}}(3.7 \times 3) = 1.54 \text{ kN/m } (107 \text{ lb/ft}),$$

*"Design of Cylindrical Concrete Shell Roofs," *ASCE Manual of Engineering Practice*, No. 31, New York, 1952.

and for a simply supported barrel, i.e., at the end support, by (17.3.3):

$$M_{\max} = 0.10 \text{ kN}\cdot\text{m/m (24 lb in./in.)}, \qquad V_0 = 0.77 \text{ kN/m (54 lb/ft)}.$$

The penetrations for the fixed and the simply supported barrels are, by (17.3.4):

$$x_{p,f} = 0.60\sqrt{Rh} = 0.60\sqrt{3 \times 0.1} = 0.33 \text{ m (1.1 ft)},$$
$$x_{p,s} = 2.40\sqrt{Rh} = 1.32 \text{ m (4.4 ft)},$$

and show that the width of the disturbed region is small compared to the barrel length $l = 12$ m; hence, the bending disturbances originating at each stiffener do not interfere with each other.

The load-carrying mechanism of the barrel may now be clearly understood. For the case of the fixed barrel, e.g., the bending action is practically nonexistent a distance 0.33 m away from the stiffener, so that the load is carried by membrane stresses over $[12 - (2 \times 0.33)]/12 = 95\%$ of the shell length. At its ends the barrel is fixed to the rigid vertical stiffeners and cannot deform; hence, the membrane strains and stresses vanish at the ends, and the load there is carried by bending and transverse shear only. The load carried in bending equals the sum of the two transverse shears V_0 at the barrel ends. The ratio r of the load carried in bending to the total load on a unit width strip of length l is, by (17.3.2), (17.3.3):

$$r = (0.38 + 0.76)\sqrt{\frac{h}{R}}\left(\frac{R}{l}\right) = 1.14\sqrt{\frac{h}{R}}\left(\frac{R}{l}\right). \tag{17.3.5}$$

In the present example:

$$r = 1.14\sqrt{\frac{0.1}{3}}\left(\frac{3}{12}\right) = 0.05 = 5\%.$$

17.4 BENDING DISTURBANCES IN TRANSLATIONAL SHELLS [12.11]

The bending disturbances at the stiffener supported, *curved* boundaries of translational shells may be approximated by cylindrical theory, ignoring the curvature *at right angles* to the stiffener. Thus, the formulas of Section 17.3 can be used, with R equal to the radius of the stiffener at the point considered and q as the total vertical load.

Example 17.4.1. Determine the maximum bending moment and shear in the elliptic paraboloid of Example 16.6.1.

The radii of the stiffeners are given by:

$$R_x = \frac{\sqrt{[1 + (dz_a/dx)^2]^3}}{d^2z_a/dx^2} = \frac{a^2}{2c_a}\sqrt{\left(1 + \frac{4c_a^2x^2}{a^4}\right)^3},$$
$$R_y = \frac{\sqrt{[1 + (dz_b/dy)^2]^3}}{d^2z_b/dy^2} = \frac{b^2}{2c_b}\sqrt{\left(1 + \frac{4c_b^2y^2}{b^4}\right)^3}. \tag{17.4.1}$$

Their maximum values occur at $x = a$, $y = b$, respectively:

$$R_{x,\max} = \frac{a^2}{2c_a}\sqrt{\left(1 + \frac{4c_a^2}{a^2}\right)^3} = \frac{9^2}{2 \times 1.75}\sqrt{\left(1 + \frac{4 \times 1.75^2}{9^2}\right)^3} = 28.6 \text{ m (94 ft)},$$
$$R_{y,\max} = \frac{b^2}{2c_b}\sqrt{\left(1 + \frac{4c_b^2}{b^2}\right)^3} = \frac{12^2}{2 \times 2.5}\sqrt{\left(1 + \frac{4 \times 2.5^2}{12^2}\right)^3} = 36.6 \text{ m (125 ft)}.$$

The maximum moment and the maximum shear occur on the stiffener parallel to y and, for a simply supported boundary, have the values obtained from (17.3.3):

$$M_{max} = 0.092 \times 3.7 \times 36.6 \times 0.1 = 1.25 \text{ kN}\cdot\text{m/m (295 lb in./in.)},$$

$$V_0 = 0.38\sqrt{\frac{0.1}{36.6}}(3.7 \times 36.6) = 2.69 \text{ kN/m (190 lb/ft)}.$$

The stresses in a circular translational shell with the same spans and rises as those of the present example are smaller because the radii R_x and R_y are smaller (see Example 16.6.2).

A very shallow elliptic paraboloid, with c_a/a and c_b/b less than $\frac{1}{20}$, behaves like a slightly curved plate. Its bending stresses, which are not boundary disturbances but pervade a large portion of the shell, are smaller than those in a flat plate of equal dimensions but may be critical in determining both the shell thickness and its reinforcement.

The bending stresses in a hyperbolic paraboloid supported on straight boundaries cannot be evaluated by cylindrical theory for lack of curvature along the supported boundary. For shallow shells with c_a/a and c_b/b less than $\frac{1}{10}$, the bending stresses pervade a large portion of the shell. For less shallow hyperbolic paraboloids the penetration of bending disturbances is substantially greater than in those with curved boundaries but is small compared to $\frac{1}{2}\sqrt{ab}$, provided the thickness h is small enough for $h/(c_a + c_b)$ to be much less than one. The bending stresses in the disturbed area near the stiffeners are of the same order of magnitude as the membrane shear stresses (16.6.9).

17.5 BENDING DISTURBANCES IN SHELLS OF REVOLUTION [12.8]

The bending disturbances in the neighborhood of the boundary of a thin rotational shell may be proved to be practically identical to those in a cylindrical shell with a radius R equal to the radius R_2 of the shell at right angles to the meridian, i.e., to the radius of the sphere tangent to the shell at the boundary, provided [Fig. 17.5.1(a)]:

1. The opening angle φ_0 of the shell is greater than 20°.
2. The shell thickness is small enough for the penetration s_1 to be much smaller than the length of the meridional arc AB.
3. The shell hugs the tangent sphere along the meridian for a length equal to at least the penetration, i.e., for a length of meridian $s_1 = R_1\varphi_1$ at least equal to $s_p = R_2\varphi_p$.

The last condition is satisfied by the toroidal shell of Fig. 17.5.1(b), but is not satisfied by the elliptical shell of Fig. 17.5.1(c).

The magnitude of the bending stresses is obtained by stating that the

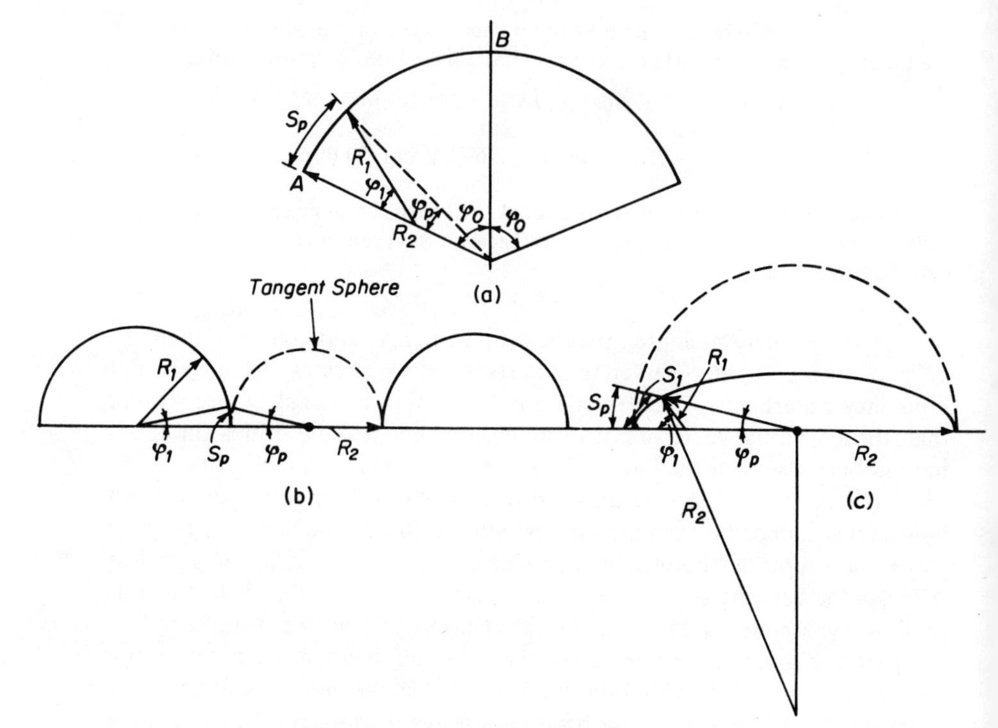

FIGURE 17.5.1

membrane displacements on the shell boundary are wiped out by equal and opposite bending displacements. A variety of commonly encountered situations are illustrated in the following examples in which $\varphi_0 = 90°$.

Example 17.5.1. A spherical concrete dome of radius $R = 15$ m (50 ft) and thickness $h = 100$ mm (4 in.) is *fixed* into a solid foundation. Its temperature increases by ΔT, while the temperature of the foundation remains unchanged. Evaluate the maximum bending moment and transverse shear in the shell.

The *outward* thermal displacement of the boundary of the shell equals:

$$\delta = -\gamma \, \Delta T \, R, \tag{a}$$

where γ is the coefficient of thermal expansion. This displacement is prevented by the foundation; hence, the bending stresses must produce an *inward* displacement equal to $+\gamma \, \Delta T \, R$, and, by (17.2.13 b, c):

$$M_0 = \frac{Eh^2}{2\sqrt{3}} \gamma \, \Delta T, \qquad V_0 = -\frac{Eh}{\sqrt[4]{3}} \sqrt{\frac{h}{R}} \gamma \, \Delta T. \tag{17.5.1}$$

In the present example, for $\Delta T = 5°\text{C}$ (9°F):

$$M_0 = \frac{23\,000}{1\,000} \left(\frac{100^2}{2\sqrt{3}} \right) \times 1 \times 10^{-5} \times 5 = 3.32 \text{ kN}\cdot\text{m/m (831 lb in./in.)},$$

$$V_0 = -\frac{1}{\sqrt[4]{3}}\left(\frac{23\,000 \times 100}{1\,000}\right)\sqrt{\frac{0.1}{15}} \times 1 \times 10^{-5} \times 5 = 0.71 \text{ kN/mm (45 lb/in.)}.$$

It is noticed that a relatively small temperature differential is capable of developing a high bending moment.

Example 17.5.2. The dome of Example 17.5.1 is acted upon by its dead load w and is *simply supported* at the boundary. Determine M_{max} and V_0.

The *outward* membrane displacement δ due to the dead load is given by (16.3.1). Hence, by (17.2.17) and (17.2.15d):

$$\begin{aligned} M_{max} &= -0.092\frac{Eh^2}{R}\frac{wR^2}{Eh} = -0.092wRh, \\ V_0 &= -\frac{Eh}{2\sqrt[4]{3}}\sqrt{\frac{h}{R}}\frac{wR}{Eh} = -0.38\sqrt{\frac{h}{R}}\,wR. \end{aligned} \tag{17.5.2}$$

For $R = 15$ m (50 ft), $h = 100$ mm (4 in.):

$$\begin{aligned} w &= 22(0.1) = 2.2 \text{ kN/m}^2 \text{ (44 psf)}, \\ M_{max} &= -0.304 \text{ kN}\cdot\text{m/m } (-72 \text{ lb}\cdot\text{in./in.}), \\ V_0 &= -1.02 \text{ kN/m } (-73 \text{ lb/ft}). \end{aligned}$$

Example 17.5.3. The dome of Example 17.5.1 is acted upon by a snow load $q = 1.5$ kN/m^2 (30 psf) and is *fixed* into its foundation. Determine M_0 and V_0.

The outward membrane displacement due to q is given by (16.3.4); hence, by (17.2.13 b, c):

$$M_0 = \frac{Eh^2}{2\sqrt{3}}\left(\frac{1}{2}\frac{qR}{Eh}\right) = 0.145qRh, \qquad V_0 = -0.38\sqrt{\frac{h}{R}}\,qR. \tag{17.5.3}$$

In this example:

$$\begin{aligned} M_0 &= 0.145 \times 1.5 \times 15 \times 0.1 = 0.33 \text{ kN}\cdot\text{m/m (73 lb in./in.)}, \\ V_0 &= -0.38\sqrt{\frac{0.1}{15}} \times 1.5 \times 15 = -0.70 \text{ kN/m } (-47 \text{ lb/ft}). \end{aligned}$$

Example 17.5.4. The spherical dome of Example 17.5.1 is *fixed* at the boundary and is acted upon by its dead load w. Determine M_0 and V_0.

By (16.3.1), the dead load w produces an outward membrane displacement $\delta = -wR^2/Eh$, while the boundary rotation α is given by (16.3.3). The bending stresses must produce equal and opposite displacement and rotation, but the influence of the rotation is small and may usually be neglected. Hence, by (17.2.13 b, c):

$$\begin{aligned} M_0 &= \frac{Eh^2}{2\sqrt{3}}\left(\frac{wR}{Eh}\right) = 0.29wRh, \\ V_0 &= -\frac{Eh}{\sqrt[4]{3}}\sqrt{\frac{h}{R}}\frac{wR}{Eh} = -0.76\sqrt{\frac{h}{R}}\,wR. \end{aligned} \tag{17.5.4}$$

In the present example:

$$M_0 = 0.29 \times 2.2 \times 15 \times 0.1 = 0.96 \text{ kN·m/m (227 lb in./in.)},$$
$$V_0 = -2.05 \text{ kN/m } (-146 \text{ lb/ft}).$$

It may be shown that the corresponding formulas taking the rotation α into account are:

$$M_0 = 0.29wRh\left(1 - 1.5\sqrt{\frac{h}{R}}\right),$$
$$V_0 = -0.76\sqrt{\frac{h}{R}}wR\left(1 - 0.76\sqrt{\frac{h}{R}}\right). \tag{17.5.4a}$$

The approximate values (17.5.4) are on the safe side and, since h/R is usually smaller than $\frac{1}{200}$, the corrections due to the rotation α are less than 10% for the moment and less than 5% for the shear.

The bending moment and shear at the boundary of rotational shells with an opening angle $\varphi_0 < 90°$ may be obtained by the following intuitive considerations.

The displacement δ in the radial direction due to a shear V_0 also in the radial direction [Fig. 17.5.2 (a)] may be approximated by cylindrical theory and, for a boundary free to rotate, is obtained from (17.2.15 d):

$$\delta = -\frac{2\sqrt[4]{3}}{Eh}\sqrt{\frac{R}{h}}RV_0. \tag{a}$$

Considering the meridian as a cantilevered beam [Fig. 17.5.2 (b)], the *horizontal* displacement δ_h due to V_0 is:

$$\delta_h = \delta \sin \varphi_0. \tag{b}$$

The *horizontal force H* per unit length capable of producing such a displacement must have a component in the radial direction equal to V_0 and, hence,

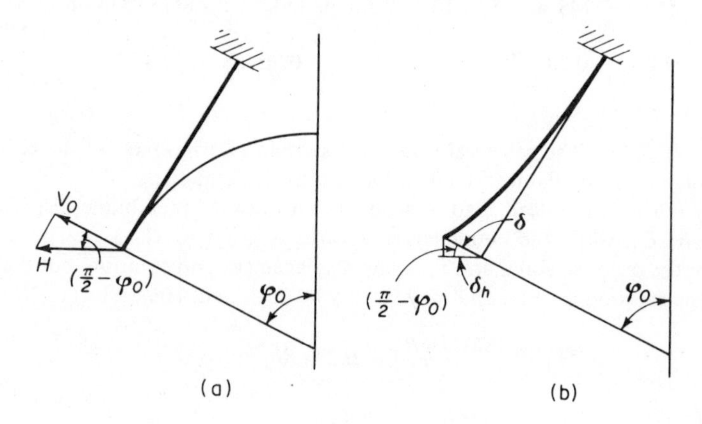

FIGURE 17.5.2

is given by:

$$V_0 = H \sin \varphi_0. \tag{c}$$

By (a), (b), and (c), δ_h becomes:

$$\delta_h = -\frac{2\sqrt[4]{3}}{Eh}\sqrt{\frac{R}{h}}RH \sin^2 \varphi_0,$$

from which:

$$H = -\frac{Eh}{2\sqrt[4]{3}}\sqrt{\frac{h}{R}}\frac{\delta_h}{R \sin^2 \varphi_0}. \tag{17.5.5}$$

In (17.5.5), H is positive when *outward* and δ_h positive when *inward.* For the case of a boundary that cannot rotate, (17.2.13 c) gives:

$$V_0 = H \sin \varphi_0 = -\frac{Eh}{\sqrt[4]{3}}\sqrt{\frac{h}{R}}\frac{\delta}{R} = -\frac{Eh}{\sqrt[4]{3}}\sqrt{\frac{h}{R}}\frac{\delta_h}{R \sin \varphi_0},$$

from which:

$$H = -\frac{Eh}{\sqrt[4]{3}}\sqrt{\frac{h}{R}}\frac{\delta_h}{R \sin^2 \varphi_0}, \tag{17.5.6}$$

and (17.2.13b) gives:

$$M_0 = \frac{Eh^2}{2\sqrt{3}}\frac{\delta}{R} = \frac{Eh^2}{2\sqrt{3}}\frac{\delta_h}{R \sin \varphi_0}. \tag{17.5.7}$$

The horizontal force H is the outward *bending thrust* on the shell boundary; the total outward thrust $\bar{H}$ on the shell boundary is the algebraic sum of the bending and the membrane thrusts, i.e., the sum of H and the horizontal component of N_φ at $\varphi = \varphi_0$ (Fig. 17.5.3):

$$\bar{H} = H - N_\varphi]_{\varphi=\varphi_0} \cos \varphi_0. \tag{17.5.8}$$

The horizontal *membrane* displacement δ_h is due either to a temperature difference ΔT or to the membrane stresses N_θ in the parallel of radius r_0 and equals either $-\gamma\, \Delta T\, r_0$, or:

$$\delta_h = \epsilon_\theta r_0 = \frac{N_\theta r_0}{Eh}. \tag{17.5.9}$$

The δ_h due to N_θ is outward (negative) or inward (positive), depending on whether N_θ is tensile or compressive. The δ_h due to ΔT is outward or inward, depending on whether ΔT is an increase or a decrease in temperature.

The following examples illustrate the application of equations (17.5.5) through (17.5.9) to shells under a variety of loads. It is essential to remember that these formulas do not apply to very flat shells with a rise of less than one-tenth of the span or an opening φ_0 smaller than 20°.

Example 17.5.5. Determine the total thrust $\bar{H}$ and the maximum moment in a concrete spherical shell of opening $\varphi_0 = 30°$, thickness $h = 100$ mm (4 in.), and radius $R = 15$ m (50 ft), if its *simply supported* boundary is restrained from moving and its temperature rises by 28°C (50°F).

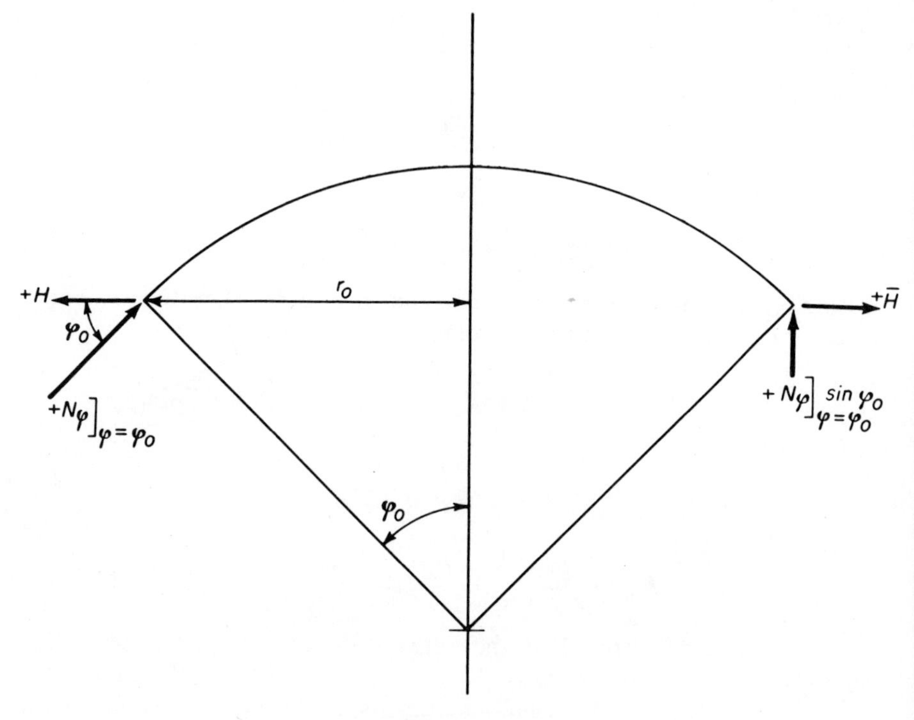

FIGURE 17.5.3

The radius of the parallel at the support is:

$$r_0 = R\sin\varphi_0, \tag{d}$$

and the *outward* horizontal displacement is given by:

$$\delta_h = -\gamma\,\Delta T\,R\sin\varphi_0.$$

The thermal expansion of the unrestrained shell does not produce membrane stresses. Hence, by (17.5.5), the thrust capable of inducing a $-\delta_h$ in the shell is:

$$\begin{aligned}\overline{H} = H &= -\frac{Eh}{2\sqrt[4]{3}}\sqrt{\frac{h}{R}}\,\frac{\gamma\,\Delta T\,R\sin\varphi_0}{R\sin^2\varphi_0}\\ &= -\frac{Eh}{2\sqrt[4]{3}}\sqrt{\frac{h}{R}}\,\frac{\gamma\,\Delta T}{\sin\varphi_0}.\end{aligned} \tag{17.5.10}$$

For $\Delta T > 0$, H is negative, i.e., is an inward thrust on the shell.
By (c):

$$V_0 = H\sin\varphi_0 = -\frac{Eh}{2\sqrt[4]{3}}\sqrt{\frac{h}{R}}\,\gamma\,\Delta T. \tag{17.5.11}$$

The maximum moment is given by (17.2.17):

$$M_{\max} = -0.092Eh^2\left(-\frac{\delta_h}{R\sin\varphi_0}\right) = -0.092\,Eh^2\gamma\,\Delta T. \tag{17.5.12}$$

M_{max} puts the outer fibers of the shell in tension. For the values of R, h, φ_0, and ΔT given in this example:

$$\bar{H} = H = -\frac{1}{2\sqrt[4]{3}}\frac{23\,000 \times 100}{1\,000}\sqrt{\frac{100}{15 \times 1\,000}}\frac{1 \times 10^{-5} \times 28}{\sin 30^\circ}$$

$$= -0.040 \text{ kN/mm } (-224 \text{ lb/in.}),$$

$$M_{max} = -0.092 \times \frac{23\,000}{1\,000} \times 100^2 \times 1 \times 10^{-5} \times 28$$

$$= -5.92 \text{ kN}\cdot\text{m/m } (-1\,325 \text{ lb in./in.}).$$

Example 17.5.6. Determine $\bar{H}$ and M_{max} due to a snow load of 2 kN/m² (40 psf) on the shell of Example 17.5.5.

By (17.5.9) and (16.2.5), the inward displacement δ_h equals:

$$\delta_h = \frac{N_\theta r_0}{Eh} = \frac{\frac{1}{2}qR\cos 2\varphi_0(R\sin\varphi_0)}{Eh} = \frac{1}{2}\frac{qR^2}{Eh}\cos 2\varphi_0 \sin\varphi_0.$$

Hence, by (17.5.6):

$$H = -\frac{Eh}{2\sqrt[4]{3}}\sqrt{\frac{h}{R}}\left(-\frac{1}{2}\frac{qR^2}{Eh}\frac{\cos 2\varphi_0 \sin\varphi_0}{R\sin^2\varphi_0}\right)$$

$$= \frac{1}{4\sqrt[4]{3}}qR\sqrt{\frac{h}{R}}\frac{\cos 2\varphi_0}{\sin\varphi_0}, \tag{17.5.13}$$

and, by (16.2.5) and (17.5.8):

$$\bar{H} = H - N_\varphi]_{\varphi=\varphi_0}\cos\varphi_0 = \frac{1}{4\sqrt[4]{3}}qR\sqrt{\frac{h}{R}}\frac{\cos 2\varphi_0}{\sin\varphi_0} - \frac{1}{2}qR\cos\varphi_0.$$

By (17.2.17), with $\delta = -\delta_h/\sin\varphi_0$:

$$M_{max} = -0.092Eh^2\left(-\frac{1}{2}\frac{qR}{Eh}\cos 2\varphi_0\right)$$

$$= 0.046qRh\cos 2\varphi_0. \tag{17.5.14}$$

With the values of q, R, h, and φ_0 of this example:

$$H = \frac{1}{4\sqrt[4]{3}}\left(\frac{2}{1\,000}\right)(15 \times 1\,000)\sqrt{\frac{100}{15 \times 1\,000}}\frac{\cos 60^\circ}{\sin 30^\circ} = 0.465 \text{ N/mm } (2.58 \text{ lb/in.}),$$

$$N_\varphi \cos\varphi]_{\varphi=\varphi_0} = \frac{1}{2}\frac{2}{1\,000}(15 \times 1\,000)\cos 30^\circ = 13 \text{ N/mm } (72.2 \text{ lb/in.}),$$

$$\bar{H} = 0.465 - 13 = -12.535 \text{ N/mm } (-835 \text{ lb/ft}),$$

$$M_{max} = 0.046\left(\frac{2}{1\,000^2}\right)(15 \times 1\,000 \times 100) \times \cos 60^\circ$$

$$= 0.069 \text{ kN}\cdot\text{m/m } (15.3 \text{ lb in./in.}).$$

Example 17.5.7. Determine $\bar{H}$ and M_0 at the boundary of the shell of Example 17.5.5, when the shell is *fixed* at the boundary and is acted upon by its dead load $w = 22(0.1) = 2.2$ kN/m² (47 psf).

By (16.2.3) and (17.5.9):

$$\delta_h = \frac{wR}{Eh}\left(\cos\varphi_0 - \frac{1}{1 + \cos\varphi_0}\right)R\sin\varphi_0, \tag{17.5.15}$$

and, by (17.5.6):

$$H = -\frac{Eh}{\sqrt[4]{3}}\sqrt{\frac{h}{R}}\left[-\frac{wR}{Eh\sin\varphi_0}\left(\cos\varphi_0 - \frac{1}{1+\cos\varphi_0}\right)\right]$$

$$= \frac{1}{\sqrt[4]{3}}wR\sqrt{\frac{h}{R}}\,\frac{\cos\varphi_0 - 1/(1+\cos\varphi_0)}{\sin\varphi_0}. \tag{17.5.16}$$

By (16.2.3):

$$\bar{H} = H - \frac{wR}{1+\cos\varphi_0}\cos\varphi_0, \tag{17.5.17}$$

and, by (17.5.7) and (17.5.15):

$$M_0 = \frac{Eh^2}{2\sqrt[4]{3}}\left[-\frac{wR}{Eh}\left(\cos\varphi_0 - \frac{1}{1+\cos\varphi_0}\right)\right]$$

$$= -\frac{1}{2\sqrt[4]{3}}wRh\left(\cos\varphi_0 - \frac{1}{1+\cos\varphi_0}\right). \tag{17.5.18}$$

With the geometry of this example:

$$H = \frac{1}{\sqrt[4]{3}}\left(\frac{2.2}{1\ 000}\right)(15\times 1\ 000)\sqrt{\frac{100}{15\times 1\ 000}}\,\frac{\cos 30^\circ - 1/(1+\cos 30^\circ)}{\sin 30^\circ}$$

$$= 1.35\ \text{N/mm}\ (8.0\ \text{lb/in.}),$$

$$\bar{H} = 1.35 - \frac{(\frac{2.2}{1\ 000})(15\times 1\ 000)}{1+\cos 30^\circ}\cos 30^\circ = 1.35 - 15.32$$

$$= -14\ \text{N/mm}\ (-83\ \text{lb/in.})\quad \text{(inward)},$$

$$M_0 = \frac{1}{2\sqrt[4]{3}}\left(\frac{2.2}{1\ 000^2}\right)(15\times 1\ 000)\times 100\times\left(\cos 30^\circ - \frac{1}{1+\cos 30^\circ}\right)$$

$$= 0.413\ \text{kN}\cdot\text{m/m}\ (98\ \text{lb in./in.}).$$

Table 17.5.1 gives a summary of the bending disturbances in cylindrical shells due to a radial boundary displacement δ, and in barrel shells due to a dead load q.

Table 17.5.1 BENDING DISTURBANCES IN CYLINDRICAL SHELLS

Type	Maximum moment	Maximum shear	Penetration
Cylinder:			
Fixed	$M_0 = 0.29Eh^2\left(\frac{\delta}{R}\right)$	$V_0 = -\left(\frac{Eh}{\sqrt[4]{3}}\right)\sqrt{\frac{h}{R}}\left(\frac{\delta}{R}\right)$	$x_p = 0.60\sqrt{Rh}$
Hinged	$M_{max} = -0.092Eh^2\left(\frac{\delta}{R}\right)$ at $x = 0.60\sqrt{Rh}$	$V_0 = -\left(\frac{Eh}{2\sqrt[4]{3}}\right)\sqrt{\frac{h}{R}}\left(\frac{\delta}{R}\right)$	$x_p = 2.40\sqrt{Rh}$
Barrel:			
Fixed	$M_0 = -0.29qRh$	$V_0 = 0.76\sqrt{\frac{h}{R}}\,qR$	$x_p = 0.60\sqrt{Rh}$
Hinged	$M_0 = 0.092qRh$	$V_0 = 0.38\sqrt{\frac{h}{R}}\,qR$	$x_p = 2.40\sqrt{Rh}$

17.6 BUCKLING OF THIN SHELLS [12.8, 12.10]

Like any thin structural element, a thin shell may buckle under compressive stresses. Although its curvatures make a shell more stable against buckling than a plate of equal thickness and span, buckling is possibly the most important limiting factor in shell design. It is therefore unfortunate that the buckling loads established by theory do not agree, in most cases, with those obtained through experiments, due to the imperfections in actual shells. In what follows, a series of semiempirical buckling formulas will be presented (without proof), which practice shows to be safe for use in preliminary design.

1. *Spherical Shells.* The theoretical buckling value of an external pressure p on a complete spherical shell of radius R and thickness h may be proved to be:

$$p_{cr} = \frac{2}{\sqrt{3(1-\nu^2)}} E\left(\frac{h}{R}\right)^2 = 1.16E\left(\frac{h}{R}\right)^2 \quad (\nu = 0). \qquad (17.6.1)$$

Experiments show that for a spherical sector the numerical coefficient in (17.6.1) is much smaller than 1.16 and does not depend essentially on the boundary conditions. The lowering of the coefficient is due to imperfections in the shell material and, above all, to deviations from a true spherical surface.

For domes with an opening angle near 90°, both nonlinear theories and experiments show that the buckling load is given approximately by:

$$p_{cr} = \alpha E\left(\frac{h}{R}\right)^2, \qquad (17.6.2)$$

with $\alpha = 0.30$. The coefficient α for *shallow* spherical shells with a rise to span ratio of the order of one-tenth or less is approximately equal to 0.15. A safety factor of from 2 to 3 is suggested in connection with any buckling formula.

The critical values of the live and dead loads are usually approximated by that of an external pressure, as shown in the following examples

Example 17.6.1. Determine the minimum safe thickness of a spherical concrete dome with $R = 30$ m (100 ft), acted upon by its own dead load.

Calling ρ kN/m³ (lb/cu ft) the unit weight of concrete, the shell weight per unit area is $w = \rho h$, and, by (17.6.2), with a coefficient of safety of 3:

$$\rho h = \frac{1}{3} 0.30 E \frac{h^2}{R^2},$$

from which:

$$h_{min} = \frac{10R^2\rho}{E}. \qquad (17.6.3)$$

With $\rho = 22\ \text{kN/m}^3$ (140 lb/cu ft), $E = 23\,000$ MPa (3.3×10^6 psi), and $R = 30$ m (100 ft), we obtain:

$$h_{\min} = \frac{10 \times (30\,000)^2 \times 22 \times 1\,000}{23\,000 \times 1\,000^3} = 8.6 \text{ mm (0.4 in.)}.$$

The actual shell thickness must be greater than $h_{\min}$ for practical reasons.

Example 17.6.2. A shallow spherical dome made out of concrete with a rise of 3 m (10 ft) and a span of 30 m (100 ft) has a thickness of 50 mm (2 in.). Check its stability under a snow load of 2 kN/m^2 (40 psf).

The total load w on the shell is:

$$w = 2 + 22\left(\frac{50}{1\,000}\right) = 3.1\ \text{kN/m}^2 \text{ (63 psf)}.$$

The radius of the shell equals, by (c) of Section 16.6:

$$R = \frac{15^2 + 3^2}{2 \times 3} = 39 \text{ m (130 ft)}.$$

With $\alpha = 0.15$, (17.6.2) gives:

$$w_{\text{all}} = \frac{0.15}{3} \times 23\,000\left(\frac{50}{39 \times 1\,000}\right)^2 = 0.001\,89\ \text{N/mm}^2 = 1.89\ \text{kN/m}^2 \text{ (35 psf)}.$$

The shell is unsafe since w_{all} is smaller than w. For $h = 100$ mm (4 in.), $w = 4.2\ \text{kN/m}^2$, $w_{\text{all}} = (100/50)^2 \times 1.89 = 7.56\ \text{kN/m}^2$, and the shell is stable.

2. *Shells of Revolution.* The buckling load for a shell of revolution is safely approximated by that of a spherical shell with a radius equal to the maximum radius R_2 of the shell.

Example 17.6.3. A conical shell of aluminum spans 15 m (50 ft), has a rise of 1.5 m (5 ft) and is 25 mm (1 in.) thick. It supports a snow load of 0.8 kN/m^2 (15 psf). Check its stability.

With a unit weight of 26.7 kN/m^3, the aluminum shell weighs 0.668 kN/m^2 and the total load on the shell is $w = 1.468\ \text{kN/m}^2$. From Fig. 17.6.1, $\tan \varphi = 0.20$, $R_2 = AD = AC/\tan \varphi = (\sqrt{7.5^2 + 1.5^2})/0.20 = 38.25$ m and, by (17.6.2), with $\alpha = 0.15$ and $E = 69\,000$ MPa:

$$w_{\text{all}} = \frac{0.15}{3} \times 69\,000\left(\frac{25}{38.25 \times 1\,000}\right)^2 = 0.001\,5\ \text{N/mm}^2 = 1.5\ \text{kN/m}^2 \text{ (30 psf)}.$$

The shell is safely stable.

3. *Synclastic Translational Shells.* The buckling load on a translational shell of principal radii R_1, R_2 is approximated by that of a spherical shell of radius $R = \sqrt{R_1 R_2}$, i.e., by:

$$p_{\text{cr}} = \alpha E \frac{h^2}{R_1 R_2}. \tag{17.6.4}$$

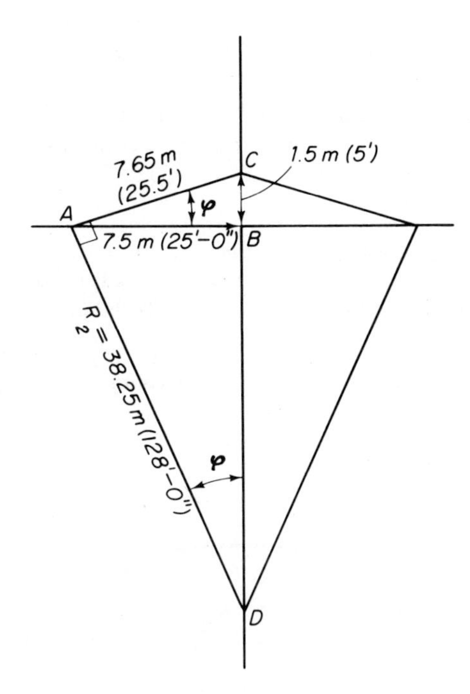

FIGURE 17.6.1

Example 17.6.4. A concrete elliptic paraboloid 75 mm (3 in.) thick covers a rectangular area of 30 m × 24 m (100 ft × 80 ft) with rises of 3 m (10 ft) and 2.4 m (8 ft) on the long and short sides, respectively. Check its stability under a snow load of 2 kN/m² (40 psf).

By (16.6.1), with $a = 15$ m, $b = 12$ m, $c_a = 3$ m, $c_b = 2.4$ m:

$$R_1 = \frac{15^2}{2 \times 3} = 37.5 \text{ m}, \qquad R_2 = \frac{12^2}{2 \times 2.4} = 30 \text{ m}.$$

With a total rise of 5.4 m the shell is not too shallow, and one may choose $\alpha = 0.20$. Hence, by (17.6.4):

$$w_{\text{all}} = \frac{0.20}{3} \times 23\,000 \frac{75^2}{(37\,500)(30\,000)} = 0.007\,4 \text{ N/mm}^2 = 7.4 \text{ kN/m}^2 \text{ (144 psf)}.$$

The total weight on the shell is:

$$w = 2 + 22\left(\frac{75}{1\,000}\right) = 3.65 \text{ kN/m}^2 \text{ (75 psf)},$$

and the shell is safely stable.

4. *Barrel Shells.* Long barrels act as beams and develop longitudinal compressive stresses in the fibers near the crown. These fibers may buckle as if they were beams on an elastic foundation: their buckling load is given

by (17.6.2), with $\alpha = 0.6$ and R equal to the largest radius of the barrel in the transverse direction.

Short barrels may theoretically buckle because of compression in the transverse direction, but it is seldom that this condition governs their preliminary design.

Example 17.6.5. An elliptical barrel (Fig. 17.6.2) of semi-axes $a = 12$ m (40 ft) and $c = 6$ m (20 ft), and thickness $h = 75$ mm (3 in.), is made out of concrete and carries a snow load of 1.5 kN/m² (30 psf). Check its stability. The total load on the shell equals:

$$w = 1.5 + 22\left(\frac{75}{1\,000}\right) = 3.15 \text{ kN/m}^2 \text{ (65 psf)},$$

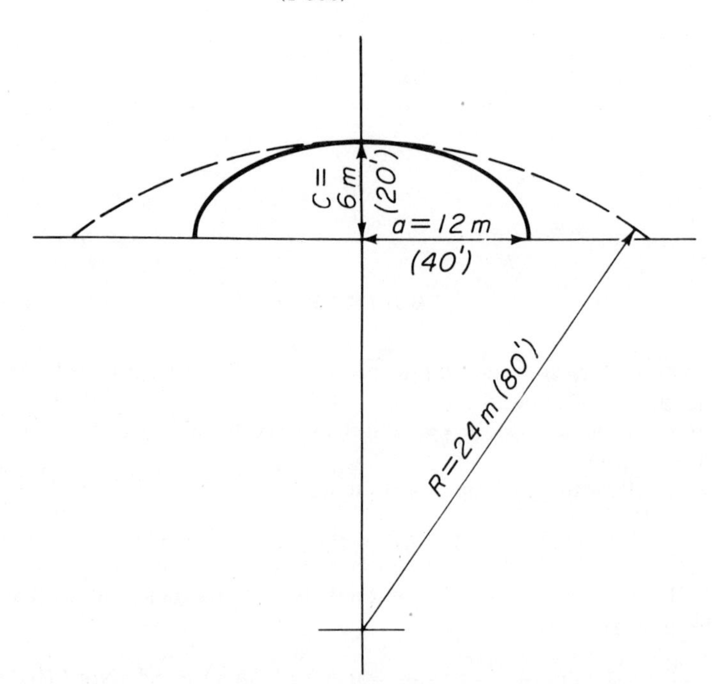

FIGURE 17.6.2

The shell cross section is flattest at the crown, where the radius may be shown to equal:

$$R = \frac{a^2}{c} = \frac{12^2}{6} = 24 \text{ m (80 ft)}, \tag{17.6.5}$$

Hence, with $\alpha = 0.6$:

$$w_{\text{all}} = \frac{0.6}{3} \times 23\,000\left(\frac{75}{24 \times 1\,000}\right)^2 = 0.044 \text{ N/mm}^2 = 44 \text{ kN/m}^2 \text{ (845 psf)}.$$

The shell is safely stable.

5. *Saddle Shells.* Saddle shells are much stronger than cylinders against buckling because the tension along the upward curved fibers stabilizes the downward, curved, compressed fibers. Hence, the buckling load for a saddle shell supported on arched end-stiffeners is safely approximated by that of a cylinder with a radius of curvature equal to that of the saddle shell at the stiffener.

A saddle surface supported on straight boundary beams has a tendency to buckle only in those areas near the center of the saddle where it is flat. The theoretical critical load w for a hyperbolic paraboloidal sector supported on straight boundaries, of sides a and b, rise c (small compared to a and b), and thickness h is given by:

$$w_{cr} = \frac{2}{\sqrt{3(1-\nu^2)}}\left(\frac{c}{b}\right)^2\left(\frac{h}{a}\right)^2 E = 1.16\left(\frac{c}{b}\right)^2\left(\frac{h}{a}\right)^2 E. \qquad (17.6.6)$$

Experiments show that the theoretical buckling value of the load given by (17.6.6) is approximately twice as large as the actual buckling load and depends on the stiffness of the supporting beams.

Example 17.6.6. A hyperbolic paraboloid made out of concrete has a downward curvature span of $2a = 15$ m (50 ft), a rise $c = 1.5$ m (5 ft), and a thickness $h =$ 50 mm (2 in.). Its total load is 4 kN/m² (80 psf). Check its stability if it is supported by stiffeners with downward curvature.

By (17.4.1), the maximum radius of the paraboloid is:

$$R_{max} = \frac{a^2}{2c_a}\sqrt{\left(1+\frac{4c_a^2}{a^2}\right)^3} = \frac{7.5^2}{2\times 1.5}\sqrt{\left(1+\frac{4\times 1.5^2}{7.5^2}\right)^3} = 23.43 \text{ m (78 ft)},$$

and, by (17.6.2):

$$w_{all} = \frac{0.6}{3}\times 23\,000\left(\frac{50}{23\,430}\right)^2 = 0.021 \text{ N/mm}^2 = 21 \text{ kN/m}^2 \text{ (395 psf)}.$$

The paraboloid is safely stable.

Example 17.6.7. A square concrete hyperbolic paraboloid supported on straight beams has sides $2a = 15$ m (50 ft), a rise $c = 1.5$ m (5 ft), and a thickness $h =$ 50 mm (2 in.). Check its stability if it is to carry, besides its dead load, a snow load of 2 kN/m² (40 psf).

The total load on the paraboloid is $22\left(\frac{50}{1\,000}\right) + 2 = 3.1$ kN/m². By (17.6.6), with a factor of $\frac{1}{2}$:

$$w_{cr} = \frac{1}{2}1.16\left(\frac{1.5}{7.5}\right)^2 \times 23\,000\left(\frac{0.05}{7.5}\right)^2 = 0.023\,7 \text{ N/mm}^2$$
$$= 23.7 \text{ kN/m}^2 \text{ (444 psf)}.$$

The coefficient of safety against buckling is $23.7/3.1 = 7.6$.

PROBLEMS

17.1 A concrete hemispherical dome of radius 15 m (50 ft) is 100 mm (4 in.) thick and carries no live load. Determine the thrust H it exerts on a ring at its base that prevents its lateral displacement but has negligible torsional rigidity.

17.2 Determine the thrust and bending moment at the base of the dome of Problem 17.1 if it is built-in into a rigid foundation.

17.3 Determine the increase in thrust in the dome of Problem 17.1 due to a live load of 1.5 kN/m^2 (30 psf).

17.4 Determine the increase in moment at the base of the dome of Problem 17.2 due to: (a) a live load of 1.5 kN/m^2 (30 psf), (b) an increase in temperature of 16°C (60°F). Establish the grade 40 steel reinforcement necessary in each case to absorb the bending moment, if the thickness is increased to 150 mm (6 in.) at the base of the dome.

17.5 Determine the penetration of the bending stresses in the dome of Problems 17.1, 17.2, 17.3, and 17.4.

17.6 A concrete barrel with simply supported ends has a span of 18 m (60 ft), a radius of 2.25 m (7.5 ft), and a thickness of 75 mm (3 in.). It carries a snow load of 2 kN/m^2 (40 psf). Determine the bending reinforcement at its ends if the thickness there is increased to 150 mm (6 in.). What percentage of the load is carried by bending stresses? Use grade 40 steel.

17.7 A barrel with the characteristics of that of Problem 17.6 has two continuous spans of 18 m (60 ft) each and is supported on three stiffeners. Determine the bending reinforcement at the intermediate stiffener. Use grade 40 steel.

17.8 An elliptic paraboloid covers a 18 m × 12 m (60 ft × 40 ft) area with boundary rises of 3 m (10 ft) and 2.4 m (8 ft) on the long and short sides. Determine the reinforcement at the boundaries, using grade 40 steel, if the thickness of 100 mm (4 in.) is increased to 150 mm (6 in.) there and if the snow load on the shell is 2 kN/m^2 (40 psf). The concrete shell is simply supported.

17.9 The spherical dome of Problem 16.1 has a ring of negligible torsional rigidity at its base. Determine the thrust H the dome exerts on the ring, and the ring reinforcement.

17.10 The spherical dome of Problem 16.2 is built-in at the boundary. Determine its maximum bending moment and its horizontal thrust.

17.11 A concrete rotational paraboloid spans 30 m (100 ft) with a rise of 9 m (30 ft). It is 75 mm (3 in.) thick and carries a 1 kN/m^2 (20 psf) snow load. Determine the bending moment around the boundary and the thrust. (Approximate the boundary displacement by those of a spherical dome of identical span, rise, and thickness, but determine the angle φ_0 by means of the equations of the paraboloid. Assume a simply supported edge.)

17.12 Determine the maximum bending moment in the rotational paraboloid of Problem 17.11 due to an increase in the temperature of 28°C: (a) when the boundary is simply supported, (b) when it is built-in.

17.13 A spherical shell of radius 36 m (120 ft) has an opening angle $\varphi_0 = 60°$. The shell is 100 mm (4 in.) thick and carries a snow load of 3 kN/m² (60 psf). Determine the thrust of the shell on a ring of negligible torsional rigidity and the maximum moment in the shell. At what distance from the shell boundary do the bending stresses vanish?

17.14 A containment dome for a nuclear reactor is a steel hemisphere with a radius of 9 m (30 ft). It is built-in at the base and must be capable of standing an internal pressure of 5 kN/m² (100 psf). Determine its thickness on the basis of membrane stresses and the increase in thickness required by the bending stresses at its base. Use A36 steel.

17.15 A submarine for deep-sea exploration consists of an A36 steel cylinder of 3 m (10 ft) diameter, closed by two hemispherical steel heads. Determine the thickness of its cylindrical and spherical portions, by membrane theory, if the submarine is to dive to a depth of 600 m (2,000 ft); also determine the maximum bending moment adjacent to the intersection between the cylinder and the hemispheres, assuming the cylinder and the hemispheres to have the same thickness (equal to the greater of the two membrane thicknesses). What thickening of the two portions is required because of the bending stresses?

17.16 What is the minimum thickness required to prevent buckling of the dome of Problem 17.1?

17.17 What is the safety factor against buckling of the dome of Problem 17.4?

17.18 What safety factor against buckling does the barrel of Problem 17.6 have?

17.19 What safety factor against buckling does the paraboloid of Problem 17.11 have?

17.20 What outer pressure could be supported by the dome of Problem 17.14 without buckling?

17.21 Are the thicknesses derived in Problem 17.15 safe against buckling? If not, what should their values be to obtain a coefficient of safety of 2?

17.22 A concrete hyperbolic paraboloid supported on straight boundaries has sides $2a = 18$ m (60 ft) and $2b = 12$ m (40 ft) and a rise of 3 m (10 ft). It is 75 mm (3 in.) thick and carries a snow load of 1 kN/m² (20 psf). Determine its coefficient of safety against buckling.

17.23 A shallow spherical shell of radius R and opening angle φ_0 is loaded by a dead load w. Determine the value of the thickness h below which buckling governs in terms of R, φ_0, E, and the allowable stress in compression, f'_c.

18. Space Frames

18.1 SPACE FRAMES [10.9]

We include in the category of space frames a large variety of flat and curved structures, built by means of straight or, at times, curved elements, and designed to cover large areas without intermediate supports. Space frames spanning hundreds of meters (feet) have been built in steel, aluminum, concrete, and wood. Their light weight and ease of construction makes them suitable for the roofing of spans, which otherwise could only be spanned (often at greater cost) by means of cables or shells. Space frames may also be used as vertical structures or as slabs in folded slab construction.

Flat space frames usually consist of two layers of bars set in parallel horizontal grids. When the bars of the two grids are connected by members lying in vertical planes, the space frame consists of intersecting truss systems and acts as a grid in which trusses replace bars. Such space frames are referred to as *truss-grids*. Figure 18.1.1 illustrates two schemes commonly used for truss-grids and shows how a bar may be common to more than one truss. At times, girders or beams are substituted for trusses in flat space frames. This leads to beam grids with a very large number of nodes, for which the methods of preliminary analysis of Sections 11.1 and 11.2 become burdensome. Their analysis may be obtained by the methods of Sections 18.2–18.5.

When the two horizontal grid systems are connected by skew bars, the space frame consists of interlocking pyramids with triangular, square, pentagonal, or hexagonal bases, and is referred to as a *space-grid* or *space frames* (Fig. 18.1.2). Space-grids are, usually, stiffer and more efficient than truss-grids.

Curved space frames may consist of one or two layers of bars and curved in one or two directions. Single-layer curved frames extend to curved

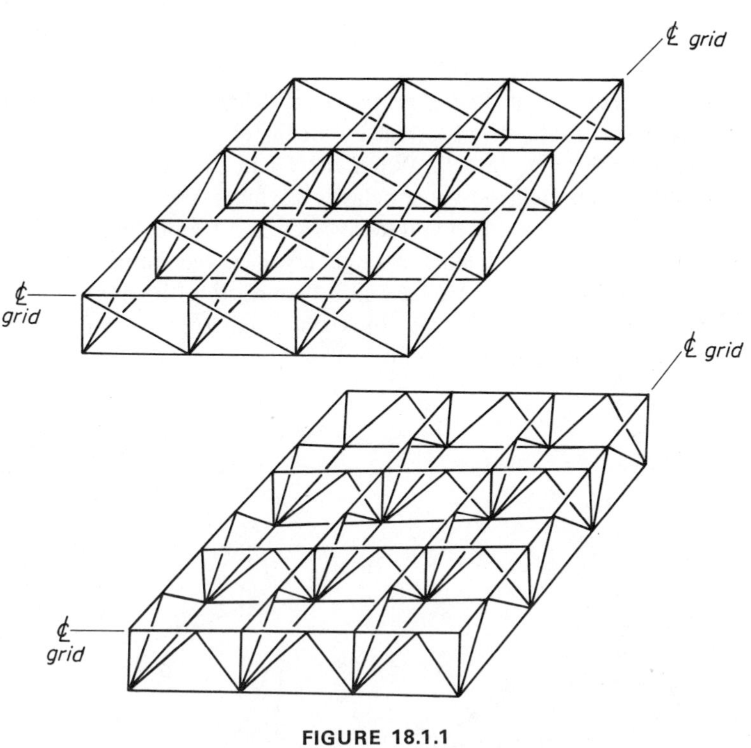

FIGURE 18.1.1

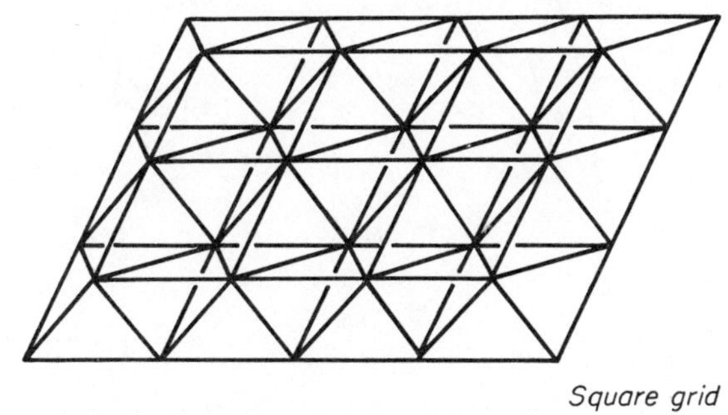

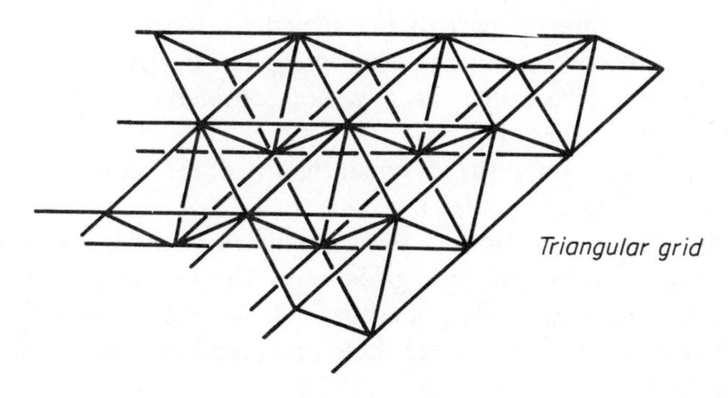

FIGURE 18.1.2

structures the principle of the flat grid system. When curved in one direction, they create barrel roofs, usually built by means of skew arches and referred to as *lamella roofs* (Fig. 18.1.3). When curved in two directions, they may take the shape of synclastic or anticlastic surfaces and be used as domes or saddle roofs (Fig. 18.1.4). Double-layer curved frames may be obtained by substituting trusses for beams in single-layer curved frames, thus leading to *curved truss-grids*, or may be created by connecting the nodes of the two *curved bar layers* by skew beams, thus obtaining true *curved space-grids*, which are structurally more efficient.

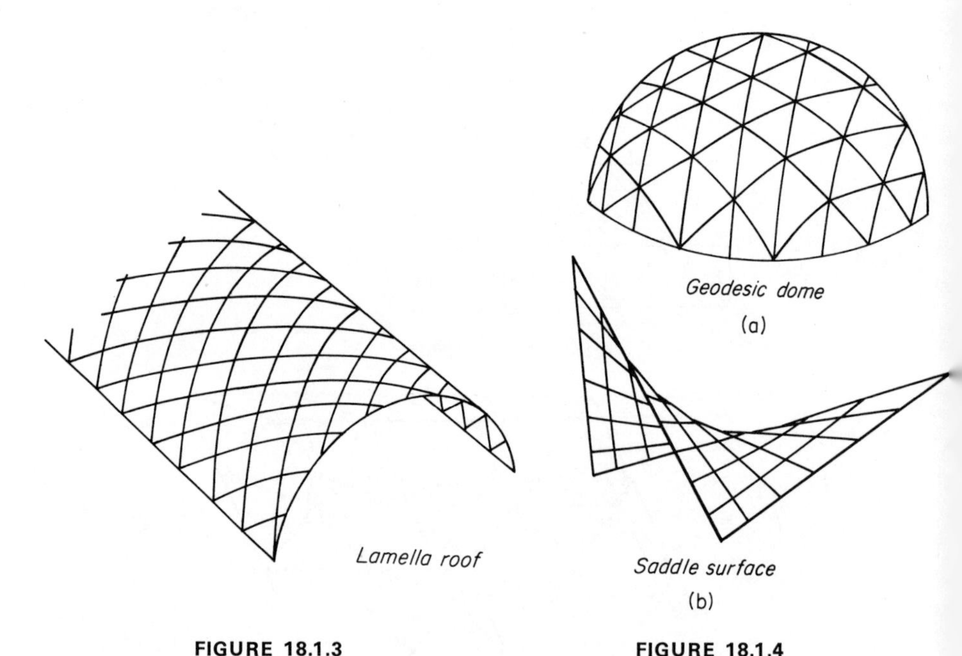

FIGURE 18.1.3

FIGURE 18.1.4

In all space frames a *skin* must connect the frame bars. At times, the skin becomes an integral part of the structural system, as when skin and bars are made of concrete and poured integrally. In space frames made out of metal or wood, the skin is usually not considered to participate in the structural action of the frame.

For all space frames, the torsional rigidity of the beams may be ignored in preliminary design.

In the following sections, a few elementary methods are presented for the preliminary analysis of truss-grids only. The rigorous analysis of large space frames and, in particular, of space-grids has been made possible by the advent of the electronic computer, but is a complex technical job to be left to the specialist.

18.2 RECTANGULAR FLAT GRIDS [10.2, 10.4]

Single-layer beam- or truss-grids with a relatively small number of nodes are analyzed by the methods of Section 11.1 and 11.2 i.e., by equating deflections at the nodal points. Although the elongation of the truss bars is the main cause of the panel point deflections in a truss, one may obtain preliminary design values for the reaction of one truss system on another by assuming trusses to be ideal I-beams, i.e., by neglecting the elongation of the verticals and the diagonals. This is analogous to neglecting shear deflections in the equivalent beam. The lesser flexibility of the trusses due to this approximation is compensated, in part, by the fact that, in reality, the truss bars are rigidly connected rather than hinged.

Example 18.2.1. Determine the stresses due to a uniform load w in the bars of the truss-grid of Fig. 18.2.1. All the trusses consist of identical bars of cross-section A.

The truss arrangement is identical to the beam arrangement of Fig. 11.1.2. Hence, the sharing of the load is given by (a) of Section 11.1, which, for:

$$l_1 = 2l_2 = 3a, \qquad I_1 = A\frac{h^2}{2} = I_2,$$

gives:

$$\frac{X}{P} = \frac{1}{1 + (\frac{1}{2})^3 \times 1} = \frac{8}{9}, \qquad \frac{P - X}{P} = \frac{1}{9}.$$

The loads on each truss are shown in Fig. 18.2.1.

The forces in the bars (Fig. 18.2.2) are obtained by equilibrium of the joints and by the section method (see Section 6.2). The diagonals D_1' and D_2' are theoretically unstressed, since the middle panels of the trusses are under pure bending due to the couples $(P/9)a$ and $(\frac{8}{9}P)(a/2)$, respectively, so that the panels B_1B_1' and B_2B_2' have zero shear.

Node A_1:

$$D_1 \sin\alpha_1 + \frac{P}{9} = 0 \quad \therefore \qquad D_1 = -\frac{P}{9\sin\alpha_1} = -\frac{P\sqrt{a^2 + h^2}}{9h},$$

Section B_1C_1:

$$\frac{P}{9}a - T_1 h = 0 \quad \therefore \qquad T_1 = \frac{Pa}{9h},$$

Node C_1:

$$-\frac{P}{9} + V_1 - D_1 \sin\alpha_1 = 0 \quad \therefore \qquad V_1 = -\frac{P}{9} + \frac{P}{9} = 0,$$

Node B_1:

$$-T_1 + T_1' = 0 \quad \therefore \qquad T_1 = T_1' = \frac{Pa}{9h} = -C_1',$$

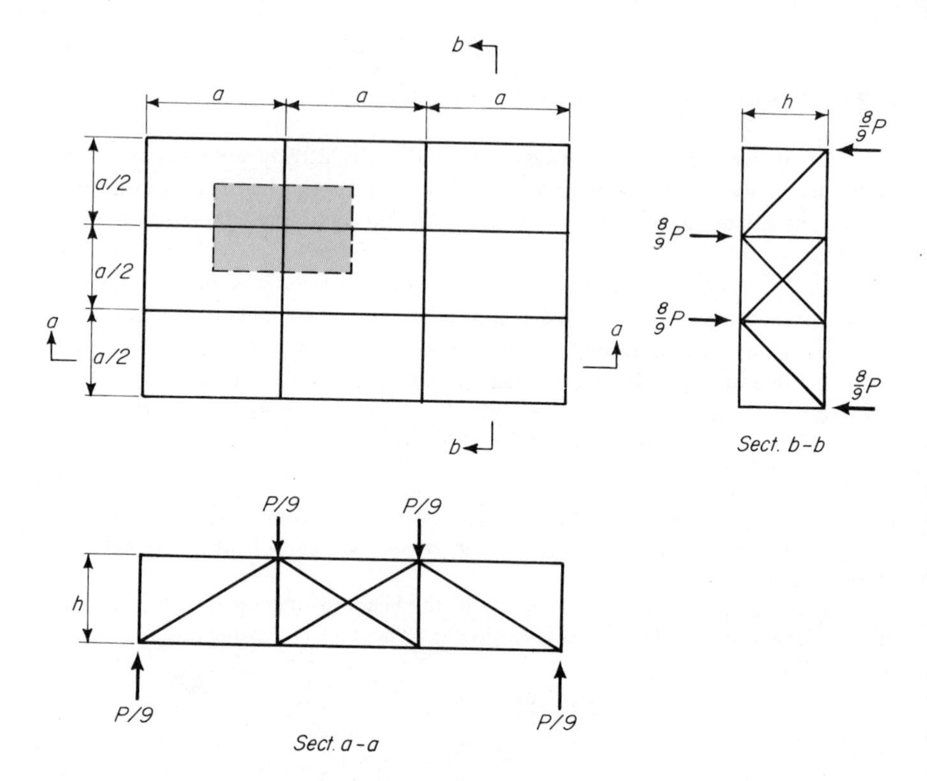

FIGURE 18.2.1

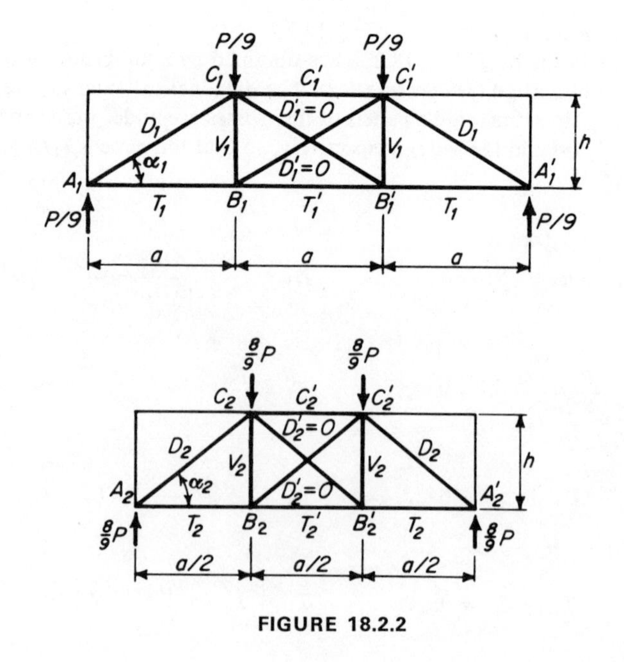

FIGURE 18.2.2

Node A_2:

$$D_2 \sin \alpha_2 + \frac{8}{9}P = 0 \quad \therefore \qquad D_2 = -\frac{8P\sqrt{(a/2)^2 + h^2}}{9h},$$

Section B_2C_2:

$$\left(\frac{8}{9}P\right)\left(\frac{a}{2}\right) - T_2 h = 0 \quad \therefore \qquad T_2 = \frac{4Pa}{9h},$$

Node C_2:

$$-\frac{8}{9}P + V_2 - D_2 \sin \alpha_2 = 0 \quad \therefore \qquad V_2 = -\frac{8}{9}P + \frac{8}{9}P = 0,$$

Node B_2:

$$-T_2 + T_2' = 0 \quad \therefore \qquad T_2 = \frac{4Pa}{9h} = -C_2'.$$

It is obvious that the analysis of a grid with hundreds of nodes, by the method of equating deflections, requires the solution of a very large number of equations, which can only be obtained by computer. In this case, simple approximate methods based on plate theory can be used for preliminary analysis, as was done for beam-grids in Section 11.1.

A plane grid, with nodes spaced at distances which are small in comparison with its lateral dimensions, behaves very much like a plate. The moments and shears in the grid elements may, therefore, be obtained by multiplying the moments and shears *per unit length* in the plate by the spacing between the grid elements.

Example 18.2.2. A rectangular, simply supported plane grid 30 × 45 m (100 ft × 150 ft) has identical ribs, parallel to the sides of the rectangle and spaced by 1.5 m (5 ft) and 2.25 m (7.5 ft), respectively (Fig. 18.2.3). Determine the maximum moments and shears in each rib due to a uniform load q_0 of 10 kN/m² (200 psf).

For a plate of aspect ratio 1.5, Table 12.2.1 gives:

$$\delta_{\max} = 0.093\frac{q_0 a^4}{Eh^3},$$

$$M_x]_{\max} = 0.081\,2\,q_0 a^2 = 0.081 \times 10 \times 30^2 = 729 \text{ kN·m/m } (162 \text{ k·ft/ft}),$$

$$M_y]_{\max} = 0.049\,8\,q_0 a^2 = 0.022 q_0 b^2 = 0.022 \times 10 \times 45^2 = 445.5 \text{ kN·m/m } (100 \text{ k·ft/ft}),$$

$$V_x]_{\max} = 0.424 q_0 a = 0.424 \times 10 \times 30 = 127.2 \text{ kN/m } (8.5 \text{ k/ft}),$$

$$V_y]_{\max} = 0.363 q_0 a = 0.242 q_0 b = 0.242 \times 10 \times 45 = 108.9 \text{ kN/m. } (7.3 \text{ k/ft}).$$

Hence, the maximum moments and shears in the ribs parallel to x are:

$$\bar{M}_x]_{\max} = 2.25 \times 729 = 1\,640 \text{ kN·m } (1{,}210 \text{ k·ft}),$$

$$\bar{V}_x]_{\max} = 2.25 \times 127.2 = 286.2 \text{ kN } (63.6 \text{ k}).$$

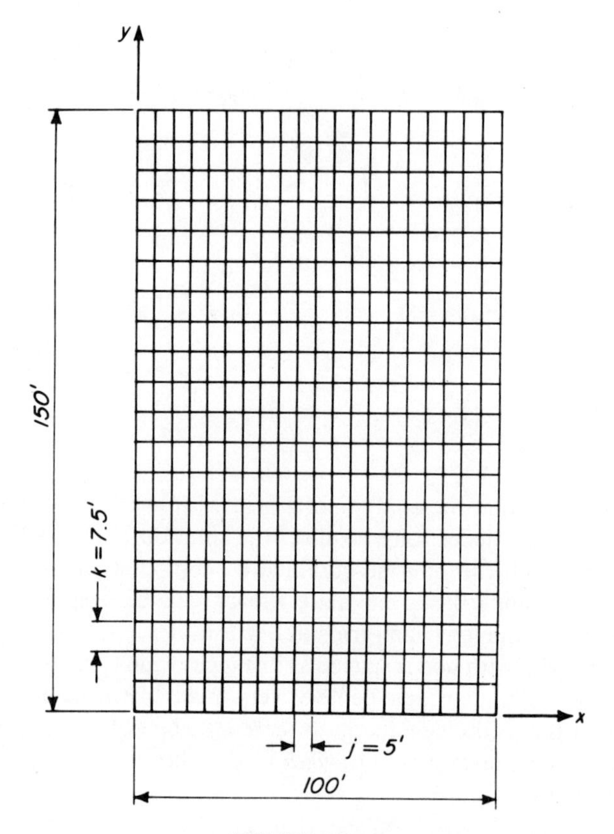

FIGURE 18.2.3

and in the ribs parallel to y:

$$\bar{M}_y]_{\max} = 1.5 \times 445.5 = 668 \text{ kN}\cdot\text{m (497 k}\cdot\text{ft)},$$

$$\bar{V}_y]_{\max} = 1.5 \times 108.9 = 163.4 \text{ kN (36.5 k)}.$$

With $F_b = 165$ MPa (24 ksi), the beams in the x-direction require moduli of:

$$S_x = \frac{1\,640 \times 1\,000^2}{165} = 9\,940 \times 10^3 \text{ mm}^3 \text{ (606 in.}^3)$$

$$\therefore \quad \text{W840} \times 298 \text{ (W33} \times 200).$$

A check on the shear stresses proves that they are low:

$$f_{s,x} = 1.5\,\frac{286.2 \times 1\,000}{38\,000} = 11.30 \text{ MPa (1.64 ksi)}.$$

The results of Example 18.2.2 are inaccurate because the plate has a finite torsional rigidity which, for all practical purposes, is nonexistent in the grid. A more accurate analysis is obtained by remembering that the load *per unit length* carried by a beam equals EI times the fourth derivative of its deflection z [see (4.1.4 c)]. Indicating by I_x and I_y the moments of inertia of the beams in the x- and y-directions, by j and k the beam spacing in the x- and y-directions (Fig. 18.2.4), and by z the deflection at a grid point, the load $q(j \times k)$ on the shaded rectangle in Fig. 18.2.4 is equilibrated by the two orthogonal beams meeting at the grid point:

$$EI_x \frac{\partial^4 z}{\partial x^4} \times j + EI_y \frac{\partial^4 z}{\partial y^4} \times k = q(j \times k). \tag{18.2.1}$$

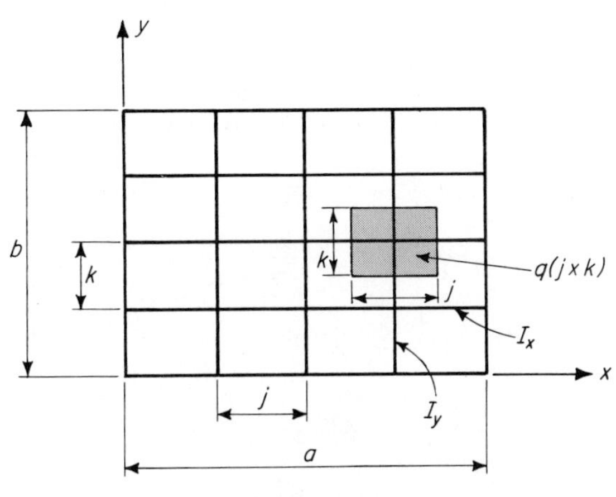

FIGURE 18.2.4

A sufficiently accurate solution of (18.2.1) for simply supported grids is obtained by letting:

$$z = z_0 \sin \frac{\pi}{a} x \sin \frac{\pi}{b} y \tag{18.2.2}$$

and by approximating the uniform load q_0 by a load:

$$q = \frac{16}{\pi^2} q_0 \sin \frac{\pi}{a} x \sin \frac{\pi}{b} y.^* \tag{18.2.3}$$

Substituting (18.2.2) and (18.2.3) in (18.2.1), one obtains:

$$z_0 = \frac{16}{\pi^6} \frac{q_0 a^4 k}{EI_x} \frac{1}{1 + (a/b)^4 (k/j)(I_y/I_x)}, \tag{18.2.4}$$

and by (18.2.2), remembering the relationships between deflections, moments,

*This approximation, like those of the following sections, is obtained by taking into account the first term only of the Fourier series expansions of z and q_0.

and shears in a beam [see (4.1.4 a) and (4.1.4 b)]:

$$
\begin{aligned}
M_x &= -EI_x \frac{\partial^2 z}{\partial x^2} = \frac{16}{\pi^4} \frac{q_0 a^2 k}{1 + (a/b)^4 (k/j)(I_y/I_x)} \sin\frac{\pi}{a}x \sin\frac{\pi}{b}y, \\
M_y &= -EI_y \frac{\partial^2 z}{\partial y^2} = \frac{16}{\pi^4} \frac{q_0 b^2 j}{1 + (b/a)^4 (j/k)(I_x/I_y)} \sin\frac{\pi}{a}x \sin\frac{\pi}{b}y, \\
V_x &= -EI_x \frac{\partial^3 z}{\partial x^3} = \frac{16}{\pi^3} \frac{q_0 a k}{1 + (a/b)^4 (k/j)(I_y/I_x)} \cos\frac{\pi}{a}x \sin\frac{\pi}{b}y, \\
V_y &= -EI_y \frac{\partial^3 z}{\partial y^3} = \frac{16}{\pi^3} \frac{q_0 b j}{1 + (b/a)^4 (j/k)(I_x/I_y)} \sin\frac{\pi}{a}x \cos\frac{\pi}{b}y.
\end{aligned}
\tag{18.2.5}
$$

Example 18.2.3. Determine the maximum deflection, moments, and shears in a square grid with $b = a$, and in a rectangular grid with $b = 2a$, for $j = k$, $I_x = I_y = I$.

By (18.2.5), for $b = a$:

$$
\begin{aligned}
z_0 &= \frac{16}{\pi^6}\frac{q_0 a^4 j}{EI}\frac{1}{1+1} = \frac{8}{\pi^6}\frac{q_0 a^4 j}{EI} = 0.008\,3\frac{q_0 a^4 j}{EI}, \\
M_x]_{\max} &= \frac{16}{\pi^4} q_0 a^2 k \frac{1}{1+1} = \frac{8}{\pi^4} q_0 a^2 k = 0.082 q_0 a^2 j = M_y]_{\max}, \\
V_x]_{\max} &= \frac{16}{\pi^3} q_0 a k \frac{1}{1+1} = \frac{8}{\pi^3} q_0 a k = 0.26 q_0 a j = V_y]_{\max}.
\end{aligned}
$$

Similarly, for $b = 2a$:

$$
\begin{aligned}
z_0 &= \frac{16}{\pi^6}\frac{q_0 a^4 k}{EI}\frac{1}{1+(\frac{1}{2})^4} = 0.015\,6\frac{q_0 a^4 j}{EI}, \\
M_x]_{\max} &= \frac{16}{\pi^4}\frac{q_0 a^2 k}{1+(\frac{1}{2})^4} = 0.153 q_0 a^2 j, \\
M_y]_{\max} &= \frac{16}{\pi^4}\frac{q_0 b^2 j}{1+(2)^4} = \frac{16}{\pi^4}\frac{q_0 (4a^2) j}{1+(2)^4} = 0.038 q_0 a^2 j, \\
V_x]_{\max} &= \frac{16}{\pi^3}\frac{q_0 a k}{1+(\frac{1}{2})^4} = 0.48 q_0 a j, \\
V_y]_{\max} &= \frac{16}{\pi^3}\frac{q_0 b j}{1+(2)^4} = \frac{16}{\pi^3}\frac{q_0 (2a) j}{17} = 0.061 q_0 a j.
\end{aligned}
$$

Example 18.2.4. Determine the maximum moments in the grid of Example 18.2.2 by beam theory.

With $b/a = \frac{3}{2}$, $k/j = \frac{3}{2}$, $I_x = I_y = I$, (18.2.5) gives:

$$
\begin{aligned}
M_x]_{\max} &= \frac{16}{\pi^4}\frac{q_0 a^2 k}{1+(\frac{2}{3})^4(\frac{3}{2})} = 0.127 q_0 a^2 k, \\
M_y]_{\max} &= \frac{16}{\pi^4}\frac{q_0 b^2 j}{1+(\frac{3}{2})^4(\frac{2}{3})} = 0.038\,3\, q_0 a b^2 j.
\end{aligned}
$$

For $q_0 = 10$ kN/m² (200 psf), $a = 30$ m (100 ft), $b = 45$ m (150 ft), $j = 1.5$ m (5 ft), $k = 2.25$ m (7.5 ft):

$$M_x]_{\max} = 0.127 \times 10 \times (30)^2 \times 2.25 = 2\,572 \text{ kN}\cdot\text{m } (1{,}900 \text{ k}\cdot\text{ft}),$$

$$M_y]_{\max} = 0.038\,3 \times 10 \times (45)^2 \times 1.5 = 1\,163 \text{ kN}\cdot\text{m } (860 \text{ k}\cdot\text{ft}).$$

These moments are larger than those of Example 18.2.2 because in that example the grid is analyzed as a plate and, hence, is incorrectly assumed to carry some load by twisting action.

18.3 SKEW FLAT GRIDS [10.3, 10.4]

By an approach analogous to that used in Section 18.2, it may be shown that the deflections, moments, and shears in a 45° skew grid of *evenly spaced identical beams* in the orthogonal directions r and s, simply supported on a rectangle of sides a, b (Fig. 18.3.1) under a uniform load q_0, are given approximately by:

$$z = \frac{16\sqrt{2}}{\pi^6} \frac{q_0 a^4 j/EI}{1 + 6(a^2/b^2) + a^4/b^4} \sin \frac{\pi}{a}x \sin \frac{\pi}{b}y, \tag{18.3.1}$$

$$\begin{aligned} M_r &= \frac{8\sqrt{2}}{\pi^4} \frac{q_0 a^2 j}{1 + 6(a^2/b^2) + a^4/b^4}\left[\left(1 + \frac{a^2}{b^2}\right) \sin \frac{\pi}{a}x \sin \frac{\pi}{b}y - 2\frac{a}{b} \cos \frac{\pi}{a}x \cos \frac{\pi}{b}y\right], \\ M_s &= \frac{8\sqrt{2}}{\pi^4} \frac{q_0 a^2 j}{1 + 6(a^2/b^2) + a^4/b^4}\left[\left(1 + \frac{a^2}{b^2}\right) \sin \frac{\pi}{a}x \sin \frac{\pi}{b}y + 2\frac{a}{b} \cos \frac{\pi}{a}x \cos \frac{\pi}{b}y\right], \end{aligned} \tag{18.3.2}$$

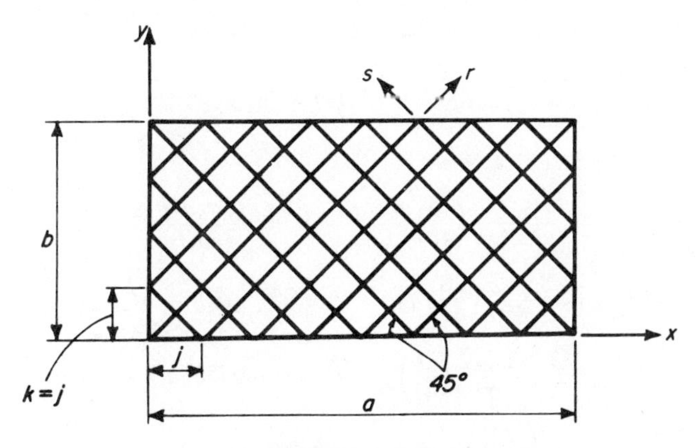

FIGURE 18.3.1

$$V_r = \frac{8}{\pi^3}\frac{q_0aj}{1+6(a^2/b^2)+a^4/b^4}\left[\left(1+3\frac{a^2}{b^2}\right)\cos\frac{\pi}{a}x\sin\frac{\pi}{b}y - \frac{a^3}{b^3}\left(1+3\frac{b^2}{a^2}\right)\sin\frac{\pi}{a}x\cos\frac{\pi}{b}y\right],$$

$$V_s = \frac{8}{\pi^3}\frac{q_0aj}{1+6(a^2/b^2)+a^4/b^4}\left[\left(1+3\frac{a^2}{b^2}\right)\cos\frac{\pi}{a}x\sin\frac{\pi}{b}y + \frac{a^3}{b^3}\left(1+3\frac{b^2}{a^2}\right)\sin\frac{\pi}{a}x\cos\frac{\pi}{b}y\right], \tag{18.3.3}$$

from which:

$$z_0 = \frac{16\sqrt{2}}{\pi^6}\frac{q_0a^4j/EI}{1+6(a^2/b^2)+a^4/b^4},$$

$$M_r]_{\max} = \frac{8\sqrt{2}}{\pi^4}\frac{1+(a^2/b^2)}{1+6(a^2/b^2)+a^4/b^4}q_0a^2j = M_s]_{\max},$$

$$M_r]_{\min} = -\frac{16\sqrt{2}}{\pi^4}\frac{a/b}{1+6(a^2/b^2)+a^4/b^4}q_0a^2j = M_s]_{\min}, \tag{18.3.4}$$

$$V_r]_{\max} = \frac{8}{\pi^3}\frac{1+3(a^2/b^2)}{1+6(a^2/b^2)+a^4/b^4}q_0aj,$$

$$V_s]_{\max} = \frac{8}{\pi^3}\frac{(a^3/b^3)[1+3(b^2/a^2)]}{1+6(a^2/b^2)+a^4/b^4}q_0aj.$$

By (18.3.4), we obtain:

1. For $a = b$:

$$z_0 = \frac{2\sqrt{2}}{\pi^6}\frac{q_0a^4j}{EI} = 0.002\,92\,\frac{q_0a^4j}{EI},$$

$$M_r]_{\max} = \frac{2\sqrt{2}}{\pi^4}q_0a^2j = 0.028\,9\,q_0a^2j = M_s]_{\max},$$

$$M_r]_{\min} = -\frac{2\sqrt{2}}{\pi^4}q_0a^2j = -0.028\,9\,q_0a^2j = M_s]_{\min}, \tag{18.3.5}$$

$$V_r]_{\max} = \frac{4}{\pi^3}q_0aj = 0.13q_0aj = V_s]_{\max}.$$

2. For $b = 2a$:

$$z_0 = \frac{16\sqrt{2}}{\pi^6}\left(\frac{16}{41}\right)\frac{q_0a^4j}{EI} = 0.009\,2\,\frac{q_0a^4j}{EI},$$

$$M_r]_{\max} = \frac{8\sqrt{2}}{\pi^4}\left(\frac{20}{41}\right)q_0a^2j = 0.057q_0a^2j = M_s]_{\max},$$

$$M_r]_{\min} = -\frac{8\sqrt{2}}{\pi^4}\left(\frac{16}{41}\right)q_0a^2j = -0.045q_0a^2j = M_s]_{\min}, \tag{18.3.6}$$

$$V_r]_{\max} = \frac{8}{\pi^3}\left(\frac{28}{41}\right)q_0aj = 0.18q_0aj,$$

$$V_s]_{\max} = \frac{8}{\pi^3}\left(\frac{26}{41}\right)q_0 aj = 0.16 q_0 aj.$$

Comparison with the results of Section 18.2 shows the advantages presented by a skew grid, particularly when $b > a$.

18.4 TRIANGULAR GRIDS

Triangular grids have beams running in three directions. One of the most commonly used is the *equilateral triangular grid*, in which the beams of the three systems make equal 60° angles between them.

By an approach analogous to that used in Section 18.2, it may be shown that the deflections, moments, and shears in a uniformly loaded equilateral triangular grid of identical beams, simply supported on a rectangle of sides a, b (Fig. 18.4.1), are given approximately by:

$$z = \frac{4\sqrt{3}}{9}\frac{16}{\pi^6}\frac{1}{(1 + a^2/b^2)^2}\frac{q_0 a^4 j}{EI}\sin\frac{\pi}{a}x \sin\frac{\pi}{b}y,$$

$$M_x = \frac{4\sqrt{3}}{9}\frac{16}{\pi^4}\frac{q_0 a^2 j}{(1 + a^2/b^2)^2}\sin\frac{\pi}{a}x \sin\frac{\pi}{b}y,$$

$$V_x = \frac{4\sqrt{3}}{9}\frac{16}{\pi^3}\frac{q_0 aj}{(1 + a^2/b^2)^2}\cos\frac{\pi}{a}x \sin\frac{\pi}{b}y,$$

$$M_r = \frac{\sqrt{3}}{3}\frac{16}{\pi^4}\frac{q_0 a^2 j}{(1 + a^2/b^2)^2}\left[\left(1 + 3\frac{a^2}{b^2}\right)\sin\frac{\pi}{a}x \sin\frac{\pi}{b}y + 2\sqrt{3}\frac{a}{b}\cos\frac{\pi}{a}x \cos\frac{\pi}{b}y\right],$$

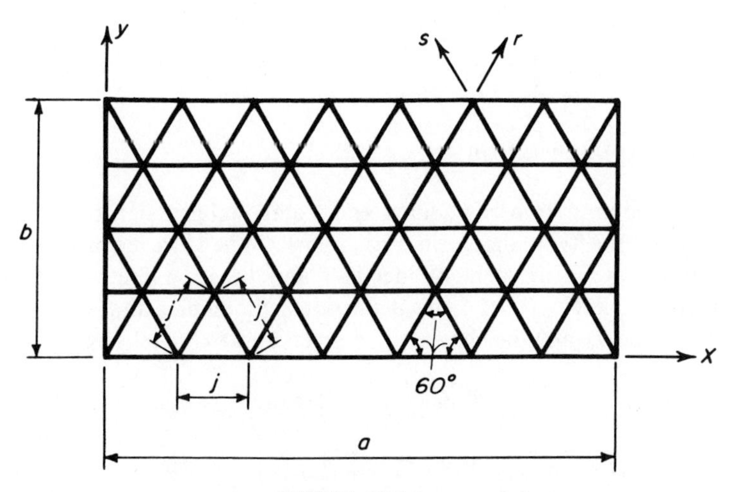

FIGURE 18.4.1

$$V_r = \frac{8\sqrt{3}}{9\pi^3}\frac{q_0 aj}{(1 + a^2/b^2)^2}\left[\left(1 + 9\frac{a^2}{b^2}\right)\cos\frac{\pi}{a}x\sin\frac{\pi}{b}y - 3\sqrt{3}\frac{a}{b}\left(1 + \frac{a^2}{b^2}\right)\sin\frac{\pi}{a}x\cos\frac{\pi}{b}y\right], \tag{18.4.1}$$

$$M_s = \frac{\sqrt{3}}{9}\frac{16}{\pi^4}\frac{q_0 a^2 j}{(1 + a^2/b^2)^2}\left[\left(1 + 3\frac{a^2}{b^2}\right)\sin\frac{\pi}{a}x\sin\frac{\pi}{b}y - 2\sqrt{3}\frac{a}{b}\cos\frac{\pi}{a}x\cos\frac{\pi}{b}y\right],$$

$$V_s = \frac{8\sqrt{3}}{9\pi^3}\frac{q_0 aj}{(1 + a^2/b^2)^2}\left[\left(1 + 9\frac{a^2}{b^2}\right)\cos\frac{\pi}{a}x\sin\frac{\pi}{b}y + 3\sqrt{3}\frac{a}{b}\left(1 + \frac{a^2}{b^2}\right)\sin\frac{\pi}{a}x\cos\frac{\pi}{b}y\right].$$

By (18.4.1), we obtain:

1. For $a = b$:

$$z_0 = 0.003\,2\frac{q_0 a^4 j}{EI},$$

$$M_x]_{\max} = 0.032 q_0 a^2 j, \qquad V_x]_{\max} = 0.10 q_0 aj, \tag{18.4.2}$$

$$M_x]_{\max} = M_s]_{\max} = 0.032 q_0 a^2 j, \qquad M_r]_{\min} = M_s]_{\min} = -0.027 q_0 a^2 j,$$

$$V_r]_{\max} = V_s]_{\max} = 0.12 q_0 aj, \qquad V_r]_{\min} = V_s]_{\min} = -0.13 q_0 aj.$$

2. For $b = 2a$:

$$z_0 = 0.008\,2\frac{q_0 a^4 j}{EI},$$

$$M_x]_{\max} = 0.081 q_0 a^2 j, \qquad V_x]_{\max} = 0.25 q_0 aj, \tag{18.4.3}$$

$$M_r]_{\max} = M_s]_{\max} = 0.035 q_0 a^2 j, \qquad M_r]_{\min} = M_s]_{\min} = -0.035 q_0 a^2 j,$$

$$V_r]_{\max} = V_s]_{\max} = 0.103 q_0 aj, \qquad V_r]_{\min} = V_s]_{\min} = -0.103 q_0 aj.$$

18.5 QUADRUPLE GRIDS

The combination of a square and a diagonal grid constitutes a quadruple grid of the type shown in Fig. 18.5.1. When the beams in the x-, y-directions have a moment of inertia I and those in the r-, s-directions a moment of inertia $I/\sqrt{2}$, the deflections, moments, and shears due to a uniform load q_0 may be shown to be given approximately by:

$$z = \frac{32}{3\pi^6}\frac{q_0 a^4 j/EI}{(1 + a^2/b^2)^2}\sin\frac{\pi}{a}x\sin\frac{\pi}{b}y,$$

$$M_x = \frac{32}{3\pi^4}\frac{q_0 a^2 j}{(1 + a^2/b^2)^2}\sin\frac{\pi}{a}x\sin\frac{\pi}{b}y,$$

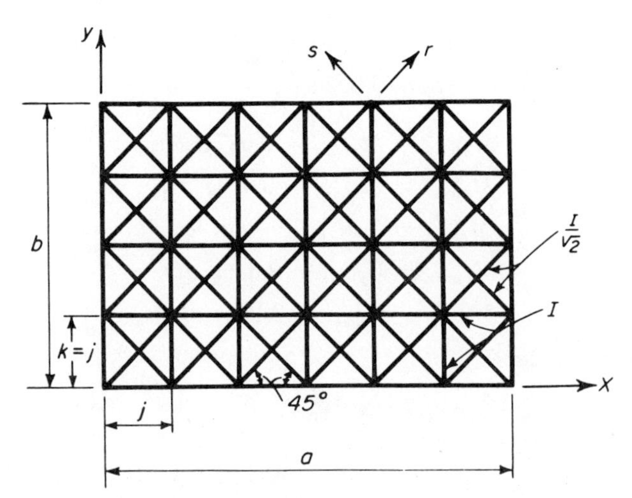

FIGURE 18.5.1

$$M_y = \frac{32}{3\pi^4}\frac{a^2}{b^2}\frac{q_0 a^2 j}{(1 + a^2/b^2)^2}\sin\frac{\pi}{a}x\sin\frac{\pi}{b}y,$$

$$M_r = \frac{8\sqrt{2}}{3\pi^4}\frac{q_0 a^2 j}{(1 + a^2/b^2)^2}\left[\left(1 + \frac{a^2}{b^2}\right)\sin\frac{\pi}{a}x\sin\frac{\pi}{b}y + 2\frac{a}{b}\cos\frac{\pi}{a}x\cos\frac{\pi}{b}y\right],$$

$$M_s = \frac{8\sqrt{2}}{3\pi^4}\frac{q_0 a^2 j}{(1 + a^2/b^2)^2}\left[\left(1 + \frac{a^2}{b^2}\right)\sin\frac{\pi}{a}x\sin\frac{\pi}{b}y - 2\frac{a}{b}\cos\frac{\pi}{a}x\cos\frac{\pi}{b}y\right], \tag{18.5.1}$$

$$V_x = \frac{32}{3\pi^3}\frac{q_0 a j}{(1 + a^2/b^2)^2}\cos\frac{\pi}{a}x\sin\frac{\pi}{b}y,$$

$$V_y = \frac{32}{3\pi^3}\frac{a^3}{b^3}\frac{q_0 a j}{(1 + a^2/b^2)^2}\sin\frac{\pi}{a}x\cos\frac{\pi}{b}y,$$

$$V_r = \frac{8\sqrt{2}}{3\pi^3}\frac{q_0 a j}{(1 + a^2/b^2)^2}\left[\left(1 + 3\frac{a^2}{b^2}\right)\cos\frac{\pi}{a}x\sin\frac{\pi}{b}y - \frac{a}{b}\left(3 + \frac{a^2}{b^2}\right)\sin\frac{\pi}{a}x\cos\frac{\pi}{b}y\right],$$

$$V_s = \frac{8\sqrt{2}}{3\pi^3}\frac{q_0 a j}{(1 + a^2/b^2)^2}\left[\left(1 + 3\frac{a^2}{b^2}\right)\cos\frac{\pi}{a}x\sin\frac{\pi}{b}y + \frac{a}{b}\left(3 + \frac{a^2}{b^2}\right)\sin\frac{\pi}{a}x\cos\frac{\pi}{b}y\right].$$

Equations (18.5.1) give:

1. For $a = b$:

$$
\begin{aligned}
z_0 &= 0.002\,8\,\frac{q_0 a^4 j}{EI},\\
M_x]_{\max} &= M_y]_{\max} = 0.027 q_0 a^2 j,\\
M_r]_{\max} &= M_s]_{\max} = 0.019 q_0 a^2 j,\\
M_r]_{\min} &= -M_s]_{\min} = 0.019 q_0 a^2 j,\\
V_x]_{\max} &= V_y]_{\max} = 0.086 q_0 a j,\\
V_r]_{\max} &= V_s]_{\max} = 0.121 q_0 a j.
\end{aligned}
\tag{18.5.2}
$$

2. For $b = 2a$:

$$
\begin{aligned}
z_0 &= 0.007\,2\,\frac{q_0 a^4 j}{EI},\\
M_x]_{\max} &= 0.070 q_0 a^2 j, \qquad M_y]_{\max} = 0.018 q_0 a^2 j,\\
M_r]_{\max} &= M_s]_{\max} = 0.031 q_0 a^2 j,\\
M_r]_{\min} &= -M_s]_{\min} = 0.025 q_0 a^2 j,\\
V_x]_{\max} &= 0.22 q_0 a^2 j, \qquad V_y]_{\max} = 0.028 q_0 a^2 j,\\
V_r]_{\max} &= V_s]_{\max} = 0.13 q_0 a^2 j.
\end{aligned}
\tag{18.5.3}
$$

18.6 TAKENAKA TRUSSES

The Takenaka truss (Fig. 18.6.1) is an assemblage of inverted tetrahedra with square bases of side a on the upper chord points U and vertices on the lower chord points V a distance h below their base, which are connected by lower chord bars V–V of length $\sqrt{2}\,a$. When the loads are applied to the upper chord points, the upper bars U–U and B–U (shown as solid lines) are in compression; the internal diagonals U–V (shown as dot-dash lines) are also in compression; the external diagonals B–V and the lower chord bars V–V (shown as dot-dash lines) are in tension. The evaluation of bar forces is obtained by space equilibrium of the truss joints.

Let us consider, for simplicity of solution, the symmetrical truss of Fig. 18.6.1 with a depth $h = a/\sqrt{2}$ and, hence, diagonals inclined at 45°, loaded by four equal loads W at the upper chord points U and supported at eight boundary points B by eight equal reactions $W/2$.

Assuming bar forces to be positive when tensile, we obtain by vertical equilibrium of the bars at a support point B_1 and with $\cos 45° = 1/\sqrt{2}$ (Fig. 18.6.2):

$$\sum F_z = -\frac{W}{2} + \frac{F_2}{\sqrt{2}} = 0 \quad \therefore \quad F_2 = \frac{W}{\sqrt{2}}.$$

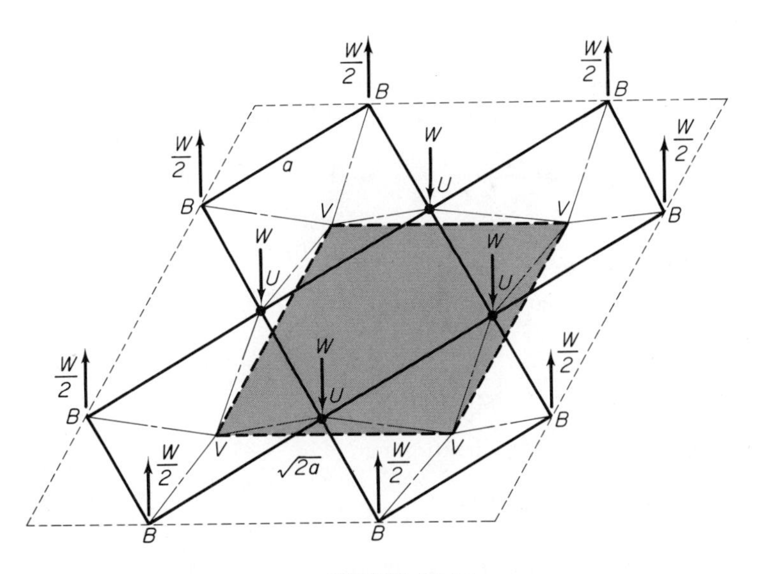

FIGURE 18.6.1

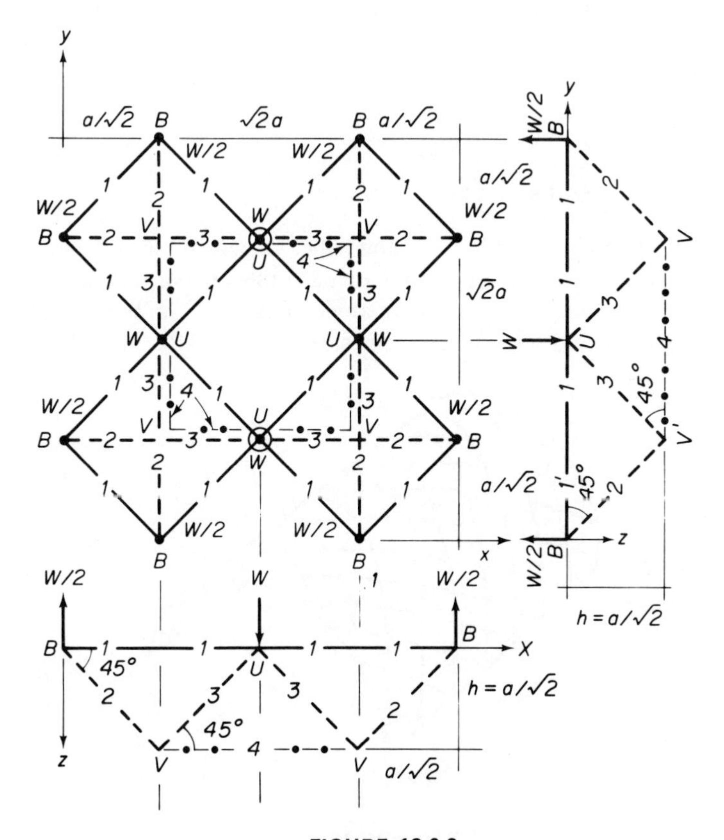

FIGURE 18.6.2

By horizontal equilibrium in the x- and y-directions at B_1:

$$\sum F_x = \frac{F_1}{\sqrt{2}} - \frac{F_1'}{\sqrt{2}} = 0 \quad \therefore \quad F_1 = F_1',$$

$$\sum F_y = +\frac{F_1'}{\sqrt{2}} + \frac{F_1}{\sqrt{2}} + \frac{F_2}{\sqrt{2}} = 0 \quad \therefore \quad F_1 = F_1' = -\frac{F_2}{2} = -\frac{W}{2\sqrt{2}}.$$

By vertical equilibrium at U, with equal forces F_3 in the diagonals U–V and U–V' because of symmetry:

$$\sum F_z = W + \frac{2F_3}{\sqrt{2}} = 0 \quad \therefore \quad F_3 = -\frac{W}{\sqrt{2}},$$

and by equilibrium in the x-direction at V':

$$\sum F_x = +\frac{F_2}{\sqrt{2}} - \frac{F_3}{\sqrt{2}} - F_4 = 0 \quad \therefore \quad F_4 = \frac{W/\sqrt{2}}{\sqrt{2}} - \frac{-W/\sqrt{2}}{\sqrt{2}} = W.$$

This last result may be checked by cutting a vertical section parallel to y through the median line U–U and taking moments, about the line U–U, of the reactions at four points B, of one of the loads and of the tensile forces in two of the lower bars 4:

$$2\frac{W}{2}(\sqrt{2}\,a) + 2\frac{W}{2}\frac{a}{\sqrt{2}} - W\frac{a}{\sqrt{2}} - 2F_4\frac{a}{\sqrt{2}} = 0 \quad \therefore \quad F_4 = W.$$

18.7 LAMELLA ROOFS [10.3]

Lamella roofs (Fig. 18.7.1) extend the concept of skew beam-grids to skew arch-grids. The ribs used in a lamella roof to cover a rectangular area of sides L–l, span the short side l at an angle of about 20° in horizontal projection to the right and left of the short side (Fig. 18.7.2). The identical ribs

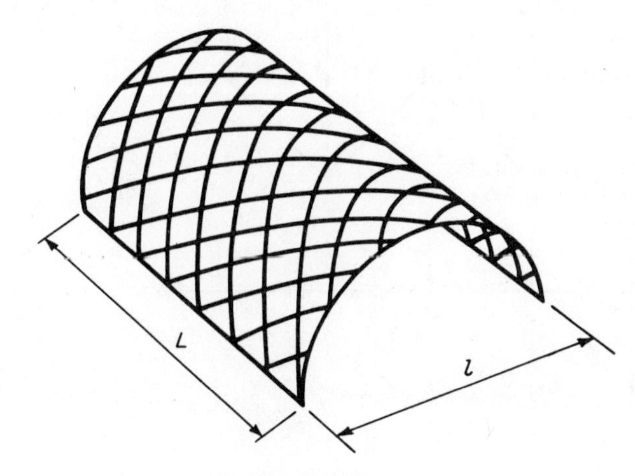

FIGURE 18.7.1

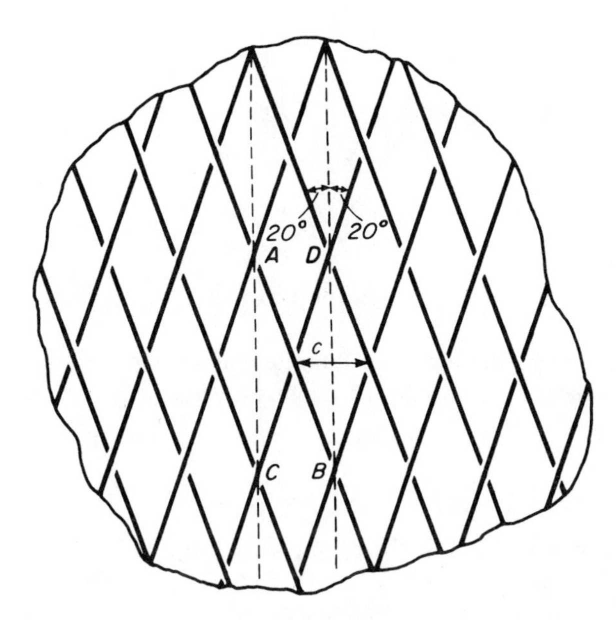

FIGURE 18.7.2

of the two intersecting systems subdivide the roof into lozenges and make an angle of about 40° between them. When the lamella roof is built of wood or steel, it is customary to build the ribs by means of straight beams twice the length of the lozenge side (Fig. 18.7.2). It is then assumed that the ribs will take thrust and bending moments, but that the bending moment will be taken only by the continuous ribs, such as AB, while the thrust will be taken by both the continuous ribs, such as AB, and the joined ribs, such as CD. The ribs are usually hinged at their springing. Lamella roofs of concrete may be poured in place or have prefabricated ribs, which are joined by welding or overlapping the reinforcement and grouting the joints. In either case, all beams take both moments and thrusts, since continuity is achieved at all joints.

Whatever the material used, roof plates usually connect the ribs rigidly transforming the roof into a monolithic structure acting as a two-hinged arch spanning the short direction l. Indicating by c the width of a lozenge, each arch of width c has an effective area A' to absorb the thrust given by (Fig. 18.7.3):

$$A' = \frac{2A}{\cos \alpha}, \tag{18.7.1}$$

and an effective modulus S' to absorb the bending moment, given by:

$$S' = \frac{S}{\cos \alpha}, \tag{18.7.2}$$

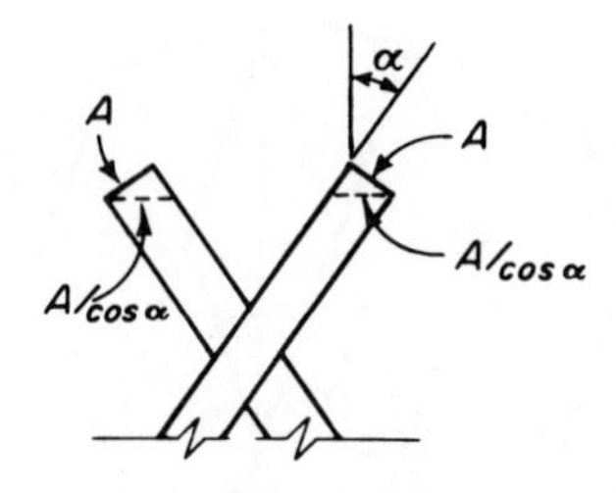

FIGURE 18.7.3

where A and S are the area and the modulus of one rib (and of the participating plate, if the roof plates are integral with the ribs). For continuous concrete ribs:

$$S' = \frac{2S}{\cos \alpha}. \tag{18.7.2a}$$

Indicating by C and M the compressive force and bending moment at a section of an arch of width c, and by F_c and F_b the allowable stresses in compression and bending, the combined compression and bending stresses must be such that:

$$\frac{C/A'}{F_c} + \frac{M/S'}{F_b} \leq 1. \tag{18.7.3}$$

The maximum moment in the two-hinged arch occurs, usually, between the sections $\frac{1}{6}$ and $\frac{1}{4}$ of the span from a support under a combination of the dead loads and $\frac{1}{2}$ or $\frac{1}{3}$ of the snow load acting on the half-arch away from the support, the other half of the arch having been cleared of snow by the wind (Fig. 18.7.4).

For a preliminary investigation, the parabolic arch of rise f may be designed as a three-hinged arch (see Section 10.3) with the dead load assimilated to a live load (Fig. 18.7.4). The parabolic arch is funicular for the uniform load. Hence, the moment M is only due to the live load w', which will be assumed equal to 40% of the code snow load.

From statics, the left reaction R_1 equals:

$$R_1 = \frac{wl}{2} + \frac{1}{4}\frac{w'l}{2} = \frac{wl}{2} + \frac{w'l}{8}.$$

The thrust H is obtained by taking moments about the center hinge O:

$$H \times f = \left(\frac{wl}{2} + \frac{w'l}{8}\right)\frac{l}{2} - \frac{w}{2}\left(\frac{l}{2}\right)^2 = \frac{wl^2}{8} + \frac{w'l^2}{16}$$

$$\therefore \quad H = \frac{wl^2}{8f}\left(1 + \frac{1}{2}\frac{w'}{w}\right). \tag{18.7.4}$$

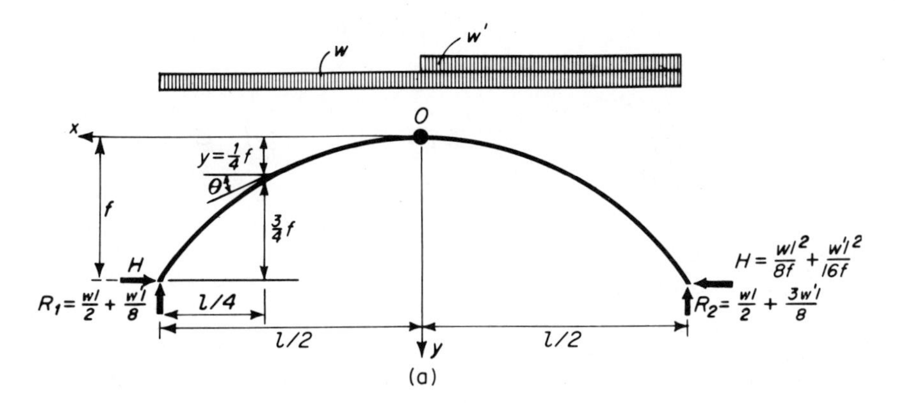

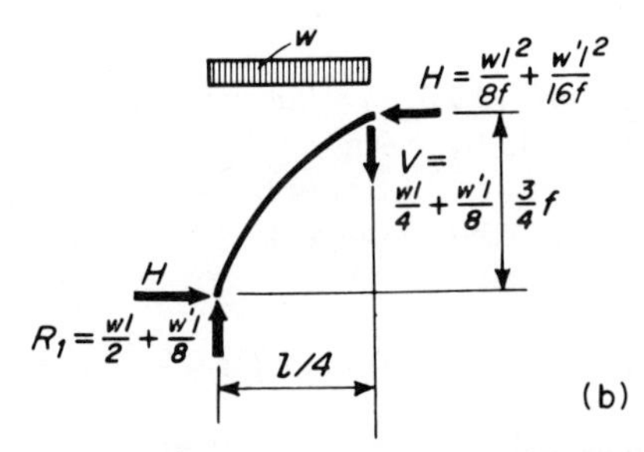

FIGURE 18.7.4

Referring the arch to the x-, y-axis of Fig. 18.7.4:

$$y = f\left(\frac{x}{l/2}\right)^2 = \frac{4f}{l^2}x^2, \tag{18.7.5}$$

and the value of y at $x = l/4$ equals $f/4$.

The slope at $x = l/4$ is given by:

$$\tan\theta = \frac{dy}{dx}\bigg]_{x=l/4} = \frac{8f}{l^2}x\bigg]_{x=l/4} = \frac{2f}{l},$$

so that:

$$\sin\theta = \frac{\tan\theta}{\sqrt{1+\tan^2\theta}} = \frac{2f}{\sqrt{l^2+4f^2}},$$
$$\cos\theta = \frac{1}{\sqrt{1+\tan^2\theta}} = \frac{l}{\sqrt{l^2+4f^2}}. \tag{18.7.6}$$

The moment at $x = l/4$ is given by:

$$M = \frac{w'l}{8}\frac{l}{4} - \frac{w'l^2}{16f}\left(f - \frac{1}{4}f\right) = -\frac{w'l^2}{64}. \tag{18.7.7}$$

The compression C at $x = l/4$ is the resultant of the shear $V = (wl/2)$

$+ (w'l/8) - (wl/4) = (wl/4) + (w'l/8)$, and of the thrust H:

$$C = H\cos\theta + V\sin\theta$$
$$= \frac{wl^2}{8f}\left(1 + \frac{1}{2}\frac{w'}{w}\right)\frac{l}{\sqrt{l^2 + 4f^2}} + \left(\frac{wl}{4} + \frac{w'l}{8}\right)\frac{2f}{\sqrt{l^2 + 4f^2}}$$
$$= \frac{wl^2}{8f}\left(1 + \frac{1}{2}\frac{w'}{w}\right)\sqrt{1 + \left(\frac{2f}{l}\right)^2}. \qquad (18.7.8)$$

For rectangular ribs of width b and depth h:

$$A = bh, \quad S = \frac{bh^2}{6}. \qquad (18.7.9)$$

Equations (18.7.7), (18.7.8), and (18.7.9), together with (18.7.1) and (18.7.2) allow the check of the stress condition (18.7.3).

Example 18.7.1. Check the stress condition (18.7.3) in a lamella roof spanning 30 m (100 ft), with $f = l/4$ and $\alpha = 20°$, built by means of 75 mm × 300 mm (3 in. × 12 in.) wood beams at a lozenge width $c = 1.2$ m (4 ft). Assume a dead load of 1 kN/m² (20 psf) and a snow load of 2 kN/m² (40 psf).

For a 1.2 m width and $\alpha = 20°$:

$$w = 1.2 \times 1 = 1.2 \text{ kN/m}, \quad w' = 0.4 \times 1.2 \times 2 = 0.96 \text{ kN/m}, \quad \frac{w'}{w} = 0.8,$$
$$\cos\alpha = 0.94; \quad b = 75 \text{ mm}, \quad h = 300 \text{ mm}, \quad A = 22\,500 \text{ mm}^2,$$
$$S = 1\,125 \times 10^3 \text{ mm}^3,$$
$$A' = \frac{2 \times 22\,500}{0.94} = 47\,870 \text{ mm}^2, \quad S' = \frac{1\,125 \times 10^3}{0.94} = 1\,197 \times 10^3 \text{ mm}^3.$$

Hence:

$$M = \frac{-0.96 \times 30^2}{64} = -13.5 \text{ kN·m } (-10 \text{ k·ft}),$$
$$C = \frac{1.2 \times 30^2}{8 \times 7.5}\left(1 + \frac{1}{2}\frac{0.96}{1.2}\right)\sqrt{1 + \left(\frac{2 \times 7.5}{30}\right)^2} = 28.17 \text{ kN } (6{,}270 \text{ lb}).$$

With $F_c = 10$ MPa (1,450 psi), $F_b = 14.1$ MPa (2,050 psi)*, (18.6.3) gives:

$$\frac{(28.17/47\,870)(1\,000)}{10} + \frac{(13\,500/1\,197)}{14.1} = 0.059 + 0.800 = 0.859 < 1.0.$$

18.8 CURVED SPACE FRAMES [10.13]

Curved space frames, single or double layered, may have the shape of any of the surfaces mentioned in Chapter 14.

The evaluation of the forces in each element of the frame, be it a bar

*See Table 2.6.1.

or a truss, is a complex task which usually requires the use of an electronic computer. However, an approximate evaluation of these forces is readily obtained when the mesh size of the space frame is small in comparison with its lateral dimensions, so that the frame may be assumed to behave like a continuous structure, i.e., like a thin shell.

The equivalence between a reticulated space frame and a thin shell is established in this section for a frame with an *equilateral* triangular grid of bars or a rectangular grid of bars with diagonals. The equivalence for other grid patterns is obtained by identical considerations.

Let us assume that the three membrane forces per unit of length N_x, N_y, and S have been evaluated (by the methods of Chapter 16) for a thin shell with the same shape and dimensions as those of the reticulated space frame under study. The bars of the frame will be assumed to be oriented in the directions x, r, and s, where r and s make angles of 60° and 120° with x (Fig. 18.8.1). Let the *compressive* forces in these bars at a point P be C_x, C_r, C_s.

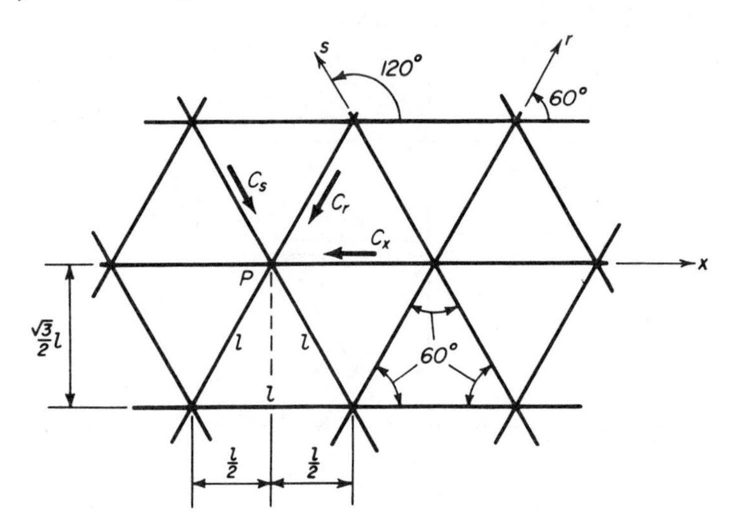

FIGURE 18.8.1

The equivalence between the shell and the frame is established by stating that there is equilibrium in the x- and y-directions between the N_x, N_y, and S unit forces applied to the sides of a shell element $ABCD$ (Fig. 18.8.2) and the forces C_x, C_r, and C_s.

Equilibrium in the x-direction [Fig. 18.8.2 (a)] requires that:

$$\frac{C_x}{2} + \frac{C_x}{2} + C_x + \frac{1}{2}C_r + \frac{1}{2}C_s = \left(2\frac{\sqrt{3}}{2}l\right)N_x$$

or:

$$4C_x + C_r + C_s = N_x(2\sqrt{3}\,l). \tag{a}$$

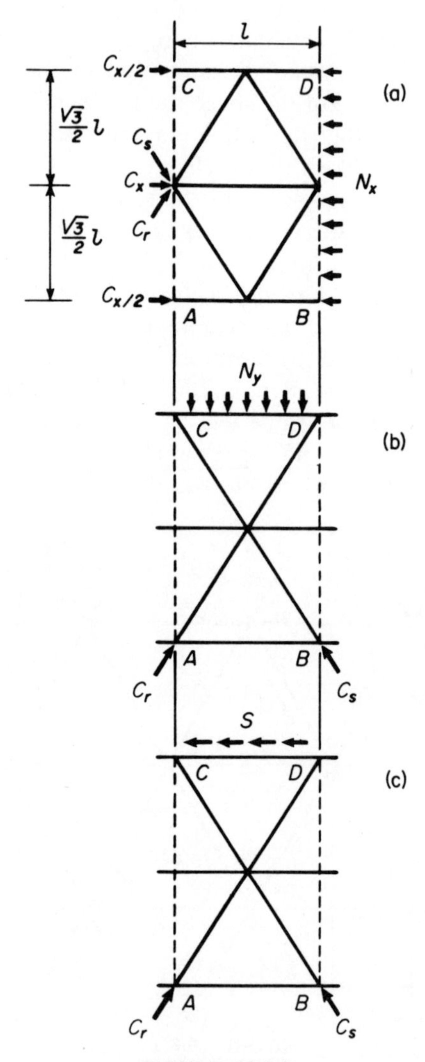

FIGURE 18.8.2

Equilibrium in the y-direction [Fig. 18.8.2 (b)] requires that:

$$\frac{\sqrt{3}}{2}C_r + \frac{\sqrt{3}}{2}C_s = N_y l$$

or:

$$C_r + C_s = N_y\left(\frac{2}{\sqrt{3}}\,l\right). \tag{b}$$

Equilibrium of shears [Fig. 18.8.2 (c)] requires that:

$$\frac{1}{2}C_r - \frac{1}{2}C_s = Sl$$

or:

$$C_r - C_s = 2Sl. \tag{c}$$

Subtracting (b) from (a) we obtain:

$$4C_x = N_x(2\sqrt{3}\,l) - N_y\left(\frac{2}{\sqrt{3}}\,l\right) \quad \therefore \quad C_x = \frac{l}{2\sqrt{3}}(3N_x - N_y). \tag{18.8.1}$$

Adding (b) to (c) we have:

$$2C_r = N_y\left(\frac{2}{\sqrt{3}}\,l\right) + 2Sl \quad \therefore \quad C_r = \frac{l}{\sqrt{3}}(N_y + \sqrt{3}\,S). \tag{18.8.2}$$

Subtracting (c) from (b) we get:

$$2C_s = N_y\left(\frac{2}{\sqrt{3}}\,l\right) - 2Sl \quad \therefore \quad C_s = \frac{l}{\sqrt{3}}(N_y - \sqrt{3}\,S). \tag{18.8.3}$$

Equations (18.8.1), (18.8.2), and (18.8.3) allow the evaluation of the forces in the bars of the curved space frame as soon as the membrane forces per unit length are known in the corresponding shell.

For a space frame with l_x by l_y rectangular meshes the compressive forces C_x, C_y, C_r in the bars parallel to x, y and the diagonal direction r are given by:

$$\begin{aligned} C_x &= N_x l_y - N_{xy} l_x, \\ C_y &= N_y l_x - N_{xy} l_y, \\ C_r &= N_{xy}\sqrt{l_x^2 + l_y^3}. \end{aligned} \tag{18.8.4}$$

Example 18.8.1. A spherical triangularly reticulated dome has a span of 90 m (300 ft) and a rise of 18 m (60 ft). It carries a total load of 2.5 kN/m² (50 psf), which is mostly a live load. Its steel bars are 3 m (10 ft) long. Determine the area of the bars.

From Fig. 18.8.3:

$$R(1 - \cos\varphi_0) = 18 \text{ m}, \qquad R\sin\varphi_0 = 45 \text{ m},$$

$$(R - 18)^2 + 45^2 = R^2 \quad \therefore \quad R = \frac{1}{2}\left(\frac{45^2}{18} + 18\right) = 65.25 \text{ m (218 ft)},$$

$$\sin\varphi_0 = \frac{4.5}{65.25} = 0.69, \qquad \cos\varphi_0 = 0.724.$$

By (16.2.5), with $N_\theta \equiv N_x$, $N_\varphi \equiv N_y$, and $S = 0$ because of symmetry, we obtain at the shell boundary where $\cos\varphi_0 = 0.724$:

$$N_x = 2.5 \times 65.25 \times (0.724^2 - 0.5) = 3.94 \text{ kN/m (264 lb/ft)},$$

$$N_y = \tfrac{1}{2} \times 2.5 \times 65.25 = 81.56 \text{ kN/m (5,450 lb/ft)},$$

and by (18.8.1), (18.8.2), and (18.8.3):

$$C_x = \frac{3}{2\sqrt{3}}(3 \times 3.94 - 81.56) = -60.4 \text{ kN } (-13{,}440 \text{ lb}),$$

$$C_y = C_s = \frac{3}{\sqrt{3}} \times 81.56 = 141.27 \text{ kN (31,470 lb)}.$$

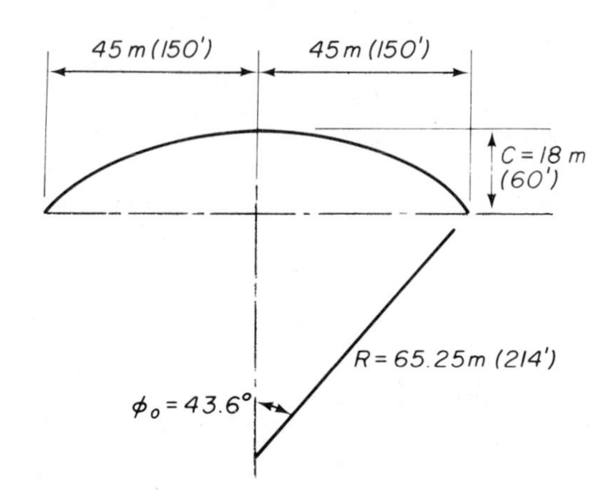

FIGURE 18.8.3

The area of the bars for $F_c = 138$ MPa (20,000 psi) is:

$$A = \frac{141.27}{138} \times 1\,000 = 1\,024 \text{ mm}^2 \text{ (1.57 sq in.)}.$$

The area of a mesh equals:

$$A_m = \frac{\sqrt{3}}{4} 3^2 = 3.9 \text{ m}^2.$$

The weight of the mesh bars is:

$$W = \frac{3}{2}(3 \times 1\,000)\,1\,024 \times \frac{77}{1\,000^3} = 0.354\,8 \text{ kN (82 lb)},$$

and the weight of the bars per unit of roof area is:

$$w_b = \frac{0.354\,8}{3.9} = 0.091 \text{ kN/m}^2 \text{ (1.9 psf)}.$$

The total dead load of the dome may be 0.5 kN/m² to 0.75 kN/m² (10 to 15 psf).

Example 18.8.2. A reticulated barrel (Fig. 18.8.4), 36 m (120 ft) long and 9 m (30 ft) wide, with a rise of 4.5 m (15 ft), is composed of steel bars with a triangular mesh size $l = 1.5$ m (5 ft). The barrel is loaded with a total load of 4 kN/m² (80 psf), which may be considered a dead load, and is supported on rigid end arches. Determine the maximum compression and the maximum tension in its bars.

By (16.5.1), (16.5.2), and (16.5.3), with $N_\varphi \equiv N_y$:

$$N_x]_{\text{max}} = 0.47\left(\frac{36}{4.5}\right)(4 \times 36) = 541.4 \text{ kN/m (36 k/ft) at midspan, at shell top,}$$

$$N_x]_{\text{min}} = -0.83\left(\frac{36}{4.5}\right)(4 \times 36) = -956.2 \text{ kN/m } (-63.5 \text{ k/ft})$$

at midspan, at shell bottom,

$$N_\varphi]_{\text{max}} = 4 \times 4.5 = 18 \text{ kN/m (1.2 k/ft) at barrel top,}$$

$$N_\varphi]_{\text{min}} = 0 \text{ at barrel bottom,}$$

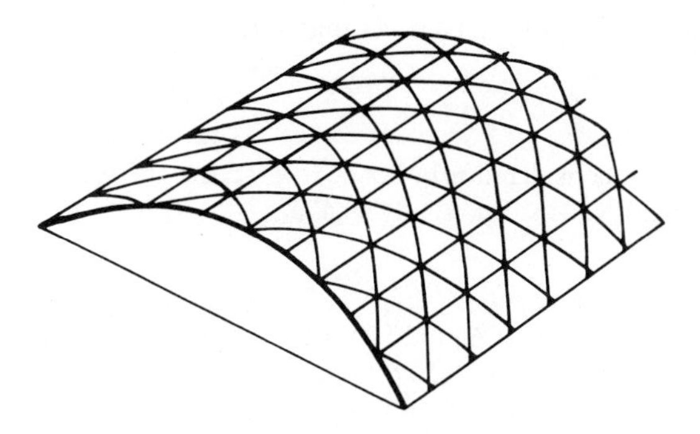

FIGURE 18.8.4

$$N_\varphi]_{\cos\varphi=2/\pi} = 4 \times 4.5 \times \frac{2}{\pi} = 11.46 \text{ kN/m } (0.76 \text{ k/ft}) \text{ at neutral axis,}$$

$$S]_{\max} = 1.10 \times 4 \times 36 = 158.4 \text{ kN/m } (10.5 \text{ k/ft}) \text{ at neutral axis, at supports,}$$

by means of which (18.8.1), (18.8.2), and (18.8.3) give:

(a) At midspan, at top of barrel:

$$C_x = \frac{1.5}{2\sqrt{3}}(3 \times 541.4 - 18) = 695.5 \text{ kN } (151 \text{ k}),$$

$$C_r = C_s = \frac{1.5}{\sqrt{3}}(18) = 15.59 \text{ kN } (3.5 \text{ k}).$$

(b) At midspan, at bottom of barrel:

$$C_x = \frac{1.5}{2\sqrt{3}}3(-956.2) = -1\,243.6 \text{ kN } (274 \text{ k}),$$

$$C_r = C_s = 0.$$

(c) At supports, at neutral axis:

$$C_x = \frac{1.5}{2\sqrt{3}}(-11.46) = -4.97 \text{ kN } (-1.12 \text{ k}),$$

$$C_r = \frac{1.5}{\sqrt{3}}(11.46 - \sqrt{3} \times 158.4) = -227.66 \text{ kN } (-50.4 \text{ k}),$$

$$C_s = \frac{1.5}{\sqrt{3}}(11.46 + \sqrt{3} \times 158.4) = 247.54 \text{ kN } (55 \text{ k}).$$

With $F_t = 138$ MPa (20,000 psi), the maximum bar area equals:

$$A = \frac{1\,243.6 \times 1\,000}{138} = 9\,012 \text{ mm}^2 \text{ } (13.70 \text{ sq in.}).$$

A triangular mesh has an area A_m given by:

$$A_m = \left(\frac{1}{2}l\right)\left(\frac{\sqrt{3}}{2}l\right) = \frac{\sqrt{3}}{4}l^2 = \frac{\sqrt{3}}{4}(1.5)^2 = 0.97 \text{ m}^2 \text{ } (10.8 \text{ sq ft}).$$

The weight of the bars of a triangular mesh is:

$$W = \tfrac{1}{2}(3l)A \times \rho = \tfrac{3}{2}(1.5 \times 1\,000) \times 9\,012 \times \frac{77}{1\,000^3} = 1.56 \text{ kN } (356 \text{ lb}),$$

and the maximum weight of the bars per unit area of roof is:

$$w_b = \frac{1.56}{0.97} = 1.61 \text{ kN/m}^2 \text{ (33 psf)},$$

or less than half the considered total load. In practice, the bar areas are adjusted to the bar forces and w_b is substantially smaller.

Example 18.8.3. A reticulated hyperbolic paraboloid spans 30 m × 60 m (100 ft × 200 ft) with a rise of 6 m (20 ft). Its triangular mesh size is 1.75 m (6 ft). The total load on it is 2.5 kN/m² (50 psf). Determine the area of its steel bars.

By (16.6.9), with $C = N_x$, $T = N_y$, $S = 0$, we obtain:

$$N_x = -N_y = 2.5 \times \frac{30 \times 60}{2 \times 6} = 375 \text{ kN/m (25.7 k/ft)},$$

by means of which (18.8.1), (18.8.2), and (18.8.3) give:

$$C_x = \frac{1.75}{2\sqrt{3}}(3 \times 375 + 375) = 758.7 \text{ kN (170.6 k)},$$

$$C_r = C_s = \frac{1.75}{\sqrt{3}} 375 = 379.3 \text{ kN (85.3 k)},$$

$$A = \frac{758.7 \times 1\,000}{138} = 5\,498 \text{ mm}^2 \text{ (8.52 sq in.)},$$

$$A_m = \frac{\sqrt{3}}{4}(1.75)^2 = 1.32 \text{ m}^2,$$

$$W = \tfrac{3}{2}(1.75 \times 1\,000) \times 5\,498 \times \frac{77}{1\,000^3} = 1.11 \text{ kN (250 lb)},$$

$$w_b = \frac{1.11}{1.32} = 0.84 \text{ kN/m}^2 \text{ (17.5 psf)}.$$

18.9 BUCKLING OF RETICULATED DOMES

Reticulated domes may buckle either locally by instability of one of the compressed bars, or as a whole, i.e., by instability of the reticulated surface as if it were a continuous shell.

The local instability is checked by the column buckling formula (5.3.6), assuming conservatively the compressed bars hinged at the ends, i.e., with a reduced length equal to l.

The buckling of the dome as a whole is checked by the formulas for the corresponding shell in which the values of the thickness and the modulus of elasticity are substituted by "equivalent" values. The derivation of these equivalent values is beyond the scope of this book, but it may be stated that reticulated domes behave like fairly thick shells (often thicker than the depth of their beams) with a very low modulus of elasticity.

For a triangulated dome with bars of length l, area A, moment of inertia I, and modulus of elasticity E, the equivalent thickness h' and the equivalent modulus E' are given by:

$$h' = 2\sqrt{3}\sqrt{\frac{I}{A}}, \tag{18.9.1}$$

$$E' = \frac{A}{3l\sqrt{I/A}}E. \tag{18.9.2}$$

For a dome with square meshes and 45° diagonals:

$$h' = 2\sqrt{3}\sqrt{\frac{I}{A}}, \tag{18.9.1a}$$

$$E' = \frac{A}{2\sqrt{3}\,l\sqrt{I/A}}E. \tag{18.9.2a}$$

Example 18.9.1. Check the local and shell buckling of the spherical dome of Example 18.8.1.

The maximum compression in the bars is 141.27 kN (31.47 k). Assuming the bars to be circular pipes of outer diameter 150 mm (6 in.) and thickness $t = 6$ mm ($\frac{1}{4}$ in.), their area and moment of inertia are [see Table 4.1.1(10)]:

$$A = \pi \times 150 \times 6 = 2\,827 \text{ mm}^2,$$

$$I = \frac{\pi}{8} \times 150^3 \times 6 = 794.81 \times 10^4 \text{ mm}^4,$$

and, with $l = 3$ m (10 ft), $E = 200\,000$ MPa (30,000 ksi), (5.3.6) gives:

$$C_r]_{cr} = \pi^2\frac{200\,000 \times 794.81 \times 10^4}{1\,000 \times 3\,000^2} = 1\,743.2 \text{ kN (436 k)}.$$

The coefficient of safety against local buckling is:

$$c_s = \frac{1\,743.2}{141.27} = 12.34 \ (13.8).$$

The equivalent thickness and modulus of the shell are:

$$h' = 2\sqrt{3}\sqrt{\frac{794.81 \times 10^4}{2\,827}} = 2\sqrt{3} \times 53 = 183.7 \text{ mm (7.35 in.)},$$

$$E' = \frac{2\,827 \times 200\,000}{3 \times 3\,000\sqrt{794.8\,1 \times 10^4/2\,827}} = 1\,185 \text{ MPa (185 ksi)},$$

by means of which (17.6.2) gives:

$$p_{cr} = 0.30 \times 1\,185\left(\frac{183.7}{65.25 \times 1\,000}\right)^2 = 0.002\,8 \text{ N/mm}^2 = 2.80 \text{ kN/m}^2 \text{ (63.2 psf)}.$$

The coefficient of safety against shell buckling is:

$$c_s = \frac{2.8}{2.5} = 1.12 \ (1.26)$$

and is obviously too low, although the bars were chosen with an area substantially greater than the area (1 024 mm²) required by strength. This result shows that shell buckling may be the governing condition in the design of a reticulated shell.

Example 18.9.2. Check the local and shell buckling of the barrel of Example 18.8.2.

With bars made out of pipes 100 mm (4 in.) in diameter and 6 mm ($\frac{1}{4}$ in.) thick, smaller than the largest pipe required by strength, we obtain:

$$A = 1\,884 \text{ mm}^2, \qquad I = 235.5 \times 10^4 \text{ mm}^4,$$

$$h' = 2\sqrt{3}\sqrt{\frac{235.5 \times 10^4}{1\,884}} = 122.5 \text{ mm (4.88 in.)},$$

$$E' = \frac{1\,884 \times 200\,000}{3 \times 1.5 \times 1\,000\sqrt{235.5 \times 10^4/1\,884}} = 2\,368.3 \text{ MPa (372 ksi)}.$$

Hence (17.6.2) with $\alpha = 0.6$ for barrel shells gives:

$$p_{cr} = 0.6 \times 2\,368.3\left(\frac{122.5}{4.5 \times 1\,000}\right)^2 = 1.053 \text{ N/mm}^2 = 1\,053 \text{ kN/m}^2 \text{ (23.3 ksf)},$$

indicating a totally safe condition against shell buckling.

The local critical compression is, by (5.3.6):

$$C_x]_{cr} = \pi^2 \frac{200\,000 \times 235.5 \times 10^4}{1\,000(1.5 \times 1\,000)^2} = 2\,064 \text{ kN (510 k)}.$$

The coefficient of safety against local buckling is also very high and indicates that, in this case, strength governs.

18.10 STEEL QUANTITIES IN RETICULATED SPHERICAL DOMES

The steel quantities in reticulated domes are not very sensitive to the orientation of the bars and, hence, can be estimated, using thin shell theory, as if their steel were uniformly distributed over the dome surface. The meridional and parallel unit forces N_ϕ and N_θ in a spherical, thin shell dome of radius R, under a dead load w are given by (16.2.3):

$$N_\phi = \frac{wR}{1 + \cos\phi}, \qquad N_\theta = wR\left(\cos\phi - \frac{1}{1 + \cos\phi}\right). \tag{a}$$

N_θ was shown to be compressive for angles $\phi \leq \bar{\phi} = 51°83$. Considering a square on the dome of unit area, with sides of length one parallel to ϕ and θ, at the colatitude ϕ, the weight of steel required to resist the meridional and parallel forces on its sides is:

$$W_s\Big]_{\phi \leq \bar{\phi}} = \gamma\left(\frac{N_\phi}{f_s} \times 1 + \frac{N_\theta}{f_s} \times 1\right) = \frac{\gamma}{f_s} wR \cos\phi,$$

where γ is the unit weight of steel and f_s the allowable steel stress. The steel weight of a parallel ring at ϕ, with a depth of $R\,d\phi$ and a circumference $2\pi r = 2\pi R \sin\phi$ (Fig. 18.10.1), and hence an area $dA = 2\pi R^2 \sin\phi\, d\phi$ equals:

$$W_s\, dA = 2\pi \frac{w}{f_s} \gamma R^3 \cos\phi \sin\phi\, d\phi. \tag{b}$$

Integrating (b) over the entire dome surface, i.e., from $\phi = 0$ to $\phi = \phi_0$, we obtain, provided $\phi_0 \leq \bar{\phi}$, the steel quantity:

$$W_s = 2\pi \frac{w}{f_s} \gamma R^3 \int_0^{\phi_0} \cos\phi \;\sin\phi\, d\phi = 2\pi \frac{w}{f_s} \gamma R^3 \left(\frac{1}{2} \sin^2 \phi_0\right), \tag{18.10.1}$$

or, since $\sin\phi_0 = r_0/R$ and $R = (r_0^2 + h^2)/(2h)$:

$$W_s = \pi \frac{w}{f_s} \gamma \left(\frac{r_0^2 + h^2}{2h}\right)^3 (r_0)^2 \left(\frac{2h}{r_0^2 + h^2}\right)^2 = \frac{\pi}{2} \frac{w}{f_s} \gamma r_0^3 \left(\frac{r_0}{h} + \frac{h}{r_0}\right). \tag{18.10.2}$$

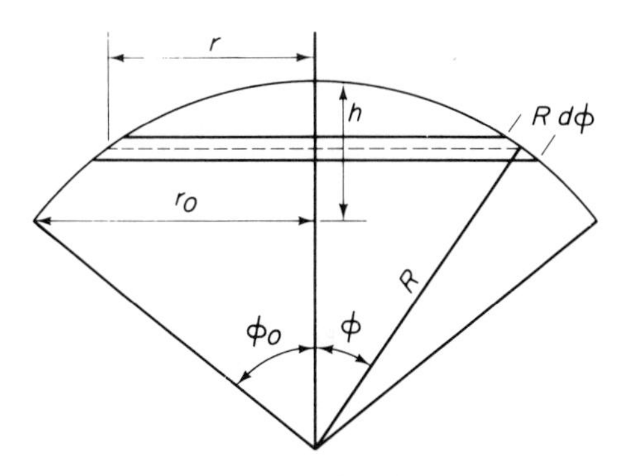

FIGURE 18.10.1

For $\phi_0 = \bar{\phi}$, (18.10.1) gives:

$$\overline{W}_s = W_s\Big]_{\phi_0 = \bar{\phi}} = 1.94 \frac{w}{f_s} \gamma R^3. \tag{18.10.3}$$

For $\phi_0 > \bar{\phi}$, the negative N_θ's from $\phi = \bar{\phi}$ to $\phi = \phi_0$ must be changed in sign to be added to the N_ϕ's in (a), obtaining:

$$W_s\Big]_{\phi_0 > \bar{\phi}} = \gamma\left(\frac{N_\phi}{f_s} \times 1 - \frac{N_\theta}{f_s} \times 1\right) = \frac{\gamma}{f_s} wR\left(\frac{2}{1 + \cos\phi} - \cos\phi\right), \tag{c}$$

and the ΔW_s to be added to $\overline{W}_s$ becomes:

$$\begin{aligned} \Delta W_s &= 2\pi \frac{w}{f_s} \gamma R^2 \left(\int_{\bar{\phi}}^{\phi_0} \frac{2 \sin\phi}{1 + \cos\phi}\, d\phi - \int_{\bar{\phi}}^{\phi_0} \cos\phi \sin\phi\, d\phi\right) \\ &= 4\pi \frac{w}{f_s} \gamma R^3 \left[\ln \frac{1.618}{1 + \cos\phi_0} - \frac{1}{4}(\sin^2\phi_0 - 0.618)\right]. \end{aligned} \tag{18.10.4}$$

For example, for $\phi_0 = 90°$, (18.10.4) gives:

$$\Delta W_s\Big]_{\phi_0=90°} = 4.85\frac{w}{f_s}\gamma R^3, \tag{d}$$

so that the total weight for a hemisphere is:

$$W_s\Big]_{\phi_0=90°} = (1.94 + 4.85)\frac{w}{f_s}\gamma R^3 = 6.79\frac{w}{f_s}\gamma R^3. \tag{18.10.5}$$

The weight $\overline{w}_s$ for unit of covered area thus becomes:

$$\overline{w}_s\Big]_{\phi_0=90°} = \frac{W_s\Big]_{\phi_0=90°}}{\pi R^2} = 2.16\frac{w}{f_s}\gamma R. \tag{18.10.6}$$

For a total load assimilated to a dead load $w = 2.5$ kN/m² (50 psf), a radius $R = 100$ m (300 ft), $f_s = 110$ MPa (16,000 psi), and $\gamma = 77$ kN/m³ (490 pcf), we obtain:

$$\overline{w}_s\Big]_{\phi_0=90°} = 2.16\frac{2.5}{110}\frac{77(100)}{1\ 000} = 0.38 \text{ kN/m}^2 \text{ (6.89 psf)}.$$

The actual value of $\overline{w}_s$ is substantially greater, since it is dictated by buckling.

PROBLEMS

18.1 Determine the stresses in the bars of the truss grid of Fig. 18.2.1 with $a = b$, when a single load P acts at one of the internal nodes. All the bars have the same cross-sectional area A, and in the cross-braced bays only the tension diagonal is active. Refer to Example 11.1.3 for sharing of the load.

18.2 A square grid similar to that of Fig. 18.1.1 (a), with twelve equally spaced square bays 1.8 m (6 ft) on a side, is loaded by a uniform load $q = 6$ kN/m² (120 psf). If the depth of the grid is 1.8 m (6 ft), determine the required area of the horizontal bars at the center and of the diagonal bars at the support for aluminum 6061-T6 (see Table 2.3.2).

18.3 Solve Problem 18.2 with eighteen 1.8 m (6 ft) bays in one direction and twelve 1.8 m (6 ft) bays in the other and a depth of 2.7 m (9 ft).

18.4 Determine the deflection at the center of the grids of Problems 18.2 and 18.3, assuming I to be computed on the basis of an ideal I-beam.

18.5 A 45° skew grid with $j = 1.5$ m (5 ft), $a = 15$ m (50 ft), $b = 7.5$ m (25 ft) consists of 300 mm × 900 mm (12 in. × 36 in.) concrete members. Determine its load-carrying capacity for $f'_c = 20$ MPa (2,900 psi).

18.6 Determine the depth required by a 45° skew concrete grid covering an area 9 m × 9 m (30 ft × 30 ft), if the width of the members is 200 mm (8 in.), $q_0 = 9$ kN/m² (175 psf), $f'_c = 30$ MPa (4,350 psi), and $j = 1.5$ m (5 ft).

18.7 Determine the maximum spacing of 200 mm × 300 mm (8 in. × 12 in.) concrete members in a square grid spanning 6 m × 12 m (20 ft × 40 ft) carrying a total load of 5 kN/m² (100 psf), if $f'_c = 20$ MPa (2,900 psi).

18.8 Determine the load-carrying capacity of a triangular grid with members, spans, and nodal points as those in Problem 18.5.

18.9 Determine the required depth of a triangular grid with members, spans, and nodal points as those in Problem 18.6.

18.10 Solve Problem 18.5 for a quadruple grid in which the width of the diagonal members is 0.7 × 300 mm = 210 mm (0.7 × 12 in. = 8.4 in.).

18.11 Check the adequacy of a parabolic lamella roof spanning 24 m (80 ft) with $f = l/4$ and $\alpha = 20°$, consisting of 75 mm × 250 mm (3 in. × 10 in.) wood members with $F_a = 10$ MPa (1,450 psi), $F_b = 14.1$ MPa (2,050 psi), spaced at a distance $c = 1.5$ m (5 ft). The dead load is 0.8 kN/m² (15 psf), and the snow load is 1.5 kN/m² (30 psf).

18.12 Determine the minimum rise of a parabolic lamella roof spanning 27 m (90 ft) and carrying an assumed dead load of 1.5 kN/m² (30 psf) and a snow load of 2 kN/m² (40 psf), if its members are 100 mm × 300 mm (4 in. × 12 in.) Douglas fir with $F_a = 10$ MPa (1,450 psi) and $F_b = 14.1$ MPa (2,050 psi), spaced at a distance $c = 1.8$ m (6 ft) with $\alpha = 20°$.

18.13 A simply supported triangulated barrel 30 m (100 ft) long and 18 m (60 ft) wide, with a rise of 9 m (30 ft), is composed of aluminum bars with a mesh size $l = 1.2$ m (4 ft). Determine the maximum tension and compression in the bars if the total load is 5 kN/m² (100 psf), and determine the bar area if $F_t = 83$ MPa (12,000 psi).

18.14 A triangulated spherical dome with a span of 60 m (200 ft) and a rise of 12 m (40 ft) consists of A36 steel bars with an average length of 2.4 m (8 ft). Determine the bar forces and total weight of the bars per unit area if the total load is 2.5 kN/m² (50 psf).

18.15 A triangulated conical shell (Fig. 16.2.3) with a span of 24 m (80 ft) and a rise of 6 m (20 ft) carries a live load of 3 kN/m² (60 psf). Determine the bar forces for a mesh size $l = 600$ mm (2 ft).

18.16 Check the local and shell buckling of the dome of Problem 18.14 if the steel bars are pipes with a nominal diameter of 150 mm (6 in.).

18.17 Check the local and shell buckling in the barrel of Problem 18.13 if the aluminum bars are pipes with a nominal diameter of 200 mm (8 in.).

18.18 A triangulated square hyperbolic paraboloid, supported on its four corners, has sides 30 m (100 ft) long and a rise of 6 m (20 ft). It carries a total load of 3 kN/m² (60 psf) and has a mesh size of 1.5 m (5 ft) in horizontal projection. Determine the cross section of its A36 steel bars and the live load it can carry.

18.19 Check the local buckling of the bars of Problem 18.18 if the steel bars are pipes 100 mm (4 in.) in outside diameter, and determine the dimensions of the compressed boundary wide flange beams.

18.20 A triangulated elliptic paraboloid covers a square area 15 m (50 ft) by the side with a total rise of 3 m (10 ft). Its mesh size in plan is 1.5 m (5 ft). It carries a total load of 3 kN/m^2 (60 psf). Determine its bar cross sections at the center of the paraboloid, at the top of its side arches, and at its corners, if the bars are steel pipes with a nominal diameter of 75 mm (3 in.).

18.21 Determine the critical value of the load for the elliptic paraboloid of Problem 18.20 if its steel pipes are 6 mm ($\frac{1}{4}$ in.) thick. Does shell buckling or local buckling govern?

Appendix A

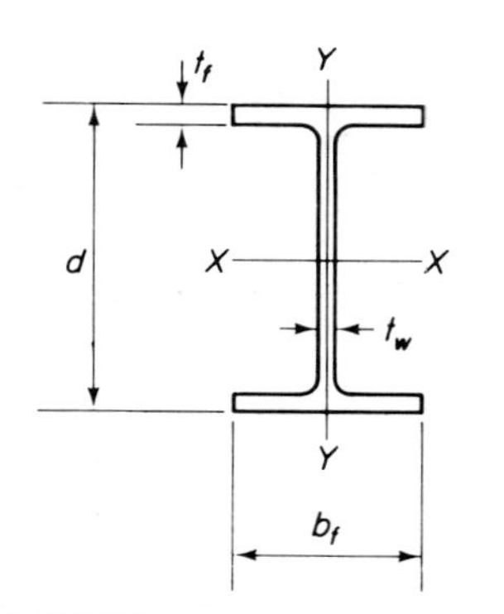

Table A.1 STEEL WIDE FLANGE SECTIONS

IN SI UNITS

Designation (nominal depth in inches and weight in pounds per linear foot)	Designation (nominal depth in millimeters and mass in kilograms per meter)	Area A, mm²	Depth d, mm	Flange		Web thickness t_w, mm	Axis X-X			Axis Y-Y		
				Width b_f, mm	Thickness t_f, mm		I, mm⁴ × 10⁴	S, mm³ × 10³	r, mm	I, mm⁴ × 10⁴	S, mm³ × 10³	r, mm
(W36 × 300)	W920 × 446	56 970	933	423	42.7	24.0	844 886	18 190	386.1	54 106	2 556	97.3
(W36 × 245)	W920 × 365	46 520	916	419	34.3	20.4	670 082	14 466	381.0	42 036	2 016	95.3
(W36 × 160)	W920 × 238	30 390	914	305	25.9	16.6	405 795	8 882	365.8	12 278	805	63.5
(W33 × 200)	W840 × 298	38 000	838	400	29.2	18.2	461 982	10 996	348.0	31 215	1 560	90.7
(W33 × 141)	W840 × 210	26 840	846	293	24.4	15.4	310 069	7 341	340.4	10 239	700	61.7
(W30 × 124)	W760 × 185	23 550	766	267	23.6	14.9	223 083	5 817	307.3	7 533	564	56.6
(W24 × 145)	W610 × 216	27 550	622	357	25.9	15.4	190 203	6 112	261.6	19 603	1 232	85.1
(W24 × 84)	W610 × 125	15 940	612	229	19.6	11.9	98 640	3 228	248.7	3 933	344	49.5

Table A-1 (cont.)

Designation (nominal depth in inches and weight in pounds per linear foot)	Designation (nominal depth in millimeters and mass in kilograms per meter)	Area A, mm²	Depth d, mm	Flange		Web thickness t_w, mm	Axis X-X			Axis Y-Y		
				Width b_f, mm	Thickness t_f, mm		I, mm⁴ × 10⁴	S, mm³ × 10³	r, mm	I, mm⁴ × 10⁴	S, mm³ × 10³	r, mm
(W21 × 68)	W530 × 101	12 900	537	210	17.4	10.9	61 598	2 294	218.4	2 693	257	45.7
(W21 × 62)	W530 × 92	11 810	533	209	15.6	10.2	55 355	2 081	216.9	2 393	228	45.0
(W21 × 55)	W530 × 82	10 450	528	209	13.3	9.5	47 447	1 803	213.4	2 010	193	43.9
(W18 × 96)	W460 × 143	18 190	461	298	21.1	13.0	69 922	3 032	195.6	9 365	628	71.6
(W18 × 64)	W460 × 95	12 190	454	221	17.4	10.2	43 701	1 934	189.5	3 155	285	50.8
(W16 × 40)	W410 × 60	7 610	406	178	12.8	7.8	21 518	1 059	168.2	1 199	135	39.6
(W16 × 36)	W410 × 54	6 840	403	178	10.9	7.6	18 604	926	165.1	1 016	115	38.6
(W14 × 150)	W360 × 223	28 450	378	394	28.7	17.7	74 500	3 933	161.8	29 259	1 485	101.4
(W14 × 142)	W360 × 211	26 970	375	394	27.0	17.3	69 505	3 720	160.5	27 469	1 396	100.8
(W14 × 61)	W360 × 91	11 550	353	254	16.3	9.6	26 678	1 508	151.9	4 453	352	62.2
(W14 × 53)	W360 × 79	10 060	354	205	16.7	9.4	22 558	1 275	149.5	2 393	234	48.8
(W14 × 22)	W360 × 33	4 190	348	127	8.5	5.8	8 241	474	140.5	291	46	26.4
(W12 × 72)	W310 × 107	13 680	311	306	17.0	10.9	24 847	1 598	134.9	8 116	531	77.2
(W12 × 65)	W310 × 97	12 320	308	305	15.4	9.9	22 183	1 442	134.1	7 284	477	76.7
(W12 × 40)	W310 × 60	7 610	303	203	13.1	7.5	12 902	850	130.3	1 835	180	49.3
(W8 × 40)	W200 × 60	7 610	210	205	14.2	9.3	6 076	582	89.7	2 039	198	51.8
(W8 × 31)	W200 × 46	5 884	203	203	11.0	7.3	4 578	449	88.1	1 540	151	51.1
(W6 × 15.5)	W150 × 23.1	2 942	152	152	6.8	6.0	1 253	164	65.3	402	53	37.1
(S7 × 15.3)	S180 × 22.8	2 890	178	92	10.0	6.4	1 527	172	72.6	110	23.6	19.5

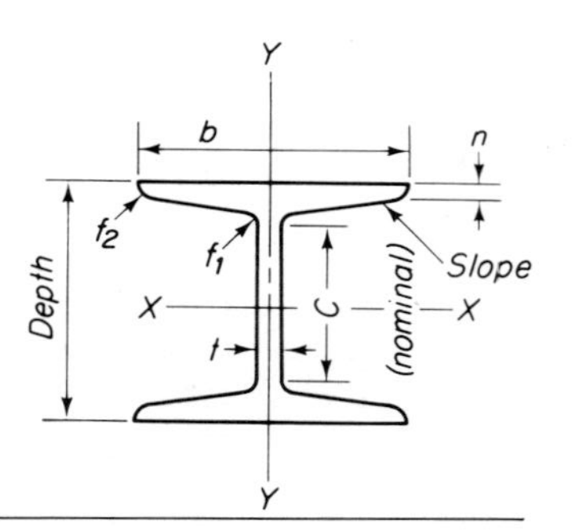

Table A.2 ALUMINUM WIDE FLANGE SECTIONS IN SI UNITS

Size											Axis X-X			Axis Y-Y		
Nominal depth and width	Mass	t	Area	Actual Depth	b	Slope	n	f_1	f_2	c	I	S	r	I	S	r
mm (in.)	kg/m (lb/ft)	mm (in.)	mm^2 ($in.^2$)	mm (in.)	mm (in.)		mm (in.)	mm (in.)	mm (in.)	mm (in.)	$\times 10^4$ mm^4 ($in.^4$)	$\times 10^3$ mm^3 ($in.^3$)	mm (in.)	$\times 10^4$ mm^4 ($in.^4$)	$\times 10^3$ mm^3 ($in.^3$)	mm (in.)
200 × 200 (8 × 8)	19 (12.99)	12.7 (0.500)	712.9 (11.05)	203 (8.00)	206.4 (8.125)	1:18.9 (1:18.9)	9.09 (0.358)	7.95 (0.313)	4.55 (0.179)	158.75 ($6\frac{1}{4}$)	5049 (121.31)	497 (30.33)	84.1 (3.31)	1519 (36.50)	150 (9.13)	46.2 (1.82)

Appendix B Answers to Selected Problems

CHAPTER ONE

1.3. 1.5 m × 1.5 m × 1.8 m (5 ft × 5 ft × 6.5 ft). **1.5.** $q = 1.4$ kN/m^2 (30 psf).
1.6. $F = 1\ 125$ kN (259 k); $M = 62\ 552$ kN·m (48,050 k·ft); $T = C = 2\ 085$ kN (480 k).
1.8. $W = 38.2$ m (135 ft).

CHAPTER TWO

2.1. $\Delta L_s = \Delta L_c = 4.2$ mm (0.17 in.); $\Delta L_w = 1.6$ mm (0.06 in.); $\Delta L_{al} = 8.4$ mm (0.34 in.).
2.3. Steel, S.F. = 2.5; Aluminum, S.F. = 2.2; Concrete, S.F. = 2.2.
2.4. $\epsilon_{al}/\epsilon_s = 2.56$; $\epsilon_c/\epsilon_s = 0.55$; $\epsilon_w/\epsilon_s = 1.03$.
2.6. $P = 1\ 836$ kN (412 k); $P_u = 5\ 100$ kN (1,147 k).

CHAPTER THREE

3.1. $W = 83$ kN (19.25 k).

Sizes of hangers:
S_2: 2∠ 64 × 64 × 5 (2∠ $2\frac{1}{2} \times 2\frac{1}{2} \times \frac{3}{16}$),
S_4: 2∠ 76 × 76 × 8 (2∠ $3 \times 3 \times \frac{5}{16}$),
S_6: 2∠ 76 × 76 × 13 (2∠ $3 \times 3 \times \frac{1}{2}$),
S_8: 2∠ 102 × 102 × 13 (2∠ $4 \times 4 \times \frac{1}{2}$),
S_{10}: 2∠ 102 × 102 × 16 (2∠ $4 \times 4 \times \frac{5}{8}$),
S_{12}: 2∠ 102 × 102 × 19 (2∠ $4 \times 4 \times \frac{3}{4}$),
S_{14}: 2∠ 152 × 152 × 14 (2∠ $6 \times 6 \times \frac{9}{16}$),
S_{16}: 2∠ 152 × 152 × 16 (2∠ $6 \times 6 \times \frac{5}{8}$),

S_{18}: 2∠ 152 × 152 × 19 (2∠ 6 × 6 × $\frac{3}{4}$),
S_{20}: 2∠ 152 × 152 × 22 (2∠ 6 × 6 × $\frac{7}{8}$),
ΔL = 45 mm (1.93 in.)

3.4. ΔL = 0.11 m (4.6 in.). **3.5.** ϕ6.35 mm ($\frac{1}{4}$ in.) rod.
3.7. The size should be increased to ϕ29 mm ($1\frac{1}{8}$ in.).
3.8. d = 31.75 mm ($1\frac{1}{4}$ in.); T = 220.5 kN (52 k).
3.9. W = 28.44 kN (6.8 k); A = 10.5 m² (120 sq ft).
3.11. Column size: 920 mm × 920 mm (36.4 in. × 36.4 in.).
3.13. 11 stories **3.15.** d = 15.88 mm ($\frac{5}{8}$ in.)
3.17. l_g = 209 m (698 ft); l_c = 563 m (1,878 ft); l_{glass} = 270 m (873 ft); l_s = 1 935 m (6,336 ft); l_a = 4 816 m (15,815 ft); l_w = 1 254 m (4,138 ft).

CHAPTER FOUR

4.1. t_b = 6 mm ($\frac{1}{4}$ in.); t_g = 10 mm ($\frac{7}{16}$ in.).
4.3. b = 614 mm (23 in.); $\Delta f_{b,\,OL}$ = 1.64%. **4.5.** b = 445 mm (17 in.).
4.6. b = 155 mm (7 in.); A_s = 954 mm² (1.64 sq. in.). **4.8.** h = 150 mm (6 in.).
4.9. δ = 18 mm (0.76 in.).
4.11. f_b = 46.5 MPa (7.08 ksi); δ = 307 mm (12.7 in.).
4.12. δ = 6.5 mm (0.24 in.); Camber: 3.3 mm (0.12 in.).
4.14. M = 512 kN·m (400 k·ft); V = 114 kN (26.4 k); no.
4.16. q = 3.7 kN/m² (80.3 psf); δ = 14 mm (0.53 in.).
4.18. M_{max} = 260 kN·m (195 k·ft); V_{max} = 173.4 kN (39 k); W460 × 82 (W18 × 55).
4.20. M_{max} = 366 kN·m (287 k·ft); M_{min} = −478.6 kN·m (−373.3 k·ft); V_{max} = 197.3 kN (46.2 k).
4.21. $M_1 = M_4 = 0$; M_2 = 712 kN·m (553 k·ft); M_3 = 462 kN·m (359 k·ft); $V_1 = V_{2L}$ = 59.3 kN (13.8 k); $V_{2R} = V_{3L}$ = 78.3 kN (32.0 k); $V_{3R} = V_4$ = 38.5 kN (9.0 k).
4.22. M_{max} = 693.18 kN·m (513.47 k·ft); M_{min} = −908.65 kN·m (−673.07 k·ft); V_{max} = 675 kN (150 k).
4.25. δ_s/δ_b = 1.67.
4.26. M = 296 kN·m (257.7 k·ft); R_{core} = 224.4 kN (53.2 k); R_{column} = 111.6 kN (23.8 k).

CHAPTER FIVE

5.1. f = 10.4 MPa (1.48 ksi) Not adequate. **5.3.** h = 302 mm (17 in.).
5.5. No.
5.6. (a) P_{cr} = 7.3 kN (1.65 k); f_{cr} = 1.9 MPa (0.28 ksi); (c) P_{cr} = 803 kN (269 k); f_{cr} = 153 MPa (30.8 ksi).
5.9. f_{max} = 7.5 MPa (1,138 psi); δ/δ_0 = 0.21. **5.10.** l_{cr} = 13.6 m (45.2 ft).
5.11. b = 70 mm (3 in.).

CHAPTER SIX

6.1. Numbering the nodes from left to right starting at the left support:

Since the truss and the loading are symmetrical about center line, the bar forces are also symmetrical about it.

$F_{1\,2} = F_{2\,4} = 520$ kN (116 k) $\quad F_{2\,3} = 0$
$F_{4\,6} = 945$ kN (210 k) $\quad F_{3\,4} = 602$ kN (134 k)
$F_{6\,8} = 1276$ kN (284 k) $\quad F_{4\,5} = -425$ kN (-95 k)
$F_{8\,10} = 1512$ kN (336 k) $\quad F_{5\,6} = 468$ kN (104 k)
$F_{10\,12} = 1654$ kN (368 k) $\quad F_{6\,7} = -331$ kN (-74 k)
$F_{3\,5} = -945$ kN (-210 k) $\quad F_{7\,8} = 334$ kN (74 k)
$F_{5\,7} = -1276$ kN (-284 k) $\quad F_{8\,9} = -236$ kN (-53 k)
$F_{7\,9} = -1512$ kN (-336 k) $\quad F_{9\,10} = 201$ kN (45 k)
$F_{9\,11} = -1654$ kN (-368 k) $\quad F_{10\,11} = -142$ kN (-32 k)
$F_{11\,13} = -1701$ kN (-378 k) $\quad F_{11\,12} = 67$ kN (15 k)
$F_{1\,3} = -735$ kN (-163 k) $\quad F_{12\,13} = -95$ kN (-21 k).

6.3. Numbering the nodes from left to right starting at the left support:

$F_{1\,3} = -34$ kN (-6 k) $\quad F_{1\,2} = 48$ kN (9 k)
$F_{3\,5} = -88$ kN (-17 k) $\quad F_{2\,3} = -48$ kN (-9 k)
$F_{5\,7} = -116$ kN (-22 k) $\quad F_{3\,4} = 29$ kN (5 k)
$F_{2\,4} = 68$ kN (13 k) $\quad F_{4\,5} = -29$ kN (-5 k)
$F_{4\,6} = 109$ kN (20 k) $\quad F_{5\,6} = 10$ kN (2 k)
$F_{6\,8} = 122$ kN (23 k) $\quad F_{6\,7} = -10$ kN (-2 k)
Other member forces may be obtained by symmetry.

6.4. Numbering the nodes from left to right starting at the left support:

$F_{1\,2} = 43$ kN (10.6 k) $\quad F_{15\,16} = 1$ kN (0.3 k)
$F_{2\,4} = 33$ kN (8.2 k) $\quad F_{2\,3} = -7$ kN (-1.6 k)
$F_{4\,6} = 24$ kN (5.8 k) $\quad F_{4\,5} = -10$ kN (-2.4 k)
$F_{6\,8} = F_{8\,10} = 14$ kN (3.4 k) $\quad F_{6\,7} = -14$ kN (-3.2 k)
$F_{10\,12} = 9$ kN (2.2 k) $\quad F_{8\,9} = 0$
$F_{12\,14} = 4$ kN (1.0 k) $\quad F_{10\,11} = 7$ kN (1.6 k)
$F_{14\,16} = -0.4$ kN (-0.2 k) $\quad F_{12\,13} = 5$ kN (1.2 k)
$F_{1\,3} = -33$ kN (-8.3 k) $\quad F_{14\,15} = 3$ kN (0.8 k)
$F_{3\,5} = -35$ kN (-8.9 k) $\quad F_{2\,5} = 12$ kN (2.9 k)
$F_{5\,7} = -27$ kN (-6.8 k) $\quad F_{4\,7} = 14$ kN (3.4 k)
$F_{7\,9} = -19$ kN (-4.8 k) $\quad F_{6\,9} = 17$ kN (4.0 k)
$F_{9\,11} = -6$ kN (-1.4 k) $\quad F_{9\,10} = -8$ kN (-2.0 k)
$F_{11\,13} = -1.7$ kN (-0.4 k) $\quad F_{11\,12} = -7$ kN (-1.7 k)
$F_{13\,15} = 2.2$ kN (0.6 k) $\quad F_{13\,14} = -6$ kN (-1.4 k).

6.6. $F_{AC} = -44$ kN (-10 k) $\quad F_{AB} = 49$ kN (11 k)
$F_{CE} = -132$ kN (-30 k) $\quad F_{BC} = -22$ kN (-5 k)
$F_{EG} = -264$ kN (-60 k) $\quad F_{CD} = 98$ kN (22 k)
$F_{BD} = 44$ kN (10 k) $\quad F_{DE} = -44$ kN (-10 k)
$F_{DF} = 132$ kN (30 k) $\quad F_{EF} = 148$ kN (33 k).

6.7. Numbering the nodes from right to left:

$F_{1\ 3} = F_{3\ 5} = -154$ kN (-33 k)
$F_{5\ 7} = -309$ kN (-67 k)
$F_{7\ 9} = -617$ kN (-133 k)
$F_{9\ 11} = F_{11\ 12} = -309$ kN (-67 k)
$F_{2\ 4} = 309$ kN (67 k)
$F_{4\ 6} = F_{6\ 8} = 463$ kN (100 k)
$F_{8\ 10} = 617$ kN (133 k)
$F_{2\ 3} = F_{4\ 5} = 90$ kN (20 k)
$F_{6\ 7} = 0$
$F_{8\ 9} = -90$ kN (-20 k)
$F_{10\ 11} = 0$
$F_{1\ 2} = 179$ kN (39 k)
$F_{2\ 5} = -179$ kN (-39 k)
$F_{4\ 7} = -179$ kN (-39 k)
$F_{7\ 8} = 179$ kN (39 k)
$F_{9\ 10} = -357$ kN (-78 k)
$F_{10\ 12} = 357$ kN (78 k).

6.9. $F_{1\ 2} = F_{5\ 7} = -723$ kN (-161 k)
$F_{2\ 4} = F_{4\ 5} = -627$ kN (-139 k)
$F_{1\ 3} = F_{6\ 7} = 623$ kN (138 k)
$F_{3\ 4} = F_{4\ 6} = 386$ kN (86 k)
$F_{2\ 3} = F_{5\ 6} = -109$ kN (-24 k)
$F_{3\ 6} = 270$ kN (60 k).

6.10. $\delta_h = -9.3$ mm (-0.37 in.); $\delta_v = 53$ mm (2.11 in.).

6.12. $\delta = 72$ mm (2.88 in.). **6.15.** $\delta = 61.5$ mm (2.46 in.).

6.16. $A_{12} = 1330$ mm² (2.09 sq in.); $A_{13} = 6019$ mm² (9.33 sq in.); Change of angle = 0.000 3 rad; $M = 2$ kN·m (0.8 k·ft); $f_b/F_t = 3\%$.

CHAPTER SEVEN

7.1. $M_{max} = 8.21$ kN·m (1827 lb·ft); $V_{max} = 14.4$ kN (980 lb); $R = 14.4$ kN (980 lb); $A_S = 325$ mm²/m (0.15 sq in./ft).

7.3. $M_{max} = -436.8$ kN·m (-336 k·ft), $V_{max} = 187.2$ kN (43.2 k);
$A_S = 3\,559$ mm² (5.65 sq in.).

7.4. Beam size: W406 × 42 (W16 × 31);
Column size: W356 × 46 (W14 × 34).

7.6. No.

7.7. $M_B = 20$ kN.m (15 k·ft); $M_C = 40$ kN·m (30 k·ft); $V_e = 4.5$ kN (1 k); $V_i = 9.0$ kN (2 k); yes.

7.8. $M_e = 15$ kN·m (11.25 k·ft); $M_i = 30$ kN·m (22.5 k·ft); member sizes are adequate.

7.9. Beam size: W530 × 101 (W21 × 68);
column size: W356 × 134 (W14 × 90).

7.11. *Axial forces in columns:*
$R_3 = \frac{1}{40}(h/l)W = 0.35$ kN (78 lb)
$R_2 = \frac{5}{40}(h/l)W = 1.75$ kN (391 lb)
$R_1 = \frac{13}{40}(h/l)W = 4.55$ kN (1,016 lb)
$R_0 = R_1 = 4.55$ kN (1,016 lb)

Shear forces in beams:
$V_3 = \frac{2}{40}(h/l)W = 0.7$ kN (156 lb)
$V_2 = \frac{8}{40}(h/l)W = 2.8$ kN (625 lb)
$V_1 = \frac{16}{40}(h/l)W = 5.6$ kN (1,250 lb)
$V'_3 = \frac{3}{40}(h/l)W = 1.05$ kN (234 lb)
$V'_2 = \frac{12}{40}(h/l)W = 4.2$ kN (938 lb)
$V'_1 = \frac{24}{40}(h/l)W = 8.4$ kN (1,875 lb)

Shear forces in columns:
$H_3 = \frac{2}{40}W = 1.4$ kN (313 lb)
$H_2 = \frac{6}{40}W = 4.2$ kN (938 lb)
$H_1 = \frac{10}{40}W = 7.0$ kN (1,563 lb)
$H_3'' = \frac{5}{40}W = 3.5$ kN (1,250 lb)
$H_2'' = \frac{15}{40}W = 10.5$ kN (3,750 lb)
$H_1'' = \frac{25}{40}W = 17.5$ kN (3,906 lb)
$H_3^{IV} = \frac{6}{40}W = 4.2$ kN (938 lb)
$H_2^{IV} = \frac{18}{40}W = 12.6$ kN (2,813 lb)
$H_1^{IV} = \frac{30}{40}W = 21.0$ kN (4,688 lb)
Axial forces in beams:
$H_3' = \frac{18}{40}W = 12.6$ kN (2,813 lb)
$H_2' = \frac{36}{40}W = 25.2$ kN (5,625 lb)
$H_1' = \frac{36}{40}W = 25.2$ kN (5,625 lb)
$H_3''' = \frac{13}{40}W = 9.1$ kN (2,031) lb
$H_2''' = \frac{26}{40}W = 18.2$ kN (4,062 lb)
$H_1''' = \frac{26}{40}W = 18.2$ kN (4,062 lb)
Moments in beams:
$M_{b3} = \frac{2}{80}Wh = 2.1$ kN·m (1,563 lb·ft)
$M_{b2} = \frac{8}{80}Wh = 8.4$ kN·m (6,250 lb·ft)
$M_{b1} = \frac{16}{80}Wh = 16.8$ kN·m (12,500 lb·ft)
$M_{b3}' = \frac{3}{80}Wh = 3.15$ kN·m (2,344 lb·ft)
$M_{b2}' = \frac{12}{80}Wh = 12.6$ kN·m (9,375 lb·ft)
$M_{b1}' = \frac{24}{80}Wh = 25.2$ kN·m (18,750 lb·ft)
Moments in columns:
$M_{C3} = \frac{2}{80}Wh = 2.1$ kN·m (1,563 lb·ft)
$M_{C2} = \frac{6}{80}Wh = 6.3$ kN·m (4,688 lb·ft)
$M_{C1} = \frac{10}{80}Wh = 10.5$ kN·m (7,813 lb·ft)
$M_{C3}'' = \frac{5}{80}Wh = 5.25$ kN·m (3,906 lb·ft)
$M_{C2}'' = \frac{15}{80}Wh = 15.75$ kN·m (11,719 lb·ft)
$M_{C1}'' = \frac{25}{80}Wh = 26.25$ kN·m (19,531 lb·ft)
$M_{C3}^{IV} = \frac{6}{80}Wh = 6.3$ kN·m (4,688 lb·ft)
$M_{C2}^{IV} = \frac{18}{80}Wh = 18.9$ kN·m (14,063 lb·ft)
$M_{C1}^{IV} = \frac{30}{80}Wh = 31.5$ kN·m (23,438 lb·ft)

7.12. *Shear forces in columns:*
$H_6 = 9.6$ kN (2.2 k)
$H_5 = 28.7$ kN (6.6 k)
$H_4 = 47.8$ kN (11.0 k)
$H_3 = 66.9$ kN (15.4 k)
$H_2 = 86.0$ kN (19.8 k)
$H_1 = 105.1$ kN (24.2 k)
Moments in columns:
$M_6 = 18.0$ kN·m (13.75 k·ft)
$M_5 = 53.8$ kN·m (41.25 k·ft)
$M_4 = 89.6$ kN·m (68.75 k·ft)
$M_3 = 125.4$ kN·m (96.25 k·ft)
$M_2 = 161.2$ kN·m (123.75 k·ft)

$M_1 = 197.1$ kN·m (151.25 k·ft)
Moments in beams:
$M'_6 = 18.0$ kN·m (13.75 k·ft)
$M'_5 = 71.8$ kN·m (55 k·ft)
$M'_4 = 143.4$ kN·m (110 k·ft)
$M'_3 = 215$ kN·m (165 k·ft)
$M'_2 = 286.6$ kN·m (220 k·ft)
$M'_1 = 358.3$ kN·m (275 k·ft)
Shear forces in beams:
$V_6 = 4.8$ kN (1.1 k)
$V_5 = 19.1$ kN (4.4 k)
$V_4 = 38.2$ kN (8.8 k)
$V_3 = 57.3$ kN (13.2 k)
$V_2 = 76.4$ kN (17.6 k)
$V_1 = 95.5$ kN (22 k)
Axial forces in columns:
$R_6 = 4.8$ kN (1.1 k)
$R_5 = 23.9$ kN (5.5 k)
$R_4 = 57.3$ kN (14.3 k)
$R_3 = 114.6$ kN (27.5 k)
$R_2 = 191.0$ kN (45.1 k)
$R_1 = 286.5$ kN (67.1 k)
Axial forces in beams:
$H'_6 = 9.6$ kN (2.2 k)
$H'_5 = 19.1$ kN (4.4 k)
$H'_4 = 19.1$ kN (4.4 k)
$H'_3 = 19.1$ kN (4.4 k)
$H'_2 = 19.1$ kN (4.4 k)
$H'_1 = 19.1$ kN (4.4 k).

7.13. $\delta = 144$ mm (5.6 in.). **7.15.** $\delta = 6$ mm (0.235 in.).

7.17. $V_1 = V_2 = 65$ kN (15 k); $C_1 = 65$ kN (15 k); $C_2 = 0$; $C_3 = 130$ kN (30 k); $H_1 = H'_1 = 70.9$ kN (16.7 k); $H_2 = 212.7$ kN (50 k); $H'_2 = 141.8$ kN (33.3 k); $H'_3 = 0$, $M_1 = M'_1 = 97.5$ kN·m (75 k·ft); $M_2 = 97.5$ kN·m (75 k·ft); $M'_2 = 195$ kN·m (150 k·ft); $M'_3 = 0$.

7.18. 24%.

CHAPTER EIGHT

8.1. $f_b = 68$ MPa (8,250 psi), $f_v = 1.27$ MPa (153 psi); $f_s = 4.9$ MPa (538 psi).

8.3. $f_{max} = 204$ MPa (22,060 psi). **8.5.** $f_{max} = 0.18$ MPa (378 psi).

8.6. $M = -2.06$ kN·m (−1,582 lb·ft); $V = 1.73$ kN (400 lb); $M_t = 1.73$ kN·m (1,308 lb·ft); Beam size: 250 mm × 250 mm (10 in. × 10 in.).

8.8. $h = 1\,080$ mm (42 in.).

8.9. The earthquake governs the design. $M = 170\,100$ kN·m (126,000 k·ft); $V =$ 4 725 kN (1,050 k); $M_t = 28\,350$ kN·m (21,000 k·ft).

8.11. $M_t = 23$ kN·m (17 k·ft). **8.12.** $M_{box}/M_I = 0.46$; $M_{t,box}/M_{t,I} = 41$.

CHAPTER NINE

9.1. $T_{max} = 42.88$ kN (99 k) at $x = 12$ m (40 ft); $T_{min} = 20.32$ kN (46.2 k) at $x =$ 3 m (10 ft).
9.3. $\delta_2 = -0.3$ m (-14 in.) up; $\delta_3 = 0.3$ m (10 in.) down.
9.6. $T_{max} = 133.62$ kN (30 k); $d = 25.4$ mm (1 in.).
9.7. (a) $T_1 = 30.13$ kN (6.04 k); $T_2 = 25.58$ kN (5.13 k); (b) $T_1 = 28.77$ kN (5.75 k); $T_2 = 20.54$ kN (4.11 k).
9.9. $P_1 = 1.87$ sec; $P_2 = 2.96$ sec.
9.10. $C = 2\,034$ kN (456 k); $T = 2\,191$ kN (491 k); $W = 1\,627$ kN (365 k).
9.13. (a) $h = 0.34$ m (0.9 ft); (b) $h = 0.25$ m (0.68 ft).
9.15. 11.87 m (43 ft).

CHAPTER TEN

10.1. 254 mm × 356 mm (10 in. × 14 in.). **10.3.** $t = 3$ mm (0.12 in.).
10.4. $h_B = 550$ mm (21 in.); $h_A = 215$ mm (8.2 in.).
10.5. (a) W920 × 446 (W36 × 300); (b) W920 × 365 (W36 × 245).
10.7. $R_A = 675$ kN (150 k); $H_A = 535.6$ kN (119 k); $M_{max} = 1\,028$ kN·m (759 k·ft).
10.8. $A = 18\,921$ mm² (30 sq in.); $M_c = 75.6$ kN·m (60 k·ft).
10.10. $h = 400$ mm (16 in.). **10.12.** $A = 15\,746$ mm² (24 sq in.).
10.13. $R_A = -2.7$ kN (-0.6 k); $R_B = 2.7$ kN (0.6 k), $H_A = 20.25$ kN (4.5 k); $H_B = 6.75$ kN (1.5 k).
10.15. $M = 11$ kN·m (1.2×10^5 lb·in.); $A_S = 83$ mm² (0.16 sq in.).
10.17. $M = 165$ kN·m (1.46×10^6 lb·in.); $A_s = 1\,250$ mm² (1.9 sq in.).
10.18. $d = 168$ mm (6.9 in.).
10.20. $H = P_0 l^2/12h$; $y = (hx/l)(3 - 4x^2/l^2),\ 0 \leq x \leq l/2$;
$y = [h(l - x)/l][3 - 4(l - x)^2/l^2],\ l/2 \leq x \leq l$.

CHAPTER ELEVEN

11.1. $x_1 = 0.882\,1P$, $x_2 = 1.092\,0P$; $x_3 = 0.835\,8P$; $x_4 = 1.128\,3P$; $R_A = 1.3\,P$; $M_A = 0.143\,1Pl$; $R_B = 1.656\,2P$; $M_B = 0.185\,0Pl$; $R_C = 0.071\,9P$, $M_C = 0.006\,5Pl$; $R_D = 0.1\,P$; $M_D = 0.009\,Pl$.
11.3. $l_2/l_1 = 1$. **11.4.** $h = 1\,310$ mm (52 in.), $\delta = 24$ mm (0.96 in.).
11.5. $I_2/I_1 = \frac{4}{9}$. **11.7.** $M_1 = 0.109\,6Pl$; $M_2 = 0.211Pl$.
11.9. $d_3/d_4 = 1.0$; $W_3 = 4wbhl$; $W_4 = 6wbhl$; $h = 1\,190$ mm (48 in.); $W_3 =$ 411 kN (96 k); $W_4 = 617$ kN (144 k).
11.10. (a) $R = P/4$, moment does not change sign; (b) $R = -P/4$, moment changes sign; (c) $R = -0.66P$, moment changes sign.
11.11. $M_{max} = 0.044\,wl^3$.
11.13. $x_1 = 0.89P$; $x_2 = 0.187P$; $x_3 = -0.044P$; $x_4 = 0.033P$.
11.14. $M_{A,max} = 0.088PR$; $M_{B,max} = 0.163PR$.
11.16. $M_{A,max} = 0.088Pl$; $M_{B,max} = 0.13Pl$; $M_{C,max} = 0.085Pl$.

11.17. $I_1/I_2 = 9$.
11.19. $f_{b,\max} = 24.2$ MPa (2,968 psi); $f_{v,\max} = 0.85$ MPa (103 psi).

CHAPTER TWELVE

12.1. $h = 8$ mm ($\frac{5}{16}$ in.). **12.2.** (b) $q = 2$ kN/m² (47 psf).
12.4. $h = 75$ mm (3 in.). **12.6.** $a = 10.7$ m (35.7 ft).
12.9. $h = 200$ mm (8 in.); $M_{c-} = 37.4$ kN·m/m (8.3 k·ft/ft); $M_{c+} = 17.9$ kN·m/m (4.0 k·ft/ft); $M_{m+} = M_{m-} = 13.0$ kN·m/m (2.9 k·ft/ft); $A_{c-} = 1\,770$ mm²/m (0.81 sq in./ft); $A_{c+} = 847$ mm²/m (0.39 sq in./ft); $A_{m+} = A_{m-} = 615$ mm²/m (0.28 sq in./ft).
12.11. $f_{\max} = 3.8$ MPa (552.5 psi); $M_1 = 0.32$ kN·m/m (73.6 lb·ft/ft).
12.12. $A_{1.5}/A_1 = 1.4$; $A_\infty/A_1 = 1.7$.
12.14. $A_{1.5}/A_1 = 1.88$; $A_2/A_1 = 3.0$.
12.15. (b) $h = 120$ mm (5 in.); $A_s = 1\,443$ mm²/m (0.65 sq in./ft).
12.16. $t = 200$ mm (8 in.); $A_s = 156$ mm²/m (1.7 sq in./ft).
12.18. $A^+_{s,a} = 332$ mm²/m (0.17 sq in./ft); $A^+_{s,b} = 221$ mm²/m (0.11 sq in./ft).
$A^-_{s,a} = 528$ mm²/m (0.26 sq in./ft); $A^-_{s,b} = 356$ mm²/m (0.18 sq in./ft).

CHAPTER THIRTEEN

13.1. $f_{s,\max} = 7.1$ MPa (1 ksi).
13.5. $f_{\max} = 495$ MPa (72 ksi); $f_{\text{actual}} = 248$ MPa (36 ksi).
13.7. $P_u = 197$ kN (43.7 k). **13.8.** $w_u = 42.36$ kN/m (2.82 k/ft).
13.10. $P_u = 112$ kN (24.8 k). **13.11.** (a) $w_u = 103$ kN/m² (1.9 ksf).
13.12. Circumscribed plate: $w_u = 103$ kN/m² (1.90 k/sq ft).
13.13. $R_1 = 0.6$ m (2 ft); $R_2 = 0.85$ m (2.83 ft).
13.14. Fails in shear.
13.15. (a) $f_{v,\max} = 3wl/(2bh)$; (c) $f_{v,\max} = 3wl^2/(2bh^2)$.
13.16. For $r/d = \frac{1}{8}$; $d = 149$ mm (5.5 in.).
13.17. (a) $P_u = 2.6\, h^2 fy$. **13.18.** (b) $W_u = 10.4\, h^2 fy$.
13.19. $M'_y = M_y[1 - (T/bhfy)^2]$.

CHAPTER FOURTEEN

14.1. (a) $z = k\sqrt{x^2 + y^2/a}$; (c) $z = k\sqrt{1 - (x^2 + y^2)/k^2}$; (f) $z = ka/\sqrt{x^2 + y^2}$.
14.2. (b) $2kx/a^2$; (d) $-k^2x/a^2z$; (e) $(k/a)\sinh(x/a)$.
14.3. Assuming the slopes are small:

Along meridian:	*At right angles:*
(a) $C_1 = 0$	(a) $C_2 = k/ax$
(b) $C_1 = 2k/a^2$	(c) $C_2 = -1/k$
(e) $C_1 = k/a^2$	(d) $C_2 = -k/a^2$
(f) $C_1 = 2ka/x^3$	(f) $C_2 = -ka/x^3$

14.4. $c_x = c/a^2$; $c_y = -c/b^2$, yes.

14.5. (c) $h/l = \frac{1}{2}$; (e) $h/l = 0.27k/a$.

14.6. (a) anticlastic, (c) developable, (e) synclastic, (g) developable.

14.7. (ad) $z = kx + c\sqrt{1 - y^2/b^2}$, developable; (bd) $z = k\sqrt{1 - x^2/a^2} + c\sqrt{1 - y^2/b^2}$; $kc > 0$, synclastic, $kc < 0$, anticlastic; (bf) $z = k\sqrt{1 - x^2/a^2} + c$, developable; (ce) $z = kx^2 - ky^2$, anticlastic; (cf) $z = kx^2 + c$, developable.

14.8. $y_2 = y_1 = y$, $z(x, y) = k[1 - (x/l)]\sqrt{1 - y^2/a^2}$, the surface is anticlastic everywhere except $y = 0$ where it is developable.

14.9. The surface is anticlastic except those on the coordinate axes where it is developable; no.

14.10. (a) $k/a = 2/r$, $m_x = 2/r$; (c) $k/l = 1/r$, $m_x = \infty$; (d) $k/a = 2/r$, $m_x = \infty$.

CHAPTER FIFTEEN

15.1. $T_{\phi,\max} = 45.47$ kN/m (2,964 lb/ft); $T_{\theta,\max} = 44.85$ kN/m (2,925 lb/ft); $A_\phi = 270$ mm² (0.4 sq in.); $A_\theta = 197$ mm² (0.3 sq in.); $A_c = 4.66 \times 10^4$ mm² (70 sq in.).

15.3. $A_0 = 33$ mm² (0.05 sq in.); $A_1 = 197$ mm² (0.32 sq in.); $A_2 = 393$ mm² (0.64 sq in.); $A_3 = 262$ mm² (0.42 sq in.).

15.5. $T_{x,\max} = 12.14$ kN/m (810 lb/ft); $T_{\theta,\max} = 18$ kN/m (1,198 lb/ft); $A_s = 400$ mm² (0.6 sq in.).

15.6. $T_{\phi,\max} = T_{\theta,\max} = 0.04$ kN/m (2.67 lb/ft); $h = 0.001$ mm (0.00004 in.).

15.8. $p = 2.3$ kN/m² (47.7 psf). **15.9.** $T = 114$ kN (25.3 k).

15.11. $\delta = 2.27$ m (7.2 ft).

15.13. $W = 3.3$ kN (784 lb); $p = 2.94$ kN/m² (62.4 psf); $h = 0.03$ mm (0.0013 in.).

15.14. $T = 0.93$ kN/m (65.5 lb/ft); $D = 172$ mm (7 in.).

15.15. $T^s_{\phi,\max} < T^p_{\phi,\max} < T^c_{\phi,\max}$; $T^p_{\theta,\max} < T^s_{\theta,\max} < T^c_{\theta,\max}$.

15.16. For $f = l/4$, S.F. $= 2.6$, the membrane will fail if $f \leq l/12.5$.

15.18. $T = 7Eh\delta^2/(12r^2)$.

15.20. $T_\phi = 10.68$ kN/m (0.71 k/ft); $T_\theta = 0$; $d_w = 4.3$ mm (0.168 in.); the point of support can be lifted 0.54 m (1.84 ft).

CHAPTER SIXTEEN

16.1. $f_{\max} = 0.78$ MPa (111 psi); $H = 62.4$ kN/m (4,278 lb/ft); no reinforcement is required.

16.3. $N_{\text{sphere}} = 26.73$ kN/m (1,856 lb/ft); $N_{\text{paraboloid}} = 28.48$ kN/m (1,978 lb/ft); $N_{\text{cone}} = 90$ kN/m (6,249 lb/ft).

16.4. $\phi = 48.7°$. **16.6.** $l_u = 3\,468$ m (11,414 ft). **16.8.** No.

16.10. If $0 \leq x \leq x_s$, $N_{x,\max} = -(wf)/6 \cos^2 \alpha$, $N_{\theta,\max} = -wf/3 \tan^2 \alpha$; if $x_s \leq x \leq x_0$, $N_{x,\max} = (4wf)/(3 \cos^2 \alpha)$, $N_{\theta,\max} = wf/3 \tan^2 \alpha$.

16.13. $\delta = 0.03$ mm (0.0128 in.).

16.15. $N_{\phi_a} = 2N_{\phi_s}$; the arch has greater displacement because it is not supported by hoop forces.

16.16. $N = 4.77$ kN/m (318 lb/ft); $S = 11.25$ kN/m (750 lb/ft); $N/N_{\phi,\max} = 17.6\%$ $< 33\%$; $S/(|N_{\theta,\max}|) = 41.6\% < 33\%$.
16.18. $N_{x,\max} = 1\,166$ kN/m (80 k/ft); $N_{x,\min} = -2\,060$ kN/m (-141 k/ft); $A_s =$ 15×10^3 mm²/m (7.0 sq in./ft).
16.20. $N_{x,\max} = 99$ kN/m (6,750 lb/ft); $S_{\max} = 198$ kN/m (13,500 lb/ft); $A_s =$ $1\,438$ mm²/m (0.68 sq in./ft); $A_{tr} = 3\,934$ mm² (6.14 sq in.).
16.22. $C = -T = 39.6$ kN/m (2.7 k/ft); $F_{Bd} = 398$ kN (90.6 k); $A_s = 2\,583$ mm² (4.05 sq in.).
16.24. $A_\phi = 2\,434$ mm² (3.8 sq in.); $A_\theta = 1\,088$ mm² (1.7 sq in.).
16.25. $A_s = 184$ mm² (0.29 sq in.).

CHAPTER SEVENTEEN

17.1. $H = -1.1$ kN/m (-77.6 lb/ft). **17.3.** $H = -0.35$ kN/m (-23.3 lb/ft).
17.5. $x_{17.1} = x_{17.3} = 2.93$ m (9.8 ft); $x_{17.2} = 0.73$ m (2.5 ft); $x_{17.4} = 0.9$ m (3 ft).
17.6. $A_s = 8.0$ mm²/m (0.004 sq in./ft); $r = 2.5\%$.
17.8. $A_{s,x} = 94$ mm²/m (0.044 sq in./ft); $A_{s,y} = 63$ mm²/m (0.030 sq in./ft).
17.9. $\bar{H} = -61.8$ kN/m ($-4{,}237$ lb/ft); $A_s = 10076$ mm² (15.89 sq in.).
17.11. $\bar{H} = -20$ kN/m ($-1{,}368$ lb/ft); $M_{\max} = -0.005$ kN·m/m (-1.0 lb·ft/ft).
17.13. $M_0 = -0.38$ kN·m/m (-86 lb·ft/ft); $\bar{H} = -56.8$ kN/m ($-3{,}865$ lb/ft); $x_{p.s.} = 4.6$ m (15.2 ft).
17.14. $h_m = 0.15$ mm (0.006 in.); $h_b = 0.24$ mm (0.009 in.).
17.16. $h = 0.83$ mm (0.0347 in.). **17.18.** S.F. = 3 807. **17.22.** S.F. = 156.
17.23. $h = (f_a R(1 + \cos\phi_0))/0.15E$.

CHAPTER EIGHTEEN

18.1.

$$C_1 = -\frac{25}{96}\frac{a}{h}\frac{P}{A} \qquad C_2 = -\frac{14}{96}\frac{a}{h}\frac{P}{A}$$

$$T_1 = -\frac{25}{96}\frac{a}{h}\frac{P}{A} \qquad T_2 = \frac{14}{96}\frac{a}{h}\frac{P}{A}$$

$$T_1' = T_1'' = \frac{2}{96}\frac{a}{h}\frac{P}{A} \qquad T_2' = T_2'' = \frac{7}{96}\frac{a}{h}\frac{P}{A}$$

$$D_1 = -\frac{25}{96}\frac{\sqrt{a^2+h^2}}{h}\frac{P}{A} \qquad D_2 = -\frac{14}{96}\frac{\sqrt{a^2+h^2}}{h}\frac{P}{A}$$

$$D_1' = \frac{23}{96}\frac{\sqrt{a^2+h^2}}{h}\frac{P}{A} \qquad D_2' = D_2'' = \frac{7}{96}\frac{\sqrt{a^2+h^2}}{h}\frac{P}{A}$$

$$D_1'' = -\frac{2}{96}\frac{\sqrt{a^2+h^2}}{h}\frac{P}{A} \qquad V_2 = -\frac{7}{96}\frac{P}{A}$$

$$V_1 = \frac{-23}{96}\frac{P}{A} \qquad V_2' = 0$$

$$V_1' = 0$$

18.2. $A_h = 1\,752$ mm² (2.68 sq in.); $A_d = 731$ mm² (1.12 sq in.).
18.4. $z_{0,18.2} = 100$ mm (4.0 in.); $z_{0,18.3} = 67$ mm (2.96 in.).
18.5. $q_0 = 47.4$ kN/m² (1,073 psf).

18.7. $j = 0.68$ m (28 in.). **18.9.** $h = 330$ mm (13 in.).

18.13. $T_{\max} = -431.3$ kN (−96 k); $C_{\max} = 228.6$ kN (51 k); $A = 5\,196$ mm² (8.0 sq in.).

18.14. $C_x = -32.2$ kN (−7,145 lb); $C_r = C_s = 75.4$ kN (16,744 lb); $w_b = 0.056$ kN/m² (1.18 psf).

18.16. Local buckling, *S.F.* = 53; shell buckling, *S.F.* = 2.16.

16.18. $A = 706$ mm² (1.0 sq in.); $q = 2.87$ kN/m² (57.6 psf).

18.19. Local buckling, *S.F.* = 18; $A_I = 6\,585$ mm² (10 sq in.).

18.20. $t_{\text{center}} = 0.75$ mm (0.029 in.); $t_{\text{arch}} = 2.25$ mm (0.086 in.); $t_{\text{corner}} = 5.2$ mm (0.2 in.).

18.21. $q_{cr} = 8.5$ kN/m² (185 psf); shell buckling governs.

Index